Neural Engineering

BIOELECTRIC ENGINEERING

Series Editor: Bin He
University of Minnesota
Minneapolis, Minnesota

Neural Engineering

Edited by

Bin He
University of Minnesota
Minneapolis, Minnesota

Kluwer Academic / Plenum Publishers
New York, Boston, Dordrecht, London, Moscow

Library of Congress Cataloging-in-Publication Data

Neural engineering/edited by Bin He.
    p.   cm. — (Bioelectric engineering)
   Includes bibliographical references and index.
   ISBN 0-306-48609-1 — ISBN 0-306-48610-5 (eBook)
    1. Biomedical engineering—Congresses. 2. Neural networks (Neurobiology)—Congresses.
3. Sensory stimulation—Congresses. 4. Imaging systems in medicine—Congresses. I. He,
Bin, 1957– II. Series.

R856.A2N47 2005
610'.28—dc22

                               2004049813

Cover image courtesy of Illinois Institute of Technology (created by Viktor Koen)

ISBN 0-306-48609-1
EISBN 0-306-48610-5

©2005 Kluwer Academic/Plenum Publishers, New York
233 Spring Street, New York, New York 10013

http://www.wkap.nl/

10 9 8 7 6 5 4 3 2 1

A C.I.P. record for this book is available from the Library of Congress

Permissions for books published in Europe: *permissions@wkap.nl*
Permissions for books published in the United States of America: *permissions@wkap.com*

Printed in the United States of America

PREFACE

Beginning centuries ago, early exploration of neural systems focused on understanding how neural systems work at the cellular, tissue, and system levels, and engineering methodologies were developed to detect, process, and model these neural signals. Recently, tremendous progress has been made in the field of neural engineering, not only understanding the mechanisms, detection, and processing of the signals, but also on restoring neural systems functions and interfacing the neural systems with external devices and computers.

The purpose of this book is to provide a state-of-the-art coverage of basic principles, theories, and methods in several important areas in the field of neural engineering. It is aimed at serving as a textbook for undergraduate or graduate level courses in neural engineering within a biomedical engineering or bioengineering curriculum, as well as a reference book for researchers working in the field of neural engineering, and as an introduction to those interested in entering this discipline or acquiring knowledge about the current state of the this rapidly developing field.

Chapter 1 deals with neural prostheses—implantable devices that mimic normal sensory-motor functions through artificial manipulation of the biological neural system using externally induced electrical currents. While these are generally separated into two classes (sensor and motor) the author provides systematic coverage of the state-of-the-art in sensory neural prostheses.

The next three chapters address neural interfacing at different levels and from different perspectives. Chapter 2 introduces the concept of interfacing neural tissues with microsystems. Microsystems technology is a rapidly developing field that integrates devices and systems at the microscopic and submicroscopic scales. Neural interfacing with microsystems provides an important basis of interfacing neural systems with a variety of artificial devices. Chapter 3 addresses another aspect of neural interfacing—brain-computer interface—which serves as a method of communication based on neural activity generated by the brain that is independent of its normal output pathways of peripheral nerves and muscles. Also reviewed are the state-of-the-art developments in this emerging field, integrating neurophysiology, signal detection, signal processing, and pattern recognition. Chapter 4 reviews the recent developments in neurorobotics, which interface directly with the brain to extract the neural signals that code for movement and use these signals to control a robotic device.

Neural stimulation is discussed in Chapter 5. Functional electrical stimulation of neural tissue can provide additional functional restoration to neurologically impaired individuals. Also covered is the fundamentals of electrical excitation introduced by electrical stimulation of neural tissue and some important applications.

Chapters 6 and 7 discuss neural signal processing and imaging. An important aspect of neural engineering is to properly analyze and interpret the neural signals—a step that plays a vital role for sensing and controlling neural prostheses and other neural interfacing devices, as well as understanding the mechanisms of neural systems. Chapter 6 provides a concise but systematic review of neural signal processing in the central nervous system; Chapter 7 teaches the basic principles and applications of electrophysiological neuroimaging. Applying electromagnetic theory and signal processing techniques, electrophysiological neuroimaging provides spatio-temporal mapping of source distributions within the brain from noninvasive electrophysiological measurements. Knowledge of such spatio-temporal dynamics of source distribution associated with neural activity would aid in the understanding of the mechanisms of neural systems and provide a noninvasive probe of the complex central nervous system.

Chapters 8, 9, and 10 focus on neural computation. Chapter 8 discusses the computational principles underlying cortical function. Recent theoretical models, presenting a range of interesting and sometimes conflicting mechanisms, are reviewed and their relationship with the underlying biology is explored. Cortical computation is an important tool for studying and understanding the mechanisms associated with processes ranging from visual, auditory, and olfactory senses to high-level brain functions such as recognition, memory, and categorization. Chapter 9 introduces nonlinear dynamics of neural systems and provides an overview of the framework to study, simulate, design, fabricate, and test biologically plausible information processing paradigms. In addition, the analog VLSI implementations of the nonlinear computational algorithms are described, providing an important link between the computational algorithms and the devices interfacing with neural systems. Chapter 10 reviews some of the important neural circuit models in order to gain a balanced understanding of the interplay between the dynamics and temporal characteristics of action potential trains and their effects on the neural information processing. Emphasis is placed on neural modeling at the cellular level and its applications for understanding the mechanisms of neural information processing.

The following two chapters emphasize neural system identification and prediction. Chapter 11 introduces important perspectives and techniques for system identification, as well as giving concrete examples of system identification strategies to study sensory processing in the central nervous system and neural control in the peripheral nervous system. An important aspect of neural engineering is not only to detect and understand signals from neural systems but to also interface with, and control, the neural systems. Chapter 12 discusses such strategies and provides an example of predicting epileptic seizures and thus allowing for proper intervention and control of the seizure.

Chapter 13 discusses retinal bioengineering. The mathematical modeling of neural responses in the retinal microenvironment as well as restoration of retinal function are reviewed. The retina has long served as a model for understanding complex parts of the nervous system, but is also simpler than other parts of the brain due to the lack of significant feedback from the brain to the retina.

This book is a collective effort by researchers and educators who specialize in the field of neural engineering. I am very grateful to them for taking time out of their busy schedules and for their patience during the entire process. It should be noted that the field of neural engineering is developing rapidly and this book can only be a part of the whole picture. Nevertheless, our intention is to provide a general overview covering important aspects of neural engineering. I am indebted to Aaron Johnson, Brian Halm, and Shoshana Sternlicht of Kluwer Academic Publishers for their great support during this project. I would also like to acknowledge the financial support from the National Science Foundation to partially support the editing effort (grants of NSF CAREER Award BES-9875344 and NSF BES-0201939/0411898) which was greatly appreciated.

Bin He
Minneapolis, Minnesota

CONTENTS

1

SENSORY NEURAL PROSTHESES

Philip R. Troyk[1,*] and Stuart F. Cogan[2]

[1]Illinois Institute of Technology, Chicago, Illinois
[2]EIC Laboratories, Norwood, Massachusetts

1.1. INTRODUCTION

The use of technology to compensate for neurological deficit or disease has long captured the imagination of researchers in neural engineering. Although the field of neural engineering, by name, is relatively young, over the past 30 years many researchers from the traditional engineering fields have devoted major portions of their careers to the development of implantable devices known as neural prostheses. These efforts are exemplary of the multidisciplinary nature of bioengineering, and have incorporated principles from a broad range of engineering fields, including electrical, mechanical, and materials engineering, as well as advanced theoretical and applied research in polymer science, electrochemistry, and neuroscience. The basic principle underlying all neural prosthetic devices is common: the artificial manipulation of the biological neural system using externally induced electrical currents with the goal of mimicking normal sensorimotor functions. However, each application requires implantable hardware systems that are specific to the desired function, and therein lay the engineering challenges.

Conceptually, one can imagine how electrical stimulation might be applied to neurons in a temporo-spatial manner that attempts to replicate the normal neuronal firing patterns. However, in practice it is extremely difficult to achieve activation of biological neural systems in a manner that approximates natural function. The major disparity between normal and artificial activation of neural systems stems from a matter of scale: Biological neural networks function by combining the firing patterns from large numbers and populations of neurons, and their associated processes, in a statistical manner. However, manipulation of these same biological neural networks by artificial electrical stimulation is limited to relatively small numbers of input/output channels, owing to the present limitations in electrode technology. Typically, metal-based electrodes implanted near electrically excitable nerve or muscle cells serve as the interface between the artificial and the biological system. The primary challenge for the neural prosthesis researcher is the development of

---

* Address for correspondence: Illinois Institute of Technology, 10 W 32nd Building E1-116, Chicago, Illinois 60616; e-mail: troyk@iit.edu.

miniature electrode structures that can be used for safe, selective, and chronic electrical stimulation.

Neural prostheses can be roughly separated into two general classes: sensory and motor. For motor prostheses, electrical stimulation of the biological neural–muscular system is used to substitute for normal control by the brain or the spinal cord. Lack of normal function can result from spinal cord or brain injury, or birth defects. Functional electrical stimulation (FES) systems have been developed for restoration of standing, walking, and hand grasp. Other systems have been investigated for diaphragmatic pacing, footdrop, urinary incontinence, assisted cough, relief of spasticity, prevention of pressure sores, sleep apnea, erectile dysfunction, and Parkinson's disease. Recently, progress has been made in developing direct cortical interfaces that can be used to mimic mouse-driven inputs to computers. It is hoped that one day such interfaces will directly control FES or mechanical prosthetic systems as a means of bridging over a spinal cord injury.

For sensory prostheses the goal is to use an artificial sensor to replace neural input that would normally come from a peripheral biological source such as an ear or eye. Owing to the importance of hearing and sight for humans, during the past three decades there have been numerous research programs for the development of sensory prostheses to substitute for normal auditory and visual function. Although sharing parallel technological developments, the auditory prosthesis, i.e. the cochlear implant, has advanced more rapidly than the visual prosthesis, with respect to clinical implementation. Cochlear implants are now an accepted clinical device for patients with acquired deafness, and research is progressing toward implantation in very young children who were deaf at birth. The cochlear implant introduces information into the peripheral nervous system, at the site of transduction of sound into neural firing—the inner ear. Other approaches are emerging for which electrodes would be implanted directly into the cochlear nuclei, within the brainstem. Similarly, visual prosthesis development has proceeded at the levels of the retina, the optic nerve, and the primary visual cortex. Although useable hearing can be obtained from 8 to 22 electrodes implanted in the cochlear nuclei, restoration of vision is a far greater challenge, because it is estimated that hundreds, or more likely thousands, of electrodes will be required to provide even minimal visual function.

For both auditory and visual prostheses, enormous technological challenges need to be overcome, and the development of the implantable devices and external hardware, as well as identifying strategies for stimulation of the neuronal system, must be accomplished by specially formed teams with combined expertise in neural signal processing, electronic design, implantable packages, electrode fabrication, surgical methods, and clinical implementation.

Although motor and sensory neural prostheses perform different functions, their designs are often quite similar. Both types artificially stimulate neural tissue with the intention of replicating absent biological stimuli. In this regard they share common elements and many of the same design principles can be applied to both classes of devices.

1.2. FUNDAMENTALS OF SENSORY NEURAL PROSTHESES

At the fundamental level, a neural prosthesis system comprises stimulating electrodes, an implanted electronic package that drives the electrodes with stimulation currents, and an extracorporeal transmitter/controller that is used to power and control the implanted device. Figure 1.1 is a generalized block diagram for a sensory neural prosthesis. Because of the

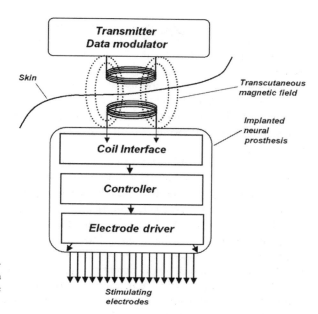

FIGURE 1.1. Block diagram of an implanted neural prosthesis. Power and data are coupled by a transcutaneous magnetic link consisting of two coils.

present-day limitations of implantable batteries, most, if not all, neural prostheses are powered by transcutaneous magnetic coupling. Unlike pacemakers, which typically produce cardiac stimulation pulses at the rate of once or twice a second using one or two electrodes, neural prostheses are required to produce hundreds of stimulation pulses each second, from many electrodes. Therefore, the power demands exceed the capacity of present-day implantable batteries. Although a pacemaker battery may last for 10 years, that same battery would probably last only a couple of days in a modern cochlear implant. For many neural prostheses, the demand for power is nearly continuous, and rather than being supplied by implanted batteries, the raw power is most often provided by an air–tissue transformer, with one coil contained within the implanted device, and one coil as part of the extracorporeal transmitter. The transmitter drives high-frequency current into the external transmitter coil, inducing current in the implanted coil via the transcutaneous magnetic field. Within the implant, the receiving-coil voltage is rectified, providing a power supply that maintains operation of the implant's electronic circuitry. Modulation of the transmitter carrier frequency allows data communication, and power transmission, to occur over the single two-coil inductive link. Similarly, a reverse telemetry signal can be communicated out of the body, by the implanted device, over a data carrier. In this way, a simple magnetic structure consisting of only two coils, one within the implant and the other within the external transmitter, serves the dual purpose of providing power to the implant and forming a two-way communication link.

Within the implanted neural prosthesis, the electronic circuitry can be functionally divided into three main sections: coil interface, controller, and electrode driver. The coil interface serves the function of establishing and maintaining the implant's power supply, establishing reference voltages and currents used by the rest of the implant's circuitry, sensing power-on reset conditions, demodulating forward telemetry command data, and modulating the coil current with reverse telemetry data. One might characterize the coil interface as a type of modem, acting as a bridge, between the controller and electrode driver, to the external transmitter, through the inductive link.

The controller decodes the command data (demodulated by the coil interface) and executes appropriate command and control functions for the implant's circuitry. Usually this section comprises mostly digital circuitry, although specialized analog functions are sometimes included. Decoding a command sent by the external transmitter, to produce a biphasic stimulation current, then sending the appropriate digital signals for pulse amplitude, pulse duration, and stimulus control is an example of a function commonly performed by the controller.

The electrode driver performs the task of using the control signals produced by the controller to generate the appropriate currents, or voltages, for a particular electrode. Producing the required biphasic current amplitude for the required pulse duration, while maintaining electrode charge balance, are functions commonly performed by the electrode driver.

The circuitry for modern neural prostheses is usually fabricated using very large scale integration (VLSI) techniques. Indeed, for obtaining complex circuit functions for a miniature-sized device while minimizing power consumption, VLSI fabrication is often the only solution. Complementary metal oxide semiconductor (CMOS) and bipolar complementary metal oxide semiconductor (BiCMOS) processes are both used to design application-specific integrated circuits (ASIC) for neural prostheses. Sometimes, a combination of more than one physical ASIC is used to optimize circuit performance for digital circuitry separately from analog circuitry.

The transducer of a neural prosthesis is the stimulating electrode. Electrical currents are passed through the electrode in order to stimulate nearby neural tissue. The basic methods of neuronal stimulation by electrical currents have not substantially changed over the past 50 years. Electrical current flowing through the electrode, and through the surrounding neural tissue, causes voltage drops across the cell membranes of nearby neurons. For any particular neuron, if the voltage drop exceeds a critical value, the neuron initiates an action potential. At observed behavioral thresholds of stimulation, a fairly large population of neurons within the vicinity of an electrode will be activated. The actual number depends on the magnitude of the stimulating current and sensitivies of the surrounding neurons, which can vary greatly (Tehovnik, 1996).

Electrodes for cortical stimulation are made from metallic conductors. The interface between the metal tip of the electrode and the biological tissue is the site of charge transfer. Within the tissue, the current flows by ionic charge carriers. The specific nature of the electrode–electrolyte interface and the manner in which charge is transferred across the interface depend on the type of metal used for the electrode. Preserving the integrity of this interface is a necessity for long-term stability of the implanted system. A well-designed neural prosthesis is approached by considering the electrode driver section of the implant circuitry and the electrode together as a system, so that the electrode and the surrounding neural tissue remain stable and healthy.

1.3. ELECTRODES FOR NEURAL STIMULATION

Establishing a chronically stable interface for exchange of information between the biological environment and the implanted electronic device is an essential starting point

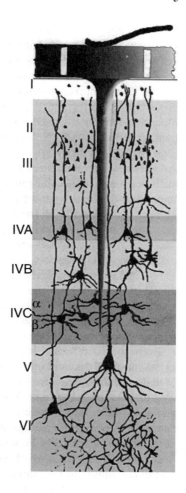

FIGURE 1.2. A microelectrode inserted into the cortex. Near the tip of the electrode, neural stimulation takes place via current that passes from the electrode tip to the surrounding neurons. (Courtesy of Cyberkinetics, Inc.)

for the development of any neural prosthesis. Although the design of neural prostheses presents numerous engineering challenges, the choice of materials and physical structures for the electrodes is often regarded as the limiting factor in prosthesis development. With present-day technology, this interface is based on the use of noncorroding metals incorporated into physical electrodes placed in close proximity to the target neurons (Figure 1.2).

Electrodes with metal stimulation sites are being used, and being considered for use, in prostheses to compensate for sensory deficits as well as for motor and therapeutic applications. Visual prostheses using electrodes implanted in the cortex (intracortical), or on the retina, have been investigated by several groups (Schmidt *et al.*, 1996a,b; Norman *et al.*, 1999; Margalit *et al.*, 2002; Troyk *et al.*, 2002) and clinical studies with retinal stimulation and implants are ongoing (Humayun *et al.*, 1999; Weiland *et al.*, 1999; Chow *et al.*, 2002). Auditory cortical prostheses using electrodes implanted over the ventral cochlear nucleus (VCN) are in clinical use (Otto *et al.*, 2002; Toh *et al.*, 2002) and implants based on microelectrodes that penetrate the VCN, to better access the tonotopic gradient of the VCN, are being actively developed (McCreery *et al.*, 2000; Rauschecker *et al.*, 2002). Intraspinal electrodes are showing promise for bladder management in spinal cord injury and may find use in managing other dysfunctions of the genitourinary system (Grill *et al.*, 1999; Jezernik *et al.*, 2002). The use of microelectrode arrays to stimulate the cortex to restore or control motor function is also being investigated (Nicolelis, 2002). Deep brain stimulation (DBS) is becoming important as a treatment for tremor, epilepsy, pain, and Parkinson's disease when pharmacological management is ineffective (Duncan *et al.*, 1991; Loddenkemper *et al.*, 2001). Nerve signal recordings using intracortical microelectrodes are being used to control extracorporeal devices that will perform functional tasks for the spinal cord injured (Maynard *et al.*, 1997). Multisite chronic recording in the cortex is becoming increasingly important for the study of cortical organization and plasticity, particularly as they relate to the development of prostheses employing cortical control or stimulation (Rousche *et al.*, 1998, deCharms *et al.*, 1999, Wessberg, 2000). The use of cortical nerve signals to control or provide feedback for functional electrical stimulation in neuromuscular prostheses is also being contemplated (Hoffer *et al.*, 1996). All of these applications share a common challenge: a chronic, safe, artificial interface with neuronal tissue using electrodes that provide a high degree of functional specificity.

The design of an implantable stimulating electrode is often a compromise between two opposing constraints: the desire for a small tip area, approaching the size of the surrounding neurons so that selective stimulation of the neurons can be obtained, and the limitation in the amount of charge per stimulation pulse that can be safely injected into the tissue without inducing deterioration of the electrode or the surrounding tissue. As the surface area of the exposed tip is reduced, the charge density, typically reported in millicoulombs per square centimeter (mC/cm^2), increases for constant stimulus charge. As the charge density increases, for a particular metal, the likelihood of irreversible damage to the metal stimulation site similarly increases. Likewise, if a fixed charge is delivered in a shorter time period, a higher current density is necessary. The current density is a measure of the rate of the charge transfer reactions, and higher current densities are more challenging to deliver without damaging the electrode. Much of the microelectrode research during the past 30 years has been directed toward the evaluation of various types of metals with regard to their individual stimulus charge-density limits.

1.3.1. *CHARGE INJECTION PROCESSES AND COATINGS*

Electrical stimulation initiates a functional response by depolarizing membrane potentials in excitable tissue. Depolarization is achieved by the flow of ionic current between two or more electrodes, at least one of which is in close proximity to the target tissue. For an electrode made from a metallic conductor, reactions at the electrode–tissue interface are required to mediate the transition from electron flow in the metal to ion flow in the tissue. These reactions can be capacitive, involving the charging and discharging of either the electrode–electrolyte double layer or a dielectric layer at the interface, or Faradaic, in which surface-confined species are reversibly oxidized and reduced. There are two strategies for increasing charge-injection capacity with capacitor electrodes. The electrochemical surface area (ESA) can be increased by roughening or by deposition of highly porous electrode coatings,

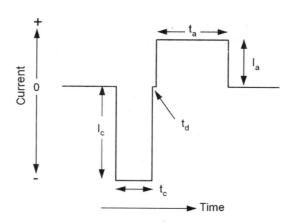

FIGURE 1.3. A biphasic, charge-balanced current pulse typical of those used in neural stimulation. For charge balance, $I_c \times t_c = I_a \times t_a$. The parameters and normal range for each value are I_c: cathodic current (1 μA–10 mA); I_a: anodic current (1 μA–10 mA); t_c: cathodic half-phase period (50 μs–10 ms); t_d: interphase dwell (0–1 ms); t_a: anodic half-phase period (50 μs–10 ms); frequency (not shown): pulses per second (10–250 Hz).

or a high-dielectric-constant film can be formed at the interface by either anodization, if the electrode is fabricated from a valve metal (titanium or tantalum), or deposition of a high dielectric-constant material. Faradaic charge injection capacity may also be increased by increasing ESA, but is most effectively increased with the use of reduction–oxidation (redox) active coatings on the electrode surface. Redox coatings, almost exclusively based on iridium oxide, are used in most intracortical and intraspinal stimulation applications and for some cardiac pacing. Presently, Faradaic electrodes provide more charge per unit geometric surface area (GSA) than do capacitive electrodes.

To avoid damage to the electrode or surrounding tissue, charge injection for neural stimulation is applied in the form of rectangular current pulses, with each pulse having cathodal and anodal components defined by current amplitudes and durations that result in an overall zero net charge (charge balance) for the pulse. A typical pulse waveform with definitions and ranges of pulse parameters is shown in Figure 1.3.

The maximum charge injection capacity of an electrode is usually defined as the charge that can be injected without polarizing the electrode beyond the potentials for reduction or oxidation of water and is based on the premise that these or other irreversible reactions would lead to byproducts that damage tissue near the electrode. Obviously, electrode dissolution reactions are also unacceptable. Other electrochemical limits, such as those for oxidation of proteins might similarly be defined. For any type of electrode, the maximum charge that can be injected within defined potential limits depends to varying degrees on the current density, which is directly proportional to the rate of the double-layer charging and redox processes, the pulse frequency, and the relative magnitudes of I_c and I_a. The geometry and porosity of the electrode also impact the uniformity and magnitude of the polarization (Posey and Morozumi, 1966; Suesserman et al., 1991). Charge injection requirements for electrodes used in some CNS stimulation applications, based primarily on neuronal activation threshold measurements, are listed in Table 1.1. Traditional noble metal electrodes, principally platinum and platinum–iridium alloys, as well as other metal electrodes including stainless steel and gold, with a maximum charge injection capability of <0.15 mC/cm², are not suitable for most of these applications (Robblee and Rose, 1990).

TABLE 1.1. Threshold Charge Injection Requirements for Neural Prostheses Electrodes in the CNS and on the Epi-retinal Surface

Application	Species	Site	Threshold charge (mC/cm^2)	Pulse width (ms)	Reference
Vision	Rabbit	Epi-retinal	0.8–5.7	0.1	Rizzo *et al.*, 1997
Vision	Human	Epi-retinal	0.16–70	2	Humayun *et al.*, 1996
Vision	Human	Epi-retinal	0.8–4.8	Not reported	Weiland *et al.*, 1999
Vision	Human	Intracortical	2–20	0.2	Bak *et al.*, 1990
Vision	Human	Intracortical	0.2–2.5	0.2	Schmidt *et al.*, 1996a,b
Hearing	Cat	Cochlear nucleus	0.06–0.09	0.15	McCreery *et al.*, 1998, 2000
Micturition	Cat	Intraspinal	4 (effective)	0.1	Grill *et al.*, 1999

Only a limited number of electrode materials or coatings are suitable for chronic stimulation at charge densities greater than 0.15 mC/cm^2. Some materials currently used for electrical stimulation in neural prostheses and pacing, with an estimate of their charge injection capacities, are listed in Table 1.2. The charge injection limits are defined as those that just avoid electrolysis of water, which occurs at potentials of approximately -0.6 and 0.8 V Ag|AgCl for reduction and oxidation, respectively, on platinum and iridium-oxide electrodes (Robblee and Rose, 1990).

Capacitive electrodes, in which no redox reactions accompany charge injection, have been evaluated for cortical stimulation (Guyton and Hambrecht, 1974; Rose *et al.*, 1985). Capacitor electrodes, which inject charge entirely by capacitive charging and discharging of the electrical double-layer, are conceptually attractive because they avoid electrolysis of water, electrode dissolution, or other irreversible electrochemical reactions that might degrade either the electrode or the surrounding tissue. However, the available charge per unit electrochemical area is small (~ 20 μC/cm^2-real at 1 V) and the use of porous electrodes and dielectric films, such as Ta_2O_5 or TiO_2, at the electrolyte interface are necessary for increasing charge capacity to physiologically useful levels. Even with surface roughening and dielectric films, the charge injection capacity of the capacitor electrodes is not adequate for small-area (<2000 μm^2) intracortical electrodes at charge and current densities necessary for stimulation (Rose *et al.*, 1985; Robblee and Rose, 1990). Capacitor electrodes based on porous titanium nitride (fractal TiN) are used extensively as large-area electrodes in cardiac

TABLE 1.2. Charge Injection Limits of Pt and Electrode Coating Materials Used or Contemplated for Applications in the CNS

Material	Mechanism	Charge limit (mC/cm^2)	Applications
Pt and Pt–Ir alloys	Faradaic	0.05–0.15	Pacing, nerve cuff electrodes, DBS
Activated iridium oxide	Faradaic	1–3.5	Intracortical stimulation
Thermal iridium oxide	Faradaic	~ 1	Cardiac pacing
Sputtered iridium oxide	Faradaic	>0.5	Limited to IDEs
Tantalum/Ta_2O_5	Capacitive	~ 0.5	Limited to animal studies
Titanium nitride	Capacitive	~ 1	Cardiac pacing

pacing (Schaldach *et al.*, 1990; Frohlig *et al.*, 1998). Titanium nitride is a metallic conductor that is deposited as a highly porous coating by reactive sputter deposition from a titanium metal target. High-porosity thin films of titanium nitride, often termed *fractal*, exhibit extremely low polarization under typical pacing conditions. These electrodes have areas >1 mm^2 and the charge injection densities in pacing are modest, about 0.5 mC/cm^2 delivered with a pulse width of >0.5 ms, compared with those required for intracortical stimulation, and are typically delivered over longer pulse widths and, consequently, lower current densities. The charge injection capabilities of fractal TiN, although considerably greater than noble metals, were less than one third of those of redox-active electrode coatings in measurements under neural stimulation conditions by Weiland *et al.* (2002).

Faradaic electrodes, based on noble metals, transfer charge by surface reduction and oxidation (redox) of a monolayer oxide film and, for Pt or Pt-alloy electrodes, H-atom plating and stripping. The contribution of double-layer capacitance to charge injection with noble metals is typically <15% of the total charge injected. Pt and Pt–Ir alloy electrodes are used extensively in cardiac pacing, deep brain stimulation, and in most nerve cuff and implantable pain management electrodes. On examining the required charge injection levels shown in Table 1.1, it is seen that the charge densities required for even threshold stimulation are well beyond the limits established for noble metal electrodes. The established alternative electrode material is Activated IRidium Oxide Film (AIROF), a Faradaic electrode coating based on a three-dimensional film of hydrated iridium oxide (Robblee *et al.*, 1983). AIROF was developed for high-charge injection electrodes under NIH funding at EIC Laboratories (Norwood, MA) (Brummer *et al.*, 1983; Robble *et al.*, 1983). AIROF is capable of injecting charge at levels appropriate for stimulating small populations of cortical neurons using small-area ($\sim$2000 μm^2) electrodes. Although extensive behavioral response studies using AIROF electrodes in the CNS are lacking, AIROF coatings have been used clinically for acute and short-term chronic (4-month) studies of intracortical stimulation of the visual cortex (Bak *et al.*, 1990; Hambrecht, 1995; Schmidt *et al.*, 1996a,b).

AIROF is formed from iridium metal by repetitive potential cycling between negative and positive limits, close to those for reduction and oxidation of water, to form a hydrated iridium oxide. The high charge capacity is obtained from a reversible Ir^{3+}/Ir^{4+} valence transition within the oxide (Brummer *et al.*, 1983). The first AIROF electrodes were fabricated from polymer-insulated iridium wire, and single-site intracortical microelectrodes formed in this way are used extensively for stimulation studies in the cortex and the spinal cord (McCreery *et al.*, 1986; Grill *et al.*, 1999; Liu *et al.*, 1999). AIROF electrodes are also used on silicon microprobe devices fabricated by thin-film silicon micro machining technology (Anderson *et al.*, 1989). For use with silicon microprobes, the AIROF is formed by activation of sputtered iridium metal films with no apparent difference in properties between AIROF formed from iridium wire or thin-film coatings.

Evaluation of AIROF charge injection electrodes usually begins with cyclic voltammetry (CV). In a CV measurement, the AIROF stimulation electrode is placed in a physiological electrolyte with a standard reference electrode and a large-area auxiliary or counterelectrode, which is usually made of platinum wire or foil. The most common reference electrodes are the saturated calomel electrode (SCE) and the Ag|AgCl (KCl) electrode, both of which are available commercially in a variety of sizes. The potential of the stimulation electrode is driven with respect to the reference electrode, whose half-cell potential remains fixed, by modulating and limiting the current flow between the stimulation and the counterelectrode.

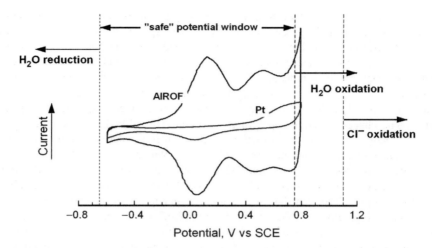

FIGURE 1.4. Typical cyclic voltammetry plots comparing a platinum electrode to an AIROF electrode. Note the higher charge capacity for the AIROF electrode. (Courtesy of EIC Laboratories)

This three-electrode measurement is usually made with an instrument called a potentiostat. The potential of the stimulation electrode is swept with a triangular waveform between two potential limits at a modest sweep rate, typically between 20 and 100 mV/s. The potential limits are usually those corresponding to the onset of water reduction and oxidation, about −0.6 and 0.8 V (vs. Ag|AgCl), respectively, for iridium oxide. The electrode potential and current are plotted as shown in Figure 1.4, where CV plots for a platinum electrode and an AIROF electrode are compared. The potential range over which no electrolysis of water occurs has been labeled the "safe" region.

The total amount of charge that can be transferred in a cathodal stimulation pulse is related to the cathodal charge storage capacity (CSC_c) of the electrode, which is typically calculated from the time integral of the cathodic current during a potential sweep from 0.8 V and −0.6 V versus Ag|AgCl (Figure 1.5). In the case of AIROF, by assuming every iridium cation in the AIROF is electrochemically active and undergoes a one-electron reduction reaction over this potential range (i.e., $Ir^{4+} + e^- \rightarrow Ir^{3+}$), the CSC_c provides an estimate of the total amount of iridium oxide at the charge injection site. A CSC_c of 1 mC is approximately equivalent to 6×10^{15} iridium atoms. Because CSC_c is measured at low current densities (<5 mA/cm^2), it is a near-equilibrium value. A stimulation current pulse is delivered at a much higher current density, often >1 A/cm^2, and only a fraction of the CSC_c can be accessed without polarizing the electrode beyond the limit for water reduction. It should be noted that the available charge depends on the starting potential of the electrode, and for cathodal-first pulses, the more positive the starting potential, in general, the more charge can be delivered in a stimulation pulse. In a physiological buffer, the maximum charge-injection limits of AIROF (with 200-μs current pulses) are generally regarded to be 1, 2, and $\sim$3.5 mC/cm^2 for cathodal, anodal, and anodically biased cathodal pulses, respectively (Beebe and Rose, 1988). For AIROF, the amount of charge that can be injected in a physiologically relevant stimulation pulse is about 5–20% of CSC_c, depending on the details of the stimulation waveform.

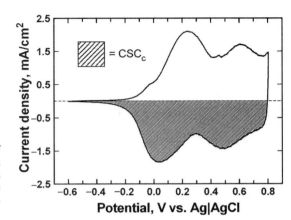

FIGURE 1.5. Cyclic voltammogram of AIROF showing the region of cathodic current for which the CSC_c is calculated. The area is directly proportional to CSC_c because the potential is related to time through the constant sweep rate.

1.3.2. FABRICATION OF NEURAL STIMULATING ELECTRODES

Electrodes for stimulation of the CNS can generally be classified as either metal wire, or metal film type. Wire-type electrodes are made from solid wire, usually iridium. Metal film type electrodes can be fabricated on a variety of substrates, including flexible substrates, with the films deposited by one of the several techniques including sputtering, adhesion of foil, or electroless deposition of a base metal with subsequent metal electroplating. For both classes, the currently preferred material for the active charge injection site is AIROF.

1.3.2.1. Metal Wire Electrodes

Discrete wire microelectrodes have been used for recording and stimulation of neural tissue for many years. The most widely used design consists of simply cutting off the end of an insulated 25- or 50-μm-diameter wire and advancing a single wire or a bundle of wires held together with glucose or carbo wax into the brain. Although this method works well for deeper structures in the brain, it does not work well for the superficial layers of the cortex, where the microelectrode is typically advanced less than 2 mm. Mechanical stability of the microelectrode tip in the brain is critical if one is trying to either record from, or stimulate, the same group of neurons consistently. Gualtierotti and Bailey (1968) were the first to describe a "neutral buoyancy" microelectrode. Their overall design constraint was that the microelectrode "float" on the brain's surface and that the wire leading from the microelectrode to a connector, usually mounted on the animal's skull, be extremely flexible. This design, although it worked fairly well, was not very practical because of the delicate fabrication techniques required to fashion the microelectrode.

The "hat pin" design, as depicted in Figure 1.6, is described by Salcman and Bak (1976). Although fairly difficult to fabricate without customized machinery, it does utilize commercially available materials and approaches Gualtierotti's (1968) "floating electrode" design. The hat pin design uses 37-μm-diameter pure iridium wire microwelded to a 25-μm-diameter gold wire lead. Iridium wire was originally selected as the electrode

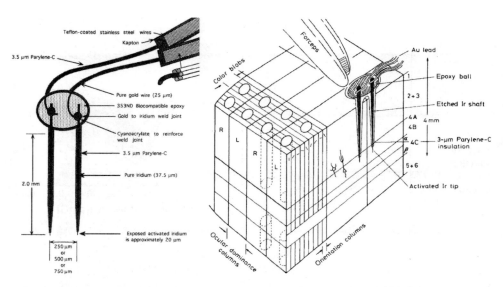

FIGURE 1.6. Dual hat pin electrode as described by Salcman and Bak showing method of hand implantation into the cortex. From Schmidt *et al.* (1996a,b).

material because of its stiffness and inert properties; however, it has the additional advantage of being easily activated to form AIROF directly on the electrode's metal surface. The iridium wire tip is sharpened to a radius of 1–5 μm, and the microweld joint is reinforced with a biocompatible epoxy. The electrode shaft is coated with about 3 to 4 μm of Parylene-C insulation. A dual-beam excimer laser is used to expose the iridium metal at the tip of the microelectrode.

In early designs, Parylene-C-coated sharpened iridium electrode tips were exposed using a high-voltage arcing technique. Although this is a commonly used technique for deinsulating electrode tips, the results can be quite variable. SEM examination of electrode tips before and after implantation consistently revealed poor adhesion of the Parylene-C insulation at the tips. The arcing process is also thought to produce tiny fractures in the Parylene-C along the shaft. Both of these phenomena are believed to result in the decreasing impedance values often seen in chronically implanted electrodes. To produce a more stable and dimensionally consistent exposed metal stimulation surface area, a technique which utilizes an excimer laser system was developed that ablates the Parylene in a precisely controlled manner, thus allowing electrodes to be consistently fabricated with precisely defined exposed tip areas. In addition, substituting the laser ablation for the arcing method allows the electrodes to be treated with surface adhesion promoters (typically silanes) before they are coated with the Parylene. Use of the surface promoters results in the Parylene-C to be more tightly bound to the iridium metal, thus reducing the risk of having the insulation push back during implantation, as well as providing a more stable metal–insulation interface during stimulation.

Other variations of this basic wire-type electrode design have evolved over the past 10 years. In an alternate design the metal wire iridium microelectrodes are fabricated from 35-μm iridium wire, with the tip of the electrode electrolytically etched to form a controlled

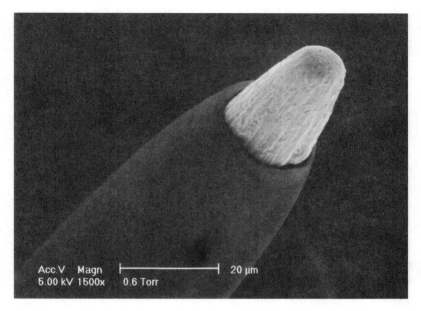

FIGURE 1.7. Discrete iridium microelectrode. The laser-ablated AIROF active tip has an exposed surface area of 2000 μm^2. (Photograph courtesy of Huntington Medical Research Institutes)

cone with an included angle of approximately 10°, and a blunt tip with a radius of curvature of 5–6 μm. In each case, the goal has been to produce electrodes with consistent tip geometries and stable AIROF coatings. Figure 1.7 shows an SEM photograph of an AIROF iridium metal wire electrode with a precision-laser-ablated tip.

1.3.2.2. Incorporation of Metal-Wire Electrodes into Arrays

Although some limited success has been reported for single metal wire stimulating electrodes implanted into the CNS, it is generally agreed that design of a sophisticated CNS sensorineural prosthesis will require larger numbers of implanted electrodes. To facilitate the implantation process, and prevent significant movement once implanted, research has been done to examine the feasibility of using groups of electrodes consolidated together within a rigid matrix. One such design is shown in Figure 1.8. The individual electrodes are cast into an epoxy matrix and connected to a multiwire cable. Implantation into the cortex is accomplished using a specialized high-speed insertion tool, at speeds of up to 1 m/s to minimize tissue damage and bleeding. Owing to this rapid insertion speed, the array can impale blood vessels with little to no bleeding. Based on this design, arrays consisting of up to 16 electrodes have been fabricated.

1.3.2.3. Metal Film Silicon Electrode Arrays

Micromachining of silicon has been used to fabricate electrode probes with multiple electrically insulated metal stimulation sites on a single probe shank. It is anticipated that

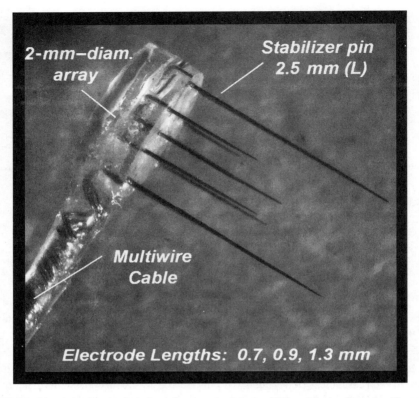

FIGURE 1.8. Stimulating intracortical electrode array containing multiple metal-wire electrodes. The stabilizer pins prevent motion of the array once implanted into the cortex. (Photograph courtesy of Huntington Medical Research Institutes)

such microprobes will be employed in neural prostheses requiring many multiple-site electrodes, and they offer one of the few approaches being developed to resolve the difficulty of connecting to the large number of neuronal sites that may be required for a clinically practical CNS prostheses. The potential of the microprobes is reflected in the extensive development effort supported by the NIH (Anderson *et al.*, 1989). Figure 1.9 shows the concept of typical micromachined probes.

Fabrication is accomplished using a modified liftoff process, in which photolithographic techniques, well known to the semiconductor industry, are employed to pattern the distribution of electrode sites, the shape of an individual electrode shank, or the combination of multiple shanks on the same silicon structure. An ongoing program of research at the University of Michigan (Wise *et al.*, 1999) continues to advance this technology, and it is anticipated that this work will culminate in the realization of extremely high density three-dimensional structures with far greater design variations and consistency than would be possible for discrete metal wire electrodes. A prototype is shown in Figure 1.10. Because the substrate for the micromachined electrode structures is silicon, including active electronic circuitry directly on the probe becomes an attractive option, and designs of this nature are currently under test. For the metal stimulation sites, AIROF is the usual

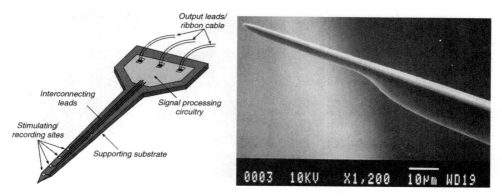

FIGURE 1.9. Micro-machined silicon probes. Left: concept drawing of a probe design. Right: Scanning electron microscope photographs of silicon probe tip. (Courtesy of University of Michigan)

FIGURE 1.10. Three-dimensional micromachined silicon probe array. (Courtesy of University of Michigan)

charge injection coating, but the probe fabrication technology also makes them suitable for use with fractal TiN (Wise *et al.*, 1999), and it is expected that they will also be suitable for use with sputtered iridium oxide (SIROF). The deposition of AIROF on the micromachined electrode structures poses unique problems, which current studies are addressing.

1.3.2.4. Stability of Activated Iridium Oxide (AIROF)

Successful and reproducible activation of iridium to AIROF depends on the surface condition of the electrode prior to activation, primarily surface cleanliness. Incompletely removed insulation or residue, from the process used to deinsulate the electrode tip (e.g., laser ablation or reactive ion etching) can occlude and mask the electrode surface, leave electroactive contamination on a surface, or lead to poor adhesion of the AIROF. This condition has been observed for the activation of both iridium shaft electrodes and thin-film electrode arrays.

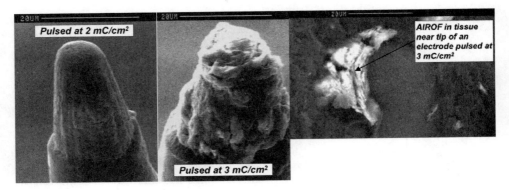

FIGURE 1.11. AIROF tip of iridium shaft electrodes pulsed for 7 h in cat cortex at 50 Hz, 400 μs/ph at the charge density indicated with 0.4 V (Ag|AgCl) anodic bias. All electrodes (8 total) pulsed at 3 mC/cm² exhibited delamination of AIROF. The third figure shows scanning electron micrograph showing delaminated iridium oxide film at the site of an electrode pulsed at 3 mC/cm². Backscatter image of histology slide. Light areas are AIROF in tissue.

The original charge injection limits for AIROF (CSC_c 28 mC/cm²), using 200-μs current pulses, were determined by visual observation of gas formation and measurement of electrochemical potential transients (Robblee *et al.*, 1983; Beebe and Rose, 1988). Visual gassing limits were subsequently found to overestimate the charge injection limits. Potential transient measurements by Beebe and Rose (1988) are more indicative of the true reversible limits, which were determined to be 1 mC/cm² for cathodal pulsing, 2.1 mC/cm² for anodal pulsing, and 3.5 mC/cm² for cathodal pulsing with anodic bias. Although these limits were previously thought to be conservative, recent studies involving *in vivo* pulsing of AIROF electrodes in cat cortex at about 3 mC/cm² found corrosion of the electrode tips and delamination of the AIROF as shown in Figure 1.11. These electrodes were evaluated at Huntington Medical Research Institutes (Pasadena, CA) as part of their ongoing studies of safe cortical stimulation. Eight electrodes pulsed at this level showed corrosion and delamination after 7 hr of pulsing.

While using AIROF coatings for neural stimulation it is important to avoid pulsing conditions that cause continued activation of the underlying iridium to AIROF. Although for thicker films the charge storage capacity of the AIROF increases, thicker films are also mechanically unstable and may delaminate. The effect of continued thickening of AIROF is shown in Figure 1.12 for an AIROF that was originally activated to a CSC_c of 28 mC/cm² and subsequently cycled at 50 mV/s between potential limits of –0.6/0.8 V for 6372 cycles followed by 6815 cycles at –0.7/0.8 V limits in phosphate-buffered saline (PBS).[1] A large increase in CSC_c is observed, from 28 to 58 mC/cm², with the formation of a pronounced oxidation peak at 0.63 V. This latter peak is associated with delamination of the AIROF, although it is not known if this is a direct result of delamination or an intrinsic evolution in voltammetric behavior with increasing AIROF thickness.

---

[1] PBS is phosphate-buffered saline having a composition of 126 mM NaCl and 22 mM $NaH_2PO_4 \cdot 7H_2O$ and 81 mM $Na_2HPO_4 \cdot H_2O$ at pH 7.3. The PBS is deaerated with a gentle flow of Ar gas to remove oxygen.

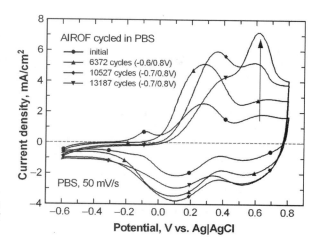

FIGURE 1.12. Effect of AIROF overactivation on voltammetric response. Note the increase in CSC and the evolution of an anodic current peak at 0.63 V.

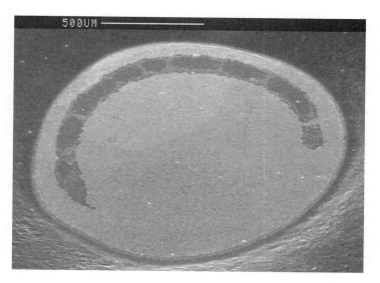

FIGURE 1.13. Delamination of AIROF at the perimeter of a thin-film iridium electrode after overactivation. Note the nonuniform current distribution during pulsing.

AIROF electrodes on silicon microprobe electrodes will also delaminate when overactivated. The delamination may result from excessive thickening of the AIROF, as in the case of iridium wire electrodes, or from activation through the full thickness of the iridium film (typically <200 nm thick). The volume increases at least by a factor of 6 when iridium metal is activated to AIROF, and if the activation advances through the iridium metal to the underlying substrate the large volume change at the interface will cause abrupt delamination. This effect is shown in Figure 1.13 for the activation of a 50-nm-thick iridium film on a 50-nm titanium adhesion layer. The AIROF was activated to a CSC_c of 60 mC/cm^2 and then abruptly delaminated around ~60% of the perimeter of the electrode where the highest current densities are expected. This result also demonstrates the importance of the nonuniform current distribution during pulsing as well as the need to avoid activating the entire thickness of an iridium film. Obviously, the use of an *in vivo* stimulation protocol that is sufficiently aggressive to activate iridium could lead to AIROF delamination. Whether long-term pulsing to the existing safe potential limits given by Beebe and Rose (1988) would continuously activate AIROF to the point of delamination is unknown, but this is a possible cause of the failure shown in Figure 1.11.

Despite the large number of potential applications, and the associated intensive efforts to develop prostheses that would require high charge capacity electrodes, only limited information on the stability of these materials for stimulation at high charge injection levels has been reported. In fact, *in vitro* stability data for iridium oxide at charge injection levels appropriate for cortical neural prostheses extends to the few reports summarized as follows:

- AIROF has charge injection limits of 1 mC/cm^2 (cathodal), 2.1 mC/cm^2 (anodal), and ~3.5 mC/cm^2 (cathodal with anodic bias) using potentials transients as a criterion for stability (Beebe and Rose, 1988). The electrodes were pulsed for a maximum time of 15 minutes to establish these limits.
- AIROF subjected to 0.25 mC/cm^2 biphasic current pulses (0.2-ms pulse width) for >1 million pulses shows no significant change in electrochemical characteristics or physical deterioration (Tanghe *et al.*, 1990).
- AIROF subjected to 1.2 mC/cm^2, 0.2 ms pulses, 50 pps for 30×10^6 pulses (167 hr) loses <10% of its initial capacity (Meyer *et al.*, 2001).

All these measurements were made at one pulse width, 0.2 ms, in electrolytes without biomolecules and did not extend to more than a few weeks of pulsing. Occasional total loss of charge injection capacity for AIROF and large changes in impedance were reported by Loeb *et al.* (1995) and were associated, in part, with handling, cleaning, and storage of the electrodes. *In vivo* evaluation of AIROF stability is limited to anecdotal observations during studies of safe stimulation with respect to tissue (Agnew *et al.*, 1986; McCreery *et al.*, 2000), and one long-term chronic study in guinea pig cortex (Weiland and Anderson, 2000). In the latter study, the stability of AIROF on silicon microprobes was evaluated for implantation periods of several weeks during which electrodes were pulsed for 10 hr (five 2-hr daily sessions) cathodally (using anodic bias) or anodally. Impedance changes in the AIROF electrodes were noted and sites on explanted electrodes that had been pulsed were often covered with an organic film whereas adjacent unpulsed sites (~150 μm separation) were uncoated. No long-term *in vitro* or *in vivo* data are available on the charge injection limits of AIROF using pulse widths that might be more suited to auditory brainstem (0.05 ms) or retinal (1 ms) stimulation.

1.3.3. REACTIONS OF NEURAL TISSUE TO STIMULATING ELECTRODES

1.3.3.1. Mechanisms of Stimulation-Induced Tissue Injury

Since 1970, the NIH has funded the development of electrode systems and studies defining protocols for safe stimulation of the cerebral cortex (with electrodes implanted on the cortical surface and with intracortical microelectrodes), cochlear nucleus, spinal cord, and peripheral nerves. Most of this work was performed by the Neurological Research Laboratory of the Huntington Medical Research Institute (HMRI), Pasadena, CA. These studies included neurophysiologic and histologic (light and electron microscopic and immunocy-tochemical) evaluations. Studies have also been performed to identify the mechanisms of electrically induced damage using disc electrodes on the brain surface. Using ion-selective microelectrodes for Ca^{2+} and K^+ these studies found evidence of a transmembrane flux of Ca^{2+} and a large increase in extracellular K^+ during damaging electrical stimulation of the cortex, either or both of which could be related to the histologic changes (shrunken and hyperchromic neurons) observed in the same animals (McCreery and Agnew, 1983). It has also been observed, using pO_2 microprobes, that ischemia was not a factor in neural injury during a damaging stimulus protocol. In nerves, autoregulation maintained pO_2 at a normal level, and pO_2 actually increased in the cerebral cortex (McCreery et al., 1990). In other studies, it was demonstrated that MK-801, a noncompetitive NMDA receptor antagonist, protected neurons during a prolonged (8 hr) intense stimulation of the cat's cerebral cortex using surface electrodes. In peripheral nerves (cat peroneal and sciatic nerve) high-frequency pulsing (50–100 Hz) delivered continuously for 8 hr results in irreversible damage to the medium-size myelinated fibers (2–5 μm diameter) axons (Agnew and McCreery, 1990; McCreery et al., 1992). However, the axonal degeneration could be completely abolished by perfusion with a local anesthetic (2% lidocaine) at the electrode site. These and other data have supported the premise that the axonal or neuronal injury in both the CNS and peripheral nervous system is related to the neuronal activity induced by prolonged, high frequency electrical stimulation (Agnew et al., 1990a,b; McCreery et al., 1997). Further studies demonstrated the feasibility of chronic intracortical stimulation of the sensorimotor cortex. (Agnew et al., 1986; McCreery et al., 1986).

1.3.3.2. Physiologic Effects of Intracortical Microstimulation

Studies have been conducted to examine the excitability changes induced in cerebral cortical neurons during prolonged microstimulation with a spatially dense microelectrode array. Arrays of 16 (AIROF) microelectrodes with interelectrode spacings of approximately 380 μm were implanted chronically into the postcruciate gyrus of cats. Beginning 90 days after implanting the arrays, neuronal responses characteristic of single-pyramidal-tract ax-ons (ULRs) were recorded in the medullary pyramid. Seven hours of pulsing of individual electrodes at 50 Hz and at 4 nanocoulombs per phase (nC/ph) induced little or no change in the ULRs' electrical thresholds. The thresholds also were quite stable when 4 of the 16 microelectrodes were pulsed on each of the 14 consecutive days. However, when all 16 microelectrodes were pulsed for 7 hr at 4 nC/ph, the threshold of approximately half of the ULRs became elevated. Recovery of excitability required 2–18 days. Prolonged sequen-tial (interleaved) pulsing of the 16 microelectrodes induced less depression of excitability than did simultaneous pulsing, but only when the stimulus amplitude was low (12 μA, 1.8 nC/ph). Stimulation at a higher amplitude (15 nC/ph) induced much more depression of excitability. These findings imply that multiple processes mediate the stimulation-induced depression of neuronal excitability. The data also demonstrate that the depression can be reduced by employing a stimulus regimen in which the inherent spatial resolution of the array is maximized (sequential pulsing at an amplitude in which there is minimal overlap of the effective current fields) (McCreery et al., 2002).

When the stimulation frequency was greater (100 or 500 Hz), the 7 hr of continuous stimulation always caused a persisting increase in the neuron's electrical threshold, and the

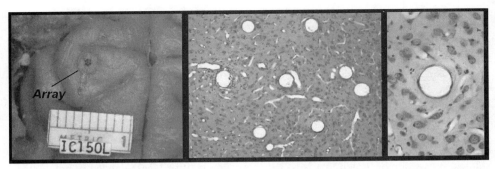

FIGURE 1.14. Seven-electrode HMRI array implanted in cat postcruciate. (Photograph courtesy of Huntington Medical Research Institute)

FIGURE 1.15. Histological section showing electrode tracks of Figure 1.2. (Photograph courtesy of Huntington Medical Research Institute)

FIGURE 1.16. Single track of Figure 1.3. (Photograph courtesy of Huntington Medical Research Institute)

severity of the increase in threshold was greater at higher pulsing rates. These frequencies exceed those needed for efficient stimulation of the visual cortex, but do not result in histologically detectable injury.

1.3.3.3. Mechanically Induced Injury from Intracortical Microelectrodes

To address the issue of mechanically induced damage during and after the implantation of intracortical microelectrodes, studies have focused on the physiological and histologic evaluation of the optimum tip configuration and insertion speed of the iridium microelectrodes comprising arrays of 7–16 electrodes. It has been demonstrated that insertion trauma, including microhemorrhages, are markedly reduced by the use of rather blunt (12-µm diameter), rather than sharp (1-µm diameter), activated iridium tips inserted at a speed of approximately 1 m/s. Figure 1.14 shows a seven-electrode array that had been implanted into a cat's left postcruciate gyrus. Figure 1.15 shows a histologic section through the seven tracks. Note the markedly, thinly ensheathed electrode tracks and the prominent neovascularization that invariably accompanies multielectrode sites. Figure 1.16 shows a high magnification of a single-electrode track. Note the few slightly flattened but otherwise normal-appearing neurons next to the track.

1.3.3.4. Histologic Evaluations of Microstimulation of the Cat Sensorimotor Cortex

To examine the effect on neural tissue of electrical stimulation directly, AIROF metal wire electrodes, implanted in cat cortex, were pulsed using a range of stimulation parameters. The implanted electrodes were pulsed with 50 Hz, 26.5 A, 150 s/ph (4 nC/ph) continuously for 7 hours on 1–14 successive days and the resulting effects on the tissue were evaluated by light microscopy (Figure 1.17). The H&E stain shows the tip of a representative pulsed electrode track, showing normal-appearing neurons within 20–25 µm of the track. Even when the stimulus frequency was increased to 500 Hz (26.5 A, 150 s/ph, 4.0 nC/ph), the neurons adjacent to the tip sites and the adjacent neuropil appeared normal, with the exception of the usual vascular hyperplasia.

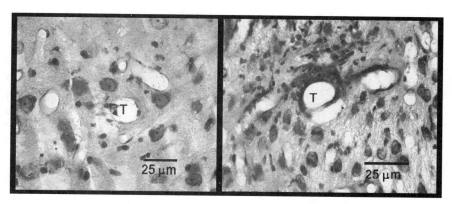

FIGURE 1.17. Electrode track resulting from 4.0 nC/ph. (Photograph courtesy of Huntington Medical Research Institute)

FIGURE 1.18. Electrode track resulting from 40 nC/ph. (Photograph courtesy of Huntington Medical Research Institute)

When the stimulus intensity was increased to 40 nC/phase and the charge density to 2 mC/cm^2, some evidence of tissue injury began to appear. The neuropil within 50 μm of the electrode tip appears somewhat spongy, and many of the nearby neurons are ragged and stellate (Figure 1.18). However, even at this very high stimulus intensity, the histologic changes extended no more than 50 μm from the electrode tip.

In summary, an established universal electronic/neural tissue interface transducer, i.e. a universal stimulating electrode, capable of chronic implantation and safe chronic electrical stimulation of neuronal tissue remains elusive at best. Capacitive-type electrodes based on insulating metal oxides lack sufficient charge capacity for the desired electrode tip sizes. Safe charge injection limits by faradaic means, using noble metals alone, are too low to be of physiological significance. AIROF is an electrochemically activated hydrated oxide film with promise to meet anticipated charge injection needs. However, the long-term stability of AIROF in the implanted environment is unknown. Chronic electrical stimulation of neuronal tissue, within the CNS, seems to be feasible, although the best data are based on studies of limited duration. In order to realize neural prostheses that include hundreds, or thousands, of subminiature stimulating electrodes technological advances beyond the current state of the art will be needed. In this regard, promise is shown by the micromachined silicon, and other high-density, electrode systems.

1.4. TRANSCUTANEOUS COUPLING OF POWER AND TELEMETRY

1.4.1. INDUCTIVE LINKS

For an implanted neural prosthesis, all power and communication is provided through a transcutaneous inductive link. In its simplest form, the inductive link is made from two coils, one contained within the neural prosthesis itself, and one contained within the extracorporeal transmitter, although some neural prosthesis systems use multiple coils. Although the link appears to be physically simple, i.e. two coils of wire, specification and design of the link is highly parametric and usually involves multiple conflicting geometric and electrical constraints. The apparent physical simplicity is deceptive, and it is not uncommon for the neural engineer to find the process of designing the inductive link to be a frustrating experience, depending heavily on a trial-and-error approach.

Transcutaneous links specific to neural prostheses have been analyzed by Donaldson and Perkins (1983), Galbraith *et al.* (1987), and Heetderks (1988), among others. Yet no one "cookbook" design procedure has evolved from this work. Probably the definitive references on coupled coils were written by Grover (1946) from work that he started at the beginning of the 20th century. For the special case of weakly coupled inductive links, Troyk and Schwan (1995), on the basis of Grover's work, presented a simplified analysis that lends itself to simple analytical solutions. Yet, each neural prosthetic design seems to defy attempts to generalize link design. Even within the electronic transformer design industry, attempts to write computer programs to automatically create the transformer and inductor design have been largely unsuccessful.

Terman, in his classic book, *Radio Engineers Handbook* (McGraw-Hill, 1943), presents comprehensive methods for designing coupled-coil systems. Indeed, most comprehensive inductive link designs use a combination of Terman's and Grover's techniques. However, much of this early information is presented in the form of design tables and graphs that display loci of coil designs. As such, the engineer often finds their use tedious when trying to adapt to modern spreadsheet-type design procedures. However, with some persistence, use of the Terman and Grover texts yield models and design formulae that are remarkably accurate for many inductive link designs commonly encountered for neural prostheses. A full discussion of these methods is beyond the scope of this book. However, there are some fundamental principles appropriate for this discussion, and they should provide the student with a first-order appreciation for the link design process.

Design of an inductive link can be broken down into several steps:

1. Specify the required implant power supply voltage and load current requirements.
2. Determine required communication data rates so that an operating frequency may be selected.
3. Identify physical constraints of both the transmitter and the implant coils so that limits on wire size, type, and number of turns may be determined. Calculate approximate range of values for coil inductances and coil equivalent series resistances.
4. Specify the range of physical separation for which the link must operate, and calculate the expected range for the link coupling coefficient.
5. Determine the expected limits of performance on the basis of models of the transmitter and implant resonant circuits.

For each step, it is necessary to use multiple design methods in order to combine together the computations of coil inductance, coil loss, coil self-resonance, link coupling coefficient, and resonant circuit responses.

Understanding the implant power supply needs, as well as the allowable transmitter power source, is an essential starting point for any inductive link design. Often, the initial expectations of performance are so unrealistic that any further consideration of the design must be accompanied by a change in the overall approach so that the performance expectations may be brought into the feasible range. Using even simplistic assumptions about how efficient the transfer of power from the transmitter to the implant can be allows the neural engineer to assess the difficulty of the design task.

Choice of the frequency of operation should be based on the combined need for small coil sizes, and adequate system bandwidth for data telemetry, especially if a simple two-coil link design is chosen for which the single link serves the dual function of power and data transmission. In general, one should choose the lowest frequency possible because coil losses dramatically increase with increasing frequency. The typical range of operating frequency for neural prosthesis systems is 100 kHz–50 MHz. At the lower end of this range, coil voltages are reduced for a constrained value of inductance because the coil voltage is directly proportional to frequency. At the upper end of this range, coil losses become severe owing to the high-frequency induced current crowding within the coil conductors.

Coil inductances are best computed using models for round flat coils as presented by both Grover and Terman. Although models for rectangular and even oddly shaped coils can be found, they offer little to a first-order model for the link. It is important to not ignore the resistive losses for the coils, because coil losses are usually the limiting factor in transfer of energy from the transmitter to the implant coil. These must be considered as frequency-dependent using not only skin-effect models, but also more importantly, proximity effect models appropriate for the desired frequency of operation. Skin-effect is a first-order model for the nonuniform distribution of current in a conductor at high frequencies; current tends to crowd near the surface of a round wire owing to eddy currents induced within the conductor. Proximity effect is an additional mechanism of current crowding caused by the proximity of one winding next to another in a magnetic field. For most inductive link designs, the proximity effect dominates the skin effect, and unfortunately proximity effect is more difficult to model. Again, a comprehensive treatment of this subject has been described by Terman. In general, it is fairly easy to design systems for operating frequencies below 1 MHz. Between 1 and 10 MHz, first-order models for coil inductances and resistances become inaccurate. Above 10 MHz, distributed computer-based models must be used.

Based on the expected separation between the two coils of the link, a coupling coefficient can be computed, and both Terman and Grover describe various methods for this computation. The coefficient of coupling, k, is a dimensionless quantity determined entirely by the geometry of the two coils and the physical separation between them. Its value $(0 < k < 1)$ is a measure of the amount of flux produced by the transmitter coil that links the implant coil. It is electrically defined as follows:

$$k = \frac{M}{\sqrt{L_p L_s}}, \tag{1.1}$$

where M = the mutual inductance between the two coils, L_p = the transmitter (primary) coil inductance, and L_s = the implant (secondary) coil inductance.

In determining k, the troublesome part of the computation is that of the mutual inductance, M. Computations of M are presented by Grover. Because k is a measure of the percentage of the magnetic flux produced by one coil that links the other coil, as the two coils are brought together, the coupling increases until it reaches a theoretical maximum of 1 (in practice $k = 1$ is not achievable because of the physical impossibility for one coil to occupy the same physical space as the other). One might think that the best performance for coupling power over a link is obtained for the maximum attainable k. Most inductive links are designed to be resonant in both the transmitter and implant circuit. Consequently, it can be shown that the secondary current reaches a maximum value at the coupling for which the reflection of the secondary resistance into the primary circuit is equal to the primary circuit resistance. This special coupling value is called the critical coupling. The critical coupling coefficient, k_c, has the approximate value

$$k_c = \frac{1}{\sqrt{Q_p Q_s}} \tag{1.2}$$

where $Q_p = \omega L_p / R_p$ (i.e. quality factor) for the primary circuit, and $Q_s = \omega L_s / R_s$ for the secondary circuit, with $R_p = $ equivalent resistance of the primary circuit, and $R_s = $ equivalent resistance of the secondary circuit.

Early designs for neural prostheses used the advantages of critical coupling as a method for link design in order to maximize the implant power supply voltage for given implant load and link spacing. The critical coupling method remains important as a first-order design approach for simple neural prostheses. Recently, for more sophisticated implant designs, operating the link at, or near, critical coupling has become less important for two reasons: 1. Modern neural prosthesis designs use active VLSI power supply regulators to compensate for changes in separation between the two coils. In some designs, the secondary circuit can even be retuned to regulate the implant coil voltage. 2. For extremely small coefficients of coupling ($k < 0.01$), characteristic of miniature implants such as injectable microstimulators, it becomes impractical to consider operating near the critical coupling point because the reflection of the implant load into the primary circuit becomes extremely small. For these systems, the only practical solution is to maintain extremely large values of transmitter coil currents (Troyk and Schwan, 1995). Nevertheless, obtaining a reasonably accurate estimate of the coupling coefficient is necessary in order to implement a circuit model for any coupled-power system.

Once the models for the primary, secondary, and coupled circuit are determined, it is easiest to use these in a computer simulation (e.g. SPICE) of the link rather than pursue a closed-form solution for the current-transfer ratio between the transmitter coil and the implant coil. The simulation model allows for rapid manipulation of the model parameters, and for assessment of the likely operation due to load variations, coil spacing, and transmitter coil deformations. One can also model the effects of metal objects (such as a wheelchair) that may come near the transmitter coil. Placement of metallic objects within the magnetic field increases the eddy current losses, and can be modeled as an increase in transmitter coil resistance. Usually a few simple measurements will give approximate values for the increase in primary circuit resistance and decrease in primary circuit inductance, caused by proximity of metal to the link.

1.4.2. GENERATING THE TRANSMITTER COIL CURRENT

Ultimately, an increased ampere–turns product of the transmitter coil must compensate for low coupling coefficients and inefficiencies in energy transfer, and it is not unusual for the required transmitter currents (necessary for maintaining powering of the implant over the expected range of coupling) to become uncomfortably high. Throughout the past 50 years, various approaches have been tried for generating large (0.2–2 A) currents in transmitter coils at radio frequencies while using lightweight batteries. Troyk and Schwan

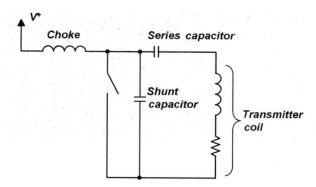

FIGURE 1.19. Topology of the Class E converter. The single transistor switch is closed at precisely the correct point in the resonant cycle so that the switching losses are reduced to near-zero.

(1992a,b) developed a closed-loop Class E transmitter driver that is capable of producing large currents (amperes) in the transmitter coil, at radio frequencies, with extremely high efficiency. In their technique, the Q of the transmitter circuit is raised to an unusually high value, 150–300. Use of the high transmitter Q reduces the transmitter power supply consumption to very low levels, even for large resonant coil currents. The basic Class E topology (Sokal and Sokal, 1975) is unique and well suited to this task because driving large currents into high-Q coils has historically been problematic when using classical push–pull (Class B) type drivers. Although one can reduce the loss of a conventional push–pull series RLC circuit by using low-loss-dielectric capacitors and litz wire for the transmitter coil, often the power loss is simply shifted from the resonant RLC circuit to the push–pull drivers. The Class E topology resolves this issue by utilizing a multifrequency resonant network. The power loss within the single-transistor active driver can be sufficiently reduced so that the low resistance of the high-Q transmitter coil becomes the dominant loss. In fact, for Class E transmitters it is desirable to use as large a circuit Q as possible.

The basic topology of the Class E oscillator is shown in Figure 1.19. The combination of the series capacitor, the shunt capacitor, and the transmitter coil forms a multifrequency network. The impedance vs. frequency plot of the multifrequency network shows a double resonance: one at a series resonant frequency, and the other at a parallel resonant frequency. Between these two frequencies can be found the Class E frequency. At this special frequency, the loss in the switching element of the converter becomes very low. The switch (usually a power FET) across the shunt capacitor is turned on at a strategic point in the resonant cycle for which the switch voltage and the derivative of the switch voltage are zero.

One serious problem with Class E circuits is maintaining operation at the precise Class E frequency. For each circuit, there is only one combination of operating frequency, duty cycle, and coil-Q that will produce the Class E conditions. Automatically maintaining the Class E frequency, for transcutaneous coil drivers, has been demonstrated through the use of voltage-mode (Zierhofer and Hochmair, 1990) and current-mode (Troyk and Schwan, 1992b, 1993) feedback. These techniques convert the Class E converter into a power oscillator whose frequency is established by the closed-loop control circuit and the characteristics of the multifrequency network.

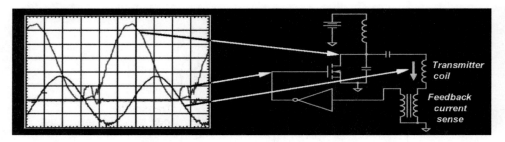

FIGURE 1.20. Waveforms associated with the operation of the current-mode Class E transmitter. The FET switch is turned on at a unique point in the resonant cycle at which the FET drain voltage is near zero, thus minimizing switching losses.

Operation and typical waveforms of a closed-loop Class E converter with current-mode control can be seen in Figure 1.20. Note that when the gate of the FET is turned on, the FET drain voltage is at the negative crest of the sinusoid. This point corresponds to the coil-current zero crossing. Because the FET is only turned on for a brief portion of the cycle, FET losses can be very low. A properly designed Class E converter can operate with efficiencies greater than 95%.

The Class E power-oscillator approach solves another problem associated with driving large currents into the transmitter coil. Mechanical deformations of the transmitter coil, or proximity to metal objects will shift the inductance and resistance of the transmitter coil. Using a classical high-Q push–pull approach, the driving current will rapidly drop as the self-resonant frequency of the RLC output circuit shifts away from the frequency of the input drive signal. For the Class E circuit, without closed-loop control, shifting of the Class E frequency away from that of the input drive signal will result in almost instantaneous destruction of the transistor driver. Using the closed-loop Class E power oscillator approach, the frequency of the transmitter automatically shifts to compensate for the changes in the transmitter coil characteristics, maintaining the low-loss Class E operation. The result is a transmitter in which transmitter-coil currents of several amperes can be produced with relatively low transmitter power-supply currents. Low transmitter supply currents are important because many neural prosthesis applications require portable extracorporeal transmitters, powered by batteries. A disadvantage to using the closed-loop Class E power oscillator is that the transmitter frequency is not constant. Because the implanted neural prosthesis's internal digital clock is typically derived from the transmitter frequency, the implant's clock frequency is also variable. Therefore, it is necessary to use the variable frequency of the transmitter as the overall system's master clock. Several neural prosthesis systems currently under development use the closed-loop Class E approach for the extracorporeal transmitter for frequencies ranging from 100 kHz to 10 MHz.

1.4.3. DATA TELEMETRY

With few exceptions, it is usually necessary to provide frequent commands to the neural prostheses in order to control the stimulus parameters. In contrast to a pacemaker, whose operation is autonomous, often the stimulus pulse amplitude and duration for a neural prosthesis are specified on a pulse-by-pulse basis. For proposed visual prosthesis systems, the required data rates can be as high as 1–2 Mbits/second. It is highly desirable to use the inductive link that has been optimized for power transfer to provide the forward command telemetry. This implies modulation of the transmitter current. Among the obvious choices for modulation techniques are amplitude-shift keying (ASK), frequency-shift keying (FSK), and phase-shift keying (PSK). Although PSK has not been extensively used for forward command telemetry of neural prostheses, ASK and FSK have both been investigated.

Forward command telemetry is accomplished by modulating the transmitter coil current and including demodulation circuitry in the neural prosthesis coil interface circuitry. Modulation of a large transmitter coil current is difficult, no matter what circuit topology is used for the transmitter. Often, the addition of modulation circuitry adds considerable loss to a particular transmitter circuit design. For Class E transmitters the high-Q transmitter coil that facilitates the design of the Class E transmitter resists rapid changes in its coil current. Two techniques have been described for ASK modulation of the Class E converter. Ziaie *et al.* (1997) and Zierhofer and Hochmair (1990) report the modulation of the transmitter's D.C. power supply in order to produce ASK of the transmitter coil current. The response time of the converter is dictated by the multifrequency network's Q and the size of the converter's input choke. Typically this method of ASK modulation produces relatively low data rates. Because within the D.C. power supply, the modulating element must carry the transmitter's full power supply current, additional power losses result. Schwan, Troyk, and Loeb (1995) report a near loss-less shifting of the Class E converter's operating point via the current mode closed-loop control circuit in order to accomplish a combined ASK/FSK modulation of the transmitter coil current. The response time of this method also depends on the size of the converter's input choke and the multifrequency network's Q; however, the additional power losses are minimal. The amplitude modulation can be used to describe two states of the transmitter current that can be digitally modulated by Manchester, pulse-width, or pulse-position encoding. For all of these ASK techniques, the level of modulation is on the order of 5% of the peak carrier, and one is confronted with bandwidth limitations caused by the natural response of the transmitter's high-Q circuitry. Therefore these techniques are used for data rates which are not much higher than 1/10 the transmitter frequency.

Schwan, Troyk, Heetderks, and Loeb (Schwan *et al.*, 1995; Troyk *et al.*, 1997) have reported an alternative to low-level Class E transmitter ASK modulation. Known as the "suspended-carrier" method, it permits the transmitter to be placed in a "suspended" state during which all of the energy contained within the converter's multifrequency network is contained in the resonant capacitors, with none in the inductor (transmitter coil), whereas in the suspended state the energy loss is minimal because no current flows. Resumption of normal operation can be near-instantaneous. In contrast to simple ASK, suspended-carrier modulation produces 100% modulation of the transmitter carrier. Using this technique the transmitter current can be turned on for as little as one cycle, and can be turned off for an arbitrary period of time; however, in order to minimize transient responses within the implant's coil it is advantageous to turn the transmitter off for whole cycles only. Suspended-carrier modulation permits forward telemetry data rates that approach the transmitter frequency, although in practice an upper data rate is constrained to about one-fourth of the transmitter frequency, owing to limitations in the implanted neural prosthesis's demodulator circuitry. Detection of the 100% modulation within the implant's coil interface section can be accomplished with minimal integrated circuitry, and does not depend on the Q of the implant coil. A disadvantage of the suspended carrier method lies in the fact that the transmitter is turned off for approximately 50% of the time, thus reducing the amount of energy transferred to

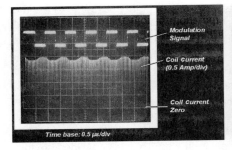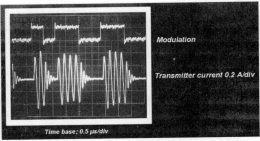

FIGURE 1.21. Oscilloscope photograph of ASK modulation of a 2-MHz Class E transmitter (left), compared to suspended-carrier modulation (right). Note the slow response time of the ASK transmitter coil current characterized by state changes occurring over several transmitter cycles. In contrast, the 100% suspended-carrier modulation occurs almost instantaneously on a cycle-by-cycle basis. (Courtesy of Illinois Institute of Technology)

the implant over the inductive link during each data-bit transfer. To compensate for this off-time, the transmitter current amplitude has to be increased, with associated efficiency penalties. Figure 1.21 shows oscilloscope waveforms for an ASK system, as compared to a suspended-carrier system.

An FSK method of modulating Class E transmitters is currently under development (EMBS 2003), and may combine rapid modulation rates, similar to those obtained for suspended carrier modulation, with constant transmitter on-times, albeit with shifting frequency. Reliable data demodulation, within the implant, can be achieved for a ±5% frequency shift about the center frequency.

1.5. TECHNIQUES FOR DRIVING ELECTRODES

1.5.1. A MODEL FOR A STIMULATING MICROELECTRODE

In order to design an electronic circuit for driving stimulating electrodes, it is useful to consider the electrode as a transducer, and develop an equivalent circuit model. A commonly used model for a metal stimulating microelectrode *in vivo* is that shown in Figure 1.22.

The model in Figure 1.22 includes impedances for the tissue and the counterelectrode, in addition to that of the stimulating microelectrode. Usually, the counterelectrode has an active surface area that is considerably larger than the microelectrode; therefore, it is a reasonable approximation to ignore the equivalent circuit of the counterelectrode relative to the microelectrode. The tissue impedance can be combined with the equivalent circuit of the microelectrode, resulting in the simplified equivalent circuit of Figure 1.23.

In this model, the leakage resistance is often ignored when deriving a first-order model, and the equivalent circuit comprises a combined access resistance and an electrode pseudocapacitance. Therefore, for purposes of electrode driver design, this series RC network is used to design the biphasic constant current driver.

To demonstrate the voltage waveform that would result if driving this idealized model with a biphasic constant-current source (assuming a very large leakage resistance), the oscilloscope waveform of Figure 1.24 is presented. In this figure it can be seen that for zero initial voltage on the capacitor, at the leading edge of the cathodic phase there is an

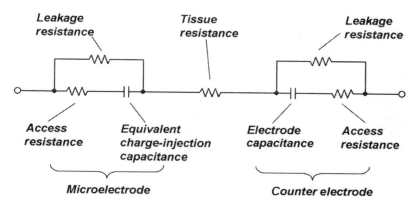

FIGURE 1.22. Equivalent circuit model for stimulating microelectrode and counterelectrode within tissue.

immediate drop in voltage caused by the *IR* drop across the access resistance of the model. Following this initial edge, the voltage ramps linearly negative according to the equation $dV/dT = C/I$, so that for a constant current, the capacitor voltage linearly increases in the cathodic direction. At the onset of the anodic phase, an *IR* drop and charging of the capacitor is once again seen, with the capacitor charging in the anodic direction. For a fixed capacitor and resistor, as well as for a balanced biphasic current, at the end of the anodic phase there should be no net charge left on the electrode, and for the model waveform in Figure 1.24, at the end of the anodic phase the electrode voltage is seen to return to 0.

Actual microelectrodes measured *in vivo* show waveforms similar to, but not identical to, the idealized model waveform of Figure 1.24. Figure 1.25 shows a computer-based oscilloscope voltage waveform for an AIROF microelectrode implanted into the visual cortex that has an active area of 500 μm². The electrode is driven with a constant-current biphasic pulse of 30 μA for a charge density of 1.5 mC/cm². Note that the waveshape during the anodic phase does not mirror that of the cathodic phase. For this particular voltage waveform, the current driver was highly balanced, with the ratio between the cathodic and anodic currents close to 1. The nonlinear waveshape during the anodic phase is a consequence of rate variations in the oxidation of Ir^{3+} to Ir^{4+} as compared to the reduction of Ir^{4+} to Ir^{3+} during the cathodic phase. It can also be seen that at the end of the stimulation,

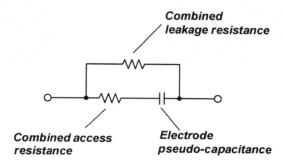

FIGURE 1.23. Simplified equivalent stimulating electrode circuit model.

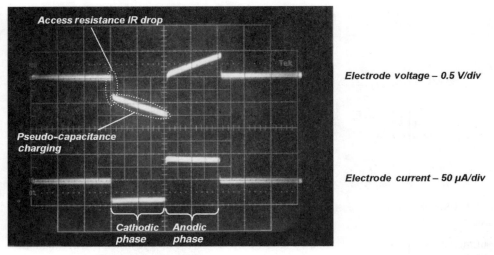

Electrode voltage – 0.5 V/div

Electrode current – 50 µA/div

Timebase – 100 µs/div

FIGURE 1.24. Oscilloscope waveform from biphasic current driver and idealized RC model of a stimulating microelectrode. Leakage resistance is assumed to be infinite.

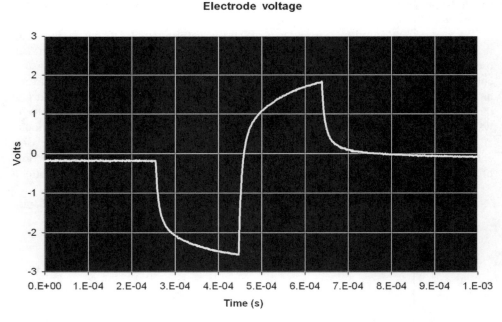

FIGURE 1.25. Computer-based oscilloscope waveform of the voltage measured on an AIROF intracortical microelectrode. Stimulation current was an *in vivo* constant-current (30-µA) biphasic pulse. Charge density was 1.5 mC/cm.

a residual anodic bias remains on the electrode. By measuring the access resistance IR drop and the subsequent rise in voltage, one can estimate the value of the access resistance and the electrode pseudocapacitance. For the electrode as measured in Figure 1.23, these values are 80 kΩ and 8 nF, respectively.

1.5.2. *IMBALANCES IN ELECTRODE CURRENT WAVEFORMS*

High-charge capacity electrodes and, in particular, ones with AIROF films require well-prescribed constant-current driving conditions. The most commonly accepted optimal driving waveform for stimulating microelectrodes is the constant-current charge-balanced biphasic waveform that has the general shape shown in Figure 1.3. It is instructive to consider why this particular waveshape is used and what the implications would be for deviations from a balanced condition.

As was pointed out in Section 1.3.1, charge injection into the biological tissue via a metal electrode involves changing the mode of current conduction from electron flow to ionic flow, i.e. within the biological tissue current is carried by ionic species. In order for the electrode to facilitate this transition without suffering deterioration it is required that the electrochemical reactions taking place at the electrode–tissue interface be reversible, and nondamaging to the electrode and to the tissue.

Consider an AIROF electrode driven under perfectly balanced conditions with a cathodal-first biphasic pulse. During the cathodal phase, charge is injected almost exclusively by the reduction reaction of Ir^{4+} to Ir^{3+}. Therefore, during the cathodal phase there is some depletion of Ir^{4+} that must be replenished during the subsequent anodic phase by oxidation of Ir^{3+}. If the charge injected during the anodic phase is less than that injected during the cathodic phase, then the electrode will be left with a net cathodic bias. For repetitive cycles, the cathodic voltage bias would theoretically rise (bias voltage would become more negative) until reaching a maximum limit determined by the electronic circuit that is driving the electrode (compliance supply voltage). In practice, for small imbalances, the cathodic bias becomes limited by the onset of chemically driven reduction reactions in the interpulse period—such as oxygen reduction—which attempt to reestablish the equilibrium potential of the electrode.

If an imbalance is present in the anodic direction, a similar situation occurs in which each cycle leaves an increasing anodic bias on the electrode. Again, for a limited imbalance, counteracting chemical reactions can limit the positive polarization of the electrode. If the polarization is extreme, however, i.e. greater than about +0.8 V Ag|AgCl, proteins and water can be oxidized, with deleterious effects on the electrode and the surrounding tissue. The situation is similar for platinum electrodes, with protein and water oxidation occurring above +0.8 V. Interestingly, the use of an anodic bias is beneficial for an AIROF electrode in that it increases the maximum charge that can be injected cathodally without polarizing the electrode to the potential for water reduction. The anodic bias increases the overall oxidation state of the AIROF, which provides more Ir^{4+} for reduction during a cathodal pulse. Oxidized AIROF, i.e. AIROF in the predominantly Ir^{4+} state, is also more electronically conductive than unbiased AIROF, which aids in facile charge injection. The optimum anodic bias depends on the type of stimulation waveform being used, but is typically between +0.4 and 0.8 V Ag|AgCl.

It is easy to calculate the average value of an imbalanced biphasic current, based on the percentage imbalance in the electrode driving circuit. Suppose a 20-µA cathodal-first biphasic pulse, with a pulse duration of 500 µs/phase at a frequency of 200 Hz has an imbalance of 1% between the two phases (in either the anodic or cathodic directions). Then for each cycle a net imbalance of 0.2 µA exists for a duty cycle of 10%. Therefore, an average current of ±20 nA will flow through the electrode–tissue system. This average current is undoubtedly large enough to cause an electrode bias, limited only by one of the reactions described above. The significance of this discussion is that electronic circuits used to drive stimulating microelectrodes either need to be unrealistically balanced to avoid damage to the electrode or tissue, or the average imbalance current needs to be controlled or blocked.

A common blocking method for dealing with the imbalance in biphasic waveforms is to place a coupling capacitor in series with the electrode driver and the electrode. Presently, there are insufficient long-term experiences with chronically implanted AIROF microelec-trodes to access the true effectiveness of using a coupling capacitor, especially when one considers that discrete capacitors do have finite leakage currents. In most cases it is likely that a combination of limits in electrode reactions, limits in the interpulse interval elec-trode driver output voltage, and limits on the absolute magnitude of the coupling capacitor leakages restricts the magnitude of the bias currents and voltages to nondestructive values. However, the relative contributions of these effects have not been quantified.

1.5.3. CONSTANT-CURRENT ELECTRONIC CIRCUITS

Constant-current circuits are usually implemented using current mirrors. The sim-plest type of current mirror is shown in Figure 1.26. For this circuit, two bipolar junction

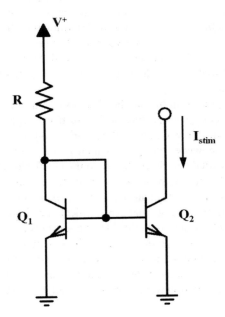

FIGURE 1.26. Current mirror composed of two bipolar junction transistors.

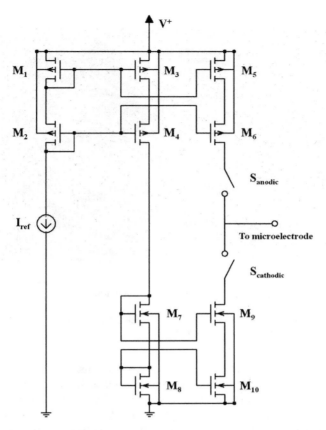

FIGURE 1.27. Practical current mirror electrode driver circuit using CMOS transistors.

transistors share a common base voltage, V_{be}. The collector current through diode-connected Q_1 is $(V^+ - V_{be})/R$. Because the base–emitter voltage of Q_2 is the same as that of Q_1, then to a first-order approximation, the collector current of Q_2 will equal the collector current of Q_1. It can therefore be said that the current in Q_2 mirrors the current in Q_1. If the collector of Q_2 is connected to an electrode as the cathodic current source, then varying either V^+ or R can control the cathodal phase of the electrode current. For this approach to function as intended, the currents through Q_1 and Q_2 need to be independent of their collector voltages, i.e. the transistor characteristics curves of I_c vs. V_{ce} need to be flat. Variations on this topology include replacing the resistor R with an active constant-current source, as well as changing the area of Q_2 relative to Q_1 so that current multiplication can be accomplished.

Because of the proliferation of CMOS fabrication processes, relative to bipolar processes, it is common to use a MOS version of current mirror circuit topology as electrode drivers. A MOS configuration for the current mirror is shown in Figure 1.27 and is based on the assumption that identical transistors whose gate-to-source voltages are identical will mirror identical drain-to-source currents. In this circuit, M_1 and M_2 are connected to a reference current generator in a cascode configuration. M_3 and M_4 mirror the reference current

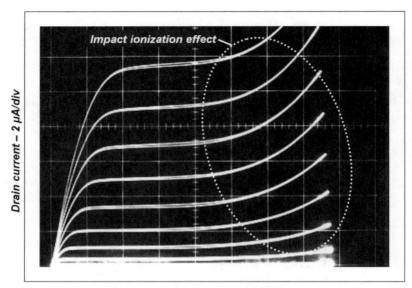

FIGURE 1.28. n-Channel MOSFET characteristic curves showing impact ionization effect. Note how the current dramatically increases for higher drain-to-source voltages.

and force the mirror of I_{ref} to flow in cascode-connected transistors M_7 and M_8. M_5 and M_6 act as the anodic phase drivers, whereas M_9 and M_{10} act as the cathodic phase drivers. The use of the cascode configuration helps to compensate for nonflat transistor characteristic curves. S_a would be closed for the anodic phase, and S_b would be closed for the cathodic phase. A noticeable advantage of this circuit is that the current through M_3, M_4, M_7, and M_8 is common. Therefore, theoretically, an identical current reference is established for both the anodic and the cathodic drivers, minimizing the degree of imbalance between the phases. Proper operation of the circuit depends on the matching between the transistors, which is usually obtained if all the transistors are on the same silicon substrate. In practice, there is some parametric variation between identically drawn CMOS transistors, and the variation between them depends on the particular process used. Increasing the size of the transistors can minimize these lithographic variations.

Despite the use of the cascode current sources, the currents in the three branches of this circuit will not match because of differences between the transistors' drain-to-source voltages. The reason for this mismatch can be seen in Figure 1.28, in which typical n-channel MOSFET characteristic curves are shown. The sharply increasing drain current that occurs for $V_{\text{ds}} > 5\,\text{V}$ is due to the impact ionization effect in which severe channel shortening causes excess carriers to be created near the drain region of the transistor. During the cathodic phase, the V_{ds} of M_9 will not match that of M_7; similarly, the voltage across M_{10} will not match that of M_8. The voltage mismatches are due to the charging of the electrode capacitance. Consequently, the current in M_9–M_{10} will not match the current in M_7–M_8. Similar mismatches for M_3 through M_6 occur during the anodic phase. The result is that despite a circuit topology designed to match the anodic and cathodic currents, they will not match.

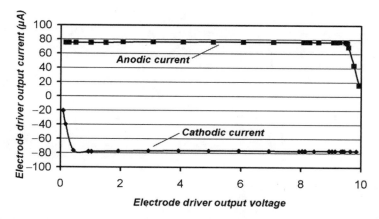

FIGURE 1.29. Anodic and cathodic characteristic curves for a 7-bit digital-to-biphasic constant-current electrode driver stage. Only the highest-bit curves are shown. The curves show that near-constant current is maintained to within ~0.5 V of the compliance voltage supply limits. For this circuit the both the microelectrode and the counterelectrodes are at 5 V during the interpulse interval.

Therefore, it can be concluded that in practice matching the anodic and cathodic currents and keeping the currents constant during the biphasic stimulation pulse are both difficult to achieve. The effects of impact ionization can be reduced by using an extended-drain transistor structure, or through feedback-controlled active cascode circuits that not only compensate for the nonflat characteristic curves but also permit near-constant current operation for electrode voltages that are very close to the compliance voltage supply limits. Anodic and cathodic characteristic curves for a digital-controlled 7-bit biphasic electrode driver can be seen in Figure 1.29. This circuit uses both extended-drain transistors for the output stages and active cascode circuits. Note the extremely flat nature and the close matching between the anodic and cathodic curves. For this circuit, the electrode is maintained at one half of the V^+ (10 V) compliance supply, with the counterelectrode also connected to this same 5 V supply.

1.6. APPLICATIONS

1.6.1. COCHLEAR IMPLANTS

A success story in the development of sensorineural prostheses is the cochlear implant. Dating from the early work of House and Urban (1973) starting in the 1960s, modern multichannel cochlear implants have become a clinically accepted method of restoring hearing, particularly for the postlingual sensorineural deaf. The cochlear implant functions by providing electrical stimulation of the hair cells in a tonotopic manner. A good description of the history and technology of the cochlear implant is given by Spelman (1999). Physically, the cochlear implant consists of an external transmitter/speech-encoding unit that transcutaneously couples power and processed sound information across an inductive link. The implanted neural prosthesis connects to a cochlear electrode that is inserted into the cochlea

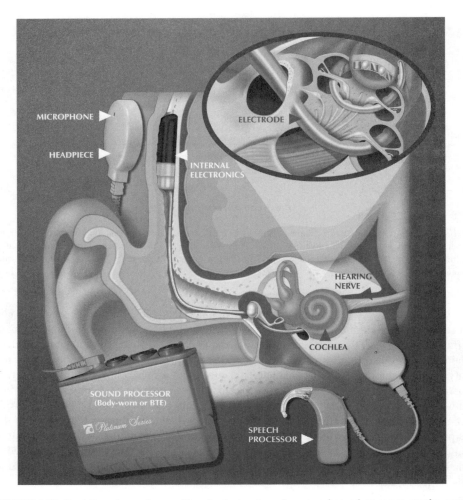

FIGURE 1.30. Depiction of a modern cochlear implant system. An external sound processor couples power and data over a transcutaneous link. In the enlarged view, the spiral cochlear electrode is shown inserted into the cochlea. The small square platinum electrodes can be seen on the inner surface. (Courtesy of Advanced Bionics Corp.)

as shown in Figure 1.30. One type of clinically available implantable device is shown in Figure 1.31.

One of the major challenges for cochlear implants is understanding how sound might be encoded, to best use a limited number of stimulating electrode sites. Clinical devices allow for 8–22 electrode channels, and debate still exists on how to best utilize them in order to most effectively provide the implanted subject with the capability to recognize spoken speech without lip reading. There are three main speech signal-processing strategies employed: compressed–analog (CA), SPEAK, and continuous interleaved sampling (CIS). In CA, the sound signal is bandpass filtered and the output of each filter is assigned to a single

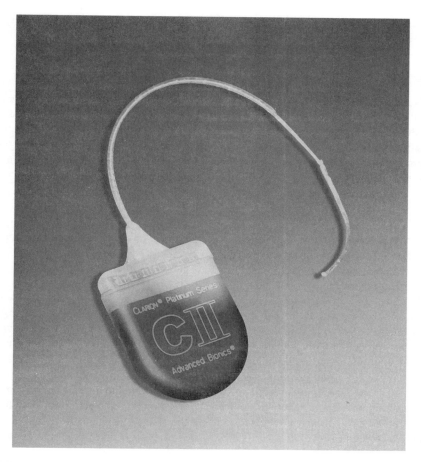

FIGURE 1.31. Modern cochlear implant. The electronic circuitry and implant coil are packaged in a ceramic case. A flexible lead connects to the cochlear electrode. (Courtesy of Advanced Bionics Corp.)

electrode. The drive to each electrode is an analog signal derived from the filter output. CA is used by Advanced Bionics for their simultaneous analog stimulation strategy. SPEAK is used exclusively by Cochlear Pty, Ltd., and is a combination of several signal-processing techniques. The sound signal is sampled then converted into a pulsed stimulation. Twenty filters are used to determine which frequency bands contain the maximum sound energy. The envelopes of the filter outputs are sampled and combined into electrode-specific biphasic stimulation pulse trains. Typically, 6 electrodes out of the possible 22 are used. CIS is a more generally used strategy developed by Wilson and other researchers at Research Triangle Institute. In the CA strategy, simultaneous stimulation produces electrode crosstalk owing to the interaction between the electric fields caused by the individual electrodes. It is felt that these interactions may distort the cochlear neural responses. CIS combats this problem by stimulating the electrodes with nonsimultaneous, nonoverlapping interleaved pulses. Only one electrode is stimulated at a time. In present-day cochlear implants, the platinum electrodes are sufficiently large so that concerns about exceeding safe charge density limits are minimal.

1.6.2. VISUAL PROSTHESES

Restoring vision by means of electrical stimulation of the visual system presents a considerably greater technical challenge than restoring hearing by means of a cochlear implant because a much larger number, at least hundreds or thousands, of parallel channels are undoubtedly required. There are two obvious places in the nervous system where it appears physically feasible to introduce dense electrode arrays: on the surface of the retina and in the visual cortex.

1.6.2.1. Retinal Prostheses

For blindness due to retinal degeneration, an array of electrodes might be implanted in the eye, on the inner surface of the retina, epiretinal (Humayun et al., 1993), or possibly on the posterior retinal surface, subretinal. The assumption is that although the retinal photoreceptors may be damaged, the ganglion cells (and other neural layers) remain functionally intact and can respond to electrical stimulation. A general advantage of the retinal approach is that the prosthesis can potentially exploit the natural retinotopic visual mapping of the retina, so that stimulation at a particular location on the retina will produce a visual percept, called a phosphene, at a known location within the visual field. There are two basic approaches, presently being researched, to devising hardware for a retinal visual prosthesis. The first uses a camera to capture the image, possibly mounted on an eyeglass frame. The camera's output is image processed and converted to a sequence of commands for controlling an array of externally powered implanted retinal electrodes (DeJaun et al., 1999). The second approach attempts to combine the image sensor and the retinal stimulator in a single structure (Peyman et al., 1998), using an array of light-activated photodiodes whose outputs connect to stimulating electrodes that are implanted within the eye and are geometrically aligned to the retina.

Combined with the photodiode approach, a subretinal array is conceptually attractive. Potentially, this technique would leave the epiretinal surface intact, and would not compromise the retinal physiology (Peachey and Chow, 1999; Zrenner et al., 1999). For one device being investigated, the photodiodes would be powered by incident light (Peyman et al., 1998). There remains unanswered engineering questions about the feasibility of this particular photodiode approach. Some researchers have suggested that the implanted device would not require any external power supply. That is, the transduction of light, via the photodiodes, to electric current would be used as the source of stimulating current for the retinal electrodes. Although the light-to-current efficiencies of future photodiodes may improve, presently the feasibility of using a photodiode to act as the sole power source for a stimulating electrode has not been demonstrated. Typical present-day photodiodes have conversion efficiencies that are on an order 10–100 times too low to be useful for stimulating retinal neurons, using known electrode technology and commonly accepted stimulation thresholds.

There are several externally powered epiretinal approaches. The general concept is shown in Figure 1.32. Most of them propose the use of an extracorporeal camera/signal processor to capture the image and convert it into a sequence of stimulation pulses. In Figure

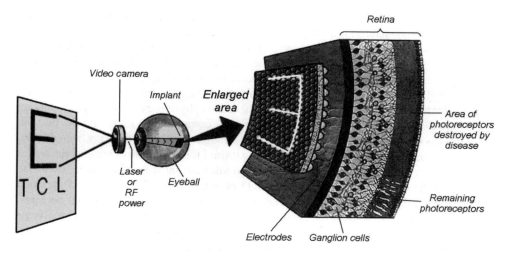

FIGURE 1.32. General approach for an epiretinal visual prosthesis. The camera captures the image and converts it into a sequence of stimulation pulses. The data stream for the electrodes and power for the implanted device is coupled into the eye using a laser beam or RF magnetic link. An electrode array is driven by a set of stimulator drivers contained on the silicon-chip implant. (Courtesy of Second Sight, Inc.)

1.29, the output of the camera modulates a power beam, implemented as either a laser or an RF magnetic link, in order to transmit power and data into the vitreous. Within the eye, an electronic chip demodulates the power beam and generates stimulation patterns for an array of epiretinal electrodes. The geometric pattern of the stimulus matches a bitmap of the original image. Stimulation currents pass into the ganglion cell layer. For all proposed retinal designs, a major engineering obstacle is the development of a flexible substrate that contains the stimulation electrode sites. Assuming that hundreds, or thousands, of electrodes may be required, designing a flexible electrode array is a significant engineering challenge.

In order to get point-specific stimulation of the retina, the electrode stimulation sites need to be small and electrically isolated. The typical charge injection needed to stimulate retinal neurons, for the diseased retina, is uncertain. However, preliminary data suggest that the required charge density will challenge the limits for AIROF, and other electroactive coatings, as discussed in Section 1.3.

For the epiretinal approaches, there is another question of physiological feasibility because of heat dissipation (Greenberg, 2000). Assuming 1000 electrodes, stimulating a collection of ganglion cells, and using typical values of stimulus pulse amplitude, duration, and frequency, a computation can be made of the range of dissipated power that would result from the resistive losses within the tissue and implant electronics. Even using conservative estimates it is possible that 0.1–0.5 W would need to be dissipated in the vitreous. The heat produced by this power dissipation is likely to produce a temperature rise within the vitreous, and presently the allowable heat dissipation within the eye is unknown. It is likely that any intraocular visual prothesis will require sophisticated packaging techniques that will challenge even the most advanced electronic packaging technology. Compared with other neural prostheses, such as cochlear implants, a retinal prosthesis would need to be 50–100 times as dense.

From the functional standpoint, it might appear that a retinal prosthesis can effectively exploit not only the natural retinotopic visual mapping, but also most of the natural neural processing that normally takes place beyond the retina. With regard to exploitation of geometry, this assumes that individual electrodes can stimulate either individual or relatively small numbers of retinal ganglion cells. To accomplish this, the electrode array needs to be in intimate contact with the epiretinal surface. It is uncertain how close the array could be placed without significant compromise of the retinal physiology, especially considering that retina disease often results in a physiologically compromised retina. At the present, microelectrodes that can penetrate the retina to the ganglion cell layer have not been demonstrated, although current research is progressing toward this goal. Use of smaller electrodes closer to the ganglion cells has the potential to alleviate the heat dissipation, owing to the likelihood of smaller stimulation threshold currents. With regard to the exploitation of natural neural processing, it has not been demonstrated that discrete spatial percepts, produced by epiretinal stimulation could have the density required to invoke higher feature extraction such as orientation or motion. Stett *et al.* (2000) have developed a model in the chicken retina to investigate multisite subretinal electrical stimulation. However, it may be that using currently available electrode technology, sensory input from retinal devices may be inherently limited to a phosphene-based image.

1.6.2.2. Optic Nerve Prostheses

Another approach to implementing a visual prosthesis involves direct stimulation of the optic nerve (Veraart, 1998). This group has investigated placing a spiral nerve cuff electrode on the optic nerve. Although still in the experimental stage, one human subject was implanted and a small collection of irregularly sized phosphenes were perceived. Whether sufficient resolution and density of visual percepts could ultimately be obtained by a spiral cuff electrode or fascicular electrodes is unknown.

1.6.2.3. Thalamic Stimulation

Electrical stimulation of the human visual system using thalamic electrodes has been reported in earlier studies by Nashold (1970) and Chapanis *et al.* (1973). In these studies spatial visual percepts were reported in subjects in whom electrodes had been implanted in a variety of thalamic locations, including the superior colliculus, the geniculocalcarine tract, lingual gyrus, and the anterior and middle portion of the optic radiation. Conceptually, the lateral geniculate nucleus (LGN) has been suggested as a obvious site for introduction of electrical stimulation of the visual system. However, the complications in surgical accessibility of this area relative to the retina and visual cortex have made it less attractive, and limited research has been directed toward this goal. Functionally, one might imagine that the LNG would provide similar opportunities for introducing visually derived information as does the retina. However, presently, a system based on implantation within the LGN remains conceptual, at best.

1.6.2.4. Cortical Visual Prostheses

Although retinal visual prostheses have received recent public attention, attempts at stimulation of the primary visual cortex has a longer history. As early as 1918, Löwenstein and Borchardt (1918) reported that while performing an operation to remove bone fragments caused by a bullet wound, the patient's left occipital lobe was electrically stimulated, and the patient perceived flickering in the right visual field. Foerster (1929) and Krause (1924) reported similar cases of visual perception caused by electrical stimulation of the visual cortex during removal of an occipital epileptic focus. The significance of their studies was that they demonstrated that the position of visual percepts within the visual field was systematically related to the area of the occipital lobe that received the stimulation. Urban (1937) inserted electrodes through an occipital burr hole 3 cm above the inion and 3 cm from the midline for the purpose of ventriculography in six patients, of which one was blind. All patients perceived spatial visual sensations of various colors and shapes. Stimulation of the human visual cortex by Penfield and Rasmussen (1950) and Penfield and Jasper (1954) revealed that visual sensations, referred to as phosphenes, could be produced by electrical stimulation. In their subjects, the visual sensations were described as stars, wheels, discs, spots, streaks, or lines.

In 1955, Shaw (Shaw, 1955) obtained a patent on a device to aid the blind through direct stimulation of the visual cortex by using percutaneous electrodes implanted into the occipital lobe that were driven by the output of a photocell. Although Shaw never performed any actual experiments, this approach was investigated, independently, by Button and Putnam (1962). In their study, three volunteers agreed to the implantation of four wires, inserted into the occipital lobe through burr holes. The percutaneous wires were connected to a crude apparatus that varied the stimulus amplitude and frequency on the basis of the output of a cadmium sulfide photocell. The electrodes were left in place for several weeks, during which a variety of experiments were carried out. The three subjects were able to scan an area, holding the photocell in their hand, and to determine the location of illuminated objects including 40–100 watt incandescent lamps and birthday candles on a cake. Owing to excessively high stimulation currents and voltages, the visual perception often filled the entire visual field. Although functionally this device did not have the capability to provide more sensory function than an audible or tactile stimulator, it did demonstrate that it was possible to electrically stimulate the visual cortex following years of acquired blindness. However, it also illustrates the naïve understanding, at that time, of the function of the visual cortex held by some researchers.

The elucidation of the retinotopic map by Hubel and Wiesel (1968) in the 1960s suggested that coherent patterns of electrically elicited sensations might be possible. The first opportunity to investigate chronic stimulation of the visual cortex resulted from experiments by Brindley and Lewin (1968) in which a 52 year-old woman received an implanted stimulation system consisting of 80 platinum electrode discs, placed on the surface of the occipital pole. Eighty associated, transcutaneously powered stimulators were implanted over most of the surface of the right cortical hemisphere, under the scalp. Approximately 32 independent visual percepts were obtained, and Brindley performed mapping studies and threshold measurements. Although some attempt was made to combine the phosphenes into crude letters and shapes, the implant did not prove to be of any practical use to the subject. Another subject received a second implant in 1972 (Brindley *et al.*, 1972; Brindley, 1973; Brindley and Rushton, 1977; Rushton and Brindley, 1977). In this subject, 79 of the 80 implanted electrodes and stimulators produced visual percepts of varied size and shape. These were meticulously mapped over 3 years, following the implantation procedure. Considering the state of the art for implantable devices at the time these studies were carried out, they were a remarkable engineering achievement. However, although these were pioneering experiments, little knowledge was gained about how electrical signals passed through such electrodes might be manipulated to introduce meaningful sensory information into the human brain. Dobelle (Dobelle and Mladejovsky, 1974; Dobelle *et al.*, 1974, 1976), Pollen (1975), and others continued to investigate stimulation of the visual cortex through surface electrodes, using relatively large electrodes placed on the pia-arachnoid surface in individuals who were totally blind, following lesions of the eyes and optic nerves. Dobelle *et al.* implanted at least three subjects with chronic cortical surface arrays. They also tested the ability of the implanted subjects to use the perceptions produced by the electrodes to "read visual Braille" (Dobelle *et al.*, 1976). Reading rates were considerably less than what could be obtained by tactile Braille. Most likely, as in Brindley's subjects, the spatial visual percepts were unsuitable for combining into meaningful patterns as would be required to identify ordinary letters or symbols. One of these subjects, who had retained the electrode array, was featured in the popular press, following an article by Dobelle in the ASAIO journal (Dobelle, 2000). A camera, connected to an improved computer-controlled image-processing system converted the image into electrical stimulus sequences. The computer output was connected to the 64-electrode array via a percutaneous connector; 21 phosphenes were obtained.

As an alternative to surface stimulation of the visual cortex, intramural and extramural studies were initiated in the early 1970s at the NIH for the systematic design, development, and evaluation of safe and effective means of microstimulating cortical tissue. By implanting floating microelectrodes within the visual cortex, with exposed tip sizes of the same order of magnitude as the neurons to be excited, much more selective stimulation can, in principle, be achieved, resulting in potentially more precise control of neuronal function. This concept is illustrated in Figure 1.33, in which an intracortical visual prosthesis is depicted. Images processed by the camera are transcutaneously communicated to an implanted visual prosthesis package with wire leads that connect to intracortical microelectrodes implanted into the primary visual cortex.

On the basis of studies by Bartlett and Doty (Barlett and Doty, 1980) and DeYoe (DeYoe, 1983) in macaques, as well as acute intracortical microstimulation studies that were performed in sighted patients undergoing occipital craniotomy (Bak *et al.*, 1990), a study was planned at the NIH to explore whether stable visual sensations could be produced from chronically implanted intracortical electrodes, and eventually lead to a short-term implantation of intracortical electrodes in a patient-volunteer with blindness. The initial questions to be answered by the short-term implant of the patient-volunteer were (i) Does the visual cortex of a person blind for a long period of time remain responsive to intracortical microstimulation? (ii) Are the visual percepts induced through intracortical stimulation stable over months of stimulation?

The short-term implantation of the visual cortex was performed in a 42-year-old woman patient-volunteer who had been totally blind for 22 years secondary to glaucoma (Schmidt *et al.* 1996a,b). Thirty-eight microelectrodes were implanted in the right visual cortex, near the occipital pole, for a period of 4 months. The electrodes were electrically accessed through percutaneous leads exiting the scalp. Confirming the earlier surface cortical stimulation work, 34 of the 38 electrodes initially produced spatial percepts with threshold currents in the range of 1.9–25 µA. Phosphene brightness, size, color, and position could be modulated by varying stimulus amplitude and frequency, and pulse duration. Repeated stimulation over a period of minutes produced a gradual decrease in phosphene brightness. The apparent size of phosphenes ranged from a "pin-point" to a "nickel" (20-mm-diameter coin) held at

FIGURE 1.33. Concept of an intracortical visual prosthesis. The camera captures the image and translates the visual information into stimulation sequences. Transcutaneous power and data are sent to the implant over a common inductive link. (Courtesy of Illinois Institute of Technology)

arm's length. When two phosphenes were simultaneously generated, the apparent distances of the individual phosphenes sometimes changed, which makes them appear at about the same distance. When three or more phosphenes were simultaneously generated, they became coplanar. Intracortical microelectrodes spaced 500 μm apart generated separate phosphenes, but microelectrodes spaced 250 μm typically did not. This two-point resolution was about five times greater than had previously been achieved with electrodes on the surface of the cortex. The 4-month study was concluded by following the informed-consent protocol, removing all extradural components.

The findings from all of these human studies, for both retinal and cortical devices, have the following significance: It appears feasible to invoke point-topographic visual precepts using retinal, surface cortical, and intracortical electrodes. When intracortical electrodes are used, visual percepts are typically smaller than those invoked by surface cortical electrodes, intracortical electrodes produce amplitude thresholds that are up to two orders of magnitude lower than those of surface cortical electrodes, they are stable over weeks to months, and varying the stimulation parameters can crudely modulate these percepts. What none of these studies demonstrated is, Can a multitude of the observed visual percepts be combined to form meaningful spatial patterns that mimic natural visual perception?

In summary, the assumption that a phosphene-based visual sensation can substitute for normal vision remains at best conceptual. Much of what is known about the function of primate visual cortex comes from neural recording studies with either single or relatively small numbers of electrodes. However, the findings from all of these human studies suggest at least the promise of one day creating a useful visual prosthesis.

ACKNOWLEDGMENT

The authors thank Douglas McCreery of Huntington Medical Research Institutes for his contributions to Section 1.3.3.

REFERENCES

Agnew, W. F., Yuen, T. G. H., McCreery, D. B., and Bullara, L. A., 1986, Histopathalogic evaluation of prolonged intracortical electrical stimulation, *Exp. Neurol.* **92:**162–185.

Agnew, W. F., and McCreery, D. B., 1990a, Considerations for safety with chronically implanted nerve electrodes, *Epilepsia* **31:**S27–S32.

Agnew, W. F., McCreery, D. B., Yuen, T. G., and Bullara, L. A., 1990b, Local anaesthetic block protects against electrically-induced damage in peripheral nerve, *J. Biomed. Eng.* **12:**301–308.

Anderson, D. J., Najafi, K., Tanghe, S. J., Evans, D. A., Levy, K. L., Hethke, J. F., Xue, X., Zappia, J. J., and Wise, K. D., 1989, Batch-fabricated thin-film electrodes for stimulation of the central auditory system, *IEEE Trans. Biomed. Eng.* **36:**693–704.

Bak, M., Girvin, J. P., Hambrecht, F. T., Kufta, C. V., Loeb, G. E., and Schmidt, E. M., 1990, Visual sensations produced by intracortical microstimulation of the human occipital cortex, *Med. Biol. Eng. Comp.* **May:**257–259.

Bartlett, J. R., and Doty, R. W., 1980, An exploration of the ability of macaques to detect microstimulation of the striate cortex, *Acta Neurobiologiae Expermentalis (Warzawa)* **40:**713–728.

Beebe, X., and Rose, T. L., 1988, Charge injection limits of activated iridium oxide electrodes with 0.2 ms pulses in bicarbonate buffered saline, *IEEE Trans. Biomed. Eng.* **35:**494–495.

Brindley, G. S., and Lewin, W. S., 1968, The sensations produced by electrical stimulation of the visual cortex, *J. Physiol. (London)* **196**(2):479–493.

Brindley, G. S., Donaldson, P. E., Falconer, M. A., and Rushton, D. N., 1972, The extent of the region of occipital cortex that when stimulated gives phosphenes fixed in the visual field, *J. Physiol. (London)* **225**(2):57P–58P.

Brindley, G. S., 1973, Sensory effects of electrical stimulation of the visual and paravisual cortex in man, In: *Handbook of Sensory Physiology, Vol. VII/3* (R. Jung, ed.), Springer-Verlag, Berlin, pp. 583–594.

Brindley, G. S., and Lewin, W. S., 1968, The sensations produced by electrical stimulation of the visual cortex, *J. Physiol. (London)* **196:**479–493.

Brindley, G. S., and Rushton, D. N., 1977, Observations on the representation of the visual field on the human occipital cortex. In: *Functional Electrical Stimulation: Applications in Neural Prostheses* (F. T. Hambrecht, and J. B. Reswick, eds.), Marcel Dekker, New York, pp. 261–276.

Brindley, G. S., Donaldson, P. E. K., Falconer, M., and Rushton, D. N., 1972, The extent of the region of occipital cortex that when stimulated gives phosphenes fixed in the visual field, *J. Physiol. (London)* **225:**57P–58P.

Brummer, S. B., Robblee, L. S., and Hambrecht, F. T., 1983, Criteria for selecting electrodes for electrical stimulation: Theoretical and practical considerations, *Ann. N. Y. Acad. Sci.* **405**:159–171.

Button, J., and Putnam, T., 1962, Visual responses to cortical stimulation in the blind, *J. Iowa Med. Soc.* **LII**(1):17–21.

Chapanis, N. P., Uematsu, B., Konigsmark, B., and Walker, A. E., 1973, Central phosphenes in man: A report of three cases, *Neuropsychologia* **11**:1–19.

Chapin, J. K., and Nicolelis, M. A., 1999, Principal component analysis of neuronal ensemble activity reveals multidimensional somatosensory representations, *J. Neurosci. Methods* **94**:121–140.

Chow, A. Y., Pardue, M. T., Perlman, J. I., Ball, S. L., Chow, V. Y., Hetling, J. R., Peyman, G. A., Liang, C., Stubbs, E. B., and Peachy, N. S., 2002, Subretinal implantation of semiconductor-based photodiodes: durability of novel implant designs, *J. Rehabil. Res. Dev.* **39**:313–321. (See also www.optobionics.com.)

DeJaun, E., Cooney, M. J., Humayun, M. S., and Jensen, P. S., 1999, Ocular surgery for the new millennium: Treatment of retinal disease in the new millennium, *Ophthalmol. Clin North Am.* **12**:539–562.

DeCharms, R. C., Blake, D. T., and Merzenich, M. M., 1999, A multielectrode implant device for the cerebral cortex, *J. Neurosci. Methods* **93**:27–35.

de Donaldson, N., and Perkins, T. A., 1983, Analysis of resonant coupled coils in the design of radio frequency transcutaneous links, *Med. Biol. Eng. Comput.* **21**:612–627.

DeYoe, E. A., 1983, An investigation in the awake macaque of the threshold for detection of electrical currents applied to striate cortex: Psychophysical properties and laminar differences. Doctoral thesis, University of Rochester.

Dobelle, W. H., 2000, Artificial vision for the blind by connecting a television camera to the visual cortex, *ASAIO J.* **46**:3–9.

Dobelle, W. H., Mladejovsky, M. G., Evans, J. R., Roberts, T. S., and Girvin, J. P., 1976, Braille reading by a blind volunteer by visual cortex stimulation, *Nature* **259**:111–112.

Dobelle, W. H., and Mladejovsky, M. G., 1974, Phosphenes produced by electrical stimulation of human occipital cortex, and their application to the development of a prosthesis for the blind, *J. Physiol. (London)* **243**:553–576.

Dobelle, W. H., Mladejovsky, M. G., and Girvin, J. P., 1974, Artificial vision for the blind: Electrical stimulation of visual cortex offers hope for a functional prosthesis, *Science* **183**:440–444.

Dobelle, W. H., Mladejovsky, M. G., Evans, J. R., Roberts, T. S., and Girvin, J. P., 1976, 'Braille' reading by a blind volunteer by visual cortex stimulation, *Nature* **259**:111–112.

Duncan, G. H., Bushnell, M. C., and Marchard, S., 1991, Deep brain stimulation: A review of basic research and clinical studies, *Pain* **45**:49–59.

Foerster, O., 1929, Beiträge zur Pathophysiologie der Sehspäre, *J. Psychol. Neurol. (Ppz.)* **39**:463–485, 477–481, 482.

Frohlig, G., Bolz, A., Strobel, J., Rutz, M., Lawall, P., Scherdt, H., Schaldach, M., and Schieffer, H., 1998, A fractally coated 1.3 mm^2 high impedance pacing electrode, *PACE* **21**:1239–1246.

Galbraith, D. C., Soma, M., and White, R. L., 1987, A wide-band efficient inductive transdermal power and data link with coupling insensitive gain, *IEEE Trans. Biomed. Eng.* BME-**34**:265–275.

Greenberg, R. J., 2000, Visual Prostheses: A Review, *Neuromodulation* **3**:161–165.

Grill, W. M., Bhadra, N., and Wang, B., 1999, Bladder and urethral pressures evoked by microstimulation of the sacral spinal cord in cats, *Brain Res.* **836**:19–30.

Grover, F. W., 1946, *Inductance Calculations*, Van Nostrand, New York.

Gualtierotti, T., and Bailey, P., 1968, A neutral buoyancy microelectrode for prolonged recording from single nerve units, *Electroencephalogr. Clin. Neurophysiol.* **25**:77–81.

Guyton, D. L., and Hambrecht, F. T., 1974, Theory and design of capacitor electrodes for chronic stimulation, *Med. Biol. Eng.* **12**:613–619.

Hambrecht, F. T., 1995, Visual prostheses based on direct interfaces with the visual system, *Bulliere's Clin. Neurol.* **4**:147–165.

Heetderks, W. J., 1988, RF powering of millimeter and submillimeter-sized neural prosthetic implants, *IEEE Trans. Biomed. Eng.* **35**:323–327.

Hoffer, J. A., Stein, R. B., Haugland, M. K., Sinkjaer, T., Durfee, W. K., Schwartz, A. B., Loeb, G. E., and Kantor, C., 1996, Neural signals for command control and feedback in functional neuromuscular stimulation: A review, *J. Rehabil. Res. Devel.* **33**:145–157.

House, W., and Urban, J., 1973, Long term results of electrode implantation and electronic stimulation of the cochlea in man, *Ann. Otol. Rhinol. Laryngol.* **82**:504–715.

Hubel, D. H., and Wiesel, T. N., 1968, Receptive fields and functional architecture of monkey striate cortex, *J. Physiol.* **195**:215–243.

Humayun, M., Propst, R., deJuan, E., *et al.* 1993, Is a functional intracular visual prosthesis feasible? *Poster Abstracts of the 24th Neural Prosthesis Workshop HIH/NINDS/NIDCD*, Bethesda, MD, p. 10.

Humayun, M. S., de Juan, E., Weiland, J. D., Dagnelie, *et al.*, 1996, Visual perception elicited by electrical stimulation of retina in blind humans, *Arch. Ophthalmol.* **114**:40–46.

Humayun, M. S., de Juan, E., Weiland, J. D., Dagnelie, G., Katona, S., Greeberg, R., and Suzuki, 1999, Pattern electrical stimulation of the human retina, *Vision Res.* **39**:2569–2576.

Jezernik, S., Craggs, M., Grill, W. M., Creasey, G., and Rijkhoff, N. J., 2002, Electrical stimulation for the treatment of bladder dysfunction: Current status and future possibilities, *Neurol. Res.* **24**:413–430.

Krause, F., 1924, Die Sehbahnen in chirurgischer Beziehung und die faradische Reizung des Sehzentrums, *Klin. Wschr.* **3**:1260–1265.

Liu, X., McCreery, D. B., Carter, R. R., Bullara, L. A., Yuen, T. G. H., and Agnew, W. F., 1999, Stability of the interface between neural tissue and chronically implanted intracortical microelectrodes, *IEEE Trans. Rehabil. Eng.* **7**:315–326.

Loddenkemper, T., Pan, A., Neme, S., Baker, K. B., Rezai, A. R., Dinner, D. S., Montgomery, E. B., and Luders, H. O., 2001, Deep brain stimulation in epilepsy, *J. Clin. Neurophysiol.* **18**:514–532.

Loeb, G. E., Peck, R. A., and Martyniuk, J., 1995, Toward the ultimate metal microelectrode, *J. Neurosci. Methods* **63**:175–183.

Löwenstein, K., and Borchardt, M., 1918, *Dtsch. Z. Nervenheilk* **58**:264–292.

Margalit, E., Maia, M., Weiland, J. D., Greenberg, R. J., Fujii, G. Y., Torres, G., Piyathaisere, D. V., O'Hearn, T. M., Liu, W., Lazzi, G., Dagnelie, G., Scribner, D. A., de Juan, E., and Humayun, M. S., 2002, Retinal prostheses for the blind, *Surv. Ophthalmol.* **47**:335–356.

Maynard, E. M., Nordhausen, C. T., and Normann, R. A., 1997, The Utah intracortical electrode array: A recording structure for potential brain–computer interfaces, *Electroencephalogr. Clin. Neurophysiol.* **102**:228–239.

McCreery, D. B., Agnew, W. F., and Bullara, L. A., 2002, The effects of prolonged intracortical microstimulation on the excitability of pyramidal tract neurons in the cat, *Ann. Biomed. Eng.* **30**:107–109.

McCreery, D. B., Agnew, W. F., Bullara, L. A., and Yuen, T. G., 1990, Partial pressure of oxygen in brain and peripheral nerve during damaging electrical stimulation, *J. Biomed. Eng.* **12**:309–315.

McCreery, D. B., Agnew, W. F., Yuen, T. G., and Bullara, L. A., 1992, Damage in peripheral nerve from continuous electrical stimulation: Comparison of two stimulus waveforms, *Med. Biol. Eng. Comput.* **30**:109–114.

McCreery, D. B., and Agnew, W. F., 1983, Changes in extracellular potassium and calcium concentration and neural activity during prolonged electrical stimulation of the cat cerebral cortex at defined charge densities, *Exp. Neurol.* **79**:371–396.

McCreery, D. B., Bullara, L. A., and Agnew, W. F., 1986, Neuronal activity evoked by chronically implanted intracortical microelectrodes, *Exp. Neurol.* **92**:147–161.

McCreery, D. B., Shannon, R. V., Moore, J. K., and Chatterjee, M., 1998, Accessing the tonotopic organization of the ventral cochlear nucleus by intranuclear microstimulation, *IEEE Trans. Rehabil. Eng.* **6**:391–399.

McCreery, D. B., Yuen, T. G., Agnew, W. F., and Bullara, L. A., 1997, A characterization of the effects on neuronal excitability due to prolonged microstimulation with chronically implanted microelectrodes, *IEEE Trans. Biomed. Eng.* **44**:931–939.

McCreery, D. B., Yuen, T. G. H., and Bullara, L. A., 2000, Chronic microstimulation in the feline ventral cochlear nucleus: Physiologic and histologic effects, *Hear. Res.* **149**:223–238.

Meyer, R. D., Cogan, S. F., Nguyen, T. H., and Rauh, R. D., 2001, Electrodeposited iridium oxide for neural stimulation and recording electrodes, *IEEE Trans. Neural Sys. Rehab. Eng.* **9**:2.

Nashold, B., 1970, Phosphenes resulting from stimulation of the midbrain in man, *Arch. Ophtalmol.* **84**:433–435.

Nicolelis, M. A., 2002, The amazing adventures of robotrat, *Trends Cogn. Sci.* **6**:449–450.

Normann, R. A., Maynard, E. M., Rousche, P. J., and Warren, D. J., 1999, A neural interface for a cortical vision prosthesis, *Vision Res.* **39**:2577–2587.

Otto, S. R., Brackmann, D. E., Hitselberger, W. E., Shannon, R. V., and Kuchta, J., 2002, Multichannel auditory brainstem implant: Update on performance in 61 patients, *J. Neurosurg.* **96**:1063–1071.

Peachey, N. S., and Chow, A. Y., 1999, Subretinal implantation of semiconductor-based photodiodes: Progress and challenges, *J. Rehabil. Res. Dev.* **36**:371–376.

Penfield, W., and Jasper, H., 1954, *Epilepsy and the Functional Anatomy of the Human Brain*, Churchill, London, pp. 116–126, 404–406.

Penfield, W., and Rasmussen, T., 1950, *The Cerebral Cortex in Man*, Macmillan, New York.

Peyman, G., Chow, A. Y., Liang, C., Chow, V. Y. F., Perlman, J. I., and Peachey, N. S., 1998, Subretinal semiconductor microphotodiode array, *Opthalmic Surg. Lasers* **29**:234–241.

Pollen, D. A., 1975, Some perceptual effects of electrical stimulation of the visual cortex in man, in: *The Nervous System, Vol. 2: The Clinical Neurosciences* (D. B. Tower, ed.), Raven Press, New York, pp. 519–528.

Posey, F. A., and Morozumi, T., 1966, Theory of potentiostatic and galvanostatic charging of the double layer in porous electrodes, *J. Electrochem. Soc.* **113**:176–183.

Rauschecker, J. P., and Shannon, R. V., 2002, Sending sound to the brain, *Science* **295**:1025.

Rizzo, J. F., and Wyatt, J., 1997, Prospects for a visual prosthesis, *Neuroscientist* **3**:251–262.

Robblee, L. S., and Rose, T. L., 1990, Electrochemical guidelines for selection of protocols and electrode materials for neural stimulation, in *Neural Prostheses: Fundamental Studies* (W. F. Agnew and D. B. McCreery, eds.), Prentice Hall, Englewood Cliffs, NJ, pp. 25–66.

Robblee, L. S., Lefko, J. L., and Brummer, S. B., 1983, Activated Ir: An electrode suitable for reversible charge injection in saline, *J. Electrochem. Soc.* **130**:731.

Rose, T. L.. Kelliher, E. M., and Robblee, L. S., 1985, Assessment of capacitor electrodes for intracortical neural stimulation, *J. Neurosci. Methods* **12**:181–193.

Rousche, P. J., and Normann, R. A., 1998, Chronic recording capability of the Utah Intracortical Electrode Array in cat sensory cortex, *J. Neurosci. Methods* **82**:1–16.

Rushton, D. N., and Brindley, G. S., 1977, Short- and long-term stability of cortical electrical phosphenes, in: *Physiological Aspects of Clinical Neurology* (F. C. Rose, ed.). Oxford: Blackwell, 123–153.

Salcman, M., and Bak, M. J., 1976, A new chronic recording intracortical microelectrode. *IEEE Trans. Bio-Med. Eng.* **14**:42–50.

Schaldach, M., Hubmann, M., Weikl, A., and Hardt, R., 1990, Sputter-deposited TiN electrode coatings for superior sensing and pacing performance, *PACE* **13**:1891–1895.

Schmidt, E. M., Bak, M. J., Hambrecht, F. T., Kufta, C. V., O'Rourke, D. K., and Vallabhanath, P., 1996a, Feasibility of a visual prosthesis for the blind based on intracortical microstimulation of the visual cortex, *Brain* **119**(Pt 1)**:507–522.

Schmidt, E. M., Bak, M. J., Hambrecht, F. T., Kufta, C. V., O'Rourke, D. K., and Vallabhanath, P., 1996b, Feasibility of a visual prosthesis for the blind based on intracortical microstimulation of the visual cortex. *Brain* **119**(Pt 2)**:507–522.

Schwan, M., Troyk, P., and Loeb, G., 1995, Suspended carrier modulation for transcutaneous telemetry links, in: *Proceedings of the 13th International Symposium on Biotelemetry*, Williamsburg, VA, pp. 27–32.

Shaw, D., 1955, Method and means for aiding the blind. U.S. Patent no. 2,721,316.

Sokal, N. O., and Sokal, A. D., 1975, Class E—a new class of high efficiency tuned single-ended switching power amplifiers, *IEEE J. Solid-State Circuits* **10**:168–176.

Spelman, F., 1999, The past, present, and future of cochlear prostheses, *IEEE Eng. Med. Biol.* **18**(3)**:27–33.

Stett, A., *et al.*, 2000, Electrical multisite stimulation of the isolated chicken retina, *Vision Res.* **40**:1785–1795.

Suesserman, M. F., Spelman, F. A., and Rubinstein, J. T., 1991, In vitro measurement and characterization of current density profiles produced by non-recessed, simple recessed, and radially varying recessed stimulating electrodes, *IEEE Trans. Biomed. Eng.* **38**:401–408.

Tanghe, S. J., Najafi, K., and Wise, K. D., 1990, A planar IrO multichannel stimulating electrode for use in neural prostheses, *Sens. Actuat.* **B1**:464–467.

Tehovnik, E. J., 1996, Electrical stimulation of neural tissue to evoke behavioral responses, *J. Neurosci. Methods* **65**:1–17.

Terman, F. E., 1943, *Radio Engineers' Handbook*, McGraw-Hill, Inc., New York.

Toh, E. H., and Luxford, W. M., 2002, Cochlear and brainstem implantation, *Otolaryngol. Clin. North Am.* **35**:325–342.

Troyk, P., and Schwan, M., 1993, Self-regulating Class E resonant power converter maintaining operation in a minimal loss region, U.S. Patent No. 5179511.

Troyk, P., and Schwan, M., 1995, Modeling of weakly-coupled inductive links, *Proceedings of the 13th International Symposium on Biotelemetry*, Williamsburg, VA, pp. 63–68.

Troyk, P. R., Heetderks, W., Schwan, M., and Loeb, G., 1997, Suspended carrier modulation of high-Q transmitters, U.S. Patent No. 5697076.

Troyk, P. R., and Schwan, M. A., 1992a, Class E driver for transcutaneous power and data link for implanted electronic devices, *Med. Biol. Eng. Comp.* **30**:69–75.

Troyk, P. R., and Schwan, M. A., 1992b, Closed-loop class E transcutaneous power and data link for microimplants, *IEEE Trans. Biomed. Eng.* **39**:589–599.

Troyk, P., Bradley, D., Towle, V., Erickson, R., McCreery, D., Bak, M., Schmidt, E., Kufta, C., Cogan, S., and Berg, J., 2002, Experimental results of intracortical stimulation in macaque IV, *ARVO 2003 Meeting*, Ft. Lauderdale, FL, May 4–9.

Troyk, P., and DeMichele, G., 2003, Inductively-coupled power and data link for neural prostheses using a Class-E oscillator and FSK modulation, in: Proceedings of *EMBS Conference*, Cancun, Mexico, September 17–21, pp. 3376–3379.

Urban, H., 1937, Zur Physiologie der Occipitalregion des Menschen, *Z. Ges. Neurol. Psychiat.* **158**:257–261.

Veraart, C., Raftopoulos, C., and Mortimer, J. T., *et al.* 1998, Visual sensations produced by optic nerve stimulation using an implanted self-sizing spiral cuff electrode, *Brain Res.* **813**:181–186.

Weiland, J. D., Anderson, D. J., and Humayun, M. S., 2002, In vitro electrical properties for iridium oxide versus titanium nitride stimulating electrodes, *IEEE Trans. Biomed. Eng.* **49**:574–579.

Weiland, J. D., and Anderson, D. J., 2000, Chronic neural stimulation with thin-film, iridium oxide electrodes at high current densities, *IEEE Trans. Biomed. Eng.* **35**:911–918.

Weiland, J. D., Humayun, M. S., and Dagnelie, G., 1999, Understanding the origin if visual percepts elicited by electrical stimulation of the human retina, *Graefes Arch. Clin. Exp. Ophthalmol.* **237**:1007–1013.

Wessberg, J., Stambaugh, C. R., Kralik, J. D., Beck, P. D., Laubach, M., Chapin, J. K., Kim, J., Biggs, S. J., Srinivasan, M. A., and Nicolelis, M. A., 2000, Real-time prediction of hand trajectory by ensembles of cortical neurons in primates, *Nature* **408**:361–365.

Wise, K. D., *et al.* 1999, Micromachined stimulating electrodes, Final Report Contract NIH-NINDS-N01-NS-5-2335, available from: www.ninds.nih.gov/ProgressReports.

Ziaie, B., Nardin, M. D., Coghlan, A. R., and Najafi, K., 1997, A single-channel implantable microstimulator for functional neuromuscular stimulation, *IEEE Trans. Biomed. Eng.* **44**:909–920.

Zierhofer, C., and Hochmair, E. S., 1990, High-efficiency coupling-insensitive transcutaneous power and data transmission via an inductive link, *IEEE Trans. Biomed. Eng.* **37**:716–722.

Zrenner, E., Stett, A., and Weiss, S., *et al.* 1999, Can subretinal microphotodiodes successfully replace degenerated photoreceptors? *Vision Res.* **39**:2555–2567.

2

INTERFACING NEURAL TISSUE WITH MICROSYSTEMS

Ph. A. Passeraub* and N. V. Thakor

Biomedical Engineering Department, Johns Hopkins University, Baltimore, Maryland

2.1. INTRODUCTION

From an engineering point of view, the nervous system, which is made of neural tissue, is a most complex biological system. All possible instruments or tools for applications related to the nervous system are part of another type of complex systems: biomedical instrument systems. As in any interacting system, "communication" can occur at the interface of these two systems. Depending on the kind of application, this communication can be either unidirectional or bi-directional. The signals found in these two systems are of very different nature:

- In the nervous system, signals carrying information over distances use evoked potentials involving a high number of ionic transport and neurotransmitter activation mechanisms. Transduction of a signal coming from the external world (e.g., chemical, optical, mechanical, magnetic) is performed by receptor cells using countless mechanisms (Bear *et al.*, 2001).
- Most signals in biomedical instrument systems are electrical, chemical, optical, mechanical, and magnetic. The nonelectric signals are usually converted at one point or another via a transducer into electrical signals in order to be compatible with the electronic computer world.

In the neural tissue the number of functions per unit volume is very high. The size of most neural cells (glial and neuronal) is typically between 10 and 50 μm in diameter (Bear *et al.*, 2001). Separated by gaps of typically 20 nm, they are densely packed together (in the range of 10^6 cells/mm³ and 10^5 neurons/mm³). Human axon diameters range from <1 to 25 μm. Their length varies from <1 mm to >1 m. Considering these values, miniaturization of biomedical instrumentation to interface efficiently to the neural tissue at such scales is crucial.

---

* Address for correspondence: Biomedical Engineering Department, Johns Hopkins University, Baltimore, Maryland; e-mail: philippe.passeraub@ieee.org

Traditionally, the primary purpose of such medical instruments and research tools has been to understand basic mechanisms of brain functions as well as neural disorders, to diagnose these disorders, and to develop new therapies. An increasing number of systems and applications have been recently demonstrated and are being developed:

- Subdural grid electrodes for neurosurgery preparation in patients with intractable focal epilepsy (Lesser *et al.*, 1998) or deep brain stimulation for tremor suppression in patients suffering from Parkinson's disease (Damian *et al.*, 2003)
- Implantable prosthetic systems for bladder control in paraplegia (Jezernik *et al.*, 2002), or for cochlear stimulation in patients with profound hearing loss (Wilson *et al.*, 2003)
- Stroke- or seizure-monitoring systems (Schneweis *et al.*, 2001; Waterhouse, 2003)
- Augmented reality applications with enhanced sensory system or brain functions (Huang, 2003), for instance in computer-assisted surgery (Vasquez-BuenosAires *et al.*, 2003)
- Advanced signal processing and computing (Zeck and Fromherz, 2001)
- Neuron-based sensors able to detect unknown dangerous biochemical agents (Mrksich, 2000)

Most of today's instruments are based on a fabrication technology that is not very compatible to the scaling down of their size. Their main drawbacks are the limited resolution of measurement or stimulation and their nonnegligible invasiveness in most cases. Recently a growing number of miniaturized microsystems interfacing the neural tissue have been demonstrated. They are based on microfabrication technologies, which allow miniaturization of highly complex systems at an unprecedented scale (Zaghloul, 2002). Microsystems are anticipated to revolutionize the engineering of interfaces between the electronic world and biological neural systems. What exactly are these microsystems? How can they be used to interface with neural tissue? What are the challenges, the state of the art, and the trends for the various types of interfaces? This chapter is an attempt to answer these very relevant questions.

2.2. NEURAL MICROSYSTEMS

2.2.1. BACKGROUND

Microsystems technology is a maturing technology that makes possible the integration of devices and systems at the microscopic and submicroscopic scales. It originates from the microelectronic batch-processing fabrication techniques. Similar and modified fabrication processes are used to micromachine mechanical structures and integrate other devices using the various properties of substrate materials and of deposited layers. For instance, miniature beams, membranes, suspended masses, or microchannels can be made (Polla *et al.*, 2000). The material properties of a monocrystalline substrate like silicon permit extraordinary mechanical strength, without hysteresis and fatigue. Sensing elements—capacitive, inductive, or piezoresistive—can be implemented to make highly sensitive and reliable sensors. Microactuators can as well be built on the same substrate. The optical properties of silicon and its semiconductor nature have been also used to make optical waveguides, various types of photosensors, and even light-emitting devices. When embedding on the same substrate

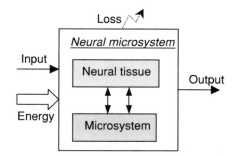

FIGURE 2.1. Neural microsystems are defined as hybrid systems comprising two interacting subsystems: the neural tissue system and the microsystem.

or in its proximity an electronic circuit of interface, a high sensitivity and a high signal-to-noise ratio can be attained (Passeraub, 1999). Nowadays, microsystems find an increasing number of applications in biomedical engineering and in particular for the development of instrument systems interfacing the neural tissue. A hybrid system consisting of a microsystem interfacing the biological system of neural tissue, either in the form of cultured neurons or of brain slices, or even as a part of an intact nervous system, can be defined as a "neural microsystem" (see Figure 2.1). The high interest in microsystems technology for neural applications resides in its modularity and its miniaturization capabilities. Potentially, this allows the matching of the very high density of functions between neural tissue and microsystems. This is of high significance for the development of simultaneous measurements of neuronal signals at a high number of locations, or to implant neural microsystems within the body.

The first reports of microfabricated systems to interface with the neural tissue can be traced back to the 1970s (Wise and Angell, 1971; Gross et al., 1977; Pine, 1980). Interestingly their application for neural tissue, and even for heart cell interfacing (Thomas et al., 1972), were among the first of the then embryonic microsystems technology. Yet for such biomedical application, this technology is currently still at an embryonic developmental stage. For other domains of applications, like in the automotive industry, microsystems have already become quite mature (e.g., accelerometer microsystems for airbags). Reasons for this slow progress can be found in the particularly complex and challenging problems faced by microsystems that exist *in vivo* or in interfaces to biological systems (Dario, 1995). A multidisciplinary approach, including knowledge and expertise in medicine and engineering, is essential for developing a successful device. Since its early development, microsystems technology has considerably progressed and offers a real potential for improving the existing neural instrumentation, and to elaborate new instruments. In the meantime, considerable new application areas have emerged in the field of biology and medicine. Many problems, such as the biocompatibility of many microsystems materials, have been successfully addressed (Kotzar et al., 2002). Lately a number of neural microsystems have even been developed commercially (MEA60 System from Multichannels Systems and Ayanda Biosystems, MED64 System from Panasonic). A growing number of research groups have reported the development of new neural microsystems in the literature. Many examples and applications are reviewed further in this chapter.

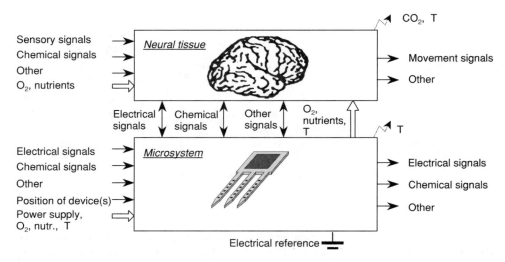

FIGURE 2.2. Schematic block representation of neural microsystems with their subsystems inputs, outputs, and interactions.

2.2.2. FUNCTION BLOCK DIAGRAM

Depending on whether it is a part of an integral nervous system, or in the form of a slice, or of cultured neurons network, subsystems made of neural tissue can have different degrees of complexity. Microsystems can also have various levels of complexity, depending on the devices and functions integrated on them. Like other complex systems the various possibilities of interactions between its subsystems and the outside world can be described using the function block diagram approach. From this simplified point of view a neural microsystem can be considered as a black box consisting of two subsystems with inputs and outputs (see Figure 2.1). Figure 2.2 depicts the typical inputs and outputs of these subsystems and their possible interactions. More details are given in the next subsections.

2.2.2.1. System Inputs and Outputs

For the "neural tissue" and the "microsystem" subsystems, the inputs coming directly from outside the neural microsystem include the following:

- Sensory signals (for seeing, hearing, touching, smelling, and tasting) acquired either by the sensory organs connected and relayed directly to the integral nervous systems, or acquired by a physical or chemical sensor and transmitted indirectly to the nervous system via electrical, chemical, or other types of signals through the microsystem.
- Chemical substances like drugs going either to the brain tissue or to the microsystem: It is delivered to the neural cells either via the vascular system by diffusion through the blood–brain barrier or via the microsystem typically with microfluidic functions in order to influence globally or locally nerves and brain functions.
- Other types of signals: For the neural tissue they can include accidental mechanical shocks, mechanical stress caused by a tumor, and magnetic signals coming from

conventional medical instruments for imaging and stimulation. For microsystems they can include mechanical and optical signals to replace a missing sensory signal, and carrying signals for stimulation and recording purposes.
- Electrical signals to communicate with the microsystem, to drive a stimulus, or to carry a signal to be modulated by the measurand (the physical parameter being quantified by measurement), or as electrical ground reference.
- The information on the microsystem position in relation to the neural tissue and on the exact location(s) of interface.
- The energy required for the biological system survival (O_2, nutrients, a stable temperature): It is generally provided directly by the body for *in vivo* applications. For the microsystem subsystem, it comprises the power supply for the embedded electronic circuitry, or the substances needed for the tissue survival delivered via the microsystem.

The outputs directly exiting the neural microsystem include the following:

- Signals from the nervous system controlling the movement of limbs.
- Other types of signals originating from the neural tissue (e.g., biochemical signals involved in the regulation of the body, as well as electrical and magnetic signals resulting from the neuronal activity and measurable with nonminiaturized medical instruments).
- Electrical signals sensed by the microsystem and reflecting the spontaneous or stimulated electrical neuronal activity, or carrying another type of signal (e.g., an electrical signal modulated by a chemical signal).
- Chemical substances either extracted by the microsystem (signals reflecting ionic exchanges plus flow as well as neurotransmitter release extracted by microdialysis or similar techniques) or delivered by the microsystem (i.e., drugs) to provide therapeutic benefit.
- Other type of signals sensed by the microsystem (e.g., optical signals like fluorescence, or mechanical signals like the flowing medium carrying chemical signals in microdialysis probing).

The by-products of the cellular respiration occurring in the brain tissue and to be eliminated from the system—mainly the CO_2—can be considered a loss. The loss for microsystems with electronic circuits is essentially a thermal dissipation resulting from its electrical power consumption. For applications requiring implantation, such loss cannot be neglected, since the neural tissue is highly sensitive to temperature.

The final system complexity depends on the application and on the number of elements integrated on it. Commonly, for each feature that is not totally integrated an additional input or output is required. For microsystems with fully embedded functions, inputs and outputs with the external world can ultimately and ideally be reduced to electrical signals plus power supply.

2.2.2.2. Subsystem Interactions

For a microsystem device, the various interfacing methods with the brain tissue include electrical, chemical, optical, and magnetic signals.

Electrical Signals

In the neural tissue, neuronal electrical signals involve the flow of a high number of different ions, positively or negatively charged, both through and along cellular membranes using various transport mechanisms. At a defined location, the flow of these charged ions reflects the neuronal activity of the system.

- *Role of the microsystem output:* to deliver electrical charges to the brain tissue through an electric conductor in order to induce an electrical field in the neural tissue, or to modify a transmembrane potential.
- *Influence on the neural system:* (a) to stimulate the tissue in order to cause a single neuron or several neurons to fire and to create an evoked potential and (b) to modify the properties of the neural tissue, for instance, to modulate an existing neuronal activity.
- *Role of the microsystem input:* to detect transmembrane potential variations in the case of intracellular recordings or field potential variations in the extracellular medium.
- *Sorts of signals:* a wide range of possible signals can be observed, ranging from single spikes (of a few milliseconds' duration) to recurrent undulation or multispike bursts, as well as to very slow potential shifts (over several minutes). Typical amplitude of the signals range from tens of microvolts to a few tens of millivolts for extracellular field. The transmembrane action potentials amplitude is stronger, typically in the range of several tens of millivolts.

Chemical Signals

In the neural tissue a chemical signal is associated with each ion involved in the ionic flow that generates electric field potentials. In addition to these ions a high number of other molecules are implicated in neuronal chemical signals, principally associated with synaptic transmission. At a given location, this ionic flow and the secretion of these molecules plus their possible chemical reactions comprise the neuronal activity of the system.

- *Role of the microsystem output:* to deliver locally or globally ions as well as molecules to modify the chemical composition of the cerebrospinal fluidic bathing medium and the extracellular medium.
- *Influence on the neural system:* to modify or inhibit a very wide range of possible mechanisms of certain regions of the brain that may be the location of injury or disorder (such as epilepsy).
- *Role of the microsystem input:* to provide measurements of nominal values and changes in neurochemical concentrations generally extra- or intracellularly at well-defined locations of the neural tissue.
- *Sorts of signals:* a high number of different molecules (e.g., single ions, amino acids, amines, peptides, and even gaseous molecules) can be observed as long as an appropriate measuring method exists. The dynamic range of the neurochemical signals is similar to neuronal electrical signals for fast signals, but can be much slower. Concentration levels, however, vary strongly and depend on the molecule of interest (e.g., typical range for calcium: from 0.2 μM to 2 mM; for glutamate: from 3 to 150 nM).

Optical Signals

Except in the retina, the neural tissue is known to be insensitive to optical signals and generates no photon. The optical properties can be modulated by various mechanisms reflecting the neural activity. For example, the so-called intrinsic optical signal is produced by changes in the transmitted or scattered light, due for instance to a volume or refractive index change in the extracellular space, comprise a type of optical signal (Duarte *et al.*, 2003). Also, using single or dual photon excitation, the changes in the possible autofluorescence or in the fluorescence of a specific dye (e.g., sensitive to a molecule or to a voltage) comprise another type of optical signal (Mainen *et al.*, 1999).

- *Role of the microsystem output:* to deliver photons to the tissue.
- *Influence on the neural system:* to transduce an optical signal into a neuronal signal for photoreceptor cells. High intensities of light can photodamage neural cells.
- *Role of the microsystem input:* to measure geometrical or refractive index changes in the tissue for a signal type, and to provide a measurable fluorescence light intensity that reflects a molecule concentration or a voltage.
- *Sorts of signals:* a light intensity.

Magnetic Signals

In the neural tissue, magnetic signals are coupled to movements of positively or negatively charged ions and reflect indirectly the neuronal electrical sources in the system.

- *Role of the microsystem output:* to induce without direct contact an electrical field in the brain tissue. With the current technology for miniaturized devices, such an output is still facing technical challenges due to thermal losses.
- *Influence on the neural system:* to stimulate the tissue in order to modulate an existing activity and to create an evoked potential.
- *Role of the microsystem input:* to detect indirectly and without contact variations of field potentials in the extracellular medium.
- *Sorts of signals:* generally short pulses of 100 μs up to a 2.0-T amplitude to induce an evoked potential magnetic field, and signals with similar dynamics to those described under electrical signals.

Oxygen, Nutrients, and Temperature

This type of interface between the neural microsystem subsystems is unidirectional. In the normal *in vivo* situation, blood flow, diffusion of oxygen and nutrients through capillary walls, and cellular respiration provide this energy to each neural cell. In extraordinary situations, for instance when harvesting neural tissue for *in vitro* preparations, the survival of the neural cells depends on replacing the normal supply by artificial means. They need to be placed in a comfortable environment at the right temperature (typically 37°C), in a fluid of specific chemical composition (matching the cerebrospinal or the cellular fluids), with sufficient oxygen and nutrients. Using microsystems technology for this purpose can be especially valuable. Here the role of the microsystem is to supply the needed energy to the neural tissue, by locally or globally delivering oxygen and nutrients or by providing temperature control to the neural tissue.

Depending on the specific applications of neural microsystems, various combinations of inputs, outputs, and interactions occur. For fundamental research, the possible combinations of function to be performed by the neural microsystem will depend on the specific biomedical question being addressed. Such systems are to be developed specifically for each type of experiment, and they are expected to provide information that no other technique is likely to permit.

2.2.3. NEURAL MICROSYSTEMS CONFIGURATIONS

In this section, various existing or envisioned applications for neural microsystems are grouped by similar block diagram configurations using the generic block diagram of Figure 2.2.

2.2.3.1. Diagnosis and Drug Discovery

On the basis of various technologies, a series of methods have already been developed:

- For diagnostic purposes, for instance to detect a possible disorder or injury in the spinal cord using evoked potentials (Sharma and Winkler, 2002).
- To screen drugs that target a specific disorder (Kupferberg, 2001) or that promote tissue survival *in vivo* and *in vitro* (Bernstein, 2001).

These types of applications require low invasiveness and a high number of repetitive tests. The drug to be potentially tested is delivered to the neural tissue either directly or through the microsystem using microchannels. Interactions at the interface can be of a chemical or electrical nature to stimulate and to record from the neural tissue. An execution of such a neural microsystem using a cerebellum slice of neonatal rats on top of an array of 60 microelectrodes is reported in Egert and Haemmerle (2002). It demonstrates that modulation of the spontaneous spike activity by dopaminergic drugs and spreading of this fast neuroelectrical activity in the neural tissue can be observed. Such neural microsystems are expected to significantly facilitate the discovery of neuroactive drugs in pharmacology.

2.2.3.2. Monitoring and Neuron-Based Sensors

Monitoring the neuroelectrical activity using grid electrodes placed directly over the cortex is a common method for preparing patients with intractable epilepsy for neurosurgery (Lesser *et al.*, 1998). The recorded signals are used to detect or relate a particular event to a certain location in the cortex. In this case the use of microsystems will be of particular interest owing to possibly lower invasiveness and an increased spatial resolution for precise cortical mapping.

For *in vitro* neural tissue preparation, the same system configuration can be used as sensor or detector. Such neural microsystems are envisioned as bio-warfare detectors for neurotoxins and other chemical agents that act against membrane channel receptors (Mrksich, 2000). Neural microsystems made of neurons grown on top of microelectrode arrays have shown regular electrophysiological behavior and stable pharmacological sensitivity for up to 9 months. Though slices of neural tissue have originally an intact neuronal circuitry, their stability for long-term recording is expected to be limited by their morphology changes. Their survival is usually shorter than cultured neurons' (Pancrazio, 2000).

2.2.3.3. Treatment of Neural Disorders

Another promising application domain for neural microsystems is targeting the therapy of a number of neural disorders. On the basis of various technologies, a number of implantable devices have already been designed, developed, and tested for tremor control caused by Parkinson's disease (Damian *et al.*, 2003) and to terminate seizures in epilepsy (Ward and Rise, 1997). They use techniques to inhibit undesired neuronal activity of the central nervous system by recording and stimulating electrical signals, as well as by infusing drugs deep in the brain, or even by stimulating the peripheral route provided by the vagus nerve to modulate brain activity (Boveja, 2001). Here the challenge is to develop a fully implantable, wireless, and autonomous system with low invasiveness, with possible drug storage, and with precise interface positioning and confinement for the optimal exchange of electrical signals and for the delivery of drug.

2.2.3.4. Neuroprostheses

From a system point of view neuroprostheses can be grouped into three main configurations corresponding to distinct types of applications.

The first one deals with restoring a sensory function (e.g., following a disease or an injury). The best existing example of such prostheses is the commercially available cochlear implant (e.g., COMBI 40+ from Medical Electronics) based on classical fabrication technology. The high density of possible interfacing locations with microsystems technology and the small size is also particularly beneficial for this type of application. For the next generation of cochlear implants, microsystems are likely to offer (a) lower resulting tissue damage, important for combining electrical and acoustic stimulation for a new generation of implants, (b) closer stimulation of the ganglion cells, which would reduce the current consumption (Wilson *et al.*, 2003), or (c) the possibility of directly stimulating the auditory nerve (Badi *et al.*, 2003), and even the auditory cortex (Rousche and Normann, 1999). The other class of sensory neuroprostheses currently in development is aimed at restoring vision for the blind. The first visual implants, based on flat 1-mm^2 subdural electrodes fabricated using classical technologies, did have strong limitations. Large currents (1–3 mA) and a 3-mm spacing between electrodes were needed to evoke distinguishable phosphenes. Microsystems technology has opened the way to overcome these limitations, allowing high numbers of microfabricated electrodes in arrays with a small stimulating surface and high selectivity (Maynard, 2001). The current neural microsystem development approaches are based on epiretinal and subretinal implants (Narayanan, 1999; Meyer, 2002) and direct optical nerve as well as visual cortex microstimulation (Normann, 1990). The main challenges for this most promising type of prosthesis are not only the technical issues, but also those encountered at the interface of neuroscience, medicine, and engineering (Maynard, 2001).

Motor function neuroprosthesis applications have a different system configuration, because sensing and stimulating devices for peripheral nerves or muscles have to replace the failing motor signal. Since the first tests on humans in the early 1960s, some unique systems using external stimulating electrodes have been developed to help paraplegic patients to stand and to walk (Graupe, 2002). Similar implantable systems based on classical technologies to restore or improve grasping functions (Popovic *et al.*, 2002), or for urinary control (Jezernik *et al.*, 2002), using electrical stimulation of peripheral nerves are commercially available (Interstim by Medtronic, FreeHand by NeuroControl). For such applications microsystems technology is highly valuable because of its miniaturization potential. Reduction of surgery pain, lower discomfort associated with the electrodes, less nerve damage, and a higher selectivity in nerve stimulation have been demonstrated recently with a microfabricated array of 100 needle microelectrodes (Normann, 1999). The task of the microsystem for such an application is to extract the command signals from the nervous system of the user directly or indirectly, to process them, and to deliver the right command signals at the proper place to control the desired limb.

The role of the third kind of neuroprosthesis application is in extracting signals coming from the brain in order to communicate or to control an artificial limb that either replaces a missing limb or complements an existing one. It is often referred to in the literature as the "brain–computer interface" (Wolpaw *et al.*, 2002). Recently, surprising experiments have shown successful control of robot arms using the electrical signals recorded from almost 100 microwire electrodes inserted in different locations in the parietal and frontal cortex of an owl monkey (Nicolelis and Chapin, 2002). This study anticipates that recordings from 500 to 700 neurons are needed to perform a one-directional hand movement with 95% precision. This number of electrodes is presumably going to increase for three-dimensional movement controls. Microsystems technology for the three-dimensional recording of neuroelectrical signals in the brain as reported in Hoogerwerf (1994) is expected to play a significant role again in the development of such applications.

2.2.3.5. Neurocomputing

Neurocomputing with hybrid microsystems—sharing electronic and biologic circuitry—is an emerging nonmedical application of neural microsystems (Mrksich, 2000). The neuronal network made of neural tissue or cells is grown on top of the microsystem electronic circuitry in a manner such as to exchange electrical pulse signals in both directions (Fromherz, 2002). In this configuration the combination of biological and electronic processes is promising not only for unraveling the nature of information processing in neuronal networks, but also for new signal-processing and computing microsystems. Noise reduction at the interface and control of neuronal network topology with synaptic connection are still big challenges in the development of these neurocomputing microsystems. The use of neural tissue slices in combination with a well-defined network could become attractive. For instance, the presumed Kalman filter property of the hippocampus could be used for probabilistic information fusion and localization (Bousquet *et al.*, 1998). Protocols to build such neural microsystems still need to be developed and standardized.

2.2.3.6. Augmented Reality

Interfacing neural tissue, and especially the brain, with the electronic world is a topic that draws considerable journalistic attention and engenders creative ideas for new applications. Neural microsystems have an equal place in such media. They are envisioned as systems able to enhance the sensory system resulting in a higher sensitivity or to complement it with additional information. Besides proposals for applications such as those that connect healthy people's brains to the Internet or that permit two persons to exchange thoughts directly from brain to brain via neural interfaces (Huang, 2003), a promising application of augmented reality can be in computer-assisted surgery. It is based on a tongue stimulator having arrays of electrodes and applying stimuli typically of 1.5 V that provides direct feedback to the surgeon for the guidance of his surgical tools (Vasquez-BuenosAires *et al.*, 2003). The day sensory neuroprostheses for rehabilitation outperform our normal sensory system (e.g., augmenting vision in patients with diseases such as macular degeneration with visual prostheses), some interesting issues will have to be faced.

2.3. GENERIC METHODS TO INTERFACE MICROSYSTEM AND NEURAL TISSUE

The first methods to interface neural tissue with microsystems have focused on the interactions achieved with the help of electrical signals and devices. In this section the existing methods that are being used, developed, or are potentially useful for neural microsystems are described.

2.3.1. HOW TO INTERFACE ELECTRICAL SIGNALS IN NEURAL MICROSYSTEMS

The focus in this and the following section is on the electrode–tissue interface, without taking into consideration the design of the microsystem and of its position relative to the tissue, which will be treated in Section 2.3.4. Contrary to conventional microwire or microneedle electrodes built from a rigid conductive material, neural microsystems mainly use nonconductive substrates (rigid or flexible) on which thin-film electrical connections and contact microelectrodes are patterned (see Figure 2.3). The working principle of electrodes and integrated microelectrodes interfacing with neural tissue has been extensively reported in the literature (Kovacs, 1994a). Electrical signals are transduced from the electronic world of the microelectrode to the ionic world of the neural tissue on the basis of two principles: capacitive coupling and charge transfer (Kovacs, 1994a). In opposition to insulated microelectrodes (Zeck and Fromherz, 2001), conductive microelectrodes make use of the chemical and electrochemical reactions taking place at the surface of the conductor bathing in the fluidic medium of the neural tissue (Neuman, 1998). Depending on the nature of the interface, the microelectrodes can be of two sorts: high pass or low pass. For applications focusing on propagated signals in the neural tissue, high-pass microelectrodes such as insulated metal microelectrodes have been found to be the most useful. When the interest is in slow extracellular signals or in membrane processes, the low-pass microelectrodes such as glass capillary microelectrodes are the best (Gesteland *et al.*, 1959). The exact nature of the interface depends on the conductor material, the surface properties of the microelectrode, as well as on the amplitude and frequency of the electrical signals involved. Integrated microelectrodes can therefore be grouped into three different categories: conductive high pass, conductive low pass, and insulated high pass. Their basic working principle,

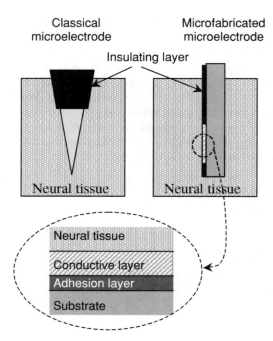

FIGURE 2.3. Schematic views (top left) of a classical microneedle electrode, and (top right and bottom) of a typical electrical microsystem–neural tissue interface.

the microfabrication methods used, and the state of the art as well as their potential for implementation in neural microsystems is described in the next subsections. Microsystems technology for electrical inputs and outputs exhibit flexibility and allow the combination of different microelectrode types on the same substrate. Thus, recording and stimulation characteristics can be optimized for a specific application.

2.3.1.1. High-Pass Conductive Microelectrodes

High-pass conductive microelectrodes are the oldest (Wise and Angell, 1971), the most used (Rutten, 2002), and probably the least understood interface between a neural device and the tissue (Dario, 1995). For this type of electrode, the transduction of signals for recording and stimulation is mainly based on capacitive coupling. Such electrodes are also polarizable (Neuman, 1998). They ideally have an electrode–electrolyte interface in which no actual charge crosses the ionic double layer when a current is applied. In normal recording and optimal stimulation conditions only displacement currents can flow through them, and no direct current flows. The electrical properties of these microelectrodes are not linear (McAdams and Jossinet, 1998). Their impedance is approximately inversely proportional to the frequency. Working with signals above 50 mV can produce harmonic distortion and/or an irreversible change to the electrode surface (Gesteland *et al.*, 1959). These parameters as well as the chemical property of the microelectrode material are important, because they determine the possible chemical and electrochemical reactions for the capacitive coupling and the charge transfer through the interface. This point is even more critical for stimulating electrodes that are generally used in the bipolar mode to keep the global transferred charge at zero, even though the electrochemical reaction involved might not be symmetrical or reversible (Neuman, 1998). The selection of inappropriate materials will cause biocompatibility or toxicity problems. It might as well increase bio-fouling of the electrode and decrease its efficiency. Noble materials have ideal-approaching polarizable properties (Neuman, 1998). The microfabrication of electrodes for neural microsystems is usually based on materials like gold (e.g., Blum, 1991), platinum (e.g., Thiébaud et al., 2000), or iridium (e.g., Borkholder et al., 1997). Special post-processing can be used to improve the desired electrical property to inject charge in the neural tissue by modifying its surface and increasing the contact area. Two design modifications are common: use of (a) platinum black (e.g., Novak and Wheeler, 1986) and (b) activated iridium, which also limits the chemical changes that could lead to tissue damage (e.g., Anderson, 1989). Other materials have also been reported. Titanium nitride, which has a high surface factor and is mechanically very stable (Egert et al., 1998); polysilicon, which is compatible with CMOS fabrication processes (Bucher et al., 1999); and indium tin oxide for transparent electrodes (Gross et al., 1985) have found unique applications as well. The deposition of thin layers (typically 50–500 nm) of these conductive materials is performed generally by evaporation (thermal or ion beam) or sputtering, on top of an adhesion layer (typically 10–50 nm of chromium or titanium). The connections and contacts are patterned using a photolithography process, followed by a lift-off or a selective etching process (e.g., Egert et al., 1998). Electrodeposition is used in some cases to notably increase the thin deposited conductive layer (Thiébaud et al., 2000). However, it can be difficult to apply because it generally requires a connecting method to set all the microelectrodes in the bath at the needed potential. The final shape of such neural microsystems with microelectrodes will depend on the application and the interface-positioning method chosen.

2.3.1.2. Low-Pass Conductive Microelectrodes

In neurophysiology, low-pass conductive microelectrodes are generally made of pulled micropipettes filled with a saline solution and a bathing chlorinated silver wire, as schematized in Figure 2.4a. They are characterized by their ability to carry direct current (DC) through them in their normal working condition. They are used to record signals of phenomena with DC shifts like spreading depression (Somjen, 2001), as reference electrodes, and in some cases also to apply an electrical field in the neural tissue (e.g., to control epileptiform activity; Gluckman et al., 2001). In this type of microelectrodes, transduction of signals is mainly based on charge transfer reactions. Their electrical properties are also not linear. Over a wide frequency (f) range their impedance is proportional to $f^{-1/4}$ (Gesteland et al., 1959). Such electrodes are also called nonpolarizable (Neuman, 1998). A few microsystem versions of nonpolarizable electrodes have been reported, mainly for the purpose of grounding chemical sensors (Bousse et al., 1986; Berg et al., 1990; Suzuki et al., 1998b). In vivo measurements of very slow neuronal signals using a nonpolarizable microelectrode, photopatterned over a cylindrical substrate (see Figure 2.4b) and implanted in the human brain, has been demonstrated (Urban, 1990). Various processes are reported to microfabricate the Ag|AgCl layer. It can be grown using electrodeposition in a solution with chloride ions, as described in Berg et al. (1990) and Kinlen et al. (1994). Deposition by electroless chemical reaction or by evaporation of AgCl is also possible (Bousse et al., 1986). When used to apply an electrical field or as reference electrode for potentiostatic measurement

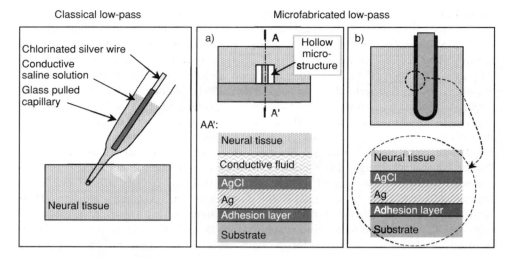

FIGURE 2.4. Schematic view of the classical low-pass microelectrode (left view), and of two typical low-pass electrical interfaces "microsystem–neural tissue" (center and right views): (a) with an indirect contact through a conductive fluid and (b) with direct contact.

(with constant-flowing direct current), the AgCl layer thickness and homogeneity will determine the life cycle of that low-pass electrode (Suzuki *et al.*, 1998a). The amount of noise in the recording system will depend on the quality of this reference electrode.

2.3.1.3. High-Pass Insulated Microelectrodes

In the case of electrodes based on classical fabrication techniques, insulating the active area generally makes little sense. The presence of an insulated layer strongly limits the efficiency and sensitivity of the electrode. However, in neural microsystems with the microelectrode separated from the neuronal cell membrane only by a very thin insulating layer, the ability to interface electrical signals rises considerably. Moreover, onboard amplifying electronics can be integrated and connected to this type of microelectrodes to improve even further the signal-to-noise ratio of recorded signals. For these high-pass insulated microelectrodes, transduction of stimulation and recording signals is exclusively based on electrical field detection and generation by capacitive coupling. Their advantage is that they do rely on a signal transduction principle that is free from electrochemical reaction, which might be particularly advantageous for long-term applications. Such a microfabricated interface with neurons has been successfully demonstrated for stimulation and for recording (Zeck and Fromherz, 2001) (see Figure 2.5). Sensing and amplification of the signal electrical field is performed in this elegant neural microsystem execution using a metal-free MOS-like transistor, whose gate is separated from the electrolyte by a thin oxide layer (typically 10 nm). A change of charges at the nearby neuron membrane will cause a change of charges at the transistor gate, resulting in the modulation of the drain-source current of the transistor.

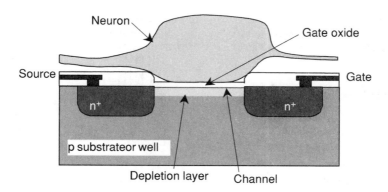

FIGURE 2.5. Schematic view of a metal-free MOS-like transistor, which channel current can be modulated by the activity of a neuron placed on top of its gate oxide.

2.3.2. HOW TO INTERFACE CHEMICAL SIGNALS IN NEURAL MICROSYSTEMS

Despite the facts that neuronal activity consists of fast and slow chemical processes, and that several transducing techniques have been developed for chemical signals, examples of microsystems exchanging chemical signals with neural tissue are still rare. Possible reasons for this can be found in the often-reported short lifetime and low selectivity of microfabricated chemical sensors and in the high performances of well-developed techniques based on standard fabrication processes to analyze chemical concentrations. Several methods to record or stimulate chemical signals in neural microsystems have been or are being developed, while others are still in the design phase. These methods include ion-selective microelectrode probes for chemical sensing and single-microchannel probes, push–pull microchannel probes, and microdialysis probes for drug delivery, sampling, or extraction. They are described in the next subsections. Detection and measurement of molecules using optical signals will be discussed in Section 2.3.3.

2.3.2.1. Selective Microelectrode Probes for the Recording of Neurochemical Signals

In neurophysiology, fast neurochemical signals include transients of species like Ca^{2+}, dopamine, and NO. Typically, the measuring system for this type of signal uses a sensor probe, placed in the region of interest, and a reference electrode, generally placed in a remote location. Depending on the neurochemical species of interest, ion-selective probes are made using two different methods: the classical metallic or carbon fiber technique (Park *et al.*, 1998) or the fluid-filled glass-pulled micropipette technique (Nicholson, 1993). Their basic working principle is illustrated in Figure 2.6. In probes made of solid materials (Figure 2.6a), an electric potential is applied to the conductive layer to cause electrons from the surrounding species to be removed. The concentration of each species involved in this electrochemical oxidation contributes directly to the current flowing through the probe. In potentiostatic recordings the current of a probe set at a fixed potential is measured. Selective layers can make the probe sensitive only to the desired species. In the case of the gaseous neurotransmitter nitric oxide (NO), selectivity in the presence of a large number

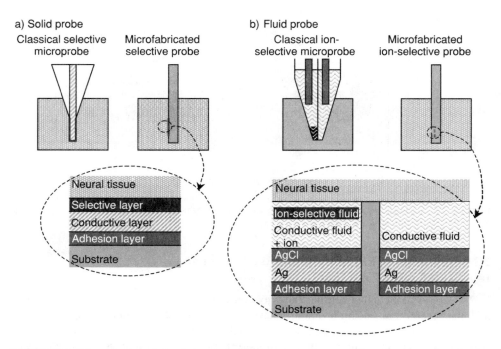

FIGURE 2.6. Schematic view of classical and microfabricated selective neural tissue interfaces: (a) using the solid microelectrode method and (b) using the fluid microelectrodes method.

of possible interfering ionic species has been achieved and demonstrated successfully *in vivo* using a carbon fiber coated with a double selective layer (Park *et al.*, 1998). Integrated versions of this type of probe have been developed to build carbon sensor arrays using sputtering (Fiaccabrino *et al.*, 1996) or even a low-tech process like screen printing (George *et al.*, 2001). To microfabricate this same selective layer, spin-coating is used to deposit the anion-blocking resin (nafion) layer. The second layer consisting of a combination of *m*-phenylenediamine and resorcinol is electrodeposited. The obtained microelectrode probes are selective and sensitive to NO at physiological levels, and are not responsive to analogous concentrations of dopamine, ascorbic acid, and nitrite (Naware *et al.*, 2003). Solid-selective probes can also be used with cyclic voltammetry, where in a defined range the potential is continuously ramped up and down to oxidize and reduce the species of interest, while measuring the resulting current signal to extract the concentration from the peak amplitude. In some cases no selective layer is even deposited. Cyclic voltammetry is then used to monitor the various peaks corresponding to different species and to determine the potential to be set for potentiostatic recording. Microfabrication here can be used to precisely control the size of the electrode, and by placing two probes in close proximity to measure in real time neurochemical as well as neuroelectrical signals, as reported in an example for the monitoring of cultured neuron (Strong *et al.*, 2001).

An interesting alternative to selective solid probes is the ion-selective fluid probe, where the fluid holds by capillary effect at the tip of the microprobe. This type of probe, based on double-barreled glass microelectrodes is well known and widely used in neurophysi-ology (Nicholson, 1993). To the best of the knowledge of the authors, no microfabricated version of the fluid probe has been reported so far. Their working principle is based on two Ag/AgCl-sensitive probes placed next to each other. One of these probes is filled with a solution containing a specific concentration of the ion of interest. The tip of this probe is filled with a fluidic ion-selective exchanger. The Nernst potential generated across the exchanger is measured with the help of the second probe (see Figure 2.6b). The integration of this sensing principle on arrays of microfabricated ion-selective probes could open the way to the imaging of the transient activity of Ca^{2+} and K^+ or of other ions in the neu-ral tissue, which cannot be observed due to the low time resolution of existing imaging techniques.

2.3.2.2. Push–Pull and Microdialysis Probes for Drug Delivery and Neurochemical Analysis

A well-accepted method of chemical exchange with the living neural tissue *in vivo* is based on fluid transport through small channels. Integration of microfluidic channels using microsystems technologies has become very common; however, their use in probe microsystems designed to interface chemical signals with the neural tissue is still very limited and in the developmental stage. A simple implementation is based on probes with single microchannels for the delivery of drug or for the sampling of the cerebrospinal fluid. Several single microchannels in parallel can be used for selective drug delivery (Chen *et al.*, 1997a). In this successful example the effective microchannel diameter is of 10 μm for a 100 pl/s flow of kainic acid and of GABA in the guinea pig inferior and superior colliculus. A problem with this single-channel approach is the difficulty of stopping the delivery of drug, due to natural diffusion occurring at the microchannel orifice. The integration of microshutters and flow meters on single-channel probes is an effective method to address this problem (Papageorgiou, 2001). Single microchannels are not only useful in *in vivo* applications, they also allow the delivery of chemical solution to cultured neurons (Heuschkel *et al.*, 1998). Before the development of microdialysis (Ungerstedt and Hallström, 1987), push–pull probes using two parallel channels or a cannula without any membrane was the most popular method in neurophysiology for direct drug delivery and fluid sampling in the neural tissue (Myers *et al.*, 1998). Microdialysis is based on push–pull probes with a semipermeable membrane separating the dialysate circulating in the probe and the fluidic medium of the neural tissue (see Figure 2.7). Differences in substance concentrations will cause diffusion from the tissue to the dialysate, or vice versa. The role of this semiper-meable membrane is to keep the dialysate free from particles and macromolecules. These membranes vary in cut-off molecular weight and chemical properties, depending on the sub-stance of interest. Different types of selectivity can be achieved. For classical fabrication processes, materials like polycarbonate, polyamide, cuprophan, and polysulphone are used to produce such membranes. Semipermeable membranes for microdialysis microsystems can be microfabricated using surface-micromachined sandwiched layers with nonoverlap-ping holes and with 30–50-nm-thick sacrificial layers between them, which once removed connect the holes by thin flat channels. Permeable polysilicon layers with pore defects of 5–20 nm can also be used for this purpose (Zahn, 2000). An alternate approach valid especially for polyimide-based flexible probes is using irradiation of heavy ions to form

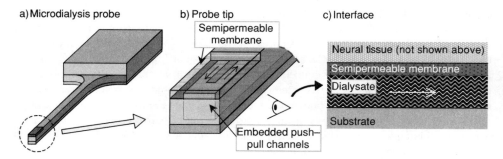

FIGURE 2.7. (a) Schematic views of a microfabricated microdialysis probe; (b) Detailed schematic view of push–pull microchannels below the semi-permeable membrane; and (c) Schematic view of interface with the neural tissue, where molecules are delivered or extracted.

traces and then chemical etching to form holes down to 10 nm diameter in the membrane (Metz, 2002).

Compared with classical microdialysis probes, microfabricated microdialysis probes are still facing some strong limitations. The smaller effective diameter of microchannels resulting from the integration can decrease the nominal flow rate by two orders of magnitude (~10 nl/min instead of 1 µl/min). With the minimal sample size required by existing methods of analysis for the collected dialysate, such as liquid chromatography, capillary electrophoresis, and mass spectroscopy (Lai *et al.*, 2003; Zhang *et al.*, 2003), the duration of measurement increases in an unacceptable manner. New analysis methods using a much smaller amount of fluid for analysis, as well as onboard chemical sensors, are essential for the introduction of microdialysis probes based on neural microsystems.

2.3.3. HOW TO INTERFACE OTHER TYPES OF SIGNALS IN NEURAL MICROSYSTEMS

As illustrated in Figure 2.2, signals interacting between neural tissue and microsystems can be of types other than electrical and chemical signals. To the best of the authors' knowledge, no neural microsystem based on other signal types have been reported yet. Nevertheless, the rapid development of microsystems technology involving optical, magnetic, and other types of signals is expected to permit the integration of promising techniques for neural microsystem recording and stimulating using neither electrical nor chemical signals. This section focuses on a few techniques, involving optical and magnetic signals, that with the development of microsystems technology are likely to become highly interesting alternatives for interfacing the neural tissue. Other types of signals are also briefly described.

2.3.3.1. Optical Signals

Phototransduction of optical signal into neuronal signals is known to occur only with rod and cone photoreceptors in the retina and with no other cell of the nervous system.

However, optical signals can be modulated by various basic mechanisms in neural tissue using fluorescence and light scattering techniques. Fluorescence techniques are based on absorption of light of one or two different wavelengths to raise an electron of the fluorescent compound to a higher energy for a very short time. When this electron returns to its ground state a photon is emitted at a different and specific wavelength. The molecule must either be autofluorescent or be attached to a nontoxic fluorescent dye. This well-accepted method, based on sophisticated microscopes, is appropriate for *in vitro* as well as *in vivo* observations of a high number of molecules. Recently, a miniaturized system based on a single fiber optic equipped with a miniaturized scanning device and a small photomultiplier has been developed and successfully used for *in vivo* two-photon laser microscopy of intracellular neuronal activity and is aimed at freely moving rats (Helmchen *et al.*, 2001). Laser can also be used for scattering and for functional imaging of the neural tissue, as demonstrated in Duarte *et al.* (2003) for the study of depolarization waves (spreading depression) in retina layers. In this technique the laser light is scattered very likely not only by the change of cell volume, but also by alteration of the heterogeneity of particle concentrations in the neural tissue. These optical techniques permit imaging with a high sensitivity. The temporal resolution of the acquired image is, however, strongly limited by the speed of scanning devices and directly dependent on the size of the image of interest. Integrated optical devices are expected to overcome such a limitation and to provide a promising solution for the recording and the study of neural activity.

2.3.3.2. *Magnetic Signals*

A number of examples show that magnetic fields as weak as the earth's can influence the nervous system in some living organisms for guidance and orientation (Wang *et al.*, 2003). Besides this identified but still not well-understood phenomenon, the interactions between neural tissue and magnetic fields are generally limited to magnetic stimulation using pulses of strong magnitude and to the measurement of the magnetic signals generated indirectly by the neuroelectrical activity in the nervous system. These indirect signals can be sensed using highly sensitive magnetic sensors, such as the SQUID sensors, which are able to measure low neuromagnetic fields in the pT range (Nowak *et al.*, 1999). With their essential bulky cooling system, the integration of SQUIDs on a microsystem is technically very challenging. Other approaches are of interest for magnetic sensing compatible with full integration, though they still lack sensitivity to very low neuromagnetic fields (Barjenbruch, 1998; Boero *et al.*, in press). Weak magnetic fields can modulate the neural firing pattern (McFadden, 2002).

Transcranial magnetic stimulation has become a well-accepted method in clinical practice since its development in 1985 (Terao and Ugawa, 2002). Besides the initial interest for cortex mapping, this technique of stimulation is now applied for therapy in psychiatry (Lisanby *et al.*, 2002) and as a potential therapy to control epilepsy (McLean *et al.*, 2001). The technology developed for this promising domain has been mainly aimed at transcranial magnetic stimulation. The stimulator consists of a bulky power source with a large capacitor, generally with a single or double air coil of 9–14 cm diameter able to generate a 100-μs pulse with up to 2 T amplitude, and a cooling system for repetitive stimuli applications. With such a large size necessary because of the 1 cm distance separating head skin and cortex, the volume of neural tissue being excited cannot go below a few cubic centimeters. For animal research, smaller coils (e.g., 2 cm) are often used (Hausmann *et al.*, 2000). Use of smaller coils for a precise stimulation of the neural tissue is a promising method, though a real technical challenge. The thermal loss of the coil, caused by a high serial resistance in smaller coils, is a main problem. Most studies conducted at low magnetic fields have used a single large coil or two Helmholtz coils. To the best of our knowledge, no neural microsystems using magnetic detection or stimulation has been developed and reported yet. The day microsystems technology for precise neuronal magnetic stimulation is available, based on advanced magnetic materials, it will provide a valuable alternative to electrical stimulation and a powerful tool to interface with neural tissue *in vivo* and *in vitro*.

2.3.3.3. Other Types of Signals

Notwithstanding their seldom use, a large diversity of other signals can be employed to interact with the neural tissue and provide interesting functions. Some of these are becoming available on microsystems, or are known to be compatible. These interactions include mechanical or acoustic signals, for instance to induce permeabilization of cell membranes to facilitate delivery of particles for therapeutic purpose (Hensley and Muthuswamy, 2002), or to detect a neurotransmitter like GABA (Zhou, 2002). They also include miniaturized NMR devices developed for minimally invasive spectroscopy that are able to provide extensive chemical information from a living tissue (Berry *et al.*, 2001) using combined magnetic and electromagnetic fields. NMR is a measuring technique that can be now integrated on a microsystem (Massin *et al.*, in press).

2.3.4. HOW TO SET THE INTERFACE POSITION IN NEURAL MICROSYSTEMS

As already mentioned, a great benefit and potential of microsystems technology lies in the possibility of interfacing to the neural tissue at multiple locations simultaneously. The question of interface location between neural tissue and microsystem is usually closely related to the final goal of the system and the type of application. The different approaches to answer this question are often a distinguishing factor between research groups developing microsystems for neural applications. The neural tissue function and structure vary depending on the region of interest and can evolve over time. For instance, the electrical recordings *in vivo* on monkeys for neuroprosthetic applications using microwire arrays have shown that over several days the neurons' properties change and that reassessment of the contribution of each interfacing electrode is necessary (Nicolelis and Chapin, 2002). In this case a large number of implanted microwires is essential so as to always have statistically sufficient contributing neurons for the model. Brain slices with their small sizes are difficult to manipulate and to position at a precise location. Also, over time they tend to become thinner and to lose their structure. Cultured neurons are generally mobile when placed and grown over a flat substrate. Formation of axons can happen in any direction. To position in a stable manner the interaction region between microsystems and neural tissue is challenging. Several techniques to connect microsystems to the extracellular and even to the intracellular region of neural cells have been or are being developed (Hanein *et al.*, 2002). The standard level of interface between microsystems and neural tissue is still at the extracellular

FIGURE 2.8. Schematic view of a flat array of microelectrodes for *in vivo* applications.

level. A selection of these techniques, grouped by application type, is presented in the next subsections.

2.3.4.1. In Vivo *Applications*

The *in situ* microsystem interface to whole brain is a delicate operation involving generally a micromanipulator setup and a stereotactic frame to hold the cranium of the animal or of the patient. The geometry of the microsystem will have a direct impact on the number of neural cells involved and on the final locations of interactions. Several variants with different geometry have been proposed, using different microfabrication processes.

Flat Arrays

The simplest kind of interface involves flat microelectrodes, as illustrated in Figure 2.8, placed for instance over the cortex like classical grid electrodes. This method is not optimal for cortical applications since its efficiency can be up to three orders of magnitude lower than that of penetrating probes with smaller and more defined interfacing regions (Rousche and Normann, 1999). For retinal stimulation, one approach is using a flat microelectrode with a 15-μm-thick circular flexible polyimide substrate especially shaped to fit the curvature of the retina in the epiretinal space (Meyer, 2002). The microfabrication process of such thin and flexible polyimide-based microsystems uses a standard silicon wafer as a temporary rigid substrate with a series of steps, including deposition of polyimide and of metallic layers, photolithography, lift-off, reactive ion etching, and separation of the thin microsystem from the temporary substrate (Meyer, 2002).

Pyramidal-Shape Needle Probe Arrays

Arrays of penetrating probes have been developed for intracortical stimulation, based on pyramidal-shape needles (see Figure 2.9). Each needle provides only one region for

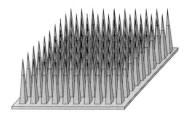

FIGURE 2.9. Schematic view of an array of pyramidal-shape micro-electrodes for *in vivo* applications.

interface. The length of the needles and the interfacing region is such that neurons in the desired layer at the right depth are stimulated (lamina 4Cb for vision) (Normann, 1990). In such an arrangement 10 × 10 arrays of 1.5-mm-long sharp needles are integrated on a 4.2 × 4.2 mm^2 silicon substrate of 120 μm thickness. The pitch distance of the needles is 400 μm. Needles have a side base of 90 μm. The active zone at the needle tip is 0.5 mm long (Campbell, 1991). Production of such microneedle electrodes involves the following microfabrication steps in succession: doping of silicon substrate, thermomigration to insulate the future needles, partial wafer sawing of parallel lines and columns to form 10 × 10 microposts, isotropic wet etching with a special holder to make the needles thin and sharp, and shadow masking for selective deposition and etching at the tip of the needles. This last step is performed by inserting needles through a thin aluminum foil at the desired depth (Campbell, 1991). The contact to the needles is from the back of the substrate, where an integrated electronic circuit for multiplexing and demultiplexing is mounted and connected (Jones, 1997).

Shank Probes (Single, Multiple, Arrays)

An alternative to pyramidal needle probes is the shank probe (Wise and Angell, 1971). Its advantage is the compatibility of integrating multiple interfacing regions on each shank and compatibility for onboard CMOS electronic circuitry for multiplexing and amplification (Olsson, 2002). Single shank was the first shape developed for neural microsystems (Wise and Angell, 1971). This design has two recording electrodes integrated on one shank. Other designs with several electrodes on the single-shank probe, as illustrated in Figure 2.10a, have been developed and successfully used to record the activity of the dorsal column nuclei in the rat spinal cord (Blum, 1991). The number of interfacing sites in this kind of neural microsystem can be significantly increased using arrangements with multiple shanks (see Figure 2.10b) and with arrays of multiple shanks (see Figure 2.10c) (Hoogerwerf, 1994). Multiple shanks with pitch distances as small as 50 μm (6 mm long) can be achieved (Xu et al., 2002). Microchannels with a diameter as small as 10 μm for drug delivery can be integrated in a shank together with electrodes (Chen et al., 1997b). Micropositioning the system to move separately implanted multiple-shank probes is being developed (Muthuswamy et al., 2002). A major drawback of shank probes based on silicon substrate is their low mechanical resistance and their brittleness. Its big advantage is precision and very high resolution due to the monocrystalline nature of the silicon. Other materials such as

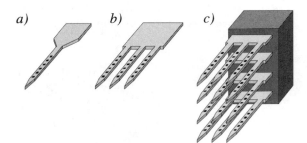

a) b) c)

FIGURE 2.10. Schematic view of (a) single-shank microelectrode arrays (MEA), (b) multiple shanks (MEA), and (c) arrays of multiple shanks (MEA), all for *in vivo* applications.

polyimide have been proposed to develop flexible and mechanically resistant single-shank probes with integrated electrodes on one or two sides (Stieglitz and Gross, 2002), and with a microtank as well as microchannels for drug delivery (Rousche *et al.*, 2001; Metz, 2002). Integrated insertion devices have been developed for the insertion of such flexible shank probes (O'Brien, 2001; Kipke, 2002). The microfabrication processes of shank probes based on silicon substrate involve bulk and surface micromachining steps that include thermal growth of masking oxide, LPCVD, metal deposition, photolithography, boron diffusion, and chemical and dry etch. The process for the microfabrication of polyimide-based shanks is simpler. It includes similar steps as for the flat microelectrode arrays. Long cylindrical substrates in Al_2O_3 is an interesting alternative for deep brain applications. They require the adaptation of microfabrication processes with a nonflat substrate. An example of such a single circular shank with eight electrodes has been successfully used in human intracerebral recordings (Urban, 1990).

Sieve Probes

The working principle of sieve probes is based on the ability of peripheral nerve fibers, or the axons, to regenerate through the array of holes in a sieve probe, as illustrated in Figure 2.11. The interesting thing about these probes is their ability to maintain over time the interface location between the neural tissue and the microsystem. Such probes have been developed for electrical signals since the early developments of neural microsystems and are still the object of recent works (Bradley *et al.*, 1997), wherein they have been successfully used for the parallel recording of evoked sensory responses from the tongue mechanoreceptors in rats. Sieve arrays with eight electrodes can be microfabricated using a silicon substrate with a local 15–20-μm-thick membrane through which 64 square holes of 90×90 μm^2 area have been etched using a process compatible with CMOS electronic circuitry (Kovacs, 1994b). An execution of sieve electrode arrays with 497 circular holes of 5 μm diameter, five of which have electrical contact around the hole, has been demonstrated to be successful for long-term chronic recording of the glossopharyngeal nerve (Bradley *et al.*, 1997). It uses a thin ribbon cable for the assembly and connection of the sieve electrode array, as well as two small tubes (not shown on Figure 2.11) to facilitate guiding and holding of the regenerating fibers. For this type of probe, flexible substrates like polyimide are being used to develop sieve electrode arrays with built-in cable and interconnection to an electronic chip (Stieglitz *et al.*, 1997). A drawback of this method is the need to cut the nerve fiber and have it regenerate through the probe. At this time the regenerating process takes time

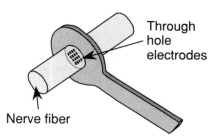

FIGURE 2.11. Schematic view of a sieve probe array for *in vivo* applications.

Through hole electrodes

Nerve fiber

and is only partial. Removal of such an implant might also cause the same problem. The improvement of this technique is likely to make this type of neural microsystem quite attractive.

2.3.4.2. In Vitro *Tissue Slice Applications*

Placed between *in vivo* applications with integral nervous systems and *in vitro* applications at the cellular level with cultured neurons, are the *in vitro* brain slice applications. Brain slice studies are of particular interest for brain research, drug development, and potentially also for computing applications using its intact network structure. Developed in the mid-1950s, brain slice preparation has become a well-accepted method for observing basic neuronal mechanisms on the cellular and on the circuit level (Colligridge, 1995). The direct access to the neural tissue overcoming the blood–brain barrier is particularly advantageous. There are two different types of brain slice preparations: acute and organotypic cultured. The neural cell structure of acute slices is closer to the living brain cell structure than are organotypic slices'. Cultured slice has been used with a planar array of 60 microelectrodes in studies lasting up to 4 weeks (Egert *et al.*, 1998), whereas in acute slices, survival is difficult (generally limited to a period of several hours to a couple of days), which hinders long-term studies. Neural microsystems based on a microelectrode array and using brain slices date back to the early 1980s (Jobling *et al.*, 1981). More recently, microelectrode arrays with 32 microelectrodes have been used to study the details of epileptiform activity propagation in hippocampal slices (Novak, 1988). The positioning of the active zones between this type of microsystem and the neural tissue depends on the relative position of the slice with the microelectrodes, and on their shape and structure. The placing of a neural tissue slice on top of a quasi-planar microsystem is generally done manually and requires some experience, as described in a study by Michael (1999). To match structures of interest in the slice, like the stratum pyramidale of the CA1, CA2, and CA3 regions in the hippocampal slice, microelectrodes can be arranged elliptically to optimize the interfacing in the desired locations (Thiébaud *et al.*, 1999). A large area of densely arranged high-resolution microelectrode arrays are likely to bypass this type of positioning problem. Figure 2.12 illustrates some of the shapes found in the literature for interfacing neural slices with microsystems as compared to the classical glass-pulled microelectrodes. Cone-shaped microelectrodes 47 μm high with 270-μm^2 active areas (Thiébaud *et al.*, 1999), and 60-μm-high pyramidal-shape microelectrodes with 40 × 40 μm^2 base (corresponding to 5000 μm^2 of active area) (Heuschkel *et al.*, 2002) have been developed and demonstrated in interfaces with hippocampal slices. Compared with flat microelectrodes, such microelectrodes have up to three times higher efficiency (Heuschkel *et al.*, 2002). Microwire-shaped microelectrodes 160 μm high with a 3.5-μm-diameter tip in doped silicon (Kawano *et al.*, 2002) have a shape similar to classical microelectrodes, and a distance that permits penetrating well inside the 300–500 μm thickness of typical slices. Such a shape might reach the efficiency of microelectrodes fabricated with classical technologies. These microsystems are microfabricated using processes similar to those used for shank probes. For neural tissue slice applications, the potential of microsystems to provide a large number of parallel regions of interaction with the tissue is of particular interest, especially when considering the restricted space surround-

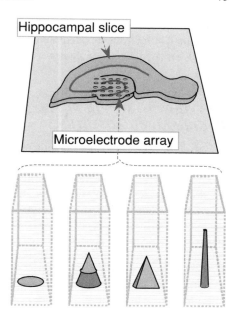

FIGURE 2.12. Schematic view of various shape variants for microelectrode contacts in tissue slice applications.

ing the slice setup, and the bulky holders for classical microelectrodes. The success of such neural microelectrodes in existing and in future applications will strongly depend on their ability to approach the performance of classical glass-pulled and needle microelectrodes.

2.3.4.3. In Vitro *Cell Applications*

In cell cultures, position and growth control of the neuron is of high significance. A number of techniques have been developed to place and maintain neurons over the active region of a microsystem, as well as to guide the growth of neuronal processes along the needed direction to build a desired network. Microsystems similar to a diving board and with integrated microelectrodes, to contact and hold a neuron cell body from the top, have been demonstrated for chronic two-way electrical connections (Regehr, 1988). Microscopic fences of 25 µm diameter and 40 µm height made of polyimide and located around the cell body can be used to immobilize neurons (Zeck and Fromherz, 2001). The placement of neurons on top of a microelectrode array and the connection between these neurons can be also obtained by placing on top of the array a three-dimensional microfluidic network made of open wells and closed microchannels (Griscom *et al.*, 2002). Chemical patterning is another approach where the microsystem substrate is prepared by photopatterning a barrier structure on a glass substrate. Neural pathways are created using selective adsorption of poly-L-lysine on glass to form a hydrophilic growth matrix. Hydrophobic regions are formed by adsorption of albumin proteins on the perfluoropolymer (Griscom *et al.*, 2002). When attached on 3–5-µm-high micropost arrays, neuronal processes can as well be guided and elongated mechanically by moving the microtool at a 36 µm/h translation speed in the desired direction (Baldi *et al.*, 2002). Cultured neurons can also be grown and maintained in microwells integrated in a shank probe with the goal to guarantee an optimal contact between electrode and neural tissue after *in vivo* implantation (Tatic-Lucic *et al.*, 1997).

2.3.5. HOW TO SUPPLY NUTRIENTS, O_2, AND STABLE TEMPERATURE IN NEURAL MICROSYSTEMS

Neural cells are totally dependent on oxygen and nutrients delivered through the cerebrospinal fluidic medium for their survival (Guyton, 1991). Choice of methods for supplying the necessary neurochemicals or drugs to a neural tissue placed in a sudden situation of need directly depends on the application. For the brain *in vivo*, such a method would be of great benefit as a therapy for acute stroke or for a disorder of neurochemical imbalance, such as Parkinson's disease. For *in vitro* preparations this situation occurs when the neural tissue is harvested from its natural environment. Oxygen, nutrients, and a stable temperature have to be supplied to assure energy provision and survival for the neural cells. In addition, for cultures of neural tissue, sterility of the environment is crucial. Perfusion chambers to deliver oxygenated artificial cerebrospinal fluid and temperature control systems play a significant role in this type of application. Typically, current supply methods for the neural tissue are based on classical instruments and devices such as container dishes or classical chambers. For neural cells cultured on top of a microsystem, the fluidic medium fills a container such as a dish in which the microsystem is placed, or a glass ring placed on top of the microsystem (Griscom *et al.*, 2002). Neurons can also be grown inside microfabricated microchannels that are used to contain and to deliver the fluidic medium (Heuschkel *et al.*, 1998). For brain slices, numerous perfusion chamber systems based on classical fabrication technologies have been developed. Their compatibility to be used with microsystems like microelectrode arrays is however low. Some attempts to assemble microdevices with brain slice perfusion chambers have recently been presented (Egert *et al.*, 1998; Heuschkel *et al.*, 2002). They usually face the problems of limited perfusion and of holding the slice in position. In some other studies (Boppart *et al.*, 1992; Thiébaud *et al.*, 1999) microelectrode arrays on perforated substrates have been used. They permit some contact between the tissue and the perfusion medium. The integration of a perfusion system for brain slices based on a microfabricated fluidic chamber has been recently reported (Passeraub *et al.*, 2003). The chamber is made of a glass substrate on top of which a thick photopolymer layer is deposited and patterned in a manner reminiscent of the Haas interface chamber (Haas *et al.*, 1979). This simple and promising microfabricated chamber is described in more detail in the next section.

The survival of neural tissue *in vitro* also depends on other tools, instruments, and pieces of equipment like tubing, fluidic connections, thermostats, thermistor, electronic systems, pumps, carbogen tank, or incubator. In the common experimental environment generally located in a laboratory, the implementation of a microsystem to interface with the neural tissue is generally limited to the adaptation of the existing dishes or chamber. Future applications such as neurocomputing are likely to associate microsystems and neural tissue outside a laboratory environment. They will require the miniaturization of a full system for neural tissue survival. Microsystems technology is expected to make such a miniaturization feasible.

2.4. EXAMPLE OF NEURAL MICROSYSTEM DEVELOPMENT

The progress of fundamental neurophysiology research is highly dependent on available tools to study the neural tissue. In this section an example of microsystem development for the study of slow extracellular electrical activity in brain slices is described.

Slow neuronal activity is believed to play a significant role and contain relevant information for the analysis of various brain disorders like epilepsy, migraine, or ischemia (e.g., Martins-Ferreira *et al.*, 2000). Brain slice preparations provide a simplified model to investigate the basic mechanisms underlying such brain disorders. The extracellular field potential signals recorded from the dentate gyrus of a rat hippocampal slice, using a zero-Ca^{2+}/high-K^+ model of epileptiform activity and a glass-pulled microelectrode, show this combination of very slow and high-frequency activity as shown in Figure 2.13. With classical glass-pulled microelectrodes, the electrical signals of only a few regions can be measured simultaneously. Such measurements based on a classical approach are limited in the number of regions that can be recorded simultaneously. It allows valuable but limited observations. Another method that permits visualization of electrical potentials in the brain tissue is based on the use of voltage-sensitive fluorescent dye that signal transmembrane potentials or intracellular Ca^{2+} and microscopy. However, due to limitations in scanning systems, such experiments face the trade-off between temporal resolution of the recorded signal versus size of acquired image. This is the case in particular when working with two-photon microscopy, a powerful tool for functional studies of brain slices at the cellular level.

Such studies can include glass-pulled microelectrode recordings and electrical stimulation to give some additional access to the global picture of the involved neural processes. It is however made difficult because of the restricted available space with the objective lens

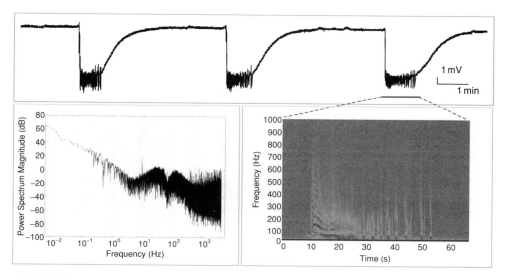

FIGURE 2.13. *Top:* experimental extracellular field potential recordings of spontaneous epileptiform activity from the dentate gyrus of a rat hippocampal slice. *Bottom left:* power spectral density of this recorded signal showing strong contribution of very slow signals. *Bottom right:* spectrogram of the recorded signals during epileptiform bursting showing high frequency contributions in the signal.

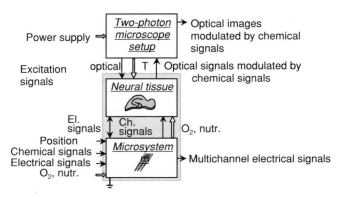

FIGURE 2.14. System configuration of experimental setup combining neural microsystem and two-photon microscopy for basic research.

positioned right on top of the brain slice. In such a case, a microsystem with onboard arrays of microelectrodes to be included in the microscopy setup is expected to be very helpful. However, this new type of experiment requires properties that have not been included in microsystems developed until now.

These properties are (a) to be able to record electrical signals down to the direct current (DC); (b) to allow optimal delivery of perfusion medium and oxygen to the brain slice by maximizing access to the slice surface; and (c) to be able to rapidly exchange various drugs infused through the perfusion medium. The microsystem presented here is a first version of a low-pass microelectrode array with an embedded perfusion system designed to address these needs. The corresponding configuration describing the role of this new microsystem interfacing a brain slice inside the setup of the two-photon microscope is described in Figure 2.14. The new microsystem is based on a microfluidic perfusion chamber newly developed and successfully tested for fast perfusion medium exchange and for brain slice studies (Passeraub *et al.*, 2003). It includes arrays of 6×6 Ag|AgCl microelectrodes. The base of the 29×11 mm^2 chamber is filled with 100×100 μm^2 microposts (see Figure 2.15). Their role is to perfuse brain slices from beneath and to uphold the tissue at the interface

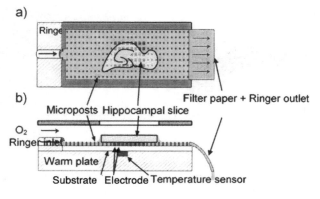

FIGURE 2.15. Schematic view from the top (a) and the side (b) of the microfabricated perfusion chamber with integrated microelectrode arrays for recordings of slow potentials.

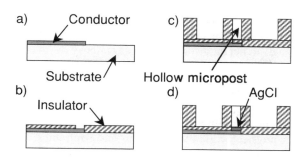

FIGURE 2.16. Microfabrication process used: (a) deposition of Ag, photolithography and lift-off of the Ag; (b) deposition of insulation layer, photolithography and patterning; (c) thick photosensitive film deposition, exposure, and development; and (d) electrodeposition of AgCl.

between perfusion medium and moisturized oxygen. These microposts with the sidewalls confine the perfusion flow within the chamber through surface tension effects. The $100 \times 100 \ \mu m^2$ microelectrodes are integrated in larger microposts. The electrical contact between the neural tissue and the low-pass microelectrodes occurs through the saline solution in a manner reminiscent of glass-pulled micropipettes, though with a $120 \times 120 \ \mu m^2$ aperture to measure field potentials from the surface of the tissue. Two Ag|AgCl reference electrodes are also integrated in the chamber. The microfabrication steps are illustrated in Figure 2.16. An adhesion layer of Ti (20 nm) and a layer of Ag (150 nm) were electron-beam-evaporated on top of a glass substrate. A lift-off process was used to define the lines, pads, and contacts. Insulation of the lines was performed using a photo-patterned layer of thick photoresist (25 μm of SU-8). The walls and the regular and the hollow microposts were produced using another layer of photo-patterned thick photoresist (300 μm of SU-8). The silver layer is to be increased to 3 μm by electrodeposition. Chlorination of silver contacts is then to be performed by electrodeposition using a 10% HCl solution. A picture of the new microfluidic system is shown in Figure 2.17 with detailed views of the microelectrode arrays.

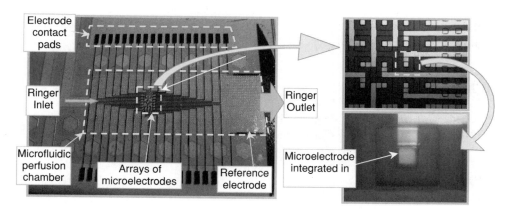

FIGURE 2.17. *Top:* Picture of a microfabricated flat perfusion chamber with integrated microelectrodes array (6×6) and with integrated reference electrodes (before chlorinating the silver microelectrodes). *Bottom left:* Detailed picture of the arrangement of the hollow microposts $(300 \times 300 \ \mu m^2)$ with microelectrodes (left on the picture). *Bottom right:* Detailed picture of a single microelectrode of $100 \times 100 \ \mu m^2$.

This new low-pass microelectrode array with embedded perfusion system, which is still in the developmental stage, is expected to contribute to the understanding of the role of slow neuronal signals in neural disorders such as epilepsy.

2.5. CONCLUDING REMARKS

Neural microsystems are highly complex systems that connect the biological world of neural tissue with the physical world of instruments and electronics. They can be described as hybrid systems consisting of two subsystems interfacing these worlds. They are currently applied in several applications for basic research and drug discovery. They are expected to notably influence the development of the newer versions of existing implants for the treatment of neural disorders, of existing neuroprostheses, and of existing tools for neuro-physiology research. Neural microsystems are also anticipated to find new applications (i.e. new types of neuroprostheses and treatments, monitoring, neuron-based sensors, neurocomputing). Each application has different system requirements and configurations. There is no universal microsystem solution to interface all forms of neural tissue. Nevertheless, based on the current developments and demonstrations, microsystems technology can be considered as highly adapted to meet the challenge of interfacing the physical world with neural tissue. Some questions are still to be clarified: What is the optimal method of interfacing a microsystem with neurons and glial cells? Is it a probe placed in the extracellular space, or rather, the intracellular space? Or is it creating an artificial synaptic connection? Or is it by developing hybrid solution using cultured neurons on the microsystem as an intermediary connection? Most probably the answers to these questions will also depend on the kind of application. For the development of *ex vivo* applications, the problem of the survival of the neural tissue outside the laboratory environment is another key issue that needs to be addressed. For implantable applications, the integration of electronic circuitry with the neural microsystem for signal processing and transmission in a suitable system is essential and has been already demonstrated in a few examples. The resolution of practical issues such as mechanical resistance or packaging will determine their implementation success for a large-scale application. In general the successful development of neural microsystems and of their applications will be directly dependent on the ability of solving these problems of highly interdisciplinary nature.

ACKNOWLEDGMENTS

This work was partially supported by the Swiss National Science Foundation, and by the NIH grants MH 062444 and MH 063159.

REFERENCES

Anderson, D. J. Najafi, K., Tanghe, S. J., Evans, D. A., Levy, K. L., Hetke, J. F., Xue, X., Zappia, J. J., and Wise, K. D., 1989, Batch fabricated thin-film electrodes for stimulation of the central auditory system, *IEEE Trans. Biomed. Eng.,* **36:**693–704.

Badi, A. N., Kertesz, T. R., Gurgel, R. K., Shelton, C., and Normann, R. A., 2003, Development of a novel eighth-nerve intraneural auditory neuroprosthesis, *Laryngoscope* **113:**833–842.

Baldi, A., Fass, J. N., De Silva, M. N., Odde, D. J., and Ziaie, B., 2002, A microtool for *in vitro* cell array manipulation, in: *2nd Annual International IEEE-EMB Special Topic Conference on Microtechnologies in Medicine & Biology*, pp. 180–183.

Barjenbruch, U., 1998, New kind of highly sensitive magnetic sensors with wide bandwidth, *Sens. Actuat. A: Phys.* **65**:136–140.

Bear, M. F., Connors, B. W., and Paradiso, M. A., 2001, *Neuroscience Exploring the Brain*, 2nd ed., Lippincott Williams and Wilkins, Baltimore.

Berg, A. V. D., Grisel, A., Vlekkert, H. H. V. D., and Rooij, N. F. D., 1990, A micro-volume open liquid-junction reference electrode for pH-ISFET, *Sens. Actuat. B* **1**:425–432.

Bernstein, K., 2001, Cogent: One slice at a time, in: *BioCentury, the Bernstein Report on BioBusiness*, BioCentury Publications, Inc., San Carlos, CA, pp. 1–2.

Berry, L., Renaud, L., Kleimann, P., Morin, P., Armenean, M., and Saint-Jalmes, H., 2001, Development of implantable detection microcoils for minimally invasive NMR spectroscopy, *Sens. Actuat. A: Phys.* **93**:214–218.

Blum, N. A., Carkhuff, B. G., Charles, H. K., Jr., Edwards, R. L., and Meyer, R. A., 1991, Multisite microprobes for neural recordings, *IEEE Trans. Biomed. Eng.* **38**:68–74.

Boero, G., Demierre, M., Besse, P.-A., and Popovic, R. S., Micro-hall devices: Performance, technologies and applications, *Sens. Actuat. A: Phys.* **106**:314–320.

Boppart, S. A., Wheeler, B. C., and Wallace, C. S., 1992, A flexible perforated microelectrode array for extended neural recordings, *IEEE Trans. Biomed. Eng.* **39**:37–42.

Borkholder, D. A., Bao, J., Maluf, N. I., Perl, E. R., and Kovacs, G. T. A., 1997, Microelectrode arrays for stimulation of neural slice preparations, *J. Neurosci. Methods* **77**:61–66.

Bousquet, O., Balakrishnan, K., and Honavar, V., 1998, Is the hippocampus a Kalman filter? in: *Pacific Symposium on Biocomputing '98*, Maui, Hawaii (USA), pp. 657–668.

Bousse, L. J., Bergveld, P., and Geeraedts, H. J. M., 1986, Properties of Ag/AgCl electrodes fabricated with IC-compatible technologies, *Sens. Actuat.* **9**:179–197.

Boveja, B. B., 2001, Apparatus and method for adjunct (add-on) therapy of partial complex epilepsy, generalized epilepsy and involuntary movement disorders utilizing an external stimulator, in: US Patent Office, USA.

Bradley, R. M., Cao, X., Akin, T., and Najafi, K., 1997, Long term chronic recordings from peripheral sensory fibers using a sieve electrode array, *J. Neurosci. Methods* **73**:177–186.

Bucher, V., Graf, M., Stelzle, M., and Nisch, W., 1999, Low-impedance thin-film polycrystalline silicon microelectrodes for extracellular stimulation and recording, *Biosens. Bioelectron.* **14**:639–649.

Campbell, P. K., Jones, K. E., Huber, R. J., Horch, K. W., and Normann, R. A., 1991, A silicon-based, three-dimensional neural interface: Manufacturing processes for an intracortical electrode array. *IEEE Trans. Biomed. Eng.* **38**:758–768.

Chen, J., Wise, K. D., and Hetke, J. F., and Bledsoe, S. C., Jr., 1997a, A multichannel neural probe for selective chemical delivery at the cellular level, *IEEE Trans. Biomed. Eng.* **44**:760–769.

Chen, J., Wise, K. D., Hetke, J. F., and Bledsoe, S. C., Jr., 1997b, A multichannel neural probe for selective chemical delivery at the cellular level, *IEEE Trans. Biomed. Eng.* **44**:760–769.

Colligridge, G. L., 1995, The brain slice preparation: A tribute to the pioneer Henry McIlwain, *J. Neurosci. Methods* **59**:5–9.

Damian, M., Davis, T. L., Konrad, P. E., Roberts, A. G., Pfister, A. A., and Charles, P. D., 2003, Deep brain stimulation: A new treatment for Parkinson's disease, *Tenn. Med* **96**:33–35.

Dario, P. C., and Carozza, M. C., 1995, Interfacing microsystems and biological systems, in: *Proceedings of the Sixth International Symposium on Micro Machine and Human Science, 1995, MHS '95.*, pp. 57–66.

Duarte, M. A., Almeida, A. C. G., Infantosi, A. F. C., and Bassani, J. W. M., 2003, Functional imaging of the retinal layers by laser scattering: An approach for the study of Leão's spreading depression in intact tissue, *J. Neurosci. Methods* **123**:139–151.

Egert, U., and Haemmerle, H., 2002, Application of the microelectrode-array (MEA) technology in pharmaceutical drug research, in: *Sensoren im Fokus neuer Anwendungen* (J. P. Baselt and G. Gerlach, eds.), Universitaetsverlag, Dresden, pp. 51–54.

Egert, U., Schlosshauer, B., Fennrich, S., Nisch, W., Fejtl, M., Knott, T., Müller, T., and Hämmerle, H., 1998, A novel organotypic long-term culture of the rat hippocampus on substrate-integrated multielectrode arrays, *Brain Res. Prot.* **2**:229–242.

Feijtl, M., 1999, The MEA—Cookbook: A guide to multichannel-recording of acute hippocampal slices, in: *Multi Channel Systems MCS GmbH*, http://www.multichannelsystems.com/download/Userguides/userguides/Cookbook.pdf.

Fiaccabrino, G. C., Tang, X-M., Skinner, N., Rooij, N. F. D., and Koudelka-Hep, M., 1996, Interdigitated micro-electrode arrays based on sputtered carbon thin-films, *Sens. Actuat. B* **35–36**:247–254.

Fromherz, P., 2002, Electrical interfacing of nerve cells and semiconductor chips, *Chem. Phys. Chem.* **3**:276–284.

George, P. M., Muthuswamy, J., Currie, J., Thakor, N. V., and Paranjape, M., 2001, Fabrication of screen-printed carbon electrode arrays for sensing neuronal messengers, *Biomed. Microdev.* **3**:307–313.

Gesteland, R. C., Howland, B., Lettvin, J. Y., and Pitts, W. H., 1959. Comments on Microelectrodes, *Proc. IRE* **47**:1856–1862.

Gluckman, B. J., Nguyen, H., Weinstein, S. L., and Schiff, S. J., 2001, Adaptative electrical field control of epileptic seizures, *J. Neurosci.* **21**:590–600.

Graupe, D., 2002, An overview of the state of the art of noninvasive FES for independent ambulation by thoracic level paraplegics, *Neurol. Res.* **24**:431–442.

Griscom, L., Degenaar, P., LePioufle, B., Tamiya, E., and Fujita, H., 2002, Techniques for patterning and guidance of primary culture neurons on micro-electrode arrays, *Sens. Actuat. B: Chem.* **83**:15–21.

Gross, G. W., Wen, W. Y., and Lin, J. W., 1985, Transparent indium-tin oxide electrode patterns for extracellular, multisite recording in neuronal cultures, *J. Neurosci. Methods* **15**:243–252.

Gross, G. W., Rieske, E., Kreutzberg, G. W., and Meyer, A., 1977, A new fixed-array multi-microelectrode system designed for long-term monitoring of extracellular single unit neuronal activity in-vitro, *Neurosci. Lett.* **6**:101–105.

Guyton, A. C., 1991, Cerebral blood flow, the cerebrospinal fluid, and brain metabolism, in: *Textbook of medical physiology*, 8th ed., Harcourt Brace Jovanovich, Inc., Philadelphia, p. 684.

Haas, H. L., Schaerer, B., and Vosmansky, M., 1979, A simple perfusion chamber for the study of nervous tissue slices in vitro, *J. Neurosci. Methods* **1**:323–325.

Hanein, Y., Böhringer, K. F., Wyeth, R. C., and Willows, A. O. D., 2002, Towards MEMS probes for intracellular recording, in: *Sensors Update*, Wiley-VCH Verlag GmbH, Weinheim, pp. 47–75.

Hausmann, A., Weis, C., Marksteiner, J., Hinterhuber, H., and Humpel, C., 2000, Chronic repetitive transcranial magnetic stimulation enhances c-fos in the parietal cortex and hippocampus, *Mol. Brain Res.* **76**:355–362.

Helmchen, F., Fee, M. S., Tank, D. W., and Denk, W., 2001, A miniature head-mounted two-photon microscope: High-resolution brain imaging in freely moving animals, *Neuron* **31**:903–912.

Hensley, A. B., and Muthuswamy, J., 2002, Ultrasound induced permeabilization of cell membranes as a therapy for cytotoxic neuronal edema, in: *Proceedings of the Second Joint Engineering Society EMBS/BMES Conference*, 2002.

Heuschkel, M. O., Guerin, L., Buisson, B., Bertrand, D., and Renaud, P., 1998, Buried microchannels in photopolymer for delivering of solutions to neurons in a network, *Sens. Actuat. B: Chem.* **48**:356–361.

Heuschkel, M. O., Fejtl, M., Raggenbass, M., Bertrand, D., and Renaud, P., 2002, A three-dimensional multi-electrode array for multi-site stimulation and recording in acute brain slices, *J. Neurosci. Methods* **114**:135–148.

Hoogerwerf, A. C., and Wise, K. D., 1994, A three-dimensional microelectrode array for chronic neural recording, *IEEE Trans. Biomed. Eng.* **41**:1136–1146.

Huang, G. T., 2003, Mind-machine merger, *Technol. Rev.* 39–45.

Jezernik, S., Craggs, M., Grill, W. M., Creasey, G., and Rijkhoff, N. J. M., 2002, Electrical stimulation for the treatment of bladder dysfunction: Current status and future possibilities, *Neurol. Res.* **24**:413–430.

Jobling, D. T., Smith, J. G., and Wheal, H. V., 1981, Active microelectrode array to record from the mammalian central nervous system *in vitro*, *Med. Biol. Eng. Comput.* **19**:553–560.

Jones, K. E., and Normann, R. A., 1997, An advanced demultiplexing system for physiological stimulation, *IEEE Trans. Biomed. Eng.* **44**:1210–1220.

Kawano, T., Kato, Y., Futagawa, M., Takao, H., Sawada, K., and Ishida, M., 2002, Fabrication and proper-ties of ultrasmall Si wire arrays with circuits by vapor–liquid–solid growth, *Sens. Actuat. A: Phys.* **97–98**:709–715.

Kinlen, P. J., Heider, J. E., and Hubbard, D. E., 1994, A solid-state pH sensor based on a Nafion-coated irid-ium oxide indicator electrode and a polymer-based silver chloride reference electrode, *Sens. Actuat. B* **22**:13–25.

Kipke, D. R., Pellinen, D. S., and Vetter, R. J., 2002, Advanced neural implants using thin-film polymers, in: *IEEE International Symposium on Circuits and Systems, 2002, ISCAS 2002*, vol. 174, pp. IV-173–IV-176.

Kotzar, G., Freas, M., Abel, P., Fleischman, A., Roy, S., Zorman, C., Moran, J. M., and Melzak, J., 2002, Evaluation of MEMS materials of construction for implantable medical devices, *Biomaterials* **23**:2737–2750.

Kovacs, G. T. A., 1994a, Introduction to the theory, design, and modeling of thin-film microelectrodes for neural interfaces, in: *Enabling Technologies for Cultured Neural Networks* (D. A. Stenger and T. McKenna, eds.), Academic Press, pp. 121–165.

Kovacs, G. T. A., Storment, C. W., Halks-Miller, M., Belczynski, C. R., Jr., Santina, C. C. D., Lewis, E. R., and Maluf, N. I., 1994b, Silicon-substrate microelectrode arrays for parallel recording of neural activity in peripheral and cranial nerves, *IEEE Trans. Biomed. Eng.* **41**:567–577.

Kupferberg, H., 2001, Animal models used in the screening of antiepileptic drugs, *Epilepsia* **42**:7–12.

Lai, L., Lin, L.-C., Lin, J.-H., and Tsai, T.-H., 2003, Pharmacokinetic study of free mangiferin in rats by microdialysis coupled with microbore high-performance liquid chromatography and tandem mass spectrometry, *J. Chromatogr. A* **987**:367–374.

Lesser, R. P., Arroyo, S., Crone, N., and Gordon, B., 1998, Motor and sensory mapping of the frontal and occipital lobes, *Epilepsia* **39**:S69–S80.

Lisanby, S. H., Kinnunen, L. H., and Crupain, M. J., 2002, Applications of TMS to Therapy in Psychiatry, *J. Clin. Neurophysiol.* **19**:344–360.

Mainen, Z. F., Maletic-Savatic, M., Shi, S. H., Hayashi, Y., Malinow, R., and Svoboda, K., 1999, Two-photon imaging in living brain slices, *Methods: Comp. Methods Enzymol.* **18**:231–239.

Martins-Ferreira, H., Nedergaard, M., and Nicholson, C., 2000, Perspectives on spreading depression, *Brain Res. Rev.* **32**:215–234.

Massin, C., Vincent, F., Homsy, A., Ehrmann, K., Boero, G., Besse, P-A., Daridon, A., Verpoorte, E., de Rooij, N. F., and Popovic, R. S., Planar microcoil-based microfluidic NMR probes, *J. Magn. Reson.* **164**: 242–255.

Maynard, E. M., 2001, Visual prosthesis, *Annu. Rev. Biomed. Eng.* **3**:145–170.

McAdams, E. T., and Jossinet, J., 1998, Non-linear transcient response of electrode-electrolyte interface, in: *20th Annual International Conference of the IEEE Engineering in Medicine and Biology*, pp. 1789–1790.

McFadden, J., 2002, Synchronous firing and its influence on the brain's electromagnetic field: Evidence for an electromagnetic field theory of consciousness, *J. Conscious. Stud.* **9**:23–50.

McLean, M., Engström, S., and Holcomb, R., 2001, Magnetic Field Therapy for Epilepsy, *Epilepsy Behav.* **2**:S81–S87.

Metz, S., Trautmann, C., Bertsch, A., and Renaud, P., 2002, Flexible microchannels with integrated nanoporous membranes for filtration and separation of molecules and particles, in: *The Fifteenth IEEE International Conference on Micro Electro Mechanical Systems, 2002*, pp. 81–84.

Meyer, J-U., 2002, Retina implant—A bioMEMS challenge, *Sens. Actuat. A: Phys.* **97–98**:1–9.

Mrksich, M., 2000, Cell-based sensors, other non-medical applications, in: *WTEC Study on Tissue Engineering Research, U. S. Review Workshop* (G. M. Holdridge, ed.), International Technology Research Institute, Baltimore, MD, pp. 133–134.

Muthuswamy, J., Salas, D., and Okandan, M., 2002, A chronic micropositioning system for neurophysiology, in: *Engineering in Medicine and Biology, 2002. 24th Annual Conference and the Annual Fall Meeting of the Biomedical Engineering Society, Proceedings of the Second Joint EMBS/BMES Conference, 2002*, pp. 2115–2116.

Myers, R. D., Adell, A., and Lankford, M. F., 1998, Simultaneous comparison of cerebral dialysis and push–pull perfusion in the brain of rats: A critical review, *Neurosci. Biobehav. Rev.* **22**:371–387.

Narayanan, N. M., 1999, Development of a silicon retinal implant: Cortical evoked potentials following focal stimulation of the rabbit retina with light and electricity, *Clin. Neurophysiol.* **110**:1545–1553.

Naware, M, Thakor, N. V., North, R., Murari, K., Passeraub, P. A., 2003, Design and microfabrication of a polymer-modified carbon sensor array for the measurement of neurotransmitter signals, In: *25th annual international conference of the IEEE Engineering in Medicine and Biology Society*, Cancun, Mexico.

Neuman, M. R., 1998, Biopotential electrodes, in: *Medical Instrumentation Application and Design*, 3rd ed. (J. G. Webster, ed.), John Wiley & Sons, Inc., New York, pp. 183–232.

Nicholson, C., 1993, Ion-selective microelectrodes and diffusion measurements as tools to explore the brain cell microenvironment, *J. Neurosci. Methods* **48**:199–213.

Nicolelis, M. A. L., and Chapin, J. K., 2002, Controlling robots with the mind, *Sci. Am.* **287**:46–53.

Normann, R. A., 1990, A penetrating, cortical electrode array: Design considerations, in: *IEEE International Conference on Systems, Man and Cybernetics, 1990*, pp. 918–920.

Normann, R. A., and Branner, A., 1999, A multichannel, neural interface for the peripheral nervous system, in: *IEEE International Conference on Systems, Man, and Cybernetics, 1999*, vol. 374, pp. 370–375.

Novak, J. L., 1988, Two-dimensional electrode array studies of propagating epileptiform neural activity in the rat hippocampal slice, in: *Electrical and Computer Engineering, University of Illinois at Urbana-Champaign*, p. 121.

Novak, J. L., and Wheeler, B. C., 1986, Recording from the aplysia abdominal ganglion with a planar microelectrode array, *IEEE Trans. Biomed. Eng.* **BME-33:**196–202.

Nowak, H., Giessler, F., Huonker, R., Haueisen, J., Rother, J., and Eiselt, M., 1999, A 16-channel SQUID-device for biomagnetic investigations of small objects, *Med. Eng. Phys.* **21:**563–568.

O'Brien, D. P., Nichols, T. R., and Allen, M. G., 2001, Flexible microelectrode arrays with integrated insertion devices, in: *The 14th IEEE International Conference on Micro Electro Mechanical Systems, 2001, MEMS 2001*, pp. 216–219.

Olsson, R. H., III., Gulari, M. N., and Wise, K. D., 2002, Silicon neural recording arrays with on-chip electronics for in-vivo data acquisition, in: *2nd Annual International IEEE-EMB Special Topic Conference on Microtechnologies in Medicine & Biology*, pp. 237–240.

Pancrazio, J. J., 2000, Cell-based sensors—electrical transduction, in: *WTEC Study on Tissue Engineering Research, U. S. Review Workshop* (G. M. Holdridge, ed.), Baltimore, Maryland, pp. 135–145.

Papageorgiou, D., Bledsoe, S. C., Gulari, M., Hetke, J. F., Anderson, D. J., and Wise, K. D., 2001, A shuttered probe with in-line flowmeters for chronic in-vivo drug delivery. In: *The 14th IEEE International Conference on Micro Electro Mechanical Systems, 2001, MEMS 2001*, pp. 212–215.

Park, J., Tran, P. H., Chao, J. K. T., Ghodadra, R., Rangarajan, R., and Thakor, N. V., 1998, In vivo nitric oxide sensor using non-conducting polymer-modified carbon fiber, *Biosens. Bioelectron.* **13:**1187–1195.

Passeraub, P. A., 1999, *An Integrated Inductive Proximity Sensor*, Hartung-Gorre Verlag, Konstanz.

Passeraub, P. A., Almeida, A. C., and Thakor, N. V., 2003, Design, microfabrication and characterization of a microfluidic chamber for the perfusion of brain tissue slices, *Biomed. Microdev.* **5:**147–155.

Pine, J., 1980, Recording action potentials from cultured neurons with extracellular microcircuit electrodes, *J. Neurosci. Methods* **2:**19–31.

Polla, D. L., Erdman, A. G., Robbins, W. P., Markus, D. T., Diaz-Diaz, J., Rizq, R., Nam, Y., Brickner, H. T., Wang, A., and Krulevitch, P., 2000, Microdevices in medicine, *Annu. Rev. Biomed. Eng.* **2:**551–576.

Popovic, M. R., Popovic, D. B., and Keller, T., 2002, Neuroprostheses for grasping, *Neurol. Res.* **24:**443–452.

Regehr, W. G., Pine, J., and Rutledge, D. B., 1988, A long-term in vitro silicon-based microelectrode–neuron connection, *IEEE Trans. Biomed. Eng.* **35:**1023–1032.

Rousche, P. J., and Normann, R. A., 1999, Chronic intracortical microstimulation (ICMS) of cat sensory cortex using the Utah intracortical electrode array, *IEEE Trans. Rehabil. Eng.* **7:**56–68. [see also *IEEE Trans on Neural Systems and Rehabilitation*]

Rousche, P. J., Pellinen, D. S., Pivin, D. P., Jr., Williams, J. C., Vetter, R. J., and Kirke, D. R., 2001, Flexible polyimide-based intracortical electrode arrays with bioactive capability, *IEEE Trans. Biomed. Eng.* **48:**361–371.

Rutten, W. L. C., 2002, Selective electrical interface with the nervous system, *Annu. Rev. Biomed. Eng.* **4:**407–452.

Schneweis, S., Grond, M., Staub, F., Brinker, G., Neveling, M., Dohmen, C., Graf, R., Heiss, W.-D., and Shuaib, A., 2001, Predictive value of neurochemical monitoring in large middle cerebral artery infarction, *Stroke* **32:**1863–1867. [editorial comment]

Sharma, H. S., and Winkler, T., 2002, Assessment of spinal cord pathology following trauma using early changes in the spinal cord evoked potentials: A pharmacological and morphological study in the rat, *Muscle Nerve* **999:**S83–S91.

Somjen, G. G., 2001, Mechanisms of spreading depression and hypoxic spreading depression-like depolarization, *Physiol. Rev.* **81:**1065–1096.

Stieglitz, T., and Gross, M., 2002, Flexible BIOMEMS with electrode arrangements on front and back side as key component in neural prostheses and biohybrid systems, *Sens. Actuat. B: Chem.* **83:**8–14.

Stieglitz, T., Beutel, H., and Meyer, J.-U., 1997, A flexible, light-weight multichannel sieve electrode with integrated cables for interfacing regenerating peripheral nerves, *Sens. Actuat. A: Phys.* **60:**240–243.

Strong, T. D., Cantor, H. C., and Brown, R. B., 2001, A microelectrode array for real-time neurochemical and neuroelectrical recording in-vitro, *Sens. Actuat. A* **91:**363–368.

Suzuki, H., Hiratsuka, A., Sasaki, S., and Karube, I., 1998a, Problems associated with the thin-film Ag/AgCl reference electrode and a novel structure with improved durability, *Sens. Actuat. B* **46:**104–113.

Suzuki, H., Hirakawa, T., Sasaki, S., and Karube, I., 1998b, Micromachined liquid-junciton Ag/AgCl reference electrode, *Sens. Actuat. B* **46:**146–154.

Tatic-Lucic, S., Wright, J. A., Tai, Y.-C., and Pine, J., 1997, Silicon cultured-neuron prosthetic devices for in vivo and in vitro studies, *Sens. Actuat. B: Chem.* **43**:105–109.

Terao, Y., and Ugawa, Y., 2002, Basic mechanisms of TMS, *J. Clin. Neurophysiol.* **19**:322–343.

Thiébaud, P., Beuret, C., Rooij, N. F. D., and Koudelka-Hep, M., 2000, Microfabrication of Pt-tip microelectrodes, *Sens. Actuat. B* **70**:51–56.

Thiébaud, P., Beuret, C., Koudelka-Hep, M., Bove, M., Martinoia, S., Grattarola, M., Jahnsen, H., Rebaudo, R., Balestrino, M., Zimmer, J., and Dupont, Y., 1999, An array of Pt-tip microelectrodes for extracellular monitoring of activity of brain slices, *Biosens. Bioelectron.* **14**:61–65.

Thomas, C. A., Springer, P. A., Loeb, G. E., Berwald-Netter, Y., and Okun, L. M., 1972, A miniature microelectrode array to monitor the bioelectric activity of cultured cells, *Exp. Cell Res.* **74**:61–66.

Ungerstedt, U., and Hallström, A., 1987, In vivo microdialysis—A new approach to the analysis of neurotransmitters in the brain, *Life Sci.* **41**:861–864.

Urban, G. A., Ganglberger, J. A., Olcaytug, F., Kohl, F., Schallauer, R., Trimmel, M., Schmid, H., and Prohaska, O., 1990, Development of a multiple thin-film semimicro DC-probe for intracerebral recordings [during surgery], *IEEE Trans. Biomed. Eng.* **37**:913–918.

Vasquez-Buenos Aires, J., Payan, Y., and Demongeot, J., 2003, An integrate study of a lingual interface into computer assisted surgery field, in: *BioMEMs & Biomedical Nanotech WORLD 2003,* Cambridge Healthtech Institute, Washington, DC.

Wang, J. H., Cain, S. D., and Lohmann, K. J., 2003, Identification of magnetically responsive neurons in the marine mollusc *Tritonia diomedea*, *J. Exp. Biol.* **206**:381–388.

Ward, S., and Rise, M., 1997, Techniques for treating epilepsy by brain stimulation and drug infusion, in: *World Intellectual Property Organization*, Patent WO9742990, Medtronic Inc. International.

Waterhouse, E., 2003, New horizons in ambulatory electroencephalography, *IEEE Eng. Med. Biol.* 74–79.

Wilson, B. S., Lawson, D. T., Müller, J. M., Tyler, R. S., and Kiefer, J., 2003, Cochlear implants: Some likely next steps, *Annu. Rev. Biomed. Eng.* **5**:207–249.

Wise, K., and Angell, J., 1971, A microprobe with integrated amplifiers for neurophysiology, in: *Solid-State Circuits Conference. Digest of Technical Papers. 1971*, IEEE International, pp. 100–101.

Wolpaw, J. R., Birbaumer, N., McFarland, D. J., Pfurtscheller, G., and Vaughan, T. M., 2002, Brain–computer interfaces for communication and control, *Clin. Neurophysiol.* **113**:767–791.

Xu, C., Lemon, W., and Liu, C., 2002, Design and fabrication of a high-density metal microelectrode array for neural recording, *Sens. Actuat. A: Phys.* **96**:78–85.

Zaghloul, M. E., 2002, MEMS, microsystems and nanosystems, in: *Proceedings of the 2002 7th IEEE International Workshop on Cellular Neural Networks and Their Applications, 2002 (CNNA 2002)*, pp. 512–514.

Zahn, J. D., Trebotich, D., and Liepmann, D., 2000, Microfabricated microdialysis microneedles for continuous medical monitoring, in: *1st Annual International Conference on Microtechnologies in Medicine and Biology*, 2000, pp. 375–380.

Zeck, G., and Fromherz, P., 2001, Noninvasive neuroelectronic interfacing with synaptically connected snail neurons immobilized on a semiconductor chip, *Proc. Natl. Acad. Sci. U.S.A.* **98**:10457–10462.

Zhang, W., Cao, X., Xie, Y., Ai, S., Jin, L., and Jin, J., 2003, Simultaneous determination of the monoamine neurotransmitters and glucose in rat brain by microdialysis sampling coupled with liquid chromatography–dual electrochemical detector, *J. Chromatogr. B* **785**:327–336.

Zhou, A., and Muthuswamy, J., 2002, Molecular recognition of neurotransmitter GABA using acoustic sensors, in: *Engineering in Medicine and Biology, 2002. 24th Annual Conference and the Annual Fall Meeting of the Biomedical Engineering Society, Proceedings of the Second Joint EMBS/BMES Conference, 2002*, pp. 2097–2098.

3

BRAIN–COMPUTER INTERFACE

Anirudh Vallabhaneni,[1] Tao Wang,[1] and Bin He[2]*

[1]University of Illinois at Chicago, Illinois
[2]Department of Biomedical Engineering, University of Minnesota, Minneapolis, Minnesota

3.1. INTRODUCTION

Human–computer interfaces (HCIs) have become ubiquitous. Interfaces such as keyboards and mouses are used daily while interacting with computing devices (Ebrahimi *et al.*, 2003). There is a developing need, however, for HCIs that can be used in situations where these typical interfaces are not viable. Direct brain–computer interfaces (BCI) is a developing field that has been adding this new dimension of functionality to HCI. BCI has created a novel communication channel, especially for those users who are unable to generate necessary muscular movements to use typical HCI devices.

3.1.1. WHAT IS BCI

Brain–computer interface is a method of communication based on neural activity generated by the brain and is independent of its normal output pathways of peripheral nerves and muscles. The neural activity used in BCI can be recorded using invasive or noninvasive techniques. The goal of BCI is not to determine a person's intent by eavesdropping on brain activity, but rather to provide a new channel of output for the brain that requires voluntary adaptive control by the user (Wolpaw *et al.*, 2000b).

The potential of BCI systems for helping handicapped people is obvious. There are several computer interfaces designed for disabled people (Wickelgren, 2003). Most of these systems, however, require some sort of reliable muscular control such as neck, head, eyes, or other facial muscles. It is important to note that although requiring only neural activity, BCI utilizes neural activity generated voluntarily by the user. Interfaces based on involuntary neural activity, such as those generated during an epileptic seizure, utilize many of the same components and principles as BCI, but are not included in this field. BCI systems, therefore, are especially useful for severely disabled, or locked-in, individuals with no

* Address for correspondence: Department of Biomedical Engineering, University of Minnesota, 7-105 BSBE, 312 Church Street, Minneapolis, Minnesota 55455; e-mail: binhe@umn.edu.

reliable muscular control to interact with their surroundings. The focus of this chapter is on the basics of the technology involved and the methods used in BCI.

3.1.2. HISTORY OF BCI

Following the work of Hans Berger in 1929 on a device that later came to be known as electroencephalogram (EEG), which could record electrical potentials generated by brain activity, there was speculation that perhaps devices could be controlled by using these signals. For a long time, however, this remained a speculation.

As reviewed by Wolpaw and colleagues (2000b), 40 years later, in the 1970s, researchers were able to develop primitive control systems based on electrical activity recorded from the head. The Pentagon's Advanced Research Projects Agency (DARPA), the same agency involved in developing the first versions of the Internet, funded research focused on developing bionic devices that would aid soldiers. Early research, conducted by George Lawrence and coworkers, focused on developing techniques to improve the performance of soldiers in tasks that had high mental loads. His research produced a lot of insight on methods of autoregulation and cognitive biofeedback, but did not produce any usable devices.

DARPA expanded its focus toward a more general field of biocybernetics. The goal was to explore the possibility of controlling devices through the real-time computerized processing of any biological signal. Jacques Vidal from UCLA's Brain–Computer Interface Laboratory provided evidence that single-trial visual-evoked potentials could be used as a communication channel effective enough to control a cursor through a two-dimensional maze (Vidal, 1977).

Work by Vidal and other groups proved that signals from brain activity could be used to effectively communicate a user's intent. It also created a clear-cut separation between those systems utilizing EEG activity and those that used EMG (electromyogram) activity generated from scalp or facial muscular movements. Future work expanded BCI systems to use neural activity signals recorded not only by EEG but also by other imaging techniques.

Current BCI-based tools can aid users in communication, daily living activities, environmental control, movement, and exercise, with limited success and mostly in research settings. A more detailed evolution of BCI systems is detailed later in this chapter. The primary users of BCI systems are individuals with mild to severe muscular handicaps. BCI systems have also been developed for users with certain mental handicaps such as autism. Basic and applied research is being conducted with humans and animals for using BCIs in numerous clinical and other applications for handicapped and nonhandicapped users.

3.2. COMPONENTS OF A BCI SYSTEM

To understand the requirements of basic research in BCI, it is important to put it in the context of the entire BCI system. The recent work of Mason and Birch (2003), which is adapted in this section, presented a general functional model for BCI systems upon which a universal vocabulary could be developed and different BCI systems could be compared in a unified framework.

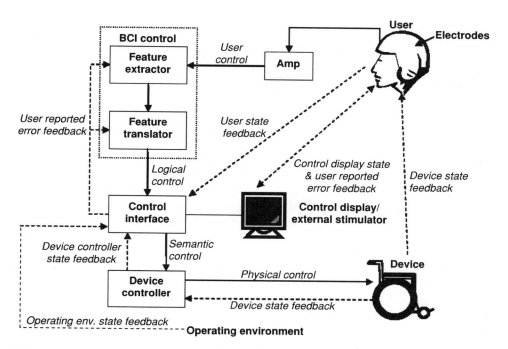

FIGURE 3.1. Functional components and feedback loops in a brain–computer interface system. The user's brain activity is measured by the electrodes and then amplified and signal-conditioned by the amplifier. The feature extractor transforms raw signals into relevant feature vectors, which are classified into logical controls by the feature translator. The control interface converts the logical controls into semantic controls that are passed onto the device controller. The device controller changes the semantic controls into physical device-specific commands that are executed by the device. The BCI system, therefore, can convert the user's intent into device action. (Revised from Mason and Birch, 2003, with permission, © 2003, IEEE)

The goal of a BCI system is to allow the user to interact with the device. This interaction is enabled through a variety of intermediary *functional components, control signals*, and *feedback loops* as detailed in Figure 3.1. Intermediary functional components perform specific functions in converting *intent* into *action*. By definition, this means that the user and the device are also integral parts of a BCI system. Interaction is also made possible through feedback loops that serve to inform each component in the system of the state of one or more components.

3.2.1. FUNCTIONAL COMPONENTS

Any BCI system is subject to the conditions in which it operates. The *operating environment* is the physical location and the surrounding objects at the location(s) in which the system is being used. This includes physical boundaries, temperature, terrain conditions, external noise, etc. Other components in the system must be able to adapt to the changing conditions in the operating environment.

A *user* is any entity that can relay its intent by intentionally altering its brain state to generate the control signals that are the input for the BCI system. The user's brain state is captured by *electrodes*, or any device that captures and converts the neural activity into detectable signals. The signal is then processed by the *Amp*, which amplifies, bandpass filters, and digitizes the signal. *User control* on the BCI Control is exerted through this signal.

The brains of the BCI system are part of the BCI control, which is responsible for processing and understanding the signal. The first part of the BCI control is the *feature extractor*. This component can handle one or more types of signals and transform the amplified signal into relevant feature values with a goal of maximizing the signal-to-noise ratio. The second part of the BCI control is the *feature translator*, which classifies the feature vectors into two or more classes. This generates a continuous *logical control* signal that is exerted on the control interface. A majority of basic BCI research is focused on creating new BCI control components and improving on existing techniques.

The *control interface* converts the logical control signal into device-dependent *semantic controls*. Control interfaces are typically context- and menu-driven so that a maximum number of semantic controls are produced utilizing a minimum number of logical controls. The *control display* (visual, aural, etc.) attached to the control interface is used to display the interpretation of the user's control signals within the device-dependent context. Some BCI systems have an *external stimulator* attached to the control interface and emits visual and/or aural stimuli for externally paced events. The release of the stimulus is controlled by the control interface and synchronized with the feature extractor for an accurate extraction of time- or phase-locked brain responses.

The *device controller* changes the semantic control signals from the control interface to *physical control* signals that are used to manipulate the device. Finally, the device itself can be any target physical object such as a wheelchair or virtual device such as a keyboard on a display screen.

3.2.2. FEEDBACK

Similar to other forms of human–computer interaction, proper use of BCI systems is extremely dependent on the adaptation of brain activity based on the feedback or response the user receives from the system. Therefore, it is critical that any BCI system provides the appropriate amount of feedback, as detailed in Figure 3.1, and adapts to the changes made by the user in response to the feedback provided by the components of the system.

From a neuroscience perspective, real-time feedback facilitates two types of corrective mechanisms. First, *continuous feedback* allows the user to control and correct errors during the execution of an action in real-time. This is in contrast to discrete or delayed feedback, which is intermittent and does not allow for real-time adaptivity. Second, the feedback that occurs after the successful completion of a command aids in gradually learning that command (Curran and Stokes, 2003).

Device state feedback is the status of the target device and is reported to the user through his or her sensory channels. It is also reported to the device controller to ensure it is synchronized with the status of the device. *Device controller state feedback* is the status of the device controller and is reported to the control interface to ensure synchronization of the semantic mapping.

The *control display state feedback* can be used to report to the user the status of the control interface and information of the status of the entire system through one or more sensory channels. This feedback loop is fed back from the control interface.

The *user-reported error feedback* is extremely important in making use of the adaptiveness of the feature extractor and translator. Through the use of a bidirectional control display, the user can report situations where intent was misclassified so that the system can adapt to the user's signal by adjusting its performance parameters or eliminating the trial from its training data.

The *user state feedback* reports the user's mental and physical state and is fed into the control interface to synchronize the system with the capabilities of the user at any particular point in time. Similarly, the *environment state feedback* is fed into the control interface to adjust system performance and functions to any changing environmental conditions.

3.3. SIGNAL ACQUISITION

As discussed before, translation of intent to action is dependent on the intent being expressed in the form of a measurable signal. Proper acquisition of this signal is important for the proper functioning of any BCI. The goal of signal acquisition methods is to detect the voluntary neural activity generated by the user through invasive or noninvasive methods. Each method of signal acquisition can be measured in terms of spatial and temporal resolution. The appropriate method to use depends on striking a balance between the feasibility of acquiring the signal in the operating environment and the resolution required for proper translation.

3.3.1. INVASIVE TECHNIQUES

Invasive signal acquisition primarily relies on electrophysiologic recordings made by neurosurgically implanting micro-electrodes inside the user's brain. The preferred site for implanting electrodes is the motor cortex. This area of the brain is more easily accessible and has large pyramidal cells, which are easier to record from. In addition, signals in this area can be generated through simple tasks such as actual or imaginary motor movements (Wolpaw *et al.*, 2000b). Other areas such as the supplementary motor cortex, subcortical motor areas, and the thalamus could also serve as potential sites for electrode implantation. Information from complementary imaging techniques such as functional magnetic resonance imaging (fMRI) can help determine potential target areas for a specific subject (Wolpaw *et al.*, 2000b). Recent developments with fMRI have allowed not only finding cortical areas of activity, but also reliable control of BCI based on fMRI imaging of changing blood oxygen levels across several cortical areas with different cognitive tasks (Weiskopf *et al.*, 2003).

Several types of electrodes have been tested on animals, but the neurotrophic cone electrode has been able to achieve a limited success in human subjects (Kennedy and Bakay, 1998). As shown in Figure 3.2, the cone contains a neurotrophic factor that causes neurites to grow into the cone and contact one of the gold wires inside the electrode which transmits the electrical signal out of the brain (Mussa-Ivaldi and Miller, 2003). Intentions are conveyed, for example, by training the patients to control a cursor with their implant, and the velocity of the cursor is determined by the rate of neural firing. The neural waveshapes are converted to pulses, and three pulses are fed as input into the BCI Control. The first and second pulses control the X and the Y position of the cursor and the third pulse serves as a mouse click or an enter command. The patients are trained using software that contains a row of icons representing common phrases (Kenney and Bakay, 1998).

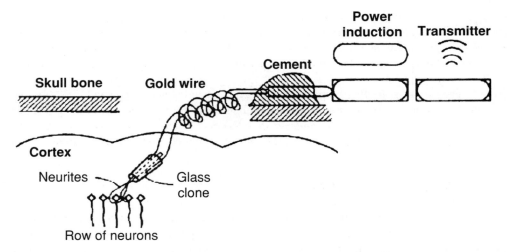

FIGURE 3.2. Recording brain activity using invasive signal acquisition methods. Cone-shaped glass electrodes are implanted through the skull and directly into specific neurons in the brain. The electrode is filled with a neurotrophic factor that causes neurites to grow into the cone and contact one of the gold wires that transmits the electrical activity to the receiver unit outside the head and then amplified and transmitted to the BCI control using a transmitter. (From Kennedy & Bakay, 1998, with permission, © 1998, IEEE)

A notable experiment has been conducted by Nicolelis and Chapin (2002) on monkeys to control a robot arm in real time by electrical discharge recorded by microwires that lay beside a single motor neuron. Various motor-control parameters, including the direction of hand movement, gripping force, hand velocity, acceleration, three-dimensional position, etc., were derived from the parallel streams of neuronal activity by mathematic models. In this system, monkeys learn to produce complex hand movements in response to arbitrary sensory cues. The monkeys could exploit visual feedback to judge for themselves how well the robot could mimic their hand movements. Refer to Figure 3.3 for a detailed description (Nicolelis, 2003).

A less invasive approach that has been well applied to epileptic patients for surgical planning is patching subdural electrode array over cortex to record electrocorticogram (ECoG) signals. Subdural electrodes are closer to neuronal structures in superficial cortical layers than electroencephalogram (EEG) electrodes placed on the scalp. It is estimated that scalp electrodes represent the spatially averaged electrical activity over a cortical area of at least several square centimeters. Several closely spaced subdural electrodes can be placed over an area of this size such that each of these electrodes measures the spatially averaged bioelectrical activity of an area very likely much smaller than several square centimeters. The advantages of subdural recordings include recording from smaller sources of "synchronized activity," higher signal-to-noise ratio than that of scalp recordings, and increased ability to record and study gamma activity above 30 Hz. Gamma activity is generated by rapidly oscillating cell assemblies composed of a small number of neurons. Consequently, gamma activity is characterized by small amplitude fluctuations that are not easily recorded with scalp electrodes (Pfurtscheller *et al.*, 2003).

Levine and coworkers (2000) have implemented a "direct brain interface" that accepts voluntary commands directly from recoding ECoG signal in epileptic patients. The subjects

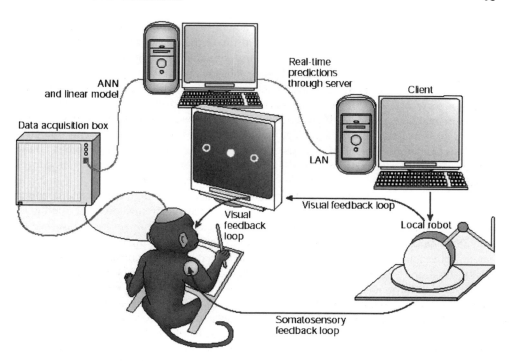

FIGURE 3.3. Experimental design used to test a closed-loop control brain–machine interface for motor control in macaque monkeys. Chronically implanted microwire arrays are used to sample the extracellular activity of populations of neurons in several cortical motor regions. Linear and nonlinear real-time models are used to extract various motor-control signals from raw brain activity. The outputs of these models are used to control the movements of a robot arm. For instance, while one model might provide a velocity signal to move the robot arm, another model, running in parallel, might extract a force signal that can be used to allow a robot gripper to hold an object during an arm movement. Artificial visual and tactile feedback signals are used to inform the animal about the performance of a robot arm controlled by brain-derived signals. Visual feedback is provided by using a moving cursor on a video screen to inform the animal about the position of the robot arm in space. Artificial tactile and proprioceptive feedback is delivered by a series of small vibromechanical elements attached to the animal's arm. This haptic display is used to inform the animal about the performance of the robot arm gripper (whether the gripper has encountered an object in space, or whether the gripper is applying enough force to hold a particular object). ANN, artificial neural network; LAN, local area network. (From Nicolelis, 2003, with permission, © 2003, *Nature*)

were instructed to make different movements of the face, tongue, hand, and foot in either a prompt-paced or a self-paced manner. Half of the ECoG recoding was used to produce an averaged ECoG segment (as "ERP templates") and the cross-correlation of templates with the continuous ECoG was used to detect ERPs that correspond to specific movements. The cortical locations of the subdural electrodes were based solely on clinical considerations relating to epilepsy surgery (as opposed to research needs). The accuracy of ERP detection for the five best subjects has hit more than 90%. In another experiment of self-paced movement study using ECoG (Pfurtscheller *et al.*, 2003), it was concluded that self-paced movement is accompanied not only by a relatively widespread mu and beta ERD, but also by a more focused gamma ERS in the 60–90 Hz frequency band.

In a different system, individual electrodes in the Utah electrode (Maynard *et al.*, 1997) are tapered to a tip, with diameters <90 μm at their base, and they penetrate only 1–2 mm into the brain. Invasive techniques cause significant amount of discomfort and risk to the patient. Researchers use them in human subjects only if it will provide considerable improvement in functionality over available noninvasive methods. A majority of the initial research, therefore, is conducted in animals, especially monkeys and rats, and is also called the brain–machine interface (BMI) (Nicolelis, 2001). Research in these animals has led to the rapid development of microelectronics that enables recording electrophysiological activities from a small group of neurons or even a single neuron. Present technology allows reliable simultaneous sampling of 50–200 neurons, distributed across multiple cortical areas of small primates, for a period of a few years (Wessberg *et al.*, 2000).

The advantage of these types of invasive techniques is the high spatial and temporal resolution that can be achieved, as recordings can be made from individual neurons at very high sampling rates. Intracranially recorded signals could obtain more information and allow quicker responses, which might lead to decreased requirements of training and attention (Sanchez *et al.*, 2004). Several issues, however, have to be considered (Lauer *et al.*, 2000). First, the long-term stability of the signal over days and years is hard to achieve. The user should be able to consistently generate the control signal reliably without the need for frequent retuning. Second is the issue of cortical plasticity following a spinal cord injury. It has been hypothesized that the motor cortex undergoes reorganization after a spinal cord injury, but the degree is unknown (Brouwer and Hopkins-Rosseel, 1997). Finally, if a neuroprosthesis that requires a stimulus to the disabled limb is used, this stimulus would also produce a significant artifact on the scalp that might interfere with the signal of interest. In such cases, BMI systems must be able to accurately detect and remove this artifact.

It is also necessary to develop a better understanding of the principles by which neural ensembles encode sensory, motor, and cognitive information (Isaacs *et al.*, 2000; Nicolelis, 2001; Serruya *et al.*, 2002). In the case of motor control, for instance, the areas of the primate brain that are involved are well known and even the physiological properties of individual neurons located in these areas have been studied well (Nicolelis, 2001). Little is known, however, about how the brain makes use of this information from neurons to generate the movements. In the movement control design, therefore, further work is needed to develop a method that can efficiently sample and accurately decode the motor signals generated by neurons so an artificial device can mimic the intended movement.

Classic experiments in primates, for example, have demonstrated that fundamental parameters of motor control emerge by the collective activation of large distributed populations of neurons in the primary motor cortex (M1). To compute a precise direction of arm movement, the brain may have to perform the equivalent of a neuronal "vote" or, in mathematical terms, a vector summation of the activity of these broadly tuned neurons. This implies that to obtain the motor signals required to control an artificial device it is necessary to sample the activity of many neurons simultaneously as well as to design algorithms that are capable of extracting motor control signals from these ensembles. Several well-established models such as linear regression, population vector, and neural network have been successfully applied to deal with large neural data to estimate the hand movement trajectory from the firing rate of motor cortex populations (Wessberg *et al.*, 2000, Taylor *et al.*, 2002, Serruya *et al.*, 2003). But these signals and models are far from providing the full range of motion that the arm can produce (Donoghue, 2002).

As mentioned earlier, experiments with humans thus far have been limited. Currently, only a few severely disabled patients have been implanted with electrodes. In some cases, success has been limited, with some patients able to communicate at a rate of only three letters per minute (Mussa-Ivaldi and Miller, 2003). Further advancements in microelectrodes, however, are required to obtain stable recordings over a long term (i.e. more than 1 year). In addition to the areas mentioned above, additional research focusing on minimizing the number of cells required for simultaneous recordings to obtain a useful signal as well as on providing feedback to the nervous system via electrical stimulation through electrodes is also essential for a potential widespread use of invasive techniques in humans. For a comprehensive review of the BMI and neurorobotic research, see Chapter 4 in this book.

3.3.2. NONINVASIVE TECHNIQUES

There are many methods of measuring brain activity through noninvasive means. Noninvasive techniques reduce risk for users since they do not require surgery or permanent attachment to the device. Techniques such as computerized tomography (CT), positron electron tomography (PET), single-photon emission computed tomography (SPECT), magnetic resonance imaging (MRI), functional magnetic resonance imaging (fMRI), magnetoencephalography (MEG), and electroencephalography (EEG) have all been used to measure brain activity noninvasively.

EEG, however, is the most prevalent method of signal acquisition for BCI. EEG has a high temporal resolution capable of measuring every thousandth of a second. Modern EEG also has a reasonable spatial resolution as signals from up to 256 electrode sites can be measured at the same time.

Practicality of EEG in a laboratory and in a real-world setting is unsurpassed. The device is portable and the electrodes can be easily placed on the subject's scalp by simply donning a cap. In addition, EEG systems have seen widespread use in numerous fields since its inception. Therefore, the techniques and technology of signal acquisition through this method have been standardized. Finally, and most important, the method is noninvasive (Wolpaw et al., 2000a).

Many EEG-based BCI systems use an electrode placement strategy suggested by the International 10/20 system as detailed in Figure 3.4. For better spatial resolution, it is also common to use a variant of the 10/20 system that fills in the spaces between the electrodes of the 10/20 system with additional electrodes (Malmivuo and Plonsey, 1995).

3.4. FEATURE EXTRACTION AND TRANSLATION

Basic research in BCI is focused on improving methods of feature extraction from the acquired signals and translating them into logical control commands for single-trial and averaged trials. A feature in a signal can be viewed as a reflection of a specific aspect of the physiology and anatomy of the nervous system (Wolpaw et al., 2000b). The goal of feature extraction methods, based on this definition, would be to obtain the specific physiological aspect of the nervous system across a specific time series. The steps involved in feature extraction and translation are detailed in Figure 3.5.

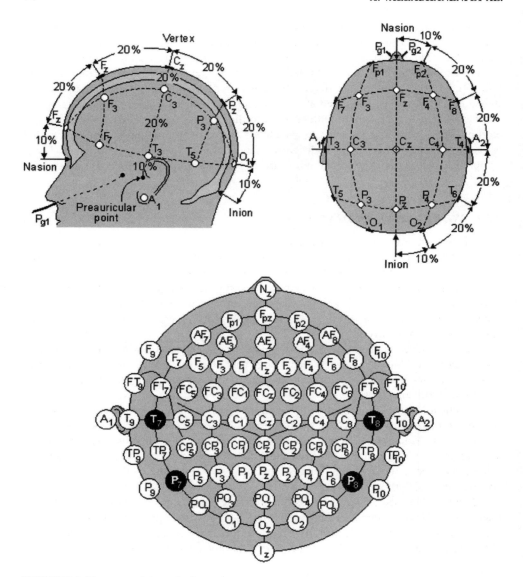

FIGURE 3.4. Placement of electrodes for noninvasive signal acquisition using an electroencephalogram (EEG). This standardized arrangement of electrodes over the scalp is known as the International 10/20 system and ensures ample coverage of all parts of the head. The exact positions for each electrode are at the intersection of the lines calculated from measurements between standard skull landmarks. The letter at each electrode identifies the particular subcranial lobe (FP, prefrontal lobe; F, frontal lobe; T, temporal lobe; C, central lobe; P, parietal lobe; O, occipital lobe). The number or the second letter identifies its hemispherical location (Z, denoting line zero refers to an electrode placed along the cerebrum's midline; even numbers represent the right hemisphere; odd numbers represent the left hemisphere. The numbers are in ascending order with increasing distance from the midline.). (From Malmivuo and Plonsey, 1995 [web edition at http://butler.cc.tut.fi/~malmivuo/bem/bembook/in/in.htm], with permission)

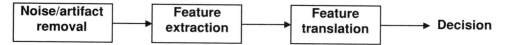

FIGURE 3.5. Processing steps required to convert user's intent, encoded in the raw signal, into device action. Signals captured through invasive or noninvasive methods contain a lot of noise. The first step in feature extraction and translation is to remove noise. This is followed by selection of relevant features through several feature extraction techniques that focus on maximizing the signal-to-noise ratio. Finally, feature translation techniques are used to classify the relevant features into one of the possible states. (From Kelly *et al.*, 2002, with permission)

3.4.1. TYPES OF SIGNALS

3.4.1.1. Spikes and Field Potentials

The brain generates a tremendous amount of neural activity. There are a plethora of signals, also referred to as components, which can be used for BCI. These signals fall into two major classes: *spikes* and *field potentials* (Wolpaw, 2003). Spikes reflect the action potentials of individual neurons and thus acquired primarily through microelectrodes implanted by invasive techniques. Field potentials, however, are measures of combined synaptic, neuronal, and axonal activity of groups of neurons and can be measured by EEG or implanted electrodes depending on the spatial resolution required. As previously mentioned, most of the BCI research is focused on using signals from EEG, and thus the most commonly used components are derived from EEG recordings.

3.4.1.2. EEG Frequency Bands

Signals recorded from EEG are split into several bands as shown in Figure 3.6. *Delta* band ranges from 0.5 to 3 Hz and the *theta* band covers the 4–7 Hz range. A majority of BCI research focuses on the *alpha* band (8–13 Hz) and the *beta* band (14–30 Hz). The beta band is sometimes considered to have an extended range of up to 60 Hz with the *gamma* band indicating all signals greater than 30 Hz.

3.4.1.3. Components of Interest

Components of particular interest to BCI can be divided into four categories: oscillatory EEG activity, event-related potentials (ERP), slow cortical potentials (SCP), and neuronal potentials.

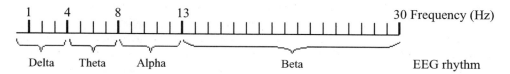

FIGURE 3.6. Different signal bands present in the EEG signal. The delta band ranges from 0.5 to 3 Hz and the theta band ranges from 4 to 7 Hz. Most BCI systems utilize components in the alpha band (8 to 13 Hz) and the beta band (14 to 30 Hz).

3.4.1.4. Oscillatory EEG Activity

Oscillatory EEG activity is caused by complex network of neurons that create feedback loops. The synchronized firing of the neurons in these feedback loops generates observable oscillations. The frequency of oscillations decreases as the number of synchronized neuronal bodies increases. The underlying membrane properties of neurons and dynamics of synaptic processes, the strength and complexity of connections in the neuronal network, and the influences from other neurotransmitter systems also play a role in determining the oscillations.

Two distinct oscillations of interest are the *Rolandic mu-rhythm*, occurring in the 10–12 Hz range, and the *central beta rhythm*, occurring in the 14–18 Hz range. Both originate in the sensorimotor cortex region of the brain. These oscillations occur continuously during "idling" or rest. During nonidling periods, however, these oscillations are temporarily modified and the change in frequency and amplitude are evident on the EEG. The amplitude of oscillations decreases as the frequency increases because the frequency of the oscillations is negatively correlated with their amplitude (Pfurtscheller and Neuper, 2001).

3.4.1.5. Event-Related Potentials

Event-related potentials (ERPs) are time-locked responses by the brain that occur at a fixed time after a particular external or internal event. These potentials usually occur when subjected to sensory or aural stimulus, mental event, or the omission of a constantly occurring stimulus.

Exogenous ERP components are obligatory responses to physical stimuli and occur due to processing of the external event but independent of the role of the stimuli in the processing of information. The random flash of a bulb, for example, will generate an exogenous component as the brain responds to the sudden flash of light regardless of the context.

Endogenous ERP components occur when an internal event is processed. It is dependent on the role of the stimulus in the task and the relationship between the stimulus and the context in which it occurred. A user trying to spell the letter R in a word, for example, will generate an endogenous ERP component if the letter R is presented since it is the event he or she is looking for. If the user is trying to spell the letter S, however, he or she will not generate an endogenous ERP component if the same letter R is presented since the relationship between the stimulus and the context in which it occurred is no longer valid.

3.4.1.6. Event-Related Synchronization/(De)synchronization

A particular type of ERP is characterized by the occurrence of an *event-related desynchronization* (ERD) and an *event-related synchronization* (ERS). Changes in the factors that control the oscillation of neuronal networks, such as sensory stimulation or mental imagery, are responsible for the generation of these event-related potentials. A decrease in the synchronization of neurons causes a decrease of power in specific frequency bands and this phenomenon is defined as an ERD and can be identified by a decrease in signal amplitude. Presence of ERD is very widespread in the alpha band, especially during tasks involving perception, memory, and judgment. Increasing task complexity or attention amplifies the magnitude of the ERD.

ERS, on the other hand, is characterized by an increase of power in specific frequency bands that is generated by an increase in the synchronization of neurons and can be identified

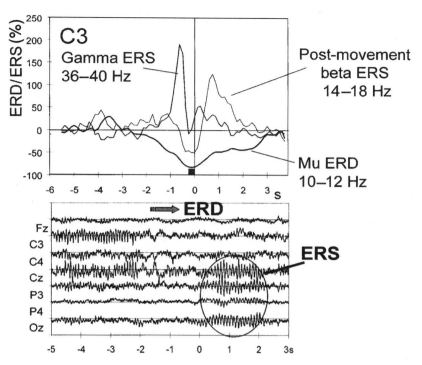

FIGURE 3.7. Evidence of event-related desynchronization (ERD) and event-related synchronization (ERS) phenomena before and after movement onset. ERD is the result of a decrease in the synchronization of neurons, which causes a decrease of power in specific frequency bands, and can be identified by a decrease in signal amplitude. ERS is the result of an increase in the synchronization of neurons, which causes an increase of power in specific frequency bands, and can be identified by the increase in signal amplitude. (From Pfurtscheller and Neuper, 2001, with permission, © 2001, IEEE)

by an increase in signal amplitude. ERD and ERS are measured relative to a baseline or reference interval, so the strength of an ERD/ERS is affected by the variance of the rhythms in this interval.

The time-locked property of ERPs is particularly evident for ERD/ERS during imagined or actual motor tasks as shown in Figure 3.7. An ERD in the mu rhythm starts 2.5 s prior to movement onset and peaks after onset of movement before recovering to baseline. A short-lived ERD in the central beta rhythm occurs prior to movement onset and is immediately followed by an ERS that peaks after movement onset. Oscillations and ERS are also found around the 40-Hz gamma band when subjected to visual stimulation owing to binding of sensory information and in motor tasks owing to sensorimotor integration. The high frequency of the gamma band works well to set up rapid coupling or synchronization between spatially separated groups of neurons (Pfurtscheller and Lopes da Silva, 1999).

3.4.1.7. Visual-Evoked Potentials

Another type of ERP commonly used in BCI is the *visual-evoked potential* (VEP), an EEG component that occurs in response to a visual stimulus. VEPs are dependent on the user's control of their gaze and thus require coherent muscular control. One frequently used VEP is the *steady-state visual evoked potential* (SSVEP).

SSVEP is an exogenous ERP component. The user visually focuses on one of two objects on a screen that flicker at different frequencies in the alpha and beta bands. The SSVEP component is amplified when the user shifts focus to the other object and then returns to baseline. The user can continue to switch focus between the two objects on the screen to generate changes in the signal (Middendorf *et al.*, 2000).

3.4.1.8. P300

The *P300* is an endogenous ERP component and occurs as part of the "oddball paradigm" (Donchin and Coles, 1988; Donchin *et al.*, 2000). In this phenomenon, users are subject to events that can be categorized into two distinct categories. Events in one of the two categories, however, are rarely displayed. The user is presented with a task that cannot be accomplished without categorization into both categories. When an event from the rare category is displayed, it elicits a P300 component, which is a large positive wave that occurs approximately 300 ms after event onset as shown in Figure 3.8. The amplitude of the P300 component is inversely proportional to the rate at which the rare event is presented. This ERP component is a natural response and thus especially useful in cases where either sufficient training time is not available or the user cannot be easily trained (Spencer *et al.*, 2001).

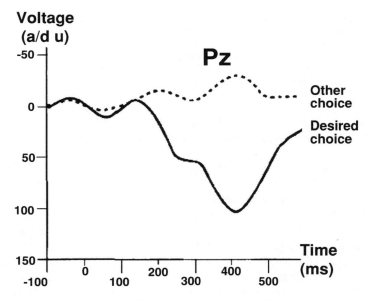

FIGURE 3.8. P300 ERP component. When the user is randomly flashed objects on a screen, the P300 component occurs when the object the user is looking for is flashed, while any of the other objects do not elicit a similar change in voltage. The amplitude of the P300 component is inversely proportional to the rate at which the object of interest is presented and occurs approximately 300 ms after the object is displayed. It is a natural response and requires no user training. (From Kubler *et al.*, 2001, with permission)

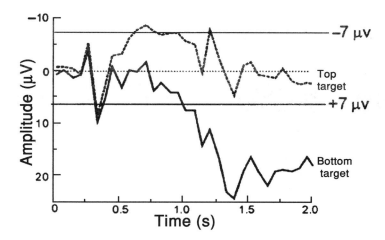

FIGURE 3.9. Different slow cortical potential (SCP) signals conveying different intents. SCPs are caused by shifts in the depolarization level of certain dendrites. It occurs from 0.5 to 10 s after the onset of an internal event and thus considered a slow cortical potential. (From Kubler *et al.*, 2001, with permission)

3.4.1.9. *Slow Cortical Potential*

A completely different type of signal is the *slow cortical potential*, which is caused by shifts in the depolarization levels of certain dendrites. Negative SCP indicates the sum of synchronized potentials, whereas positive SCP indicates reduction of synchronized potentials from the dendrites. As behavioral and cognitive performance of the user improves, so do the synchronized potentials, resulting in an increase of negativity of SCP. Since this cortical potential occurs anywhere from a 0.5 to 10 s after the onset of an internal event, as shown in Figure 3.9, it is referred to as the slow cortical potential (Birbaumer *et al.*, 1999, 2000; Wolpaw *et al.*, 2000b).

3.4.1.10. *Neuronal Potential*

Neuronal potential is a voltage spike from individual neurons as shown in Figure 3.10. This potential can be measured for a particular neuron or a group of neurons. The signal is a measure of the average rate, correlation, and temporal pattern of the neuronal firing. The central nervous system presents information on the firing rate of each neuron. Therefore, learning can be measured through changes in the average firing rate of neurons located in the cortical areas associated with the task.

Neuronal potential is extremely useful since it can achieve two-dimensional controls for the BCI by identifying the location of the neurons which are firing and also their rate of firing (Wolpaw, 2003). Research in neuronal potentials has been limited to animals until very recently because of the invasive procedures required to implant the electrodes as well as a lack of electrodes that generate stable recordings over a long period of time. The limited work, however, helps prove that better machine control is achievable by isolating signals with better spatial resolution (Wolpaw *et al.*, 2002; Moxon, 2005).

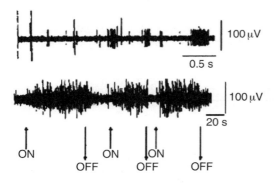

FIGURE 3.10. Neuronal potentials. These signals are voltage spikes and can be measured from individual neurons or groups of neurons. They are reflective of the average firing rate, correlation, and temporal pattern of neuronal firing. The top frame indicates signal recorded from idling neurons and the bottom frame from active neurons. (From Kennedy, 1998, with permission, © 1998, IEEE)

3.4.2. TRAINING

According to the review by Curran and Stokes (2003) and other research (Kostov and Polak, 2000; Laubach *et al.*, 2000), which is adapted here, the effectiveness of BCI is dependent on the capacity of the user to willingly and consistently control their EEG activity. Unlike motor tasks, control of brain activity is harder to achieve since the user can neither identify nor discern the activity. The user can only comprehend their EEG activity through the feedback received from the components in the BCI system.

The goal of training, therefore, is to have users voluntarily produce detectable EEG signals that can be altered to achieve a specific result. From the definition of BCI, it should be evident that the components produced by the user must be voluntary. Although the user might not be aware of how and when the signals are generated, the signal generation process can only be activated by voluntary actions from the user. BCI systems, however, differ in whether these voluntary signals must be produced through conscious mental activity (e.g., adding numbers) (Birbaumer, 1999) or as an automatic response to the situation that requires minimal conscious effort (e.g., riding a bicycle).

3.4.2.1. Cognitive Tasks

Most training methods require the user to perform specific cognitive tasks. These methods focus on developing the user's ability to generate EEG components through voluntary and conscious mental activity. *Motor imagery* (MI) tasks have been among the most widely used cognitive tasks. In each trial the user imagines or plans one of several motor movements (i.e. left or right hand movement) based on visual or aural cues. Research has shown that this generates signals from the sensorimotor cortex of the brain and can be detected by EEG (Annett, 1995; Jeannerod, 1995). After several training sessions, the user is able to control the amplitude and frequency of the required component (Babiloni *et al.*, 2000).

Other commonly used cognitive tasks do not involve motor imagery. Rather, they require the user to perform actions such as arithmetic (addition of a series of numbers), visual counting (sequential visualization of numbers), geometric figure rotation (visualization of rotation of a 3D object around an axis), letter composition (nonvocal letter composition), and baseline (relaxation). Research has shown that these tasks produce discernable components detectable by EEG (Pfurtscheller *et al.*, 1993; Penny and Roberts, 1999; Babiloni *et al.*, 2000; Birbaumer *et al.*, 2000; Penny *et al.*, 2000).

3.4.2.2. Operant Conditioning

In contrast, the *operant conditioning* approach does not require the user to perform specific cognitive tasks. The focus of this method is on helping the user gain automatic control of the device by thinking about anything. The feedback provided by the system serves to condition the user to continue to produce and control the EEG components that have achieved the desired outcome. With continuous practice, the user is able to gain control of the device without necessarily being aware of the specific EEG components being produced. It is important to note, however, that operant conditioning method often uses motor imagery tasks to initially acclimate users to the concept that brain waves can be controlled.

3.4.2.3. Factors That Affect Training

Both methods of training, cognitive tasks and operant conditioning, are subject to numerous external factors. Some of the most common factors are concentration, distractions, frustration, emotional state, fatigue, motivation, and intentions. It is important to counteract these factors during training by providing ample feedback and varying the duration or frequency of the training sessions.

In addition, the EEG components produced by cognitive tasks are vulnerable to the amount of direction provided to the user. Motor imagery, for example, is subject to issues such as first/third-person perspective, visualization of the action versus retrieving a memory of the action performed earlier, imagination of the task as opposed to a verbal narration, etc. Research has yet to prove whether users can effectively control such fine details to produce significant change in the components they produce.

Because the focus of BCI is to provide a means of communication for the disabled, it is possible that some users have suffered from mentally debilitating diseases that do not allow them control of all areas of the brain. The left hemisphere of the brain, for example, is the center of activity for tasks involving language, numbers, and logic, whereas the right hemisphere is more active during spatial relations and movement imagery. Users need to be paired with the cognitive tasks that best suit their capabilities.

As indicated earlier, it is possible to discern different cognitive tasks based on the EEG components generated when the task is achieved. When using a combination of cognitive tasks during training, overlap of EEG signals can occur if the tasks require similar skills or cortical areas. It is important to choose tasks with contrasting EEG components for easy discrimination.

Another factor to consider during training is the particular EEG component to use. Slow cortical potentials, for example, are a natural response and thus require less training time than for users trying to control their mu rhythm. As mentioned above, choosing contrasting cognitive tasks accelerates training. It is also important to maintain consistent training regimens to ensure subjects retain their ability to control their EEG components.

The tasks used in training a user carry forward into general BCI usage. The method of training, therefore, determines the method of signal acquisition. Neuronal activity generated by cognitive tasks is restricted to certain areas of the brain. This allows signal acquisition to occur over a few electrodes that encompass the specific region. The operant conditioning method, however, can only work on a BCI that uses all or unspecific electrode locations since the mental activity used to control the objective is not defined.

3.4.3. SIGNAL PROCESSING AND FEATURE EXTRACTION TECHNIQUES

The user is able to voluntarily generate detectable signals to convey his or her intent. Signal acquisition methods, however, capture noise generated by other unrelated activity in or out of the brain. Appropriate features need to be extracted by maximizing the signal-to-noise ratio.

The goal of all processing and extraction techniques is to characterize an item by discernable measures whose values are very similar for those in the same category but very different for items in another category. Such characterization is done by choosing relevant features from the numerous choices available. This selection process is necessary, because unrelated features can cause the translation algorithms to have poor generalization, increase the complexity of calculations, and require more training samples to attain a specific level of accuracy.

3.4.3.1. Artifact/Noise Removal

Because signals are often captured across several electrodes over a series of time, existing methods concentrate on either spatial domain processing or temporal domain processing, or both. In addition, research has shown that a lot of noise captured in EEG is generated by non–central nervous system (CNS) activity, especially muscular movements in the facial muscles (Wolpaw *et al.*, 2002). To counteract this noise, another set of techniques focus on non-CNS artifact removal.

To minimize noise in the signal, it is important to understand its sources. Noise can be captured through neural sources when components not related to the target signal are captured. Noise can also be generated by nonneural sources such as muscular movements, particularly of the facial muscles. This type of noise is especially important, because signals generated by muscular movements are overpowering and can be mistaken for the target signal. The problem is further complicated when the frequency or amplitude of the noise and the target signal are similar.

Typically more prominent than EEG signals, non-CNS artifacts are the result of unwanted potentials from eye movements, scalp-recorded EMG activity, and other such nonneural sources. Simple instructions to the user to not use facial muscles or to disregard the trials that contain artifacts can be used, but are not always adequate to remove this noise. Mathematical operations such as linear transformations and component analysis are also used for artifact removal (Makeig *et al.*, 2000; Müller *et al.*, 2000).

3.4.3.2. Characteristics of Feature Extraction Methods

Blum and Langley (1997) create an analogy of feature selection or extraction algorithms as heuristic search techniques that process large amounts of irrelevant data to find and extract a few relevant features. They further characterized algorithms designed for feature extraction based on four criteria.

The first criterion is the definition of a starting point(s) that will also determine the direction of the search as well as the operators to decide the succeeding states. Algorithms could start with an empty set of features and successively add features based on a scoring function. This method is called forward selection. Another option is to start with all available features and remove certain features based on a scoring function. This method is known as backward elimination. Some algorithms even apply a combination of forward selection followed by backward elimination or vice versa.

The second criterion is the organization of the search. Because it is not efficient to do a comprehensive search of the entire feature space, algorithms use techniques such as greedy selection, stepwise addition and elimination, or best-search to select the next feature that will improve the score over the current set.

The third criterion is the strategy used to evaluate all possible subsets of features. Most algorithms tend to use a scoring function that reflects a feature's ability to discriminate among the different classes. Many algorithms score features on the basis of information theory or contribution to the classification accuracy.

The fourth criterion is the terminating condition for the search. Some feature extraction algorithms stop when successive iterations fail to improve the score of the feature set above a certain threshold. Others continue to search as long as there is no decrease in the score or accuracy of the feature set. Another option used is to sort each of the features on the basis of some scoring function and selecting a breakpoint at which all features above this point are automatically selected.

3.4.3.3. Types of Feature Extraction Methods

Also discussed by Blum and Langley (1997), feature extraction techniques can be divided into three categories (Yom-Tov & Inbar, 2002). The first category is called embedded algorithms, wherein the feature selection is a part of the translation, also called classification, method. The feature selection procedure adds or removes features to counter prediction errors as new training data is introduced. Embedded algorithms, however, are of little use when there is a high level of interactions between relevant features.

The second category is called filter algorithms, which select specific features prior to, and independent of, the translation process. They work by removing irrelevant features (those providing redundant data or contaminated by noise) prior to training the translation technique. One way of filtering involves calculating each feature's correlation with the target function and then selection of a fixed number of features with the highest scores. Another filtering method involves the derivation of higher-order features based on features from the raw data and sorting these higher-order features on the basis of the amount of variance they explain and selecting a fixed number of highest scoring features.

The final category is called wrapper algorithms, which select features by utilizing the translation algorithms to rate the viability or quality of a feature set. Rather than selecting a feature set on the basis of the results of the classification, these algorithms utilize the translation algorithm as a subroutine to estimate the accuracy of a particular subset of features. This type of algorithm is unique to a translation algorithm and particularly useful with limited training data.

In certain occasions, existing signals are not enough for high accuracy feature extraction. Some methods introduce more signals to capture additional information about the state of the brain, for example, by using 56 electrodes where only 2 were previously used. This increased spatial data can be processed to derive common spatial patterns. This is achieved by projecting the high-dimensional spatiotemporal signal onto spatial filters that are designed such that the most discriminative information is inherent in the variances of the resulting signals (Ramoser *et al.*, 2000).

3.4.3.4. Spatial and Temporal Domain Processing

Spatial filtering techniques are useful for extracting features with a specific spatial distribution (McFarland *et al.*, 1997; Muller-Gerking *et al.*, 1999). In BCI systemss that utilize mu or alpha rhythms, the selection of spatial filters can greatly affect the signal-to-noise ratio. A high-pass spatial filter such as the *bipolar derivation* calculates the first spatial derivative and emphasizes the difference in the voltage gradient in a particular direction. The *surface Laplacian* (Hjorth, 1975; Perrin *et al.*, 1987; He and Cohen, 1992; Le *et al.*, 1992; Nunez *et al.*, 1994; Babiloni *et al.*, 1996; He, 1999; He *et al.*, 2001) also acts as a high-pass filter and can be approximated by subtracting the average of the signal at four surrounding nodes from the signal at the node of interest (Hjorth, 1975). It is the second derivative of the spatial voltage distribution and as the distance to the surrounding nodes increases, its sensitivity to higher spatial frequencies decreases, whereas a decreasing distance from the surrounding nodes increases its sensitivity to higher spatial frequencies (Wolpaw *et al.*, 2002).

Temporal domain processing techniques are also useful in maximizing the signal-to-noise ratio. These methods work by analyzing the signal across a period of time. Some temporal domain-processing methods such as Fourier analysis require significantly long signal segments, whereas others such as band-pass filtering or autoregressive analysis can work on shorter time segments. Though all temporal domain-processing methods work well during offline BCI analysis, some of them are not as useful as spatial domain-processing methods during online analysis because of the quick responses required (Wolpaw *et al.*, 2002).

3.4.3.5. Extracting ERD/ERS Features

ERD/ERS components can serve as an ideal example of how to extract relevant features from a raw EEG signal, and several procedures exist to calculate these ERP components, which have been covered in detail by Kalcher and Pfurtscheller (1995) and Pfurtscheller and Lopes da Silva (1999) and shown in Figure 3.11. Since EEG signals are recorded from multiple channels that are referenced to a common electrode, the raw data is reference-dependent and must be dereferenced or, in other words, converted into reference-free data. This can be done through using methods such as common average reference, surface Laplacian, or local average reference.

Computation of the time course of ERD/ERS can be done using the classical ERD method, also known as the power method, and requires the following steps. (1) The raw EEG signal from each trial, where $x(i,j)$ is the jth sample of the ith trial, needs to be bandpass filtered ($x_f(i,j)$). (2) The amplitude samples need to be squared to obtain power

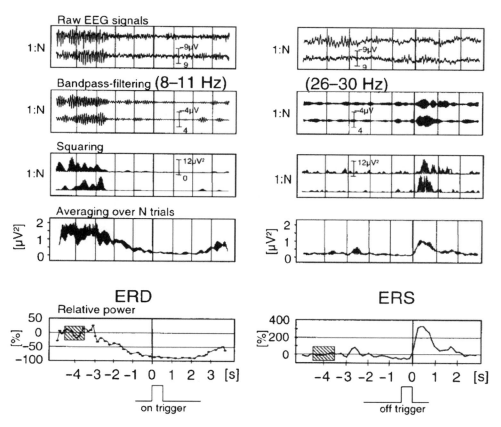

FIGURE 3.11. Techniques required to extract ERD and ERS from raw EEG signal. First, the raw EEG signal from each trial is bandpass filtered. Second, the amplitude samples are squared to obtain the power samples. Third, the power samples are averaged across all trials. Finally, variability is reduced and the graph is smoothed by averaging over time samples. (From Pfurtscheller and Lopes da Silva, 1999, with permission from Elsevier)

samples ($x_f^2(i,j)$). (3) The power samples need to be averaged across all the trials. (4) The variability must be reduced by averaging over time samples. This calculation of the instantaneous power is summarized in Eq. (3.1).

$$\bar{P}_{(j)} = \frac{1}{N} \sum_{i=1}^{N} x_f^2(i,j) \qquad (3.1)$$

Though ERD is known to occur in the alpha band and ERS is known to occur in the beta band, the range of frequencies to use for bandpass filtering needs to be more accurate. This range can be calculated by comparing two short-term power spectra to detect the most reactive frequency band or utilizing the continuous wavelet transform method or using the mean peak center of gravity frequency as the basis to adjust frequency bands individually.

The signal processing method to compute the time course of ERD/ERS described above produces a time course of phase-locked and non-phase-locked power changes and band power values. To discern between phase-locked and non-phase-locked power changes, the same procedure can be followed when substituting step 2 with a calculation of point-to-point intertrial variance ($IV_{(j)}$) prior to averaging over time, thus replacing Eq. (3.1) with Eq. (3.2).

$$IV_{(j)} = \frac{1}{N-1} \sum_{i=1}^{N} \{x_f(i, j) - \bar{x}_f(j)\}^2 \qquad (3.2)$$

This variant procedure is useful in cases with lower frequency components where the non-phase-locked ERD, characterized by a decrease in power, can be hidden by a phase-locked increase in power caused by a different ERP.

ERD/ERS is typically expressed as a percentage of change in power compared to baseline power measures taken prior to onset of event. These values can be calculated by taking the ratio of the difference between the power or intertrial variance at each sample point or an average of sample points within the frequency band of interest during the period after event onset (A) in the jth channel and the baseline power or intertrial variance prior to the event averaged over k samples (R) (Eqs. (3.3) and (3.4)).

$$ERD\% = [(A_{(j)} - R)/R] \times 100 \qquad (3.3)$$

$$R = \frac{1}{k} \sum_{j=n_o}^{n_0+k} A_{(j)} \qquad (3.4)$$

As the number of extraction methods created or adapted for BCI increases, it is difficult to compare their relative effectiveness in isolating the required features. An r^2 measure has been used as the scoring method. The r^2 score is a measure of the proportion of the total variance in the features generated by the method that is accounted for by the user's intent.

3.4.4. TRANSLATION TECHNIQUES

Translation techniques are algorithms developed with the goal of converting the input features (independent variable) into device control commands (dependent variable) (Wolpaw *et al.*, 2002). Translation techniques used widely in other areas of signal processing are adapted to BCI.

Discussed by Wolpaw *et al.* (2002), effective BCI techniques have three levels of adaptation. First, the technique must be able to adapt to the uniqueness of each user's signal features. Second, the technique must be able to reduce the impact of spontaneous variations that occur during regular use by making periodic online adjustments. Finally, the technique must be able to accommodate and engage the adaptive ability of the brain through increasing levels of feedback to encourage stronger feature signal generation.

Wolpaw and coworkers suggested that the success of a translation technique is determined by three criteria. The first criterion is the appropriateness of the selection of features. In other words, from all the features extracted, how well is the translation technique able to select those features that accurately convey the user's intent. The second criterion is the level at which the technique can assist the user's control of signal features through its adaptive capacity. The final criterion is the effectiveness of the method in translating command into logical control.

There are numerous types of feature translation algorithms. Some utilize simple characteristics such as amplitude or frequency. Others utilize single features while advanced algorithms utilize a combination of spatial and temporal features produced by one or more physiological processes (Bianchi and Babiloni, 2003). Algorithms currently in use include, but are not limited to, linear classifiers (Babiloni *et al.*, 2001; Wang and He, 2004), Fisher discriminant (Blankertz *et al.*, 2002) CSSD and Fisher discriminant (Wang *et al.*, 2004), Mahalanobis distance based classifiers (Cincotti *et al.*, 2002), neural networks (NN) (Penny and Roberts, 1998; Peters *et al.*, 1998; Robert *et al.*, 2002; Deng and He, 2003), support vector machines (SVM) (Vallabhaneni and He, 2004), and hidden Markov models (HMM) (Obermaier *et al.*, 2001).

3.4.5. *EXTRACTION AND TRANSLATION IN ACTION: A CASE STUDY ON CLASSIFICATION OF MOTOR IMAGERY TASKS*

Some BCI systems are based on classification of motor imagery tasks through recognition of mental states. In this study, a common data set of EEG recordings, made available by Dr. Osman (Osman and Robert, 2001; Sajda *et al.*, 2003) from University of Pennsylvania (http://liinc.bme.columbia.edu/competition.htm), is used to investigate three (Methods A, B, and C) different combinations of feature extraction and translation techniques offline.

3.4.5.1. *Experiment Setup*

Subjects were seated in front of a display screen and instructed to wear an EEG cap with 59 electrodes placed according to the International 10/20 system. The subjects were then trained to synchronize an indicated response base within 100 ms of a timed cue.

Each subject was put through several trials, with each trial epoch lasting 6 s as shown in Figure 3.12. Every trial started out with a blank screen displayed for 2 s. This was immediately followed up with a fixation point displayed for 500 ms to indicate that the trial has begun. This was replaced by either an E or an I for 250 m to instruct the subject

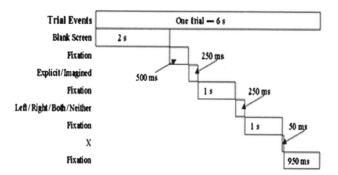

FIGURE 3.12. Onset of cues during one trial epoch. Subjects were seated in front of a monitor. For each trial, the display changed based on the timing sequence indicated above.

whether to perform an explicit or imagined hand movement when cued. The fixation point was returned to the screen for another 1 s.

This fixation point was followed by the display of L, R, B, or N for 250 ms to instruct the subject to employ the left, right, both, or no index finger when movement is cued. An X was displayed for 50 ms to cue the subject to perform the instructed action and was preceded by the fixation point for 1 s and was also followed up with the fixation point for 950 ms.

3.4.5.2. Data Preparation

For each trial, EEG data was recorded from all 59 electrodes. A final data set was created with 180 trials from each subject that contained only the imagined left or right finger movements. Half of these trials were labeled as left or right to use for training purposes.

All three methods redefined the 0th second to be when the L/R cue was displayed. All time data was referenced to this point. Method A used data from 1.75 s before to 2.25 s after the 0th second from only 9 centroparietal electrodes (F3, Fz, F4, C3, Cz, C4, P3, Pz, P4). Method B used data from 0.10 s before to 2 s after the 0th second from FC3, FC1, C3, C1, FC2, FC4, C3, and C4. Method C used data channel combinations—2-pair (C1/2, C3/4) and 4-pair (FC1/2 and FC3/4).

3.4.5.3. Feature Extraction

For all three methods, the first step in data processing utilized spatial filters, a very popular method in BCI, for accentuating localized activity and reducing diffused activity. Particularly, the surface Laplacian method was used. As previously described, it is the second spatial derivative of the instantaneous voltage distribution. With the assumption that distance between every adjacent channel is approximately equal, the surface Laplacian (Eq. 3.5) can be estimated by the difference between the potential V_j, at the jth channel of interest and the average value of the set of surrounding channels S_j (Hjorth, 1975).

$$V_j^{Lap} = V_j - \frac{1}{4} \sum_{k \in S_j} V_k \qquad (3.5)$$

Spline algorithms have been developed for computing the surface Laplacian from the recorded potentials (Perrin *et al.*, 1987; Nunez *et al.*, 1994; Babiloni *et al.*, 1996, 1998; He, 1999; He *et al.*, 2001), whereas the finite difference algorithm (Hjorth, 1975) was adopted because of the computational efficiency.

Because motor imagery tasks are known to cause an ERD/ERS phenomenon in the mu and beta rhythms, the appropriate frequency ranges were isolated. Methods A and B focused only on the standard mu-rhythm-associated components. Therefore, they used a fifth-order Butterworth bandpass filter with appropriate filter coefficients to extract components in the 8–13 Hz frequency range as shown in Figure 3.13.

Method C, however, recognized that ERD/ERS calculation requires an exact frequency band based on analysis of the signals (Wang and He, 2004). So this method followed a more precise technique to determine the most optimal frequency bands for each subject as they are affected by individual differences. For each trial, the frequency range from 5 to 25 Hz

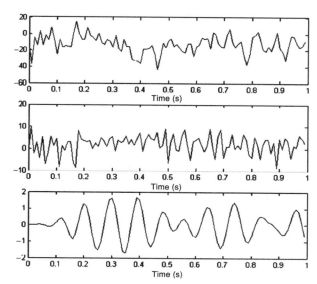

FIGURE 3.13. Results of concurrent steps of feature extraction on raw EEG signal in Methods A and B. It is difficult to detect a coherent component in the raw EEG signal depicted in the top frame as it is filled with noise. The second frame shows the signal after being processed through a surface Laplacian filter to get the spatial signal distribution. The signal is then bandpass filtered for 8–13 Hz, as shown in the third frame, to isolate the mu-rhythm ERD features noticeable around 0.5 s.

(covering mu and beta bands) was isolated and then divided into twenty bins, each with a 2 Hz bandwidth and a 50% frequency overlap with adjacent bins.

This collection of narrow bandpass filters decomposed the EEG signals into different frequency components, resulting in a rhythmical component of EEG, i.e., a temporal signal with amplitude variations that modulated the rapid oscillations. These can be approximately expressed as (Eq. 3.6).

$$x(t) = a(t) \cos(2\pi f_0 t + \varphi(t)) \tag{3.6}$$

where the enveloped portion, $a(t)$, should include the time-locked features. In order to extract $a(t)$, we can first calculate the Hilbert transform of the $x(t)$ using Eq. (3.7).

$$x_h(t) = a(t) \sin(2\pi f_0 t + \varphi(t)) \tag{3.7}$$

Eq. (3.8) can then be derived by combining Eqs. (3.6) and (3.7) (Papoulis, 1977).

$$z(t) = x(t) + jx_h(t) = a(t) \exp(2\pi f_0 t + \varphi(t)) \tag{3.8}$$

where $z(t)$ is termed the analytical signal. Thus $a(t)$ can be approximated using Eq. (3.9).

$$a(t) = |z(t)| = \sqrt{x^2(t) + x_h^2(t)} \tag{3.9}$$

The ERD/ERS phenomenon was then isolated through a grand average approach over all the envelopes because each envelope contains the relevant information for instantaneous power estimations for that particular narrow frequency band. In addition, the nonstationary

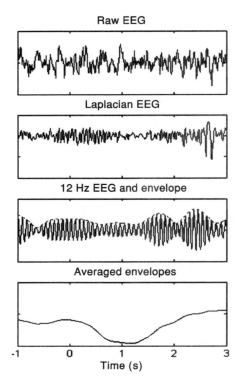

FIGURE 3.14. Results of concurrent steps of feature extraction on raw EEG signal in Method C. It is difficult to detect a coherent component in the raw EEG signal depicted in the top frame as it is filled with noise. The second frame shows the signal after being processed through a surface Laplacian filter to get the spatial signal distribution. The signal is then bandpass filtered, as shown in the third frame, to isolate the frequencies of interest. The features become evident in the fourth frame as they are extracted by utilizing a grand averaging method over a fixed bin or window size.

nature of the EEG signal would alter the dynamic region of the signal amplitude, and since spatial noise is amplified owing to the surface Laplacian filter, amplitude is normalized to counter these effects. Figure 3.14 shows the affect of each processing step on the raw EEG signal.

Returning to Method A, feature selection was done by extracting the amplitude values with a resolution of 1 Hz for the 8–13 Hz frequency range from a standard power spectral density (PSD) calculation. PSD, derived using the Welch's averaged modified periodogram method, describes how the power (or variance) of a time series is distributed with frequency and is obtained by the fast Fourier transform (FFT) of the autocorrelation function (ACF) of the time series.

The PSD calculations derived six features, 1 for each 1 Hz division, from each of the nine electrodes and a total of 54 features for each trial. Concurrently plotting the features from each channel, C3, C4, and Fz were chosen as the features from these channels, which exhibited prominent differences between trials labeled left and right (Deng and He, 2003).

Following the Butterworth filter, in Method B (Vallabhaneni and He, 2004), the data is squared, as recommended in the classic ERD calculation method, and feature extraction is performed using spatiotemporal principal components analysis (PCA). PCA is commonly used in data analysis for extracting the most significant parts of any data without changing the meaning of the data itself. This process reduces the dimensionality of a given data set to the few principal components that explain the majority of the variance in the data.

FIGURE 3.15. Representation of the three-layer supervised artificial neural network. PSD values from 90 trials (45 left and 45 right) from the three channels of interest were used for training and fed into the first layer with 18 neurons (one for each feature), which were mapped to a hidden layer with 3 neurons and the final output layer with 1 neuron with 2 possible values (0 for left, 1 for right). A separate set of 90 trials were used for testing the prediction accuracy of the neural network. (From Deng and He, 2003, with permission, © 2003, IEEE)

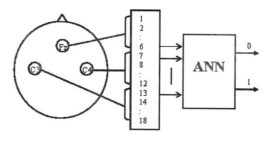

The steps involved in PCA are the following:

1. Center the data matrix by subtracting the mean from each dimension.
2. Calculate the covariance matrix of the centered data.
3. Calculate the eigenvectors and eigenvalues of the covariance matrix.
4. Select the eigenvectors with the highest n corresponding to the eigenvalues as they are the principal components of the data set.
5. Final data set = [principal components]$^T \times$ [centered data]^T

Spatiotemporal PCA involves, for each trial, performing the standard PCA along the spatial dimensions or channels to find spatial factors that explain the spatial variance and indicate interchannel signal correlation and patterns. This is followed by performing PCA along the time series to identify temporal features. Principal components were chosen based on a scree test threshold of 95, which means that the sum of the percentage of variance in the data explained by the chosen principal components was at least 95%. The final dimensions of the feature space, therefore, varied for each subject.

3.4.5.4. Feature Translation

All three methods utilized different supervised learning methods as feature translation techniques. Method A used a three-layer supervised artificial neural network (ANN) as shown in Figure 3.15 (Deng and He, 2003). This method requires a desired output to learn and creates a model that correctly maps the input to the output so that trials with unknown output can be predicted.

The input layer with 18 neurons (i.e. one neuron for each of the six features from the three channels of interest) was followed by the hidden layer with three neurons and the output layer with one neuron that had two possible values of 0 (left) and 1(right). Ninety trials were used to train the ANN, of which 45 were left-hand trials and 45 were right-hand trials. A separate set of 90 trials was used to test the prediction accuracy of the ANN.

Feature translation in Method B was performed using a linear kernel-based SVM provided by OSU SVM Classifier Matlab Toolbox (http://www.eleceng.ohio-state.edu/~maj/osu_svm/). SVMs, as a kernel-based statistical learning method, perform especially well in a high dimensional feature space as they efficiently avoid overfitting of the data. SVMs are also supervised learning methods that use knowledge from training data to classify unknown trials with similar features. It should theoretically be possible to find a hyperplane across an input space that completely separates the trials in the right and left classes. In such cases, a kernel-based method would be unnecessary. This data set, however, is not perfectly separable across the input space, and such data sets are rare. The only possible way to generate a separating hyperplane is by mapping the data onto a higher dimensional feature space as defined by the kernel. As the dimensionality increases, it should be possible to create a feature space that can completely separate the two classes in any data set. The risk of artificially projecting the data onto a high dimensional feature space, however, is to subject the SVM to the possibility of finding trivial solutions that overfit the data.

If there are n trials for each subject, let each of the m dimensions in the final data matrix be a feature vector in an m-dimensional input space. The kernel, $K(X,Y)$, that is used to measure the similarity between the two trials X and Y is the dot product of their input spaces, as shown in Eq (3.10).

$$K(X, Y) = X \cdot Y \qquad (3.10)$$

A collection of support vectors are generated using this kernel function. These support vectors are used in the classification decision.

In a two-class problem, a linear SVM has the decision function, as shown in Eq (3.11), where w is a weight vector, b is a constant bias, and x is a particular support vector.

$$f(x) = w \cdot x + b \qquad (3.11)$$

Both w and b are automatically chosen to maximize the margin between the decision hyperplane and the class. The class of a particular trial is determined by the sign of f.

Prior to translation, Method C used PCA to reduce the dimensionality of the feature set and increase computational efficiency. Feature vectors projecting onto the largest three principal components were retained for translation.

For translation, Method C used a Bayesian linear classifier as the basic technique. This method maximizes the ratio of interclass variance to the intraclass variance in any particular data, thereby guaranteeing maximal margin of separation.

A discrete linear classifier, with discriminant function denoted as h_{ij}, was created for each of the frequency bands, i, in each channel, j, in training stage. The accuracy values $a_{i,j}$ of each of these classifiers reflected the adaptation of every frequency band and channel. A normalized frequency weight, $w_{i,j}$, is defined by

$$w_{i,j} = \begin{cases} (2a_{i,j} - 1)^m, & a_{i,j} > 0.5 \\ 0, & a_{i,j} \leq 0.5 \end{cases} \qquad (3.12)$$

where m is the control parameter used to further emphasize those bands with larger accuracy values.

On the basis of the frequency weights, classification of the trial at each channel could be made using the channel discriminant function,

$$g_j(v) = \sum_{i=1} w_{i,j} \, \text{sgn}(h_{i,j}(v)) \qquad (3.13)$$

TABLE 3.1. Classification Accuracies for Methods A, B, and C for Three Subjects

| | Method A (%) | | Method B (%) | | Method C (%) | | | |
| | | | | | 2 pairs | | 4 pairs | |
	Training	Testing	Training	Testing	Training	Testing	Training	Testing
Subject 1	90.0	70.0	n/a	74.4	83.0	70.0	89.0	70.6
Subject 2	85.6	64.4	n/a	91.1	93.0	85.6	94.8	90.6
Subject 3	84.4	82.2	n/a	82.8	88.6	82.2	95.0	92.2
Average	86.7	72.2	n/a	82.8	88.2	79.6	92.9	84.5

The correct classification accuracy for the jth channel of all frequency bands was obtained using this method. So the decision for v was made according to the sign of $g_j(v)$, and its absolute value measures the likelihood of this decision. The logical way of synthesizing all the channels is by summarizing the likelihood together as the final discriminant function, i.e.

$$q(v) = \sum_{j=1}^{M} g_j(v) \tag{3.14}$$

Therefore, the final classification decision was made by synthesizing all frequency bins and channels according to their classification contribution as rated by training.

3.4.5.5. Results

All three feature extraction and translation methods were tested on identical data sets from three different subjects using n-fold cross-validation. The classification accuracy of all the methods, as detailed in Table 3.1, is promising. These methods may be further applied in an online feature extraction and translation method to measure their performance in a real-time environment.

It is evident from the methods described above and the resulting accuracies that it is possible to achieve high classification rates using different methods for a small set of subjects. It is important to note that signal variation does exist between subjects on a consistent basis across the three methods. This is because of signal variations in each subject and also their individual capacity to train and produce the required signal components. The adaptive capabilities of the extraction and translation algorithms must allow to accommodate such differences in users for successful application. Note also that Method C was demonstrated to provide good performance for a larger set of subjects (Wang and He, 2004).

3.5. TYPICAL BCI SYSTEMS

With the growing combinations of signals, feature extraction, and translation techniques, the number of different BCI systems is rapidly growing. Basic research is typically started on offline BCI systems, where signal acquisition is followed by feature extraction and translation as a separate step. This type of BCI allows researchers to refine and test extraction and translation algorithms before utilizing them in applied research. Ultimately, however, a BCI technique needs to be tested online for assessing its performance.

Another important categorization of BCI systems is *external* versus *internal*. External BCI systems, also known as *exogenous* BCI systems, classify on the basis of a fixed temporal context to an external stimulus not under the user's control. These systems utilize components evoked by external stimuli such as VEP. These BCI systems do not require extensive training but do require a controlled environment and stimulus. Internal BCI systems, also known as *endogenous* BCI systems, on the other hand, classify based on a fixed temporal context to a timed event or internal stimulus. These systems utilize components evoked by tasks such as motor imagery and do require significant user training (Wolpaw *et al.*, 2002).

In addition, BCI systems vary in the use of specific or unspecific signal acquisition methods. Specific BCI systems use signals recorded at well-chosen positions where effect is expected such as with specified cognitive tasks. Unspecific BCI systems, meanwhile, use signals recorded from electrodes all over the brain such as with operant condition method.

Over the past 20 years, researchers have created models of several working BCI systems. One such BCI system was developed by Farwell and Donchin. They created a BCI that could be used to type out words by selecting letters, words, and commands from a display. The system used an externally paced method to flash letters randomly on a 6×6 matrix on a screen while the user thought about the next letter he or she wanted. When the expected letter flashed on the screen, the user would generate a detectable P300.

Because the system focused on detecting only P300s, signal acquisition was done at specific electrodes. Feature extraction generated 36 feature vectors, one for each square on the screen. As the P300 is time-locked to the stimulus, when a particular row or column was flashed, a 600-ms window of signal was added to each of the corresponding feature vectors. The translation was done by continuously ranking all the features using various methods. The letter, word, or command corresponding to the highest-ranked feature vector was classified as the user's intent (Farwell and Donchin, 1988).

Kerin and Aunon developed a BCI system that allowed users with severe physical disabilities to communicate with their surroundings by spelling specific codewords that were predefined commands. Depending on the cognitive task performed by the user, the system could detect differences in lateralized spectral power levels. Because the cognitive tasks were not defined, EEG signals were collected from electrodes covering the parietal and occipital regions. Feature extraction involved generating two feature vectors using fast Fourier transform (FFT) and auto-regressive (AR) spectral estimation methods and then running them through a bandpass filter in four frequency bands. A third vector was generated using AR coefficients of the signals in the four frequency bands. The Bayesian quadratic classifier performed feature translation based on the power or AR coefficients of the features (Kerin and Aunon, 1990).

Wolpaw and coworkers developed a BCI system that allowed users to control prosthetic devices by moving a cursor up or down on the screen to select one of two icons. His team also used internally paced events. The EEG was recorded at specific position as the user actively controlled the power of their mu rhythm. EEG signals were collected from specific bipolar electrode locations around C3 (refer to Figure 3.4). The EEG power spectrum was calculated using FFT with a 3-Hz resolution to generate the feature vector. A power value centered at 9 Hz was used as the amplitude of the mu rhythm. Feature translation was done using linear discriminant analysis on the basis of the power of the mu rhythm divided into five possible levels. Each of the five levels defined a direction and magnitude for the cursor, which eventually helped the cursor reach its targets on the top and bottom of the screen (Wolpaw et al., 1991).

Like Farwell and Donchin's system, Sutter also presented a BCI system for locked-in individuals to communicate by selecting letters, words, and commands from a screen. This BCI, however, used SSVEP generated by flashing alternating symbols on a display rather than a P300 component. Signal acquisition for most users was done by EEG, through electrodes over the occipital cortex. In one user, however, an intracranial electrode array was utilized. Since SSVEP is time-locked to the change in stimulus, the feature vector was the averaged EEG signal for a specific period after stimulus onset. These feature vectors were overlaid onto 64 template responses, corresponding to a particular letter, word, or command, and the user's intent was classified on the basis of a predefined threshold between the calculated feature vector and the template (Sutter, 1992).

Pfurtscheller and coworkers developed a BCI system that used mu rhythm EEG recordings measured over the sensorimotor cortex. This BCI, however, used ERD in the mu rhythm rather than the amplitude used by Wolpaw and coworkers. The raw EEG signals were filtered to the alpha band (8–12 Hz) and then squared to estimate the instantaneous mu power. Five consecutive mu power estimates during ERD were combined to create a five-dimensional feature vector that was classified using one-nearest-neighbor (1-NN) classifier with reference vectors generated using the learning vector quantization (LVQ) method (Birch and Mason, 2000). LVQ is a type of vector quantization method where the high-dimensional input space is divided into different regions, with each region having a reference vector and a class label attached. During feature translation, an unknown input vector is classified using the 1-NN classifier, where it is assigned to the class label of the reference vector to which it is closest (Pfurtscheller et al., 1993).

A few more BCI systems, along with revisions to existing systems were also presented. Pfurtscheller and coworkers introduced the concept of using signals from the sensorimotor cortex generated from imagined motor movements (Pfurtscheller et al., 1994, 1997). Wolpaw and McFarland revised the cursor-based BCI system with predefined directional movements to utilize recordings from multiple channels and allow simultaneous two-dimensional cursor control by using the sum of the feature vector power levels as a magnitude of vertical movement and the difference of the power levels as a magnitude of horizontal movement. This allowed the user to select from four icons, one at each corner of the screen (Wolpaw and McFarland, 1994).

Birbaumer and coworkers introduced a BCI system that utilized SCPs. This system also enabled users to create text messages by selecting letters from a virtual keyboard displayed on a screen. The EEG signals were recorded from electrodes over the frontal cortex. The feature vector was represented by the amplitude of the SCP waveform averaged over a predefined sliding window. Feature translation was done using a linear classification with a heuristic threshold customized to the user. The feature vectors were classified into two states, move up or move down, which moved a cursor up or down on the screen (Birbaumer et al., 1999).

Kennedy and coworkers created a new version of the spelling device that used the firing rate of particular neural groups. The user would make selections by actively controlling the firing rate of neurons through imagined movements, which was recorded by two electrodes implanted within the area of the primary motor cortex that controlled hand movements. The user was presented with a cursor on a screen with a dynamic matrix of letters, words, or commands. The feature vector was a reflection of the firing rate recorded by both electrodes. Increasing neuronal firing rate resulted in increasing output levels, which defined the horizontal and vertical cursor speed. A particular cell could be selected on the basis of a dwell time threshold, a predefined amount of time that the user would leave the cursor on the cell to indicate that it was their choice (Kennedy *et al.*, 2000).

The systems described above are only a small sampling of the work that has been done in BCI. It is difficult to objectively compare the wide array of BCI systems that utilize a variety of signals and are aimed at different applications. The goal of every BCI system, however, is to communicate the user's intent accurately. As the number of possible choices increases, accuracy alone becomes a weak scoring methodology. For communication systems the traditional unit of measure is the amount of information transferred for a unit of time. For BCI systems, therefore, the performance measure can be indicated by bits per trial and bits per minute. This provides a tangible measure for making intrasystem and intersystem performance comparisons.

The bit rate for a BCI system can be calculated easily. Let there be N possible choices for each trial, where each choice has the same probability of being the desired choice. If the probability, P, that the desired choice will be selected is constant and the probability of each of the undesired choices being selected is equal, then the bit rate per trial, B, can be defined by Eq. (3.15) (Wolpaw *et al.*, 2000b).

$$B = \log_2 N + P \log_2 P + (1 - P) \log_2[(1 - P)/(N - 1)] \qquad (3.15)$$

As the number of possible choices increases, an equivalent bit rate can be achieved even with a lower accuracy when compared to systems with smaller number of possible choices as shown in Figure 3.16. It is also important to note that for any BCI, the relationship between accuracy and rate of information transfer is not linear. Increasing the accuracy, for example, from 80 to 90% in a two-choice system nearly doubles the information transfer rate.

3.6. BCI DEVELOPMENT

Basic and applied research in BCI has advanced rapidly, especially over the past 20 years. There are dozens of active research groups around the world involved in the development of BCI technology. Though BCI has emerged from theory to reality with systems now being introduced into commercial applications, the field still holds tremendous untapped potential.

Considering the current rate of development, advancing the proliferation of BCI technology requires an increase in the number of people and funding involved. BCI requires increasing cooperation between various fields such as computer science, neuroscience, engineering, psychology, medical imaging, etc. Establishing multidisciplinary research teams is essential.

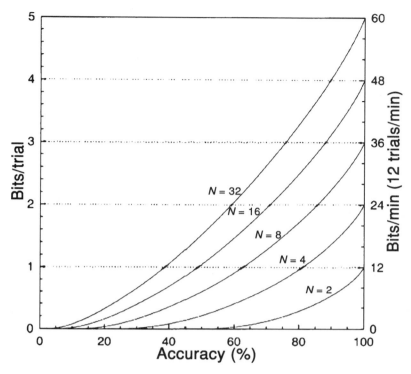

FIGURE 3.16. Information transfer rate for different values of N. For a trial with N possible selection and the conditions (1) each selection has an equal probability of being desired, (2) probability (P) that the desired selection will be selected is always the same, and (3) each of the other undesired selections has the same probability of selection, the bit rate can be defined as $B = \log_2 N + P \log_2 P + (1 - P) \log_2[(1 - P)/(N - 1)]$. For each N, bit rate is shown for accuracy levels greater than chance ($\geq 100/N$). (From Wolpaw *et al.*, 2000b, with permission, © 2000, IEEE).

Factors such as insurance coverage, for example, can aid in faster commercial applications of BCI devices. If users can be reimbursed for the expenses incurred in operating the BCI device, they would be more willing to utilize them. Increasing interest from the industry would not only standardize manufacturing and lower the unit cost, but also increase research funding. This would require a larger target audience and development of BCI technology for applications beyond assistance of patients with disabilities.

Future work in BCI technology should focus on basic and applied research. New usable features are required for more precise and faster user control. New algorithms are required for feature extraction and translation. Advancements in signal acquisition methods such as invasive techniques also potentially hold great promise in increasing spatial and temporal resolution. Novel applications of EEG inverse solutions, which convert the smeared scalp EEG onto source signals in the brain or over the brain surface (Mosher *et al.*, 1992; Babiloni *et al.*, 1997; He *et al.*, 2001, 2002a, 2002b; He and Lian, 2002, 2004), promise to overcome the limitations of scalp EEG and may lead to high resolution noninvasive BCI (Qin *et al.*, 2004). Better and easier training methods to make BCI devices easier to use will make BCI systems usable by extremely locked-in users. Finally, increasing awareness of the potential of BCI technology will be essential in generating public interest and channeling government funding.

ACKNOWLEDGMENTS

The authors would like to thank Dr. Osman of the University of Pennsylvania for making his data available, and Jie Deng for useful discussions. This work was supported in part by grants NSF BES-0218736, NSF BES-0411898, NSF BES-9875344, and NIH R01EB00178 (BH).

REFERENCES

Annett, J., 1995, Motor imagery: Perception or action, *Neuropsychologia* **33**(11):1395–1417.

Babiloni, F., Babiloni, C., Carducci, F., Fattorini, L., Onorati, P., and Urbano, A., 1996, Spline Laplacian estimate of EEG potentials over a realistic magnetic resonance-constructed scalp surface model, *Electroencephalogr. Clin. Neurophysiol.* **98**(4):363–73.

Babiloni, F., Babiloni, C., Carducci, F., Fattorini, L., Anello, C., Onarati, P., and Urbano, A., 1997, High resolution EEG: A new model-dependent spatial deblurring method using a realistically—shaped MR—constructed subject's head model, *Electroenceph. Clin. Neurophysiol.* **102**:69–80.

Babiloni, F., Carducci, F., Babiloni, C., and Urbano, A., 1998, Improved realistic Laplacian estimate of highly-sampled EEG potentials by regularization techniques, *Electroencephalogr. Clin. Neurophysiol.* **106**(4):336–343.

Babiloni F., Cincotti, F., Bianchi, L., Pirri, G., Millan, J., Mourino, J., Salinari, S., and Marciani, M. G., 2001, Recognition of imagined hand movements with low resolution surface Laplacian and linear classifiers, *Med. Eng. Phys.* **23**:323–328.

Babiloni, F., Cincotti, F., Lazzarini, L., Millán, J., Mouriño, J., Varsta, M., Heikkonen, J., Bianchi, L., and Marciani, M. G., 2000, Linear classification of low-resolution EEG patterns produced by imagined hand movements, *IEEE Trans. Rehabil. Eng.* **8**(2):186–188.

Bianchi, L., and Babiloni, F., 2003, Comparison of different feature classifiers for brain computer inerfaces, *Proc. 1st Int. IEEE Conf. Neural Eng.* 645–647.

Birbaumer, N., 1999, Rain Man's revelations, *Nature* **399**(6733):211–212.

Birbaumer, N., Ghanayim, N., Hinterberger, T., Iversen, I., Kotchoubey, B., Kübler, A., Perelmouter, J., Taub, E., and Flor, H., 1999, A spelling device for the paralysed, *Nature* **398**(6725):297–298.

Birbaumer, N., Kübler, A., Ghanayim, N., Hinterberger, T., Perelmouter, J., Kaiser, J., Iversen, I., Kotchoubey, B., Neumann, N., and Flor, H., 2000, The Thought Translation Device (TTD) for completely paralyzed patients, *IEEE Trans. Rehabil. Eng.* **8**(2):190–193.

Birch, G. E., and Mason, S. G., 2000, Brain–computer interface research at the Neil Squire Foundation, *IEEE Trans. Rehabil. Eng.* **8**(2):193–195.

Blankertz, B., Curio, G., and Müller, K., 2002, Classifying single trial EEG: Towards brain computer interfacing, *Adv. Neural Inf. Proc. Systems* **14**:157–164.

Blum, A. L., and Langely, P., 1997, Selection of relevant features and examples in machine learning, *Artif. Intell.* **97**:245–271.

Brouwer, B., and Hopkins-Rosseel, D., 1997, Motor cortical mapping of proximal upper extremity muscles following spinal cord injury, *Spinal Cord* **35**:205–212.

Cincotti, F., Mattia, D., Babiloni, C., Carducci, F., Bianchi, L., Millan, J., Mourino, J., Salinari, S., Marciani, M., and Babiloni F., 2002, Classification of EEG mental patterns by using two scalp electrodes and Mahalanobis distance based classifiers, *Method Inform. Med.* **41**:337–341.

Cincotti, F., Scipione, A., Timperi, A., Mattia, D., Marciani, M. G., Millan, J., Salimari, S., Bianchi, L., and Babiloni, F., 2003, Comparison of different feature classifiers for brain computer interfaces, *Proc. 1st Int. IEEE EMBS Conf. Neural Eng.* 645–647.

Curran, E. A., and Stokes, M. J., 2003, Learning to control brain activity: A review of the production and control of EEG components for driving brain–computer interface (BCI) systems, *Brain Cognition* **51**:326–336.

Deng, J., and He, B., 2003, Classification of imaginary tasks from three channels of EEG by using an artificial neural network, *Proc. 25th Ann. Int. Conf. IEEE EMBS.* [CD-ROM]

Donchin, E., and Coles, M. G. H., 1988, Is the P300 component a manifestation of context updating? *Behav. Brain Sci.* **11**:355–425.

Donchin, E., Spencer, K. M., and Wijesinghe, R., 2000, The mental prosthesis: Assessing the speed of a P300-based brain–computer interface, *IEEE Trans. Rehabil. Eng.* **8**(2):174–179.

Donoghue, J., 2002, Connecting cortex to machines: Recent advances in brain interfaces, *Nature Neurosci. Suppl:*1085–1088.

Ebrahimi, T., Vesin, J., and Garcia, G., 2003, Brain-computer interface in multimedia communication, *Signal Process. Mag.* **20**(1):14–24.

Farwell, L. A., and Donchin, E., 1988, Talking off the top of your head: Toward a mental prosthesis utilizing event-related brain potentials, *Electroencephalogr. Clin. Neurophysiol.* **70**(6):510–523.

He, B., 1999, Brain electric source imaging: Scalp Laplacian mapping and cortical imaging, *Crit. Rev. Biomed. Eng.* **27**:149–188.

He, B., and Cohen, R., 1992, Body surface Laplacian ECG mapping, *IEEE Trans. Biomed. Eng.* **39**(11):1179–1191.

He, B., Lain, J., and Li, G., 2001, High-resolution EEG: A new realistic geometry spline Laplacian estimation technique, *Clin. Neurophysiol.* **112**(5):845–852.

He, B., and Lian, J., 2002, Spatio-temporal Functional Neuroimaging of Brain Electric Activity, *Critical Review of Biomedical Engineering*, **30**:283–306.

He, B., and Lian, J., 2005, Electrophysiological Neuroimaging, *In* He (Ed): *Neural Engineering*, Kluwer Academic Publishers.

He, B., Lian, J., Spencer, K. M., Dien, J., and Donchin, E., 2001, A Cortical Potential Imaging Analysis of the P300 and Novelty P3 Components, *Human Brain Mapping*, **12**:120–130.

He, B., Zhang, X., Lian, J., Sasaki, H., Wu, S., and Towle, V. L., 2002a, Boundary Element Method Based Cortical Potential Imaging of Somatosensory Evoked Potentials Using Subjects' Magnetic Resonance Images, *NeuroImage* **16**:564–576.

He, B., Yao, D., Lian, J., and Wu, D., 2002b, An Equivalent Current Source Model and Laplacian Weighted Minimum Norm Current Estimates of Brain Electrical Activity, *IEEE Trans. on Biomedical Engineering* **49**:277–288.

Hjorth, B., 1975, An on-line transformation of EEG scalp potentials into orthogonal source derivations, *Electroencephalogr. Clin. Neurophysiol.* **39**(5):526–530.

Isaacs, R. E., Weber, D. J., and Schwartz, A. B., 2000, Work toward real-time control of a cortical neural prosthesis, *IEEE Trans. Rehabil. Eng.* **8**(2):196–198.

Jeannerod, M., 1995, Mental imagery in the motor context, *Neuropsychologia* **33**(11):1419–1432.

Kalcher, J., and Pfurtscheller, G., 1995, Discrimination between phase-locked and non-phase locked event-related EEG activity, *Electroencephalogr. Clin. Neurophysiol.* **94**:381–384.

Keirn, Z. A., and Aunon, J. I., 1990, A new mode of communication between man and his surroundings, *IEEE Trans. Biomed. Eng.* **37**(12):1209–1214.

Kelly, S., Burke, D., de Chazal, P., and Reilly, R., 2002, Parametric models and spectral analysis for classification in brain-computer interfaces, *Proc. 14th Int. Conf. Digit. Sign. Process.* **1**:307–310.

Kennedy, P. R., and Bakay, R. A., 1998, Restoration of neural output from a paralyzed patient by a direct brain connection, *NeuroReport* **9**:1707–1711.

Kennedy, P. R., Bakay, R. A. E., Moore, M. M., Adams, K., and Goldwaithe, J., 2000, Direct control of a computer from the human central nervous system, *IEEE Trans. Rehabil. Eng.* **8**(2):198–202.

Kostov, A., and Polak, M., 2000, Parallel man–machine training in development of EEG-based cursor control, *IEEE Trans. Rehabil. Eng.* **8**(2):203–205.

Kubler, A., Kotchoubey, B., Kaiser, J., Wolpaw, J., and Birbaumer, N., 2001, Brain–computer communication: Unlocking the locked in, *Psychol. Bull.* **127**(3):358–375.

Laubach, M., Wessberg, J., and Nicolelis, M. A. L., 2000, Cortical ensemble activity increasingly predicts behavior outcomes during learning of a motor task, *Nature* **405**(6786):567–571.

Lauer, R. T., Peckham, P. H., Kilgore, K. L., and Heetderks, W. J., 2000, Applications of cortical signals to neuroprosthetic control: A critical review, *IEEE Trans. Rehabil. Eng.* **8**(2):205–208.

Le, J., Menon, V., and Gevins, A., 1992, Local estimate of surface Laplacian derivation on a realistically shaped scalp surface and its performance on noisy data, *Electroenceph. Clin. Neurophysiol.* **92**:433–441.

Levine, S. P., Huggins, J. E., BeMent, S. L., Kushwaha, R. K., Schuh, L. A., Rohde, M. M., Passaro, E. A., Ross, D. A., Elisevich, K. V., and Smith, B. J., 2000, A direct brain interface based on event-related potentials, *IEEE Trans. Rehabil. Eng.* **8**(2):180–185.

Makeig, S., Enghoff, S., Jung, T. P., and Sejnowski, T. J., 2000, A natural basis for efficient brain-actuated control, *IEEE Trans. Rehabil. Eng.* **8**(2):208–211.

Malmivuo, J., and Plonsey, R., 1995, *Bioelectromagnetism—Principles and Applications of Bioelectric and Biomagnetic Fields*, Oxford University Press, New York.

Mason, S. G., and Birch, G. E., 2003, A general framework for brain-computer interface design, *IEEE Trans. Neural Syst. Rehabil. Eng.* **11**(1):70–85.

Maynard, E., Nordhausen, C., and Normann, C., 1997, The Utah intracortical electrode array: A recording structure for potential brain–computer interfaces, *Electrencephalogr. Clin. Neurophysiol.* **102**:228–239.

McFarland, D. J., McCane, L. M., David, S. V., and Wolpaw, J. R., 1997, Spatial filter selection for EEG-based communication, *Electroencephalogr. Clin. Neurophysiol.* **103**:386–394.

Middendorf, M., McMillan, G., Calhoun, G., and Jones, K. S., 2000, Brain–computer interfaces based on steady-state visual evoked response, *IEEE Trans. Rehabil. Eng.* **8**(2):211–214.

Mosher, J. C., Lewis, P. S., and Leahy, R. M., 1992, Multiple dipole modeling and localization from spatio-temporal MEG data, *IEEE Trans. Biomed. Eng.* **39**:541–557.

Moxon, K. A., 2004, Neurorobotics, in: *Neural Engineering* (He, ed.), Kluwer Academic Publishers, 2005.

Müller, K., Kohlmorgen J., Ziehe, A., and Blankertz, B., 2000, Decomposition algorithms for analyzing brain signals, in: *Adaptive Systems for Signal Processing, Communications and Control* (S. Haykin, ed.), pp. 105–110.

Muller-Gerking, J., Pfurtscheller, G., and Flyvbjerg, H., 1999, Designing optimal spatial filters for single-trial EEG classification in a movement task, *Clin. Neurophysiol.* **110**(5):787–798.

Mussa-Ivaldi, F. A., and Miller, L. E., 2003, Brain–machine interfaces: Computational demands and clinical needs meet basic neuroscience, *Trends Neurosci.* **26**(6):329–334.

Nicolelis, M., 2001, Actions from thoughts, *Nature* **409**:403–407.

Nicolelis, M., 2003, Brain–machine interfaces to restore motor function and probe neural circuits, *Nat. Rev. Neurosci.* **4**(5):417–422.

Nicolelis, M., and Chapin, J., 2002, Controlling robots with mind, *Sci. Am.* **287**(4):46–53.

Nunez, P., Silberstein, R., Cadusch, P., Wijesinghe, R., Westdorp, A., and Srinivasan, R., 1994, A theoretical and experimental study of high resolution EEG based on surface Laplacians and cortical imaging, *Electroencephalogr. Clin. Neurophysiol.* **90**(1):40–57.

Obermaier, B., Guger, C., Neuper, C., and Pfurthscheller, G., 2001, Hidden Markov models for online classification of single trial EEG data, *Pattern Recogn. Lett.* **22**:1299–1309.

Osman, A., and Robert, A., 2001, Time-course of cortical activation during overt and imagined movements, in: *Proceedings of the Cognitive Neuroscientists Annual Meetings*, New York.

Papoulis, A., 1977, *Signal Analysis*, McGraw-Hill Book Company, New York.

Penny, W. D., and Roberts, S. J., 1998, Bayesian neural networks for detection of imagined finger movements from single-trial EEG, *Neural Networks* **12**:877–892.

Penny, W. D., and Roberts, S. J., 1999, EEG-based communication via dynamic neural network models, *Proc. Int. Joint Conf. Neural Networks*. [CDROM]

Penny, W. D., Roberts, S. J., Curran, E. A., and Stokes, M. J., 2000, EEG-based communication: A pattern recognition approach, *IEEE Trans. Rehabil. Eng.* **8**(2):214–215.

Perrin, F., Bertrand, O., and Pernier, J., 1987, Scalp current density mapping: value and estimation from potential data, *IEEE Trans. Biomed. Eng.* **34**:283–288.

Peters, B. O., Pfurtscheller, G., and Flyvbjerg, H., 1998, Mining multi-channel EEG for its information content: An ANN-based method for a brain–computer interface, *Neural Networks* **11**:1429–1433.

Pfurtscheller, G., and Neuper, C., 2001, Motor imagery and direct brain–computer communication, *Proc. IEEE* **89**(7):1123–1134.

Pfurtscheller, G., Flotzinger, D., and Kallcher, J., 1993, Brain–computer interface: A new communication device for handicapped persons, *J. Microcomp. App.* **16**:293–299.

Pfurtscheller, G., Flotzinger, D., and Neuper, C., 1994, Differentiation between finger, toe and tongue movement in man based on 40-Hz EEG, *Electroencephalogr. Clin. Neurophysiol.* **90**(6):456–460.

Pfurtscheller, G., and Lopes da Silva, F. H., 1999, Event-related EEG/MEG synchronization and desynchronization: Basic principles, *Clin. Nuerophysiol.* **110**(11):1842–1847.

Pfurtscheller, G., Neuper, C., and Flotzinger, D., 1997, EEG-based discrimination between imagination of right and left hand movement, *Electroencephalogr. Clin. Neurophysiol.* **103**(6):642–651.

Pfurtscheller, G., Neuper, C., Guger, C., Harkam, W., Ramoser, H., Schlögl, A., Obermaier, B., and Pregenzer, M., 2000, Current trends in Graz brain–computer interface (BCI) research, *IEEE Trans. Rehabil. Eng.* **8**(2):216–219.

Qin, L., Ding, L., and He, B., 2004, Motor imagery classification by means of source analysis for brain computer interface applications, *Journal of Neural Eng.* **1**:135–141.

Ramoser, H., Muller-Gerking, J., and Pfurtscheller, G., 2000, Optimal spatial filtering of single trial EEG during imagined hand movement, *IEEE Trans. Rehabil. Eng.* **8**(4):441–446.

Robert, C., Gaudy, J., and Limoge, A., 2002, Electroencephalogram processing using neural networks, *Clin. Neurophysiol.* **113**:694–701.

Sajda, P., Gerson, A., Muller, K., Blankertz, B., and Parra, L., 2003, A data analysis competition to evaluate machine learning algorithms for use in brain-computer interfaces, *IEEE Trans. Neural Syst. Rehabil. Eng.* **11**(2):184–185.

Sanchez, J. C., Carmena, J. M., Lebedev, M. A., Nicolelis, M. A., Harris, J. G., and Principe, J. C., 2004, Ascertaining the importance of neurons to develop better brain–machine interfaces, *IEEE Trans. Biomed. Eng.* **51**(6):943–953.

Serruya, M., Hatsopoulos, N., Paninski, L., Fellows, M., and Donoghue, J., 2002, Instant neural control of a movement signal, *Nature* **416**:141–142.

Serruya, M., Hatsopoulos, N., Paninski, L., Fellows, M., and Donoghue, J., 2003, Robustness of neuroprosthetic decoding algorithms, *Biol. Cybern.* **88**(3):219–228.

Spencer, K. M., Dien, J., and Donchin, E., 2001, Spatiotemporal analysis of the late ERP responses to deviant stimuli, *Psychophysiology* **38**(2):343–358.

Sutter, E. E., 1992, The brain response interface: Communication through visually-induced electrical brain responses, *J. Microcomp. App.* **15**:31–45.

Vallabhaneni, A., and He, B., 2004, Motor imagery task classification for brain computer interface applications using spatio-temporal principle component analysis, *Neurol. Res.*, **26**(3):282–287.

Vidal, J., 1977, Real-time detection of brain events in EEG, *Proc. IEEE* **65**:633–664.

Wang, T., and He, B., 2004, An efficient rhythmic component expression and weighting synthesis strategy for classifying motor imagery EEG in brain computer interface, *J. Neural Eng.* **1**(1):1–7.

Wang, Y., Zhang, Z., Li, Y., Gao, X., Gao, S., and Yang, F., 2004, An algorithm based on CSSD and FDA for classifying single—trial EEG, *IEEE Trans. Biomed. Eng.* **51**(6):1081–1086.

Weiskopf, N., Veit, R., Erb, M., Mathiak, K., Grodd, W., Goebel, R., and Birbaumer, N., 2003, Physiological self-regulation of regional brain activity using real-time functional magnetic resonance imaging (fMRI): Methodology and exemplary data, *Neuroimage* **19**(3):577–586.

Wessberg, J., Stambaugh, C., Kralik, J., Beck, P., Laubach, M., Chapin, J., Kim, J., Biggs, S., Srinivasan, M., and Nicolelis, M., 2000, Real-time prediction of hand trajectory by ensembles of cortical neurons in primates, *Nature* **408**:361–365.

Wickelgren, I., 2003, Neuroscience: Tapping the mind, *Science* **299**(5606):496–499.

Wolpaw, J. R., 2003, Brain–computer interfaces: Signals, methods, and goals, *Proc. 1st Int. IEEE EMBS Conf. Neural Eng.* **1**:584–585.

Wolpaw, J. R., Birbaumer, N., McFarland, D. J., Pfurtscheller, G., and Vaughan, T. M., 2002, Brain–computer interfaces for communication and control, *Clin. Neurophysiol.* **113**(6):767–791.

Wolpaw, J. R., Birbaumer, N., Heetderks, W. J., McFarland, D. J., Peckham, P. H., Schalk, G., Donchin, E., Quatrano, L. A., Robinson, C. J., and Vaughan, T. M., 2000b, Brain–computer interface technology: A review of the first international meeting, *IEEE Trans. Rehabil. Eng.* **8**(2):164–173.

Wolpaw J. R., and McFarland, D. J., 1994, Multichannel EEG-based brain–computer communication, *Electroencephalogr. Clin. Neurophysiol.* **90**(6):444–449.

Wolpaw, J. R., McFarland, D. J., Neat, G. W., and Forneris, C. A., 1991, An EEG-based brain–computer interface for cursor control, *Electroencephalogr. Clin. Neurophysiol* **78**(3):252–259.

Wolpaw, J. R., McFarland, D. J., and Vaughan, T. M., 2000a, Brain–computer interface research at the Wadsworth Center, *IEEE Trans. Rehabil. Eng.* **8**(2):222–226.

Yom-Tov, E., and Inbar, G. F., 2002, Feature selection for the classification of movements from single movement-related potentials, *IEEE Trans. Neural Syst. Rehabil. Eng.* **10**(3):170–176.

4

NEUROROBOTICS

Karen A. Moxon*

School of Biomedical Engineering, Drexel University, Philadelphia, Pennsylvania
Department of Neurobiology and Anatomy, College of Medicine, Drexel University,
Philadelphia, Pennsylvania

4.1. INTRODUCTION

Images from Hollywood suggest that by directly communicating with the brain it may be possible to control human behavior (Terminal Man) or provide a new reality far more interesting than what we currently experience (The Matrix). Unfortunately, Hollywood has always been a bit ahead of science and our ability to directly interface with the brain is at its infancy. There are, however, some clear examples of successful neural prosthetic devices that suggest the possibility of restoring function after injury. For example, over 30,000 auditory prostheses have been successfully implanted in patients with sensorineural hearing loss (Rubenstein and Miller, 1999). These devices bypass normal signaling mechanisms in the ear by translating sounds into patterns of stimulation and directly activate nerve cells to improve hearing in a broad range of patients. Another example of successful neural prosthetics is the technique for electrically stimulating either the muscles or nerves that innervate them to restore some function after paralysis. Over 150 functional electrical stimulation (FES) devices have been implanted into patients. These devices have been used to assist in breathing, bladder control, posture, and locomotion. There are now commercially available neural prosthetic devices (Smith *et al.*, 1987; Peckham *et al.*, 2000) that restore hand grasp function by stimulating muscles through electrodes. The electrodes are controlled by movement of the shoulder or neck and they stimulate nerves in the arm or wrist to restore grasping function in patients who have suffered loss of function in their arms or hands.

The examples mentioned above involve recording and stimulating in the periphery. However, signals recorded from the brain have also been used for neuroprosthetic devices. Studies recording fluctuations in the electrical activity of the brain using electrodes placed on the surface of the scalp [electroencephalography (EEG)] showed that these signals were modulated depending on the state of the person and the type of sensory input. The signals

---

* Address for correspondence: Assistant Professor, School of Biomedical Engineering, and Department of Electrical and Computer Engineering, 3141 Chestnut Street, Philadelphia, Pennsylvania 19104-2875; e-mail: karen.moxon@drexel.edu.

recorded in EEGs reflect mostly large synaptic currents from across large areas of the brain. Although these signals are not able to discriminate precise, local patterns of neural activity, they can show conditions of synchronous activity from large numbers of cells. The most obvious example is the changes in synchronous activity that occurs during sleep. Moreover, EEGs can indicate sensory stimulation by aggregate activity in a given area in response to the specific sensory stimulus, known as an evoked response. This kind of evoked response, referred to as an evoked potential, can be recorded from human subjects under different conditions and can be used to control a computer interface (Middendorf *et al.*, 2000).

Several groups have utilized evoked response activity as part of a brain–computer interface in a clinical setting, which is discussed in more detail in chapter 3 (Vallabhaneni *et al.*, 2005). For example, visual evoked cortical potentials have been used to control a speech synthesizer (Sutter, 1992) in patients who have lost their ability to communicate in a normal fashion. The visual evoked cortical potentials that are normally generated in response to flashes of light are used to determine where on the computer screen a patient is looking. The computer is divided into blocks with alphanumeric symbols or commonly used words written on them. The subject looks at the block he wants to select while random flashes of light are emitted. The computer uses the visual evoked cortical potentials to determine which block the subject is looking at. The system then selects the appropriate block in the matrix. The advantage of using this method is that it can be used on severely disabled persons that only have control of their gaze and for whom few alternatives exist.

An alternative and promising development from the Wadsworth Center (Wolpaw *et al.*, 2000) is based on learning to control the amplitude of mu (8–12 Hz) or beta (13–28 Hz) rhythms sent to EEG-based brain–computer interfaces that allow for movement by disabled individuals. The work focuses on developing a new brain function using the EEG as an outlet, by analyzing the possibility that a person could control the EEG rapidly and accurately. Individuals can learn to control the cursor when the EEG output is in the appropriate spectral band. Visualizing the movement of the cursor on the computer screen serves as feedback that helps the patient learn to control the signals. Ultimately, EEG-based brain–computer interfaces can provide cursor-based menu selection and operate a robotic device to assist with a host of daily functions. Recently, brain–computer interfaces that rely on slow wave potentials have been used on patients who suffer from complete paralysis (Birbaumer *et al.*, 1999). These patients, who lack sustained voluntary control of the musculature, can successfully control the movement of a cursor and use a spelling program that allows them to communicate. These are just a few examples of the types of brain–computer interfaces being implemented. However, these technologies have several limitations. The response times of the EEG signals, currently on the order of seconds, is comparatively slow. There are also long training periods and limited spatiotemporal resolution of EEG as compared with signals recorded from multiple single neurons, which is the subject of the remainder of this chapter.

Recent developments show that it is possible to interface directly with the brain, extract the neural signals that code for movement, and use these signals to control a robotic device (Figure 4.1). This emerging field of neurorobotics may eventually be a viable therapeutic method to overcome paralysis and restore sensorimotor function. The general approach is to record neural signals that code for the intention to move, from microelectrodes implanted into the brain. This signal is then used to control an external device, which creates a new line of communication for the brain. Although much work still needs to be done to build a device

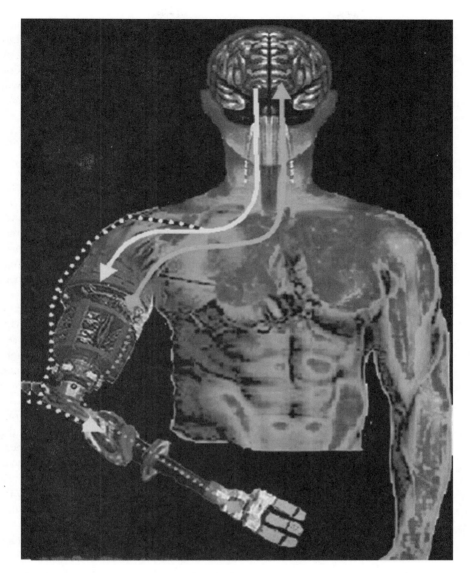

FIGURE 4.1. The vision of neurobotic control. Although current technology allows single neuron action potentials recorded from the brain to be used to control a robotic device or a cursor on the screen, the ultimate goal would be to record these neural signals and have them either control a prosthetic limb that replaces a lost limb or stimulate the nerves and muscle of the patient's body to restore functional output. The yellow arrow indicates signals from the brain to the prosthetic, and the blue arrow indicates sensory information being returned to the brain.

for use in humans, in the short term, neurorobotic devices are proving to be an excellent tool to understand how ensembles of neurons code for motor commands, the role of plasticity in these circuits, and, most important, how these circuits in the brain are modified after injury to the spinal cord, peripheral nerves, muscles, or limbs.

This chapter will focus on recent advances in the field of neurorobotics and examine how it is being used as a tool by neuroscientists to probe brain functions. An example of how these techniques have been used in a clinical setting will be reviewed as well as an overview of the electrode and signal-processing issues in the design of a functional neurorobotic system will be provided. Finally, a look at what the future is likely to hold for the field will be discussed.

4.2. DIRECTLY INTERFACING WITH THE BRAIN

Perhaps the most characteristic qualities of brain function are its richness and complexity. This complexity can be seen at all levels of analysis from the molecular to the psychological. Rules that govern the functioning of the brain at one level affect and influence functioning at other levels. From an anatomical point of view, the brain is composed of many different structures that rely on and interact with each other through a complex system of feedforward and feedback pathways. Information, in its most basic form, is represented in the nervous system by coordinated patterns of activity involving large numbers of neurons interconnected across multiple structures.

There are over 100 billion neurons in the brain and the operation of these neural systems must be interpreted in terms of these vast numbers of divergent–convergent synaptic junctions between large populations of neurons. Individual neurons typically receive 10–10,000 synaptic inputs from other nearby neurons as well as neurons located throughout the brain. Therefore, it is necessary to understand some of the basic features of neural communication, including convergence, divergence, and summation. Furthermore, evidence suggests that neural signals can be processed in both a parallel and a serial manner.

Communication between groups of neurons can be divergent in the sense that one projecting neuron typically activates several hundred receiving cells, and convergent in the sense that any one neuron also typically receives input from hundreds of sender cells. These signals are summed together such that a typical neuron must receive numerous simultaneous inputs in order to reach threshold and fire an action potential. Then, the firing of this cell will have physiological significance only if its firing is correlated in time and space with the coordinated firing of other cells projecting to a common set of receiving cells. In this way, a signal can be propagated through the brain.

Neural signals can be processed in a serial manner (e.g., from the retina to visual regions of the thalamus to the primary visual cortex) or in a parallel fashion (e.g., coordination of motor plans in the motor cortex, basal ganglia, and cerebellum). To properly grasp the dynamics of these systems one needs to develop a working model that is capable of representing this extensive parallel as well as serial integration of signals over space and time. Without such models it is difficult to develop methods to use neural signals to do useful work such as control a robotic device.

4.2.1. REPRESENTATION OF INFORMATION IN THE BRAIN

Several principles of brain function emerged in the early 20th century as a result of the application of techniques to record single neurons. These theories underlie the working models that form the foundation for how neurorobotic devices operate. The first of these principles is based on the early work of Sherrington (1906) on network activity in the spinal cord. He identified excitatory states in the central nervous system, initially referred to as "circularities", that examined the transmission of signals from one place to another (Kubie, 1930). These "circularities" are now generally known as recurrent loops. This work suggested for the first time the possibility that large numbers of neurons acted in concert and influenced each other.

To try to understand how and where these signals represented information (or memories in this case), Lashley (1950) investigated the ability of animals to retain the ability to perform a learned task after lesions were made in various parts of the brain. These studies suggested that information is diffusely represented in composite cortical regions. Lashley further suggested that any part of the region associated with a learned task was equally effective in storing the memory for that task, otherwise known as the theory of equipotentiality. Furthermore, Lashley postulated that the more tissue devoted to a task, the better the system could perform, also known as the theory of mass action. For example, rats use their whiskers as their main tactile organ and the largest part of their primary somatosensory cortex is devoted to whiskers, whereas humans use their thumbs to master skills associated with using tools and a disproportionate amount of human cortex is devoted to the representation of the thumb as compared to other digits.

In 1949, Donald Hebb (Hebb, 1949) presented a revolutionary theoretical construct that described how information could be represented and stored in the brain by groups of neurons. The theory consisted of two parts. The first part was a mechanism describing how the efficacy of synaptic function could be modulated by use. This rule, known as the Hebbian rule, suggested that simultaneous activity between a "sender" (presynaptic) cell and a "receiver" (postsynaptic) cell would increase the strength, or efficacy, of the synapse and make this pathway more likely to respond to future stimuli. These changes in synaptic strength are generally referred to as synaptic plasticity.

The second was a biologically plausible mechanism describing how individual neurons did not, by themselves, convey information, but rather worked together as assemblies of activity to represent information. His theory of cell assemblies is probably the most influential proposal describing how large populations of neurons underlie brain processing. More important for our understanding of how the brain functions, Hebb gave a description of how individual neurons could simultaneously participate in different cell assemblies and be involved in multiple functions and representations of information.

John (1972) expanded on the work of Hebb and Lashley and presented a theory of information representation in the brain based on the idea that memory traces are stochastic, diffuse, redundant and primarily related to function rather than anatomy. His "statistical versus switchboard" theory postulated that spatiotemporal patterns of neural activity may retain their coherence while moving freely across anatomical regions of the brain. These ideas have been implemented in several computational models (Abeles, 1991; MacGregor, 1991; Moxon et al., 2003a,b) and form the underlying theoretical basis of the most advanced neurorobotic experiments being performed today.

Each of these major theoretical developments, discussed in more detail elsewhere (Abeles, 1991; MacGregor, 1993), had profound effects on the way scientists explore and interpret data recorded from the brain. Moreover, they demonstrated that "information" is represented in the brain as large-scale neuronal interactions among widely distributed and interconnected populations of neurons, which are now generally referred to as neural ensembles. This theoretical construct is essential to our understanding of how sensory information is processed, converted into percepts, stored into memory, and then used to generate behaviors.

4.2.2. CODING STRATEGIES OF ENSEMBLES OF SINGLE NEURONS

The main advantage of brain–computer interfaces based on EEG recordings is that they are obtained through scalp electrodes, and are therefore noninvasive. The major disadvantage is that the ability of the patient to change the signal for cursor control is slow and error prone. An alternative is microelectrodes implanted directly into the brain tissue that can record signals directly from specific, identifiable populations of neurons from within the brain itself. Typically, these electrodes can record from a single neuron or can record field potentials from discrete populations of neurons.

The electrical potential recorded with an extracellular electrode inserted into the brain tissue is the sum of electrical activity around the recording site of the electrode (Jack *et al.*, 1975) (Figure 4.2). This electrical activity consists of the electrical fluctuations (or changes in membrane potential) of neurons near the recording electrode. These electrical fluctuations derive from changes in membrane potential of a population of cells, whereas the

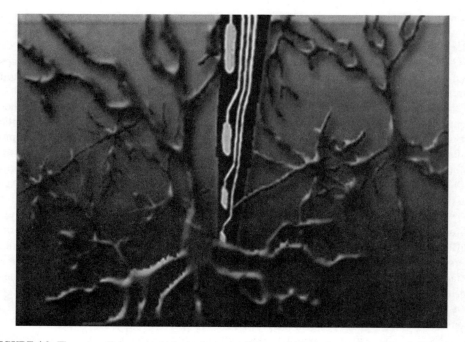

FIGURE 4.2. The extracellular space in the brain surrounding a multichannel recording electrode. The signals for a neurorobotic system are the action potentials generated by neurons in the brain. These signals, commonly referred to as spikes are signals picked up by extracelluar recording electrodes that are within about 100 μm of the cell body.

characteristic output signal of individual neurons consists of action potentials. An action potential is caused by the rapid movement of ions across the nerve cell membranes and is triggered when the influx of positively charged ions reach a critical value, known as the threshold. Action potentials can generally be recognized or discriminated from the electrical potentials recorded. When this occurs, a brief (1–2 ms) but substantial change in the permeability of the membrane to positively charged ions occurs, which creates a relatively large (50–100 μV) change in the potential recorded by the extracellular electrode if the electrode is close enough to the cell (within about 50 μm). These generally look like spikes in the analog signal recorded, and the firing of action potentials by neurons are often referred to as spikes.

As single-neuron electrophysiological techniques matured, investigators began to understand that regions of the brain contained single neurons that had response properties to a single modality (i.e., vision, touch, etc.). Furthermore, single cells also respond to a relatively small subspace of the modality known as the receptive field (Mountcastle, 1957, the sensorimotor cortex; Hubel and Wiesel, 1959, the visual cortex; and Rosenblith, 1957, the auditory cortex). These investigators and others also perfected techniques to examine the anatomical structure of the cortex and discovered that cells were grouped in patterns that repeated across large areas of the cortex. This is generally referred to as the modular organization of the cortex. Using the basic ideas of Hebb, Lashley, and John described in the previous section, and of these abovementioned pioneering electrophysiological and anatomic studies, several investigators (Szentagothai, 1975; Eccles, 1981) developed hypotheses to explain how the response properties of single neurons could combine, in awake-behaving animals, to represent information.

4.2.3. DECODING THE NEURAL SIGNAL

Very quickly investigators developed the capacity to record and analyze the simultaneous activity of tens of neurons from a local area within the brains of rats and monkeys (Gerstein and Perkel, 1978). Successful experiments led to insights into neural information processing. For example, Gerstein and Perkel (1969) used simultaneously recorded neurons to examine connectivity among neurons. By measuring the correlation of the time between action potentials, they examined how the functional connectivity between neurons could represent information. This work highlighted the need for the simultaneous recording of large numbers of neurons. The ultimate goal, which is being able to record from "enough" neurons such that you can view multiple "functional states" of the neurons, is starting to be realized and forms the basis of neurorobotic control in its present state. Nevertheless, each of the various recording techniques widely used today have advantages and limitations, which will be discussed below.

The development of an implantable multiple-microelectrode system in which each electrode was individually adjustable (Humphrey, 1970) maximized the number of recorded neurons. In subsequent experiments, this microelectrode device was used with tungsten microelectrodes to record multiple single neurons simultaneously in the motor cortex of awake monkeys while the animals performed a variety of arm movements (Humphrey *et al.*, 1970). Recently, investigators have been able to chronically implant arrays of microelectrodes and record for weeks to months and sometimes years in rats and monkeys, respectively. This ability to record chronically allowed investigators to record the neural activity from the same neurons while the animal performed a variety of tasks on consecutive days and even after neural injury. For the past 30 years, the technology to record multiple single neurons has advanced to the point where it is now possible to record up to 50 neurons in rats and hundreds in monkeys (Kralik *et al.*, 2001). Investigators now chronically implant multiple electrodes and record from awake-behaving animals to investigate how ensembles of sensory (Freeman, 1983; Eggermont, 1993; Nicolelis *et al.*, 1995; Moxon *et al.*, 1999) and motor (Evarts, 1974; Mountcastle *et al.*, 1975; Nicolelis *et al.*, 1993; Wilson and McNaughton, 1996a,b) neurons code information. The results of Humphreys and his colleagues showed for the first time that information about a movement is carried not simply in the frequency of spikes (or action potentials) from individual neurons but to a significant extent in the temporal relations between spikes (temporal coding vs. rate coding). This work of Humphreys has been the basis for several of the neurorobotic studies in monkeys.

More recently, Georgopoulos *et al.* (1986) postulated a population coding scheme for the cortical control of arm movements. This influential theoretical construct suggested that single neuron activity in the motor cortex was tuned to respond to the movement of a limb in a particular direction by a cosine function that related firing rate to direction. This algorithm has successfully been used for neurorobotic control in monkeys.

Another critical concept to emerge from the animal work is that the average response of a single cell over many trials is well correlated to a sensory stimulus. Furthermore, averaging many cells over one trial can be used to classify sensory input and thereby investigate how ensembles of neurons code for the sensory input (Rolls *et al.*, 1997, 1998; Ghazanfar *et al.*, 2000; Foffani and Moxon, in press). The technological developments for chronic, multiple single neuron recording in awake-behaving animals combined with the concepts of distributed, stochastic coding and directional tuning created a viable basis to begin investigation of neurorobotic interfaces.

The seminal studies of neurorobotic control began with a few simple questions: (1) how can we transform the neural spike patterns from large numbers of single neurons into a signal that codes for limb movement (i.e., a population function), (2) can these "motor codes" be used to generate a "neuronal population function" to control a robotic arm in real time, (3) is it possible to develop an interface with sufficient accuracy in real time in the absence of actual limb movement, and (4) will the activity of recorded ensembles of neurons adapt to the neurorobotic controller?

4.3. NEUROROBOTIC CONTROL

4.3.1. FEASIBILITY OF NEUROROBOTIC CONTROL

The pioneering work of Chapin *et al.* (1999) showed that signals produced by an ensemble of neurons could be recorded from the brain while the animal was performing a motor task and be used to substitute for the animal's motor behavior to control a robotic arm. These studies have been repeated in nonhuman primates and in humans. The studies have also corroborated that such interfaces are also feasible in these species, and can be used to control more complex movements. Furthermore, this body of work has demonstrated several important features of the neural coding of movement.

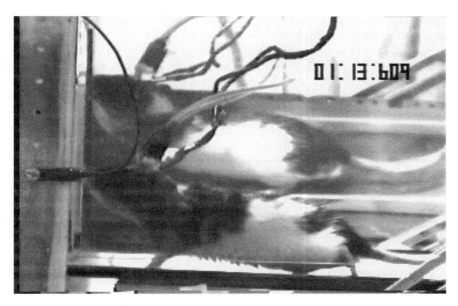

FIGURE 4.3. A typical chronic recording from a rat freely moving within the recording chamber. Notice the headstage device on the rat's head and the wires that tether the rat to the recording equipment. There is a mirror behind the rat that allows a view of the opposite side of the animal. This animal was implanted with 32 microwires and recorded for 2 months. The number in the upper right-hand corner allows videotaped analysis of the animal's behavior to be time-locked to the animal's movements.

In the first study of neurorobotic control in rats (Chapin *et al.*, 1999) arrays of microwires were implanted into the forelimb motor region of the thalamus and cortex to record the activity of large numbers of single neurons simultaneously (Figure 4.3). The rats were trained to obtain a water reward from a robot arm by pressing a lever to proportionally move the robot arm to and from the water source. As the lever was depressed, the robot arm was extended to reach a water dropper; when the lever was released, the robot arm returned to the rat, delivering a water reward. During the experiment, rats initially controlled the lever with their forepaw while the neural signals were continuously recorded. The neural signals from each cell were combined into a neural population signal that increased and decreased as the rat's forelimb moved the lever. This neural population signal was well correlated to the limb movement. Then, after about 5 min of activity, control of the robot arm was switched to the brain-derived neural population signal. The real-time neural population signal replaced the lever movement and successfully moved the robot arm to the water-drop position to obtain the water reward.

Immediately after switching to neural control mode, animals initially continued to press the lever down to the original threshold position for obtaining the water for several trials. However, during subsequent trials, the animals quickly realized that limb movement was no longer a requisite for robot arm movement. On subsequent trials, the normally high correlation between limb movement and robot arm movement declined such that even though the animal continued to make sporadic lever movements, usually following failures, the animal was generally able to obtain water without complete lever movement. This loss of overt limb movement was replaced with a high correlation between the neural population signal and robot arm movement that allowed the rat to receive its reward. Despite the absence of limb movements, the magnitude and timing of the individual neuron signals did not significantly change and, therefore, the neural population signal remained sufficient for neurorobotic control.

Several characteristics of neural coding were elucidated by this study. First, as has been shown previously, most single neurons are well correlated, on average, to a particular phase of the movement. However, a single neuron by itself is not a good predictor of movement on a single trial. Second, the neural activity from tens of neurons can reliably code for limb movement in individual trials. Third, this neural activity is largely predictive and the best correlations between neural population activity and limb movement occur during a short interval immediately preceding the limb movement. Finally, this neurorobotic control does not require overt movement of the limb, suggesting that the neural activity generated in the brain to signal limb movement can be activated independent of actual limb movement. This result is critical for use of neurorobotic control in paralyzed patients.

4.3.1.1. Development of Neural Population Functions and Implementation in Real Time

Although the outcome of these experiments was predicted by the theories outlined in the previous section, the implementation of these experiments depended on several developments. It was necessary to develop a method to extract information from the individual neurons related to the limb movement, to implement this algorithm in real time and use it to actuate the robot arm on a single trial. Although 90% of the neurons recorded from the forelimb motor regions of the thalamus and cortex exhibited some statistically significant activity related to the timing of forelimb movement, the activity of single cells was unreliable for predicting limb movement on a single trial. Despite the high correlation of neural activity with the downward movement of the lever, the information capacity of individual neurons was insufficient for this task because the activity of a single neuron cannot produce a signal that approached the positioning accuracy and smoothness of native limb movements. This is because each individual neuron has a very low firing rate. Moreover, there is too much variability in the response of a single neuron during a single trial to be reliable. For example, in the 100-ms period around the onset of the pressing movement, the spike count of a single neuron varies from 0 to 3 spikes. Overall, correlation of single neurons with the lever movement on a single trial was poor (about 0.3) and thus was not able to provide trial-to-trial reliability or the temporal specificity that would be required for controlling a motor device.

Because it was known that the information about limb movement is distributed across a neuronal ensemble and should be manifested as patterns of correlated activity among neurons associated with motor output, the activity from many cells on a single trial should be sufficient to code for the limb movement. To extract the information from the populations of neurons, principal components analysis was used (Chapin and Nicolelis, 2000). Effectively, principal components analysis extracts the maximal amount of covariance of the signal by treating the spikes recorded from each neuron as a series of random variables (Jackson, 1991). From these random variables one can create a correlation matrix that describes the coordinated activity between the spike times of each neuron. From the correlation matrix, one calculates the eigenvector that maximizes the eigenvalue, thereby describing the greatest amount of covariance among the neural activity. This eigenvector describes a set of weighting coefficients that can be used to transform the neural activity into a population function that is well correlated to the movement of the limb.

There are two major advantages to using principal components analysis to create the neural population function. The first is that this method involves a linear combination of the activity of the cells recorded and is therefore computationally efficient. This is important for the real-time analysis of the neural spike trains. The second is that principal components analysis does not rely on *a priori* knowledge of the limb movement but is still surprisingly accurate. In fact, many other methods currently being used for neurorobotic control rely on knowing in advance the response properties of the neurons to different components of the movement (i.e., tuning properties of the cells; refer to Georgopoulos, 1996, discussed above). Application of these techniques would be difficult for patients who cannot move their limbs because the tuning properties could not be identified. In other neural systems, principal components analysis has been shown to optimally condense the salient information of the ensemble to a single population function (Chapin and Nicolelis, 1999).

4.3.1.2. Implementation of the Neural Population Function in Hardware

Though each neuron has a unique but partially overlapping response to the movement of the forelimb, averaging the activity of these neurons using weights defined by the first principal component (PC1) creates a smooth and distinct output function that predicts the rat's limb movement. This is accomplished by using a multichannel electronic device to record the individual spike times of each neuron and then averaging the neuronal activity by applying the weights defined by PC1. The resulting output is a single analog voltage signal used to position the robot arm. Robot arm control could be arbitrarily switched from the lever (actual forelimb movement) to this brain-derived neuronal population signal.

However, the signal derived from principal components analysis does not encode the movement trajectory as a direct real-time image but instead codes for the movement in a short period just prior to movement onset (Chapin and Nicolelis, 2000). This result could be explained by the fact that the activity of these cells is best correlated with the onset of movement of the downward pressing of the lever. This neuronal activity precedes onset of wrist-flexor/elbow-extensor EMG activity by up to 90 ms, and onset of detectable downward movement by up to 100 ms. Therefore, the neural activity best correlated with limb movement is completed before limb movement directly related to lever movement, and hence robot movement. As the rat initiates limb movement, the robot arm under neural control moves out, obtains a drop of water, and returns to the rat before the rat completes the lever press. During the course of a single experiment, as described above, the rat quickly figures out that limb movement is not required, and the correlation between neural activity and robot arm movement is maintained whereas the correlation between neural activity and limb movement declines.

To understand this premovement activity and to optimize the neural population signal to bring the neural signal into register with the actual limb movement, a delay was needed to replicate the limb movement in time as well as space. The first principal component was "temporally stretched" by dividing it into five parallel signals lagged at successive 100-ms delays. Linear techniques were then applied to this tapped delay line signal to predict the limb movement. The results were more highly correlated to the limb movement than the first principal component alone but resulted in a significant number of false positives. These data suggested that the linear transformation techniques were not sufficient to reject neural signals relating to non–lever-pressing behaviors.

To try to minimize these false positives, nonlinear techniques were used. Because principal components analysis is known to be mathematically equivalent to the general learning rule used by artificial neural networks, artificial neural networks were a likely candidate for a neural population function. Artificial neural networks are based on the work of Hopfield and Herz (1995). Hopfield suggested that networks of simple integrate-and-fire neurons could be used to perform pattern recognition (Hopfield, 1995). This could be accomplished by increasing the synaptic strength between neurons, and is based on the ideas first postulated by Hebb (1949). Hebb suggested that the "strength" of connections between neurons would increase as the probability of their firing together increased. Hopfield used this principle in models of artificial neural networks. In an artificial neural network connections or "synapses" between "artificial neurons" are updated according to this "Hebbian" learning rule. These synapses are represented by weights in the network, which are modified so that the output can correctly classify an input into an appropriate category.

Several different classes of artificial neural networks now exist depending on the type of problem to be solved. One type of network, known as a recurrent artificial neural network, has the potential to match the firing patterns of neurons to limb movement. This type of artificial neural network uses the output of the network to directly modulate the input through the use of recurrent connections within the network architecture. These recurrent connections encode temporal information in the data used to train the network. This eliminates the need for the temporal stretching of the neural population signal implemented with the first principal component (Principe *et al.*, 2000). However, an important issue to consider is that an artificial neural network cannot utilize the raw neural data from each individual neuron recorded as inputs to the network because there are too many cells. Furthermore, the information provided by each cell on a single trial is sparse, since neurons do not fire often. This makes it very difficult for the network to converge on a solution. Therefore, there is always a preprocessing step, such as principal components analysis, on the data that combines the activity from many cells into a few dense functions that retain most of the covariance among the neurons.

In the neurorobotic rat study, dynamic backpropagation learning was used with the first principal component (PC1) as the input. In this case, the learning rule adjusted the weightings within the network to optimally transform the PC1 input signal into an output function that closely matched the limb movement. Interestingly, the recurrent artificial neural network learned to recognize distinct features of the PC1 signal to predict the timing and magnitude of the limb movement. When a recurrent artificial neural network was used to transform the raw PC1 signal, the output was highly correlated to the movement trajectory in both space and time and the number of false positives significantly reduced. In particular, the onset and termination of lever presses were found to be well predicted on the basis of the steepness of the slope during the rising and falling phases of the population function defined by PC1. This result suggests that the neural encoding of the population function may be less related to the overall magnitude of neural population activity than to highly distinct spatiotemporal patterning of neuronal activity within the populations (Chapin and Nicolelis, 2000). Since this work, several investigators have instituted various artificial neural network approaches to decode neural signals (Ghazanfar *et al.*, 2000).

4.3.1.3. Feasibility of Neurorobotic Control after Injury

Although the results of this study in healthy rats clearly show the feasibility of neuro-robotic control, therapeutic use in a clinical setting will require interfacing with a system that is damaged. The potential problems of recording from paralyzed patients, namely that cortical representation of the limb might be lost or perhaps not functionally useful, could severely limit the effectiveness of such devices. Neurophysiological studies in primate cortex have shown that the somatotopic representation of the body undergoes profound changes after limb amputation or nerve damage such that the area of the cortex that previously represented the affected limb now codes for a different part of the body. However, recent work in monkeys and the work of Kennedy and colleagues in humans (Kennedy and Bakay, 1998; Kennedy *et al.*, 2000) suggested that this may not be a concern because the same plasticity in the cortex that allows reorganization after injury may be used to reorganize circuits in the brain to take advantage of alternative forms of communication.

In order to perform recordings of single neurons in human patients with severe neuromuscular disorders, Kennedy developed a novel electrode device called a cone electrode (Kennedy, 1989), which is essentially a microwire electrode surrounded by a solution of neurotrophic factors in a glass tube. This electrode was implanted into patients who have lost the ability to move any part of their body except make small movements of the face (either lifting an eyebrow or closing the eyelids). The goal was to implant the electrode into what is usually considered to be the motor area for the hand in order for single neurons to grow into the cone to produce a tight junction between the recording electrode and the neuron (Kennedy *et al.*, 1992).

After implantation, neural activity appeared in about 3 weeks and robust recording began at about 3 months. The goal of the study was to use the single neuron recorded through the cone electrode to move a cursor on a computer screen. During these sessions, the patient was able to listen to the neural activity amplified over loudspeakers to gain auditory feedback about the activity of the recorded cells. Essentially, the patient thinks about moving the cursor, and the goal is that the movement of the cursor should become correlated to the neural activity over time.

Initially it was presumed that since the electrode was implanted into the hand region of the brain, if the patient was asked to imagine making hand movements neural activity would increase and could then be used to move the cursor. However, reliable activity in the firing patterns of recorded cells was not seen under these conditions. Nor was there any consistent neural activity in response to sensory stimulation of the hand or manipulation of the limb. After 5 months of implantation, movements of the eyebrow reliably activated the neural signals being recorded. These responses were consistent with amputation studies in monkeys that showed that the facial area of the cortex innervated the hand area after the amputation.

By 6 months, the patient was able to lie quietly, with no apparent movement during neural signal activation and drive the cursor toward a target on the computer screen. There was an increase in recorded spikes just before the cursor reaches the target and then the signal stopped when the cursor reached the target. The patient was able to communicate that when he drove the cursor, he was not thinking about moving any of his body parts, but rather was concentrating directly on moving the cursor. It is important to note that muscle activity was suppressed during this period. Eventually the patient could gain independent control over the neural firing and the muscle movement in the eyebrow. This was exploited by allowing the neural activity to drive the cursor horizontally across the screen and using the muscle activity from the eyebrow to drive the cursor vertically up and down the screen. This provided the patient with rapid, 2D control of the cursor (Kennedy and King, 2000).

Further training improved the ability of the neural activity to move the cursor across a row of icons with different symbols that represent different commands without any EMG activity. Although the rate was slow—approximately 30 seconds to move the cursor to a particular target (the subject was able to move the cursor faster but this resulted in increased errors)—the fact that the patient was only thinking about moving the cursor suggests that when this new communication line was made available to the brain, the brain was able to adapt quite readily, take advantage of this novel output, and create a cursor-related cortex pattern of firing in the cortex. This data further suggests neither the precise location nor the amount of reorganization of the cortex after injury or disease as was previously thought. The same plastic potential that permits cortical reorganization after injury is still available for the cortical reorganization in response to new channels. The fact that the monkey studies also showed that neural activity was better able to control a cursor directly than through the use of the subject's own limb (Taylor *et al.*, 2002) suggests that if enough neurons could be recorded from the human subjects, a faster, more efficient activation of the cursor might have been achieved.

4.3.2. NEUROROBOTIC CONTROL AS TOOL FOR INVESTIGATING NEURAL CODING STRATEGIES

These techniques for neurorobotic control were quickly adapted for use in primates (Wessberg *et al.*, 2000). The primate model is a valuable system to study the possibilities of neurorobotic control because their nervous system anatomy and physiology are similar to humans, and has a similar level of complexity. To adapt the results of neurorobotic control from the rat to the monkey a deeper understanding is required of how the brain performs motor control. Neurorobotics is also an ideal tool to investigate strategies used by the brain to control a robotic device. Studies to date have focused on how activity in the motor areas of the cortex code for forelimb movement. The results of these studies have led to several important advances in our understanding of the neural control of movement, namely, that (1) many motor areas of the cortex contributed to the neural code for movement (Wessberg *et al.*, 2000; Carmena *et al.*, 2003); (2) neural activity is remarkably adaptive; (3) a great deal of versatility is evident in a relatively small (about 100) number of neurons; specifically, the same neural population functions predict movement in a variety of directions and placements in space; and (4) functional accuracy of neurorobotic control is enhanced by feedback.

4.3.2.1. Plasticity

Several investigators have successfully shown that the neural activity recorded from the cortex of awake-behaving monkeys can be used to control the movement of a cursor in a 3D space (Serruya, 2002; Taylor *et al.*, 2002; Carmena *et al.*, 2003) and, more important, can be used to control a robot arm in 3D space (Wessberg *et al.*, 2000; Carmena *et al.*, 2003). The results confirmed the work done in rats and showed that the neural activity preceded limb movement and that neural control could continue in the absence of limb movement. In addition, it was shown that many motor areas of the cortex contributed to the neural code for movement (Wessberg *et al.*, 2000; Carmena *et al.*, 2003). These areas include dorsal pre–motor cortex, supplemental motor area, medial intraparietal area of the posterior parietal cortex, primary motor cortex (MI), and the primary somatosensory cortex. It is likely that other areas not yet explored also contribute to the coding of neural control signals.

More important, these studies showed that the neural activity is remarkably adaptive and as the monkey is allowed to use the neurorobotic device in place of limb movement, significant changes in the neural activity take place. For example, significant dynamic changes in the coupling between neuronal activity and movement and other nonstationary influences were found to be highly significant, suggesting that the neural ensembles may be continuously reacting to the experimental paradigm. If neuronal weights used to generate the neural population signal that controls the robot are continuously updated, significant improvement can be made in the ability to predict hand trajectories (Taylor *et al.*, 2002).

4.3.2.2. *Multiple Brain Areas Code for Movement*

These experiments led to important insights into how neural ensembles might function. For example, large numbers of neurons were recorded from several different brain regions as described above (Wessberg *et al.*, 2000), and neuronal population functions (neuronal weightings) were generated from data recorded while the monkey made hand movements in one direction. The versatilities of these population functions were tested to determine if the same weightings could be used to predict hand movements in another direction. Not only were the population functions trained on reaches to the left able to predict reaches made to the right, but population functions trained on proximal movements were able to predict distal movements and vice versa. This suggests a great deal of versatility from a relatively small (about 100) number of neurons. In addition, similar to the rat studies, no *a priori* information about the tuning properties of these neurons was used to create the neuronal population functions. This feature suggests that a random population of neurons could be used for long-term control of a prosthetic device.

Because such experiments have included recordings from several regions of the cortex, it is possible to evaluate the relative contribution of these different areas to the prediction of the hand trajectories. The neurons from each region of the cortex were analyzed separately and a correlation was made between the number of neurons used for a particular prediction and the accuracy of the prediction (Carmena *et al.*, 2003). From these correlation studies, one could extrapolate, for each region of the brain, the number of neurons that would be required for 90% accuracy. If enough neurons were recorded (around 300 for each region), each area could independently track hand position with better than 90% accuracy. These results are consistent with the idea that motor control signals for arm movements appear concurrently in large areas of the frontal and parietal cortices and that, in theory, each of these cortical areas individually could be used to generate hand trajectory signals in real time. Because it has been previously shown that posterior parietal and primary motor cortex are influenced by motor parameters other than hand position, the inclusion of signals from large areas of the cortex may provide a better representation of hand movement than any one area alone. In this way, the relative contributions of these cortical areas may change according to other events or the demands of the particular motor task, improving the overall effectiveness of the neurorobotic control.

4.3.2.3. Feedback

Additional studies in primates showed that visual feedback could enhance the ability of the neuronal population to control a robotic device. Furthermore, physical movement of the monkey's paw was not necessary for the neural signals to control the movement of virtual object(s) rendered in 3D (Taylor *et al.*, 2002). These studies were performed in a way similar to the previously described traditional EEG-based brain–computer interface studies done on humans except that the cursor could move in a 3D virtual environment. The cursor was moved from a central-start position to one of eight targets located radially at the corners of an imaginary cube. Monkeys were first trained to move a joystick to control the cursor. The goal was for the monkey to get the cursor to hit a target that appeared on the screen at one of the corners of a cube. During the task, single neuron activity was recorded from several cells while hand movement controlled the cursor and the data were analyzed off-line to understand the relationship of their firing patterns to the arm movements. Movement of the hand was predicted from this data on the basis of the fixed tuning properties of the cells evaluated during this hand-controlled cursor movement part of the experiments based on algorithms first developed by Georgopoulos (1986). The movement of the cursor was then switched to real-time brain control. A neuronal population function based on these tuning curves and the neuronal population signal was able to move the cursor to the target. Because this off-line analysis of the cell's tuning properties would not be possible to be performed in movement-impaired individuals, the cell's tuning properties were also calculated from on-line data as the subject attempted 3D brain-controlled cursor movement with their limbs restrained. The tuning properties identified during brain-controlled cursor movements for each cell was substantially different from the tuning properties under hand control of the cursor. In fact, the brain-controlled cursor movements became increasingly more accurate at predicting the trajectory of hand movement compared with the hand-controlled activity.

These data suggest that the visual feedback to the monkey regarding the position of the target and the cursor and an adaptive algorithm are sufficient to create neuronal responses that can control the movement of an external device (Taylor *et al.*, 2002). Electromyographic (EMG) activity during brain-controlled sessions showed suppression of the EMG activity as the ability to track the target improved. This result is similar to the previously described studies performed in humans and is a further indication that it is possible to develop effective brain control in the absence of physical limb movements or muscle activation.

4.3.3. NEUROROBOTIC CONTROL AS A THERAPEUTIC DEVICE

The work on neurorobotic control in primates, as described above, has changed the nature of the debate over invasive neurorobotic devices and EEG-based brain–computer interfaces and, for the first time, put together many of the necessary components for a functioning neurorobotic controller. These components include the ability of the same neurons to encode for multiple movements, the ability of the neurons to adjust to the dynamics of actuating a physical device and the demonstration that neurons now directly code for the robotic device. For the first time neurorobotics may be a viable alternative to EEG-based brain–computer interfaces. If enough neurons are recorded, fast reliable signals may be extracted from patients with severe injury to directly and efficiently control a robotic device.

The neurorobotic studies presented in the previous section require electrodes to be implanted into the brain tissue. There are several advantages of internal electrodes as compared with brain–computer interfaces that use EEG recordings from the scalp. First, the temporal response of the signal recorded from the electrode is on the order of tens of milliseconds, and is therefore quicker than noninvasive methods. Second, the quality of the signal is higher, and internal electrodes are able to track in real time either the actual movement of a limb in 3D space (Wessberg *et al.*, 2000) or the "desire" to move the limb if the limb is constrained (Taylor *et al.*, 2002). Translating these techniques that utilize invasive recording techniques for use in the human brain in order to provide a communication channel to an external device is a radical alternative to EEG-based brain–computer interfaces. To justify the increased risk of the surgical procedure to implant the electrodes into the brain the resulting communication link must be significantly better than that achieved through noninvasive methods. Although this appears to be the case when normal animal subjects are implanted, only limited studies have been performed on injured subjects (see Section 4.3.1).

The application of neurorobotics in paralyzed patients who are unable to effectively deliver neural signals for the intention to move their muscles could dramatically change the quality of life for millions of patients. These include patients who have the necessary brain function to formulate commands for movement but the means to enact the motor intent are gone.

Examples of such circumstances include patients with spinal cord injuries and neuromuscular disorders, such as amyotrophic lateral sclerosis. Initially, neurorobotic devices could take the form of devices that control a cursor on a screen much the same way that EEG-based brain–computer interface devices work (Donoghue, 2002) using signals recorded from motor cortex structures. Ultimately, these devices should be able to extract voluntary motor commands in real time from the electrical activity of large populations of cortical and subcortical neurons and used to either actuate an external robotic device (Schmidt, 1980) or enact motor function in the patient by directly stimulating the patient's own musculature (Nicolelis, 2003). The full repertoire of movements, in particular, movements of individual fingers, has not yet been shown, despite the fact that the speed and accuracy of neurorobotic control signals from multiunit implanted electrodes is much better than EEG-based brain–computer interfaces.

To test the ability of a single population of neurons to code for multiple motor outputs, a recent study in monkeys has shown that, if enough neurons are recorded, it is possible to have the same population of neurons code for reaching and grasping (Carmena *et al.*, 2003). In this study both reaching (hand position and velocity in three dimensions) and grasping were accurately predicted by neural signals recorded simultaneously from the frontoparietal cortex in real time. Most important, the monkeys learned to reach and grasp virtual objects using the same population of cells. This was demonstrated by having the monkey perform the tasks serially so that the neural signal was able to accurately move the robot arm in space to reach the target followed by changes in the force component of the signal to predict the grasp. In addition, the ability of the neural signals to code for reaching and grasping was performed even in the absence of overt arm movements. These results emphasize that by recording from large numbers across many regions of the brain, ensembles of neurons may be able to code for multiple motor outputs.

Finally, control of reaching and grasping was accomplished through the use of a robot arm. Unlike the previous studies where the cursor movement was directly controlled by the neural signal, in this study, the neural signal controlled a robot arm whose output moved the cursor. This setup required the monkeys to adjust to the dynamics of the artificial actuator that must physically move in space. When the robot arm was introduced there was an immediate drop in performance, which the animals were able to overcome with additional training.

These studies also showed changes in the tuning properties of cells that support the earlier studies in humans (see Section 4.3.1). Long-term neurorobotic control leads to the development of novel tuning parameters though a process of cortical plasticity. These include the parameters that represent the artificial actuator dynamics, which are distinctly different from those for hand movement. In this way, neurons directly code for actuator dynamics much like the cursor-related cells recorded in human subjects. This could allow patients to have the perception that the robotic device is now an integral part of their own bodies allowing for the possibility of more efficient control (Carmena *et al.*, 2003; Nicolelis, 2003). Of course, there are many issues that need to be resolved before the success of such neurorobotic applications can be claimed in humans. First, the electrode interface must be improved. The reliability of recordings from single neurons needs to be improved so that every surgery can consistently yield good data. Second, the equipment that is used to record, filter, amplify, and discriminate 100 single neurons is presently too big and cumbersome for daily use by a human. The technology must be miniaturized and made more efficient in order to fit on a small chip that can be implanted under the skin. Telemetry methods must be improved to then send the signals from the subject to the robotic device or, ideally to a neural prosthetic stimulator implanted in the subject's own body to restore the natural movements of the patient. These technologies are in their infancy but rapidly improving. Some of the issues and advance in electrode technology and brain–machine interface (BMI) technology are discussed in the remaining sections of this chapter.

4.4. HARDWARE REQUIREMENTS FOR NEUROROBOTIC CONTROL

In order for neurorobotics to be a viable interface that allows paralyzed patients to restore mobility or communication, systems must be designed that can record the neural signals from hundreds of neurons simultaneously in real time, combine these neural signals into a useful output signal, and transmit them to a robotic device (Table 4.1). Design challenges exist at each of these stages. Multi-channel electrodes must be inserted into

TABLE 4.1. Neurorobotic Control Device Subassemblies

ELECTRODE SUBASSEMBLY	• Neural tissue interface
	• Electrode interface
SIGNAL CONDITIONING	• Pre-amplifier circuits
	• Noise sources
	• Single neuron detection
	• Algorithms for decoding the neural signal
SIGNAL ACQUISITION	• Telemetry subsystem
	• Device packaging constraints
	• Bandwidth limitations
	• Power source constraints

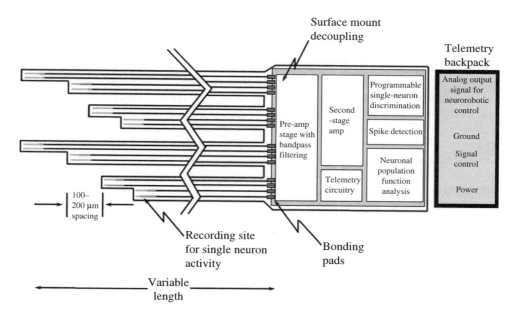

FIGURE 4.4. Schematic of the components for a neurorobotic control device. As outlined in the text, the device must consist of arrays of electrodes, whose signal is preamplified and filtered near the source. On a chip small enough to reside under the skin, circuits must be constructed that amplify the signal sufficiently to discriminate single neurons and store spike-detection algorithms. The resulting spike times from the individual cells will be weighted according to some neural population function and combined into an analog output signal that is then telemetered to the robotic device.

the neural tissue and consistently record single-neuron action potentials for years. The signals from approximately 100 channels must be filtered and amplified and the single neuron spike times discriminated from the analog signal. The resulting spike times must be combined in a meaningful way to create a neuronal population function capable of controlling a robotic device. These signal-processing requirements must be performed in real time and must fit on a device that resides under the skin (Figure 4.4). Power must be supplied to run the system and the signals must be sent telemetrically to the robotic device to avoid the need for wires to tether the patient to the robot. The system is made up of several subsystems that must work together to achieve the desired output. These subsystems include (1) microelectrodes, (2) signal-conditioning device, (3) neurorobotic control algorithms, and (4) packaging and telemetric devices. Although substantial progress has been made in these areas and an exhaustive review of current technologies is beyond the scope of this chapter, a brief overview of the issues and recent developments in the realization of a real-time neurorobotic controller are discussed in the following section.

4.4.1. THE NEURAL INTERFACE

Several issues regarding the design of recording electrodes must be resolved in order to obtain accurate, chronic, multisite recordings necessary for a viable brain–machine interface. These include the integrity of the implanted electrodes, the materials comprising the electrodes, the impedance of the recording electrodes, the number of recording sites, and the ability to precisely place electrodes. The signal measured by extracellular recording electrodes reflects the voltage change between the electrode tip and a reference electrode (Moxon, 1999b). The reflection of the action potential in the external environment is carried by the external membrane resistance. However, in the CNS, neurons are surrounded by glia and an aqueous solution rich in salts. Therefore the electrode interface consists of a metal in contact with a salt solution (Wise and Weissman, 1971; White and Gross, 1974; Prohaska et al., 1986). This can be a very corrosive environment and multisite chronic electrodes must be insulated with a material that can maintain its integrity in this environment (Wise et al., 1970; White et al., 1983; Bement et al., 1986; Blum et al., 1991; Nordhausen et al., 1996; Moxon et al., 2000, 2004a,b).

In addition, the metal–electrolyte interface forms a half-cell with a DC offset potential (Wise et al., 1970). The half-cell potential is a result of the buildup of a charge layer between the tissue and an adjacent layer on the metal surface, and the magnitude of the potential is dependent on the metal used in the electrode and the size and shape of the recording site. This half-cell potential is typically several hundred millivolts and is nonstationary due in part to the relative displacement of the electrode with respect to the neural tissue during the respiratory cycle or cardiac output cycle. Because discriminating the action potential—itself only a few hundred millivolts—from this background activity is the focus of the recording, this DC offset potential must be reduced or eliminated during the recording (see Section 4.4.2 below). At the interface, a double layer of electrical charge is formed, creating a high impedance interface (Lui et al., 1999).

To date, the most successful chronic, multiunit recordings have been obtained with low impedance microwires (see NB Labs, Dennison, TX). Microwire electrodes generally consist of platinum or stainless steel wire coated with an insulating material, such as Teflon, with a single recording site at the tip (Blum et al., 1991). Microwires have the advantage that they can be used to reach deep structures in the brain, and recordings from as many as 32 and over 100 wires in the rat and primates, respectively, are now commonplace (Kralik et al., 2001). However, it is difficult to precisely control the interelectrode spacing with arrays of microwires because the tips of the wires can readily be displaced as the electrode is inserted into the tissue.

One solution to this problem of interelectrode spacing has been addressed by the development of micromachined arrays of single-contact electrodes (Rousche et al., 1999). For example, the Utah array (Nordhausen and Maynard et al., 1996) consists of an array of 100 silicon needles micromachined from a monothlithic piece of silicon with precise interelectrode distance and platinum recording sites at the tip of each needle. The Utah array has the advantage that the spacing between recording sites can be precisely determined. Although microwires and the Utah array are sufficient for many applications, certain design features impose inherent limitations. First, the overall density of the recording sites is limited by the fact that both the microwires and the Utah array electrodes have only one recording site at their tips. Second, the development of a therapeutic neurorobotic device requires not only the ability to perform stable recordings from large numbers of single neurons for a period of weeks without the need to move the electrode into fresh tissue, but also the ability to record from deep structures within the brain with precise spacing between recording sites.

A. 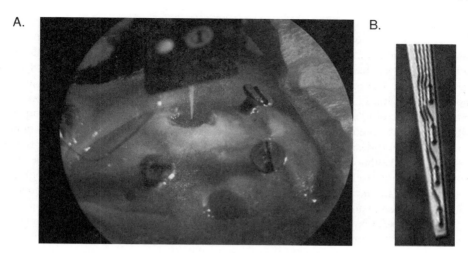 B.

FIGURE 4.5. Photograph of a single, four-site, ceramic electrode array connected to a connector for chronic recording in a rat. The electrodes can be made of varying lengths to reach deep tissue in the brain. (A) The electrode is placed over an opening in the brain just prior to inserting into the tissue. Metal screws provide grounding and help anchor the electrode for chronic recording. (B) An expanded view of the tip showing four individual recording sites. This design allows a maximum of eight recording sites on a single array about the size of a single microwire electrode.

To address these issues, several labs have employed standard thin-film technologies from the semiconductor industry to manufacture multisite recording probes on an insulated substrate (thin film probes), including silicon substrates (silicon probes) (Wise and Weissman, 1971; Pochay *et al.*, 1979; Eichmann *et al.*, 1986; Prohaska *et al.*, 1986; Drake *et al.*, 1988; Blum *et al.*, 1991; Hoogerwerf and Wise, 1994), or substrates doped with boron (boron-diffused silicon probes) (Wise *et al.*, 1994). Each silicon probe is about the size of a single microwire or smaller (approximately 50 µm), with multiple recording electrodes patterned along the length of the silicon-based device generating more recording sites for the same volume of electrode. Recently, the boron-diffused probes have been used to record from multiple sites in the cerebral cortex of rats for at least 18 weeks (Vetter *et al.*, 2004).

In addition to silicon, ceramic has been used as an alternative to provide enhanced insulating properties and stiffness to allow the electrode to reach deep structures in the brain (Figure 4.5) (Moxon *et al.*, 2004a). Ceramic substrates offer a promising alternative to silicon substrates and to the several existing electrode designs. Ceramic substrates are suitable for chronic recordings from large numbers of neurons, like the Utah array and the microwires, but, in addition, also have multiple recording sites along the shaft, similar to the boron-diffused microelectrodes. Each multisite ceramic array can consistently record single-neuron action potentials from four recording sites equally spaced along the length of the array for at least 3 weeks and arrays generally continue to record at least one single neuron for months. With the ceramic substrate devices, it is possible to increase the number of recording sites per array to as many as eight sites and still keep the overall dimensions of the array as small as a single microwire.

Furthermore, in addition to the increased number and precision placement of recording sites, ceramic-based electrodes are impermeable to solutions of electrolytes, even during long-term chronic implant. This property is critical because even tiny breaches in the insulation that protects the whole electrode (outside the recording sites) will severely degrade the recordings. The unique properties of the ceramic also allow it to be used to insulate the conducting lines as well as act as a substrate, effectively encasing the whole electrode, except the recording sites, in a hard ceramic coating that is impervious to electrolyte infiltration, which could allow the system to effectively record single neurons indefinitely. The ceramic array design also has several additional advantages. First, the ceramic is rigid, reducing the likelihood of buckling during surgical implantation, especially as the arrays become longer to reach deep brain structures. Second, the strength of the ceramic substrate is greater than similar-sized probes made of silicon, decreasing the likelihood of breakage during surgical implantation.

Despite these breakthroughs, there is still a clear need to increase the longevity of the recordings. In many cases, the duration of recording ability is not only limited by the technical liabilities of individual electrode design, but also on the response of the brain to foreign bodies and surgical damage. When the electrodes are inserted into the neural tissue, they create damage that eventually builds a scar around the electrode, preventing further recording (for an overview see Moxon *et al.*, 1999b). Much work is currently underway to reengineer the neural response to eliminate this scar and increase the biocompatibility. Enhanced surgical procedures (Vetter *et al.*, 2004) and surface modifications (Moxon *et al.*, 2004b) suggest that we may be able to record single neurons from throughout the brain, including deep structures, indefinitely.

4.4.2. SIGNAL CONDITIONING

A critical component of a neurorobotic control device for clinical applications is an integrated circuit device that can condition the signals recorded from the electrode. Current hardware for amplifying and filtering the neural signals includes headstage devices for current amplification, racks with signal-conditioning boards for bandpass filtering and amplifiers for each electrode all connected to one or more desktop PCs for data storage. The solution for making this equipment small enough to fit on a patient is clearly to use custom-designed mixed-signal (analog and digital) very large scale integrated circuits (VLSI) to acquire and process the electrical signals recorded from the electrodes (Moxon *et al.*, 2000). Because we are only interested in the time the cell fires an action potential (see Section 4.4.3 below), the goal is to filter and amplify the signal sufficiently to discriminate single-neuron action potentials on-chip, and monitor when an action potential occurs. The spike times and not the analog signal can therefore be transmitted telemetrically to a robotic device. This approach eliminates many of the bandwidth and connector problems that have plagued earlier designs (Takahashi and Matsuo, 1984; Bement *et al.*, 1986; Wise and Najafi, 1991; Ji and Wise, 1992; Bai and Wise, 2001; Kim and Kim, 2000; Bai *et al.*, 2000).

However, regardless of how the resulting neural spikes are processed the problems associated with conditioning the biological signal recorded from the brain to discriminate single neurons are the same and have been examined in detail (Smith *et al.*, 1967; Wise *et al.*, 1970; Takahashi and Matsuo, 1984; Bement *et al.*, 1986). The quality of the neural signal recorded from a microelectrode is poor. The amplitude of the action potential is small, generally less than 500 μV peak to peak. It is corrupted with both low and high frequency noise from biological sources that can exceed 500 μV. These include neural activity from far-field neurons and muscle activity from chewing and eye blinks. Moreover, as stated above (Section 4.4), there is a significant DC offset potential. Finally, the interface of the electrode with the neural tissue is a high impedance interface, often greater than 1 MΩ and the frequency of an action potential is generally in the neighborhood of 1 kHz. With conventional solid-state electronics, these signals have been easily handled using field effect transistors or op-amps (Chapin and Nicolelis, 2000), followed by bandpass filtering and amplification. However, for large numbers of recording channels, these devices can become too large to be feasible for applications involving humans. Therefore, many investigators have used microfabrication techniques to mount signal-conditioning devices directly on the silicon-based electrode or build independent very large scale integrated (VLSI) circuits on a small chip to reject noise and select action potentials.

The signal-conditioning device must maximize the ratio of the action potential to the background noise amplitude or the signal-to-noise by (1) rejecting the DC offset potential and low frequency biological noise, (2) transforming the high input impedance to a low impedance output to match impedance for the bandpass filter input stage, and (3) minimizing the inherent electrical noise of the device itself. All this must be performed on a device small enough to fit on the microelectrode array. To maximize the signal-to-noise ratio, the signal-conditioning circuitry should reside as close to the source (i.e., the microelectrode) as possible. The design of a signal-conditioning unit requires that the electrode signal-to-noise ratio be large enough to exceed the equivalent input electrical noise of the devices themselves (Szabo and Marczynski, 1993; Guillory and Normann, 1999).

Noise sources for the signal-conditioning devices can be classified into two categories—inherent noise and ambient noise. The inherent noise sources are caused by the physical properties of the active and passive electrical components. Inherent noise sources include thermal noise from excitation of carriers within the substrate, shot noise due to the DC bias current, and flicker, which is inversely proportional to the frequency and often referred to as $1/f$ noise (Grey and Meyer, 1993). Inherent noise can be minimized by the selection of material used for the device. While most signal-conditioning units for neural probes are CMOS devices (Ji and Wise, 1992; Kim and Kim, 2000; Bai and Wise, 2001), which can have significant noise in the bandpass range of 100 Hz–kHz, recent designs have used an NMOS transistor for highpass filtering (Mohseni and Najafi, 2004).

The ambient or remote noise sources can dominate for single-ended electronics and find their way into the neural signal recording through connectors or lead wires from the electrodes. This common-mode noise is usually most dominant and caused by electromagnetic interference from surrounding electronic devices. This noise can be 60-Hz noise from AC power, radio signals, or noise from cell phones. For devices that utilize an on-board digital clock, broadband or white noise can interfere with the recorded signal through the metal conductor lines on an integrated circuit. This induced noise is capacitively, or inductively, coupled into the signal lines. If the design is such that the signal-conditioning, population-function-generator, and telemetry circuits are all combined into a single chip, this source can be mostly eliminated.

Several groups have developed procedures to embed circuit devices directly onto the silicon substrate of the microelectrode. This approach clearly maintains the signal-conditioning circuitry close to the signal source but is technically challenging because it does not allow for any off-chip components. For example, one approach uses a diode capacitor highpass filter (Ji and Wise, 1992), while Bai and Wise (2001) used a unity-gain op-amp. The most common method for rejecting low-frequency signals involves combining the capacitance of the electrode with a low-input impedance at the source to produce a high-pass filter (Szabo and Marczynski, 1993; Mohseni and Najafi, 2004). However, this technique requires the amplifiers to be matched to a particular type of electrode and may require trimmable devices to control input impedance (Gray and Meyer, 1993). An alternative and perhaps more cost-effective design than that where the circuit devices are embedded on the microelectrode substrate is a single, monolithic VLSI chip mounted on the electrode. This allows the electrodes and the signal-conditioning circuitry to be tested independently, and only high quality devices of both types to be combined. The device has a variable bandpass gain that effectively enhances the signals of interest (single-neuron action potentials) but requires a single off-chip capacitor. This approach has lead to commercially available VLSI chips (Plexon Inc., Dallas, TX) that increase the recording capability of many single-neuron recording applications.

4.4.3. NEUROROBOTIC CONTROL ALGORITHMS

Once the signal recorded from the electrode is properly conditioned, the time at which each recorded cell fires an action potential must be determined on-line, in real-time. Multiple neurons recorded from a single microelectrode are generally identified by the shape of their waveform, and in this way can be discriminated from other neurons and from the analog signal (Wheeler *et al.*, 1999). The typical parameters defining an individual neuronal waveform include the amplitude and duration of the spike. For neurobiological experiments, it is common to record the analog signal and use sophisticated algorithms to detect and sort action potentials off-line (Lewicki, 1998; Letelier and Webber, 2000; Hulata *et al.*, 2002; Pouzat *et al.*, 2002; Kim and Kim, 2003; Nguyen *et al.*, 2003) and then reconstruct the spike times of individual cells off-line to ensure a precise and accurate sorting of the waveforms (Bar-Gad *et al.*, 2001). This is obviously not suitable for a neurorobotic controller. A neurorobotic control device requires an on-line spike-sorting algorithm that can discriminate an action potential waveform from the analog signal in real-time (Figure 4.6). In general, the method requires saving a segment of data and analyzing the waveform off-line. A template is built off-line, and this template captures features of the action potential that distinguishes it from the background signal and other cells. These captured features are usually the amplitude and duration of the spike. This template is then used on-line, during neurorobotic control, to select spike times from the analog signal (Wheeler, 1999; Chapin and Nicolelis, 2000; Moxon *et al.*, 2000; Kralik *et al.*, 2001).

In order to implement this type of procedure, one would envision that during off-line spike discrimination, the patient would reside near the spike-sorting processor. Fully sampled neural signals would be transmitted to the spike-sorting processor and templates would be created. These templates would then be downloaded to the neurorobotic control device for on-line spike sorting. Because neurons drift and the electrode may move within the brain, it is important to re-discriminate neurons and create new templates periodically.

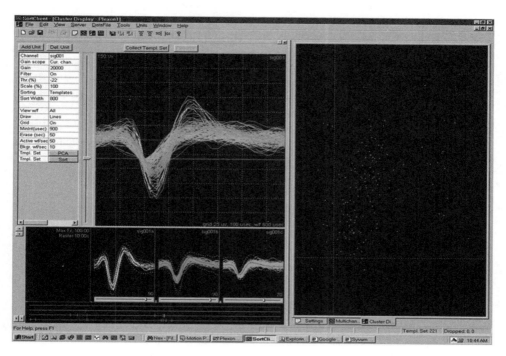

FIGURE 4.6. On-line discrimination of multiple single-neuron action potentials recorded from a single microwire implanted in the animal pictured in Figure 4.3. On this wire three distinct single neurons could be recorded simultaneously. The panel on the right displays the shape of the action potential in principal components space. This is used to discriminate spikes originating from the same recording site.

This is clearly cumbersome for the patient and, depending on the design of the spike-sorting processor, may require the patient to travel to a medical facility. Ideally, this type of neurorobotic control will require devices that can automatically update the discrimination parameters and are also small enough to reside on the noninvasive part of the electrode itself. This will allow the signals to be continuously adapted to optimize single neuron detection.

Several lines of evidence suggest that it may be possible to simplify this process considerably. One possibility is that it will not be necessary to discriminate single cells recorded from the same microelectrode from each other. It has been shown that multiunit data, or the combined spikes from more than one cell recorded from the same microelectrode, have many of the same response properties as precisely discriminated single cells (Fanselow *et al.*, 2002; Carmena *et al.*, 2003). Moreover, the ability to decode sensory stimuli is only slightly impaired when using multiunit data as compared to single-unit data. It is possible that the neural population function can compensate for this loss in decoding and that multiunit signals are used instead of single units. If the process of neuron discrimination can be reduced to distinguishing any spike from background noise, the problem is greatly reduced, and a simple amplitude detector will likely suffice. This is also comparatively easy to implement in hardware. A threshold could be set automatically based on the variance of the signal. A general rule could be set that any signal that exceeds the variance by a certain amount will be considered a spike. Spikes on each channel would be sent to the neural population algorithm without identifying whether or not they came from a single neuron or multiple units. During neurorobotic control, the neural population algorithm would weight the spikes coming from each channel depending on the neural information on that channel's participation in the task. There are several algorithms that have been successfully used to develop these weights, which are discussed in more detail in Section 4.3. These include principal components analysis, cosine tuning functions, and artificial neural networks. It is likely that different algorithms will work better for different patients and for different types of tasks. The system may also require an "off" function to ensure attempts at neurorobotic control are not performed when the patient is engaged in another task.

4.4.4. PACKAGING AND TELEMETRY

For a neurorobotic control device that can be used in humans the packaging of the device is critical. Everything must be small enough to reside under the skin and the signal that passes telemetrically to the robotic device must conform to bandwidth and power constraints. Several designs have utilized on-board channel selection, to reject electrodes for which the signal is poor and multiplexing to reduce the number of connectors and limit the bandwidth. Another alternative is to include the spike-detection algorithm on the chip and pass only the spike times to the robotic device. This will minimize amplification of the neural signal (which reduces power) and minimizes the bandwidth of the transmitted signal. Regardless of the signal transmitted, connecting to each individual electrode creates the possibility for noise sources through pin inductance. Furthermore, with tens or even hundreds of connectors, space problem becomes a critical issue as well. There is a clear need for improved packaging technologies to increase the number of input/outputs per square millimeter and permit an integrated circuit device. Packaging improvements will be needed to increase the number of channels per unit area of the neurorobotic circuit device and ultimately improve the quality of the neural control signal (Bakoglu *et al.*, 1990).

Another important feature of the overall package is the ability to telemeter the signal away from the patient. A telemetric signal acquisition system can be divided into two parts: (1) a device for converting the neural signal for transmission and (2) a radio-frequency transceiver on the robotic device. As mentioned above, the neural signals recorded from the electrodes are amplified and filtered, and single-neuron action potentials are discriminated from the analog signal. One solution is to transform the spike times into a single analog control signal based on the neural control algorithm and transmit this control signal to the robotic device. Ideally, this should be done telemetrically, without attaching wires directly to the patient, because wires that cross the skin boundary will always entail a risk of infection, irritation, and mechanical damage. The telemetry design must consider the transmission frequency, the modulation method, and the data-encoding scheme (Moxon *et al.*, 2000).

The general guidelines for picking a transmission frequency for the low-power, short-transmission neurorobotic application are based on the bandwidth, or data-transfer rate, of the channel. A general rule of thumb is to use a transmission frequency at least 20 times greater than the bandwidth. The overwhelming constraint involves minimizing power consumption in the implanted transmitter and a second important and related constraint is the weight of the final device, including the power source. These are functions of the number and types of signal transmitted. For example, radio-telemetry systems have been designed for freely-moving animals as small as insects (Takeuchi and Shimoyama, 2004) and as large as monkeys (Obeid *et al.*, 2004). For typical signal transfer the upper end of the spectrum for frequency selection is generally set by the frequency limitations of the semiconductor technology and by the desired power consumption. Generally, the higher the frequency, the more the power required to generate the signal. If only spike times are transmitted, or a single analog neural control signal that will likely contain only low-frequency components, the power consumption can remain low.

Many schemes exist for encoding data for serial transmission on a radio-frequency data link. A digital frequency-modulated method of frequency shift keying is one low-power encoding scheme that uses a small number of off-chip components. Standard frequency shift keying communication protocol can be implemented using a voltage-controlled oscillator to modulate the signal for transmission (Towe, 1986). The radio-frequency link receiver hardware resides on the robotic device and consists of the receiving filter and antenna and the demodulator. If standard frequency shift keying modulation is used to send the signal, then on the receiving side, a phase-locked loop can be used to track the incoming frequency changes and to directly demodulate the signal.

Finally, any implantable or wearable system will need to be powered by a battery. Because battery life and weight are critical, power consumption per channel must be as small as possible (Vittoz *et al.*, 1993). Most implantable commercial devices use primary cells for energy storage. Commercial manufacturers of implantable devices usually custom design the battery cells for each application of their devices. This allows them to optimize the voltage, current drain characteristics, and the shape of the battery to make the best use of space in the implanted device. For this emerging field of neurorobotics, more efficient batteries and circuits will be required to build a realistic implantable neurorobotic control device.

Rechargeable batteries have been suggested but such devices may have to deal with the problems of transcutaneous wires to deliver the power to the implant or inductive power transfers that recharge the battery through the skin. The advantages of VLSI technology and novel packaging trends that are being developed will allow the integration of mixed analog/digital signals on a single chip. New designs are increasing our ability for discrete, single-channel analog functions, which include the preamplifiers, bandpass filters, and analog-to-digital (A/D) converters. The goal is to be able to integrate small devices with more channels, lower noise, lower power, and packaged into smaller and lighter-weight hybrid circuit subassemblies.

4.5. NEW DIRECTIONS FOR A NEUROROBOTIC CONTROL

The theoretical constructs first elaborated in the early part of the 20th century combined with more recent advances in technology for simultaneously recording from large numbers of neurons suggest that it is not only possible to perform neurorobotic control to replace limb movement but that these efforts may be more efficient than the more traditional BCI devices that rely on signals recorded from the surface of the scalp. Recent experiments using humans with sensorimotor damage have demonstrated that they are capable of successful neurorobotic control of a cursor on a computer screen. These data suggest that the adult injured nervous system can adapt to new avenues of communication and establish effective control of a cursor. However, much needs to be done to further our understanding of these plastic responses of the injured brain and to develop the hardware necessary to making neurorobotic control a viable alternative for a broad range of disorders of the nervous system.

The issues and problems associated with the development of a neuroprosthetic control device outlined above suggest the need for a multidisciplinary design team consisting of biomedical, electrical, and chemical engineers as well as neuroscientists and medical professionals, while taking into consideration the patients' needs. State-of-the-art developments are still required in order to design an optimized and efficient hardware system that can successfully interface with the brain, record neural signals for decades, and produce an effective neuroprosthetic command signal. These developments include improved materials for neural biocompatibility, mixed-signal (analog and digital) VLSI, novel power-consumption technologies, and telemetric device-packaging technologies.

However, integrated telemetric subassemblies are currently being developed, using the technologies described above, including on-chip spike discrimination algorithms (Obeid et al., 2003). These telemetric subassemblies can then be used to transmit the brain-derived command signals that control the patient's limbs. Artificial neural network algorithms (Principe et al., 2000) that can be downloaded to the neuro-chip are being developed. We expect these devices to be small enough to reside subdurally and eventually provide consistent neural signals that can last for decades.

ACKNOWLEDGMENTS

The author would like to thank Dr. Maria Gulinello for her help in editing the manuscript and bringing it to its final form and Drs. Ronald MacGregor and John Chapin for their inspiration and mentorship, without which this manuscript would not have been possible.

REFERENCES

Abeles, 1991, *Corticonics*, Academic Press, Boston, MA.

Bai, Q., and Wise, K. D., 2001, Single-unit neural recording with active microelectrode arrays, *IEEE Trans. Biomed. Eng.* **48**(8):911–920.

Bai, Q., Wise, K. D., and Anderson, D. J., 2000, A high-yield microassembly structure for three-dimensional microelectrode arrays, *IEEE Trans. Biomed. Eng.* **47**(3):281–289.

Bakoglu, H., Baldwin, G., Li, Z., Tsai, C., and Zhang, J., 1990, *Circuits, Interconnections and Packaging for VLSI*, Addison-Wesley, Boston, MA.

Bar-Gad, I., Ritov, Y., Vaadia, E., and Bergman, H., 2001, Failure in identification of overlapping spikes from multiple neuron activity causes artificial correlations, *J. Neurosci. Methods* **107**(1–2):1–13.

Bement, S. L., Wise, K. D., Anderson, D. J., Najafi, K., Drake, K. L., 1986, Solid-state electrodes for multichannel multiplexed intracortical neuronal recording, *IEEE Trans. Biomed. Eng.* **33**(2):230–240.

Birbaumer, N., Ghanayim, N., Hinterberger, T., Iversen, I., Kotchoubey, B., Kubler, A., Perelmouter, J., Taub, E., and Flor, H., 1999, A spelling device for the paralysed, *Nature* **398**(6725):297–298.

Blum, N. A., Carkhuff, B. G., Charles, H. K., Edwards, R. L., and Meyer, R. A., 1991, Multisite microprobes for neural recordings, *IEEE Trans. Biomed. Eng.* **38**(1):68.

Bragin, J., Hetke, C. L., Wilson, D. J., Anderson, J. E., Jr, and Buzsaki, G., 2000, Multiple site silicon-based probes for chronic recordings in freely moving rats: Implantation, recording and histological verification, *J. Neurosci. Methods* **98:**77–82.

Carmena, J. M., Lebedev, M. A., Crist, R. E., O'Doherty, E., Scatucci, D. M., Dimitrov, D. F., Patil, P. G., Henriquez, C. S., and Micolelis, M. A. L., 2003, Learning to control a brain–machine interface for reaching and grasping by primates, *PLOS Biol.* **1**(2):1–16.

Carter, R., and Houk, J. C., 1993, Multiple single-unit recordings from the CNS using thin-film electrode arrays, *IEEE Trans. Rehabil. Eng.* **1:**3–18.

Chapin, J. K., Moxon, K. A., Markowitz, R. S., and Nicolelis, M. A. L., 1999, Realtime control of a robot arm using simultaneously recorded neurons, *Nat. Neurosci.* **2**(7):1–7.

Chapin, J. K., and Nicolelis, M. A. L., 2000, Brain control of sensorimotor prosthesis, in: *Neural Prostheses for Restoration of Sensory and Motor Function* (J. K. Chapin and K. A. Moxon, eds.), CRC Press, Boca Raton, pp. 45–74.

Chapin, J. K., and Nicolelis, M. A. L., 1999, Principal component analysis of neuronal ensemble activity reveals multidimensional somatosensory representations, *J. Neurosci. Methods* **94:**121–140.

Donoghue, J. P., 2002, Connecting cortex to machines: Recent advances in brain interfaces, *Nat. Neurosci.* **4**(Suppl.):1085–1088.

Drake, K. L., Wise, K. D., Farraye, J., Anderson, D. J., and Bement, S. L., 1988, Performance of planar multisite microarrays in recording extracellular single-unit intracortical activity, *IEEE Trans. Biomed. Eng.* **35:**719–732.

Eccles, J. C., 1981, The modular operation of the cerebral neocortex considered as the material basis of mental events, *Neuroscience* **6:**1839–1859.

Eggermont, J. J., 1993, Functional aspects of synchrony and correlation in the auditory nervous system, *Concepts Neurosci.* **4:**105.

Eichman, H., and Kuperstein, M., 1986, Extracellular neural recording with multichannel microelectrodes, *J. Electrophysiol. Tech.* **13:**189.

Evarts, E. V., 1974, Precentral and postcentral cortical activity in association with visually triggered movement, *J. Neurophysiol.* **37**(2):373.

Foffani, G., and Moxon, K. A., 2004, PSTH-based classification of sensory stimuli, *J. Neurosci. Methods* **135:**107–120.

Freeman, W. J., 1983, The physiological basis of mental images, *Biol. Psych.* **18:**1107–1125.

Gerstein, G. L., Perkel, D. H., and Subramanian, K. N., 1978, Identification of functionally related neural assemblies, *Brain Res.* **140:**43–62.

Gerstein, G. L., and Perkel, D. H., 1969, Simultaneously recorded trains of action potentials: Analysis and functional interpretation, *Science* **164**(881):828.

Georgopoulos, A. P., Schwartz, A. B., and Kettner, R. E., 1986, Neuronal population coding of movement direction, *Science* **233:**1416–1419.

Ghazanfar, A. A., Stambaugh, C. R., and Nicolelis, M. A., 2000, Encoding of tactile stimulus location by somatosensory thalamocortical ensembles, *J. Neurosci.* **20:**3761–3775.

Gray, P. R., and Meyer, R. G., 1993, *Analysis and Design of Analog Integrated Circuits*, John Wiley & Sons, Inc., New York.

Guillory, K. S., and Normann, R. A., 1999, A 100-channel system for real time detection and storage of extracellular spike waveforms, *J. Neurosci. Methods* **91:**21–29.

Hebb, D. O., 1949, *Organization of Behavior*, McGraw-Hill, New York.

Hoogerwerf, A. C., and Wise, K. D., 1994, A three-dimensional microelectrode array for chronic neural recording, *IEEE Trans. Biomed. Eng.* **41:**1136–1146.

Hopfield, J. J., and Herz, A. V., 1995, Rapid local synchronization of action potentials: Toward computation with coupled integrate-and-fire neurons, *Proc. Natl. Acad. Sci. U.S.A.* **92:**6655–6662.

Hopfield, J. J., 1995, Pattern recognition computation using action potential timing for stimulus representation, *Nature* **376:**33–36.

Hubel, D. H., and Wiesel, T. N., 1959, Receptive fields of single neurones in the cat's striate cortex, *J. Physiol.* **148:**574–591.

Humphrey, D. R., 1970, A chronically implantable multiple micro-electrode system with independent control of electrode position, *Electroencephal. Clin. Neurophysiol.* **29:**616.

Humphrey, D. R., Schmidt, E. M., and Thompson, W. D., 1970, Predicting measures of motor performance from multiple cortical spike trains, *Science* **170**(959):759.

Hulata, E., Segev, R., and Ben-Jacob, E., 2002, A method for spike sorting and detection based on wavelet packets and Shannon's mutual information, *J. Neurosci. Methods* **117**(1):1–12.

Jack, J. J., Noble, B. D., and Tsien, R. W., 1975, *Electric Current Flow in Excitable Cells*, Oxford Unversity Press.

Jackson, J. E., 1991, *A User's Guide to Principal Components*, John Wiley and Sons, Inc., New York, pp. 1–25.

John, E. R., 1972, Switchboard versus statistical theories of learning and memory, *Science* **177**:850–864.

Jin, J., and Wise, K. D., 1992, An implantable CMOS circuit interface for multiplexed microelectrode recording arrays, *IEEE Trans. Biomed. Eng.* **27**(3):433–443.

Kennedy, P. R., and Bakay, R. A. E., 1998, Restoration of neural output from a paralyzed subject by a direct brain connection, *NeuroReport* **9**:1707.

Kennedy, P. R., *et al.*, 2000, Direct control of a computer from the human central nervous system, *IEEE Trans. Rehabil. Eng.* **8**:198–202.

Kennedy, P. R., 1989, The cone electrode: A long-term electrode that records from neurites grown onto its recording surface, *J. Neurosci. Methods* **29**:181–193.

Kennedy, P. R., Bakay, R. A. E., and Sharpe, S. M., 1992, Behavioral correlates of action potentials recorded chronically inside the cone electrodes, *Neuroreport* **2**:605.

Kennedy, P. R., and King, B., 2000, Dynamic interplay of neural signals during the emergence of cursor related cortex in a human implanted with the neurotrophic electrode, in: *Neural Prostheses for Restoration of Sensory and Motor Function* (J. K. Chapin and K. A. Moxon, eds.), CRC Press, Boca Raton, pp. 45–74.

Kim, K. H., and Kim, S. J., 2000, Noise performance design of CMOS preamplifier for the active semiconductor neural probe, *IEEE Trans. Biomed. Eng.* **47**(8):1097–1105.

Kim, K. H., and Kim, S. J., 2003, Method for unsupervised classification of multiunit neural signal recording under low signal-to-noise ratio, *IEEE Trans. Biomed. Eng.* **50**(4):421–431.

Kralik, J. D., Dimitrov, D. F., Krupa, D. J., Katz, D. B., Cohen, D., and Nicolelis, M. A., 2001, Techniques for long-term multisite neuronal ensemble recordings in behaving animals, *Methods* **25**(2):121–150.

Kreiter, A. K., Aertsen, A. M., and Gerstein, G. L., 1989, A low-cost single-board solution for real-time, unsupervised waveform classification of multineuron recordings, *J. Neurosci. Methods* **30**(1):59–69.

Kubie, L. S., 1930, A theoretical application to some neurological problems of the properties of excitation waves which move in closed circuits, *Brain* **53**:166–177.

Lashley, K. S., 1950, In search of the engram, *Symp. Soc. Exp. Biol.* **4**:454–482.

Letelier, J. C., and Weber, P. P., 2000, Spike sorting based on discrete wavelet transform coefficients, *J. Neurosci. Methods* **101**(2):93–106.

Lewicki, M. S., 1998, A review of methods for spike sorting: The detection and classification of neural action potentials, *Network* **9**(4):R53–R78.

Lin, Y., Tsai, C., Huang, H., Chiou, D., and Wu, C., 1999, Preamplifier with a second-order high-pass filtering characteristic, *IEEE Trans. Biomed. Eng.* **46**:609–612.

Liu, X., McCreery, D. B., Carter, R. R., Bullara, L. A., Yeun, T. G. H., and Agnew, W. F., 1999, Stability of the interface between neural tissue and chronically implanted intracortical microelectrodes, *IEEE Trans. Rehabil. Eng.* **7**:315.

MacGregor, R. J., 1991, Sequential configuration model for firing patterns in local neural networks, *Biol. Cyber.* **65**:339–349.

MacGregor, R., 1993, *Theoretical Mechanics of Biological Neural Networks*, Academic Press, Boston.

Middendorf, M., McMillan, G., Calhoun, G., and Jones, K. S., 2000, Brain–computer interfaces based on steadystate visual evoked response, *IEEE Trans. Rehabil. Eng.* **8**:211–213.

Mohseni, P., and Najafi, K., 2004, A fully integrated neural recording amplifier with DC input stabilization, *IEEE Trans. Biomed. Eng.* **51**(5):832–837.

Moxon, K. A., 1999, Multichannel electrode design: Considerations for different applications, in: *Methods for Simultaneous Neuronal Ensemble Recordings* (M. A. L. Nicolelis, eds.), CRC Press, Boca Raton, FL, pp. 25–45.

Moxon, K. A., Gerhardt, G. A., Bickford, P. C., Rose, G. M., Woodward, D. J., and Adler, L. E., 1999, Multiple single units and populations responses during inhibitory gating of hippocampal auditory response in freely-moving rats, *Brain Res.* **825**:75–85.

Moxon, K. A., Kalkhoran, N. M., Markert, M. A., Sambito, M. A., McKenzie, J. L., and Webster, J. T., 2004b, Nanostructured surface modification of microelectrodes to enhance biocompatibility for a direct brain machine interface, *IEEE Trans. Biomed. Eng.* **1**(6):881–889.

Moxon, K. A., Leiser, S. C., Gerhardt, G. A., Barbee, K., and Chapin, J. K., 2004a, Ceramic based multisite electrode arrays for electrode recording, *IEEE Trans. Biomed. Eng.* **51**(4):647–656.

Moxon, K. A., Morizio, J., Chapin, J. K., Nicolelis, M. A. L., and Wolf, P. D., 2000, Designing a brain–machine interface for neuroprosthetic control, in: *Neural Prostheses for Restoration of Sensory and Motor Function* (J. K. Chapin and K. A. Moxon, eds.), CRC Press, Boca Raton, pp. 45–74.

Moxon, K. A., Gerhardt, G. A., Gulinello, M., and Adler, L. E., 2003a, Inhibitory control of sensory gating in a computer model of the CA3 region of the hippocampus, *Biol. Cyber.* **88**(4):247–264.

Moxon, K. A., Gerhardt, G. A., and Adler, L. E., 2003b, Dopaminergic modulation of the P50 auditory evoked potential in a computer model of the CA3 region of the hippocampus: Its relationship to sensory gating in schizophrenia, *Biol. Cyber.* **88**(4):265–275.

Moxon, K. A., Leiser, S. C., Gerhardt, G. A., Barbee, K., and Chapin, J. K., 2004, Ceramic based multisite electrode arrays for electrode recording, *IEEE Trans. Biomed. Eng.* **51**(4):647–656.

Mountcastle, V. B., 1957, Modularity and topographic properties of single neurons of cat's somatic sensory cortex, *J. Neurophysiol.* **20**:408–434.

Mountcastle, V. B., Lynch, J. C., Georgopoulus, A., Sakata, H., and Acuna, C., 1975, Posterior parietal association cortex of the monkey: Command functions for operations within extrapersonal space, *J. Neurophysiol.* **38**(4):871.

Najafi, K., and Wise, K., 1986, An implantable multielectrode array with on-chip signal processing, *IEEE J. Solid-State Circuits* **21**:1035–1044.

Nguyen, D. P., Frank, L. M., and Brown, E. N., 2003, An application of reversible-jump Markov chain Monte Carlo to spike classification of multi-unit extracellular recordings, *Network* **14**(1):61–82.

Nicolelis, M. A., 2003, Brain–machine interfaces to restore motor function and probe neural circuits, *Nat. Rev. Neurosci.* **4**(5):417–422.

Nicolelis, M. A., and Fanselow, E. E., 2002, Thalamocortical optimization of tactile processing according to behavioral state, *Nat. Neurosci.* **5**(6): 517–523.

Nicolelis, M. A. L., Lin, R. C. S., Woodward, D. J., and Chapin, J. K., 1993, Dynamic and distributed properties of many-neuron ensembles in the ventral posterior medial thalamus of awake rats, *Proc. Natl. Acad. Sci. U.S.A.* **90**:2212.

Nicolelis, M. A. L., Baccala, L. A., Lin, R. C. S., and Chapin, J. K., 1995, Sensorimotor encoding by synchronous neural ensemble activity at multiple levels of the somatosensory system, *Science* **268**:1353.

Nordhausen, C. T., Maynard, E. M., and Normann, R. A., 1996, Single unit recording capabilities of a 100 microelectrode array, *Brain Res.* **726**:129.

Obeid, I. M., Morizio, J. C., Moxon, K. A., Nicolelis, M. A. L., and Wolf, P. D., 2003, Two multichannel integrated circuits for neural recording and signal processing, *IEEE Trans. Biomed. Eng.* **50**(2):255–258.

Obeid, I., Nicolelis, M. A., and Wolf, P. D., 2004, A multichannel telemetry system for single unit neural recordings, *J. Neurosci. Methods.* **133**(1–2):33–38.

Peckham, P. H., Kilgore, K. L., and Keith, M. W., 2000, Advances in upper extremity functional restoration employing neuroprostheses, in: *Neural Prostheses for Restoration of Sensory and Motor Function* (J. K. Chapin and K. A. Moxon, eds.), CRC Press, Boca Raton, pp. 45–74.

Pochay, P., Wise, K. D., Allard, L. F., and Rutledge, L. T., 1979, A multichannel depth array fabricated using electron-beam lithography, *IEEE Trans. Biomed. Eng.* **26**(4):199–206.

Pouzat, C., Mazor, O., and Laurent, G., 2002, Using noise signature to optimize spike-sorting and to assess neuronal classification quality, *J. Neurosci. Methods* **122**(1):43–57.

Principe, J. C., Euliano, N. R., and Lefebvre, W. C., 2000, *Neural and Adaptive Systems*, John Wiley & Sons, New York.

Prohaska, O. J., Olcaytug, F., Pfundner, P., and Draguan, H., 1986, "Thin-film multiple electrode probes: Possibilities and limitations," *IEEE Trans. Biomed.Eng.* **33**(2):223–229.

Rolls, E. T., Treves, A., Robertson, R. G., Georges-Francois, P., and Panzeri, S., 1998, Information about spatial view in an ensemble of primate hippocampal cells, *J. Neurophysiol.* **79**:1797–1813.

Rolls, E. T., Treves, A., and Tovee, M. J., 1997, The representational capacity of the distributed encoding of information provided by populations of neurons in primate temporal visual cortex, *Exp. Brain Res.* **114**:149–162.

Rosenblith, W. A., 1957, Relations between auditory psychophysics and auditory electrophysiology, *Trans. N. Y. Acad. Sci.* **19**(7):650–657.

Rousche, R. J., and Norman, R. A., 1998, Chronic recording capability of the Utah intracortical electrode array in cat sensory cortex, *J. Neurosci. Methods* **82**:1–15.

Rousche, P. J., Petersen, R. S., Battiston, S., Giannotta, S., and Diamond, M. E., 1999, Examination of the spatial and temporal distribution of sensory cortical activity using a 100-electrode array, *J. Neurosci. Methods* **90**:57.

Rubenstein, J. T., and Miller, C. A., 1999, How do cochlear prostheses work? *Curr. Opin. Neurobiol.* **4**:399–404.

Schmidt, E. M., 1980, Single neuron recording from motor cortex as a possible source of signals for control of external devices, *Ann. Biomed. Eng.* **8**(4–6):339–349.

Schmidt, E. M., 1999, Electrodes for many single neuron recordings, in *Methods for Neural Ensemble Recordings* (M. A. L. Nicolelis, ed.), CRC Press, New York.

Serruya, M. D., Hatsopoulos, N. G., Paninski, L., Fellows, M. R., and Donoghue, J. P., 2002, Instant neural control of a movement signal, *Nature* **416**(6877):141–142.

Sherrington, C. S., 1906, *The Integrative Activity of the Nervous System*, Yale University Press, New Haven.

Smith, B., Tang, Z., Johnson, M. W., Pourmehdi, S., Gazdik, M. M., Buckett, J. R., and Peckham, P. H., 1987, An externally powered, multichannel, implantable stimulator-telemeter for control of paralyzed muscle, *IEEE Trans. Biomed. Eng.* **45**(4):463–475.

Sutter, E. E., 1992, The brain response interface: Communication through visually-induced electrical brain responses, *J. Microcomp. Appl.* **15**:31–45.

Szabo, I., and Marczynski, T. J., 1993, A low-noise preamplifier for multisite recording of brain multi-unit activity in freely moving animals, *J. Neurosci. Methods* **47**:33–38.

Szentagothai, J., 1975, The "module-concept" in cerebral cortex architecture, *Brain Res.* **95**:475–496.

Takahashi, K., and Matsuo, T., 1984, Integration of multi-microelectrode and interface circuits by silicon planar and three-dimensional fabrication technology, *Sens. Actuat.* **5**(1):89–99.

Takeuchi, S., and Shimoyama, I., 2004, A radio-telemetry system with a shape memory alloy microelectrode for neural recording of freely moving insects, *IEEE Trans. Biomed. Eng.* **51**(1):133–137.

Taylor, D. M., Tillery, S. I. H., and Schwartz, A. B., 2002, Direct cortical control of 3D neuroprosthetic device, *Science* **296**:1829–1832.

Towe, B., 1986, Passive biotelemetry by frequency keying, *IEEE Trans. Biomed. Eng.* **33**.

Vallabhaneni, A., Wang, T., and He, B., 2005, Brain-Computer Interface, In He (Eds): Neural Engineering, Kluwer Academic Publishers.

Vetter, R. J., Williams, J. C., Hetke, J. F., Nunamaker, E. A., Kipke, D. R., 2004, Chronic neural recording using silicon-substrate microelectrode arrays implanted in cerebral cortex, *IEEE Trans. Biomed. Eng.* **51**(6):896–904.

Vittoz, E., Borel, J., Gentil, P., Noblanc, J., Nouailhat, A., and Verdone, M., 1993, Design of low-voltage low-power IC's, in: *Proceedings of the 23rd European Solid State Device Research Conference*, p. 927.

Wessberg, J., Stambaugh, C. R., Kralik, J. D., Beck, P. D., Laubach, M., Chapin, J. K., Kim, J., Biggs, S. J., Srinivasan, M. A., and Micolelis, M. A. L., 2000, Real-time prediction of hand trajectory by ensembles of cortical neurons in primates, *Nature* **48**:361–365.

Wheeler, B. C., 1999, Automatic discrimination of singe units, in: *Methods for Neural Ensemble Recordings* (M. A. L. Nicolelis, ed.), CRC Press, New York, p. 61.

White, R. L., and Gross, T. J., 1974, An evaluation of the resistance to electrolysis of metals for use in biostimulation microprobes, *IEEE Trans. Biomed. Eng.* **21**:487.

White, R. L., Roberts, L. A., Cotter, N. E., Kwon, O. H., 1983, Thin-film electrode fabrication techniques, *Ann. N.Y. Acad. Sci.* **83**:183–190.

Williams, J. C., Rennaker, R. L., and Kipke, D. R., 1999, Long-term recording characteristics of wire microelectrode arrays implanted in cerebral cortex, *Brain Res. Prot.* **4**:303–313.

Wilson, M. A., and McNaughton, B. L., 1996a, Dynamics of the hippocampal ensemble code for space, *Science* **261**:1055.

Wilson, M. A., and McNaughton, B. L., 1996b, Reactivation of hippocampal ensemble memories during sleep, *Science* **265**:6761.

Wise, K., 1998, Micromachined interfaces to the cellular world, *Sens. Mater.* **10**:385–395.

Wise, K., and Angell, J., 1975, A low-capacitance multielectrode probe for use in extracellular neurophysiology, *IEEE Trans. Biomed. Eng.* **22**:212–219.

Wise, K. D., Angell, J. B., Starr, A., 1970, Integrated circuit approach to extracellular microelectrodes, *IEEE Trans. Biomed. Eng.* **17**(3):238–246.

Wise, K. D., and Najafi, K., 1991, Microfabrication techniques for integrated sensors and microsystems, *Science* **254**:1335–1342.

Wise, K. D., Najafi, K., and Drake, K. L., 1994, A multichannel microprobe for intracortical single-unit recordings, *Proc. IEEE/NSF Symp. Biosens.*, 87–89.

Wise, K. D., and Weissman, R. H., 1971, Thin films of glass and their application to biomedical sensors, *Med. Biol. Eng.* **9**:339–350.

Wolpaw, J. R., McFarland, D. J., and Vaughan, T. M., 2000, Brain–computer interface research at the Wadsworth Center, *IEEE Trans. Rehabil. Eng.* **8**:222–225.

5

ELECTRICAL STIMULATION OF THE NEUROMUSCULAR SYSTEM

Dominique M. Durand,[*] Warren M. Grill,[#] and Robert Kirsch

Neural Engineering Center, FES Center, Department of Biomedical Engineering, Case Western Reserve University, Cleveland, Ohio
[#] Department of Biomedical Engineering, Duke University, Durham, North Carolina

5.1. INTRODUCTION

Patients with paralysis or disease of the nervous system can have severe functional deficits. Although rehabilitation and neural regeneration can provide some improvement (Grill and Kirsch, 2000; McDonald and Sadowsky, 2000), there is still a large gap to close in order to restore function. Functional electrical stimulation (FES) of neural tissue can be successfully applied to provide additional functional restoration to neurologically impaired individuals. By placing electrodes within excitable neural tissue and passing current through these electrodes, it is possible to activate pathways to the brain or to muscles. The activated pathways can then excite or inhibit their intended target. Neural prostheses refer to applications for which electrical stimulation is used to replace a previously lost or damaged neural function.

Electrical stimulation has been applied to restore neural function in several neural systems (see review by Hambrecht, 1979; Grill and Kirsch, 2000). The most successful neural prosthesis is the cochlear prosthesis, and electrical stimulation of the auditory nerves can restore hearing in deaf patients (Clark, 1990). Stimulation of the phrenic nerve of patients with high-level spinal cord injury generates diaphragm contractions and can restore ventilation (Glenn *et al.*, 1984; Schmit and Mortimer, 1999). Electrical stimulation of the visual cortex produces visual sensations called phosphenes (Brindley and Lewin, 1968). Electrical stimulation of the retina to restore vision is also studied (Margalit *et al.*, 2002) and a visual prosthesis for the blind is currently being tested. Restoration of both upper extremities hand function and lower extremities gait can be partially achieved. There are also several applications of electrical stimulation in the central nervous system. For example, deep brain stimulation of thalamic nuclei has been found particularly efficient at decreasing tremor in patients with Parkinson's disease (Dostroski and Lozano, 2002). Similarly, electrical

---

[*] Address for correspondence: Neural Engineering Center, Department of Biomedical Engineering, Case Western Reserve University, Cleveland, Ohio 44106; e-mail: dxd6@cwru.edu.

stimulation of the vagus nerve has been shown capable of reducing seizure frequency in patients with epilepsy (Rutecki, 1990; George *et al.*, 2002).

Activation of the neural tissue can also be generated by magnetically induced electric fields. Magnetic stimulation of the nervous system is completely noninvasive and has been effective in stimulating brain tissue because the magnetic field can easily penetrate the skull. Moreover, it is reported to cause less pain than electrical stimulation. However, this stimulation method is not efficient, requires a large amount of power, and its effect is difficult to localize. The mechanism of action of magnetic stimulation shares similarities with electrical stimulation (Roth and Basser, 1990). However, there are several important differences (Nagarajan and Durand, 1993) that are not reviewed in this chapter.

Electrical excitation results from the interaction between electric fields and neural excitable tissue. This interaction involves the determination of the propagation of the voltage inside cylindrical neural structures such as axons and dendrites and can be derived from the cable equation. The interaction also involves the determination of the voltages and electrical fields generated by the electrode inside the volume conductor. These voltages can be measured or calculated using Maxwell's equations. The relationship between the applied extracellular field and intracellular transmembrane voltage is described by the source term of the cable equation and is derived below (Section 5.2.1). The effect of waveform of the stimulation pulse, the electrochemistry at the electrode interface, and the tissue damage are reviewed (Section 5.2.2). In the last section (5.2.3), some of the fundamental principles and the electrodes used for neuromuscular prostheses are reviewed. Lower extremities, upper extremities, and bladder prostheses are discussed.

5.2. MECHANISMS OF EXCITATION OF APPLIED ELECTRICAL FIELDS

5.2.1. ANATOMY AND PHYSIOLOGY

Axonal excitation by applied current is effected when the transmembrane current generated by the electrode depolarizes the membrane sufficiently to activate the sodium channels located in high densities at various locations in the cell. Once activated, the sodium current will further increase until the membrane reaches an unstable fixed point. This is the point at which a full action potential ($\sim$100 mV) will develop. Once started, the action will propagate unattenuated either smoothly along the unmyelinated fibers or discretely at the nodes of Ranvier of the myelinated fibers (Fig. 5.1A).

Action potentials can be generated artificially by placing an electrode either inside or outside the cell. Because axons are very small (several micrometers in diameter), it is not possible currently to place electrodes inside several axons without damaging them. Therefore, electrodes used to activate the nervous systems are placed in the extracellular space. For example, surface electrodes are placed directly on the skin whereas epimysial electrodes, small disk electrodes, are placed inside the body but on the surface of the muscle. Intramuscular electrodes are small wires placed inside the muscle with needles. All three types of electrodes generate muscle activation by stimulating not the muscle itself but rather the nerve fibers. Because the threshold activation of muscles fibers is much higher than the threshold of single axonal fibers, these electrodes activate the nerves going to the muscles and are placed as close as possible to the nerve. Electrodes can be placed directly on the nerve (cuff electrodes) or inside the nerve (intrafascicular electrodes). The current threshold for these types of electrodes is significantly lower and these electrodes have varying degrees of selectivity and potential for causing damage to the nerve.

Selectivity is a particularly important issue in the design of neural protheses. Selectivity is defined as the ability of a stimulation system to activate any chosen set of axons. In order to generate a movement, small fibers connected to small muscle units are activated first by the nervous system. If more force is required, additional larger units will be recruited. Because large fibers have a lower threshold of activation compared to small fiber, the recruitment order is reversed and the recruitment curve is very steep, making control of the movement difficult.

Although the dynamics of membrane channels play a major role in the excitation properties of nerves, the sodium channels are almost completely inactive at the resting potential. Therefore, the passive properties of the membrane contribute significantly to the determination of the membrane voltage along the axons and to the site of excitation. The transmembrane voltage is the difference between the intracellular voltage and the extracellular voltage. This extracellular voltage is generated by the current of the electrode and can be estimated with a reduced set of Maxwell equations, the quasi-static formulation.

5.2.2. ELECTRIC FIELDS IN VOLUME CONDUCTORS

The volume conductor surrounding the neural excitable tissue is usually assumed to be purely resistive, with resistivity varying between 50 and 500Ω·cm. The resistivity of the volume conductor is not constant either spatially or directionally, and is defined at each point of the volume by a resistivity tensor. If the resistivity is the same in all directions, the volume is isotropic. If the resistivity is the same at all points of the volume, then the volume is homogeneous.

Although the volume conductor clearly contains material with high dielectric constant (cell membrane for example), its capacitive properties can be neglected when the frequency is below 10 kHz (Plonsey, 1969). Similarly, the inductive properties can be neglected at these low frequencies. However, when the amplitude of the change in time of the magnetic field (dB/dt) is large, as in the case of magnetic stimulation, induction can generate electric fields capable of exciting axons. In the case of electrical stimulation, the current amplitudes involved are too small to generate significant induced fields. Therefore, for most applications of electrical stimulation of the nervous system, the quasi-static formulation of Maxwell's equations can be applied.

5.2.2.1. Quasi-Static Formulation of Maxwell Equations

The electric fields generated by currents applied to the extracellular space of neural tissue can be calculated by solving Maxwell equations. For frequencies under 10 kHz, both the capacitive and inductive properties can be neglected and a simplified set of equations known as the quasi-static formulation can be used (Plonsey, 1969):

$$\text{Conservation of charge}: \quad \boldsymbol{\nabla} \cdot \boldsymbol{J} = 0 \tag{5.1}$$

$$\text{Gauss's law}: \quad \boldsymbol{\nabla} \cdot \boldsymbol{E} = \rho/\varepsilon \tag{5.2}$$

$$\text{Ohm's } law \text{ } for \text{ } conductors : \quad \boldsymbol{J} = \sigma \boldsymbol{E} \tag{5.3}$$

$$\text{Electric field}: \quad \boldsymbol{E} = -\boldsymbol{\nabla} V \tag{5.4}$$

where $\boldsymbol{E}$ is the electric field (V/m) defined as the gradient of the scalar potential V, $\boldsymbol{J}$ the current density (mA/m^2), σ the conductivity (inverse of resistivity) (in s/m), ρ the charge density in C/m^3, ε the permittivity of the medium .

5.2.2.2. *Equivalence between Dielectric and Conductive Media*

At any point P in the volume conductor, the current flowing could be generated by a source at that point (J_s) or could be the ohmic current generated by a distant source (Plonsey, 1969). Therefore, the current density $\boldsymbol{J}$ at any point is the sum of a source term $\boldsymbol{J}_s$ and an ohmic term $\boldsymbol{J}_\Omega$.

$$\boldsymbol{J} = \boldsymbol{J}_\Omega + \boldsymbol{J}_s = \sigma \boldsymbol{E} + \boldsymbol{J}_s \tag{5.5}$$

Using Eq. (5.1)

$$\boldsymbol{\nabla} \cdot \boldsymbol{J} = \boldsymbol{\nabla} \cdot (\sigma \boldsymbol{E}) + \boldsymbol{\nabla} \cdot \boldsymbol{J}_s = 0 \tag{5.6}$$

Assuming a homogeneous volume conductor, $\boldsymbol{\nabla} \cdot (\sigma \boldsymbol{E}) = \sigma (\boldsymbol{\nabla} \cdot \boldsymbol{E})$; therefore

$$\boldsymbol{\nabla} \cdot \boldsymbol{E} = -\boldsymbol{\nabla} \cdot \boldsymbol{J}_{s}/\sigma \tag{5.7}$$

Because $\boldsymbol{E} = -\boldsymbol{\nabla} V$

$$\nabla^2 V = \boldsymbol{\nabla} \cdot \boldsymbol{J}_{s}/\sigma = -I_{v}/\sigma \tag{5.8}$$

where I_v is a volume current in A/m^3 and ∇^2 is the Laplacian operator. The volume current I_v can be calculated from the knowledge of the distribution of sources in the volume conductor. This equation is the equivalent of the Poisson equation derived for dielectrics:

$$\nabla^2 V = -\rho/\varepsilon \tag{5.9}$$

derived for dielectric media. Using the following equivalence,

$$\rho \Leftrightarrow I_v$$
$$\varepsilon \Leftrightarrow \sigma$$

the solution of the Poisson equation for dielectric problems can be applied to the calculation of the current and voltage distribution in volume conductors.

5.2.2.3. *Monopole Point Source*

The voltage and electric field generated by a point source can be obtained by using the known solution of Maxwell equations. However, for the point source, the solution can be obtained by using spherical symmetry. The current density J at a point P located at a distance r from the source is equal to the total current crossing a spherical surface with radius r divided by the surface area:

$$J = \frac{I}{4\pi r^2} u_r \tag{5.10}$$

where u_r is the unit radial vector and r the distance between the electrode and the measurement point. The electric field is then obtained from Eq. (5.3):

$$E = \frac{I}{4\pi \sigma r^2} u_r \tag{5.11}$$

The electrical field is the gradient of the potential. In spherical coordinates,

$$E = -\frac{d\phi}{dr} u_r \tag{5.12}$$

The potential at point P referred to a reference electrode located at infinity is

$$\phi = \frac{I}{4\pi \sigma r} \tag{5.13}$$

The potential along an axon located 1 cm below a monopolar anodic electrode passing 1 mA is plotted in Fig. 5.1B. Both the current density J and the electric field E have a radial distribution with an amplitude inversely proportional to the square of the distance to the source. The potential decay is inversely proportional to the distance. At $r = 0$, the potential goes to infinity and this singularity can be eliminated if the electrode is modeled as a sphere with a radius a. Equation (5.13) is then valid on the surface of the electrode $r = a$ and for $r > a$ (Nunez, 1981).

Assuming the medium to be homogeneous and linear, and using superposition, Eq. (5.13) can be generalized to n monopolar electrodes with a current I_i located at a distance r_i from the recording point. The voltage is then given by

$$\phi = \frac{1}{4\pi \sigma} \sum_n \frac{I_i}{r_i} \tag{5.14}$$

For an axon located in the volume conductor as shown in Figure 5.1, the voltage along the axon located 1 cm away from an anode with a 1-mA current source is given by the following equation and is plotted in Figure 5.1B:

$$\phi = \frac{I}{4\pi \sigma \sqrt{d^2 + x^2}} \tag{5.15}$$

A

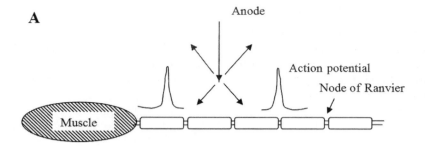

B

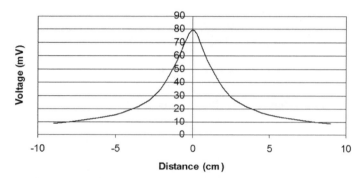

Distance (cm)

FIGURE 5.1. (A) Electrical stimulation of a myelinated fiber. An electrode is located near the axon and an anodic stimulus is applied to the electrode. The current flows outside and inside the axons, generating polarization of the membrane. Action potentials are generated underneath the electrode and propagate both orthodromically and antidromically. (B) Extracellular voltage generated by 1 mA and measured along an axon located at 1 cm from the electrode (resistivity of the medium is assumed to be 100 $\Omega \cdot$cm).

5.2.2.4. Bipolar Electrodes and Dipoles

Using Eq. (5.14), it is possible to calculate the voltage generated by a current source and a current sink separated by a known distance d. The potential at point P (assuming that the voltage reference is at infinity) is given by

$$\phi = \frac{I}{4\pi\sigma}\left(\frac{1}{r_1} - \frac{1}{r_2}\right) \tag{5.16}$$

where r_1 and r_2 are the distances between the measuring point and the electrodes (Figure 5.2). When the distance d between the two electrodes is small compared to the distance r, the equation for the potential reduces to

$$\phi = \frac{Id\cos\theta}{4\pi\sigma r^2} \tag{5.17}$$

where the angle θ is defined in Fig. 5.2. The current distribution for a bipolar electrode and

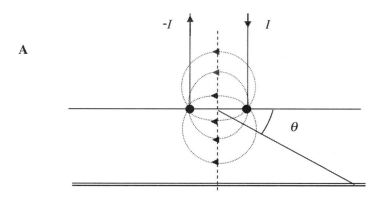

A

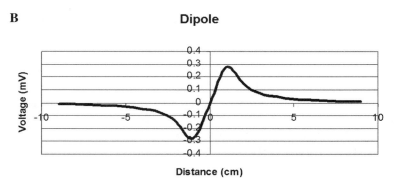

B

Dipole

Distance (cm)

FIGURE 5.2. Dipole stimulation. (A) Two electrodes with opposite polarity located next to each other generate a set of current loosely approximated by circles. (B) Voltage along an axon located 1 cm from a dipole with two electrodes (1 mA and −1 mA) separated by 0.1 cm. Note that the voltage decay is larger than from monopoles and the amplitude of the voltage is significantly lower.

for a dipole are shown in Figure 5.2A. The current distribution is no longer symmetrical as in the case of a monopolar electrode. There is an equipotential line ($\phi = 0$) passing between the two electrodes (Figure 5.2A). Therefore, an axon or nerve located near or on this line (transverse excitation) has a very high threshold for excitation. The voltage generated by a dipole is inversely proportional to the square of the distance between the source and the recording site. Because the voltage decays rapidly compared with that for a monopole, the dipole is a much more selective method of stimulation. However, the current threshold for a dipole is significantly higher than that for a monopole because the ratio of the monopole voltage to the dipole voltage is proportional to r/d.

5.2.2.5. Inhomogenous Volume Conductors

Biological volume conductors are highly nonhomogenous, and the complexity of the volume conductors requires in most cases numerical solutions using finite element or finite boundary methods. However, the effect of the boundary between two layers of various

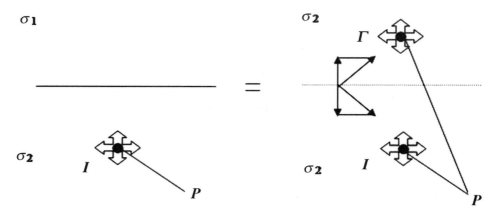

FIGURE 5.3. Method of images. The method of images can be used to calculate the voltage generated by an electrode in a homogenous volume conductor. The two semi-infinite volume conductors with conductivities σ_1 and σ_2 (A) are replaced by a single semi-infinite volume conductor with conductivity σ_1 and an additional image electrode (B).

conductivities can be studied in a simple configuration: the two-layer problem. Consider two volume conductors with conductivities σ_1 and σ_2 separated by an infinite plane as shown in Figure 5.3. A monopolar electrode is placed in region 1 and potentials are recorded in the same region. Solving for the boundary conditions at the interface, it can be shown that the inhomogenous volume conductor can be replaced by a homogenous volume by adding a mirror current source (Fig. 5.3) with an amplitude equal to (Nunez, 1981)

$$I' = \frac{\sigma_1 - \sigma_2}{\sigma_1 + \sigma_2} I \qquad (5.18)$$

The voltage at point P is then given by Eq. (5.14) for voltage in the region 1 only. In the case where layer 1 is a volume conductor such as the body and 2 is air ($\sigma_2 = 0$), the mirror source I' is equal to I. The boundary condition that the current density normal to the surface is zero is satisfied as illustrated in Figure 5.3. If the stimulation electrode is located at the interface, the zero-current boundary condition is satisfied by simply doubling the amplitude of the current. Similarly, if the recording electrode is located on the surface, the zero-conductivity layer will double the size of the recorded potentials. If layer 2 is a perfect conductor ($\sigma_2 = \infty$), the current density must be normal at the boundary. The condition is satisfied if $I' = -I$. This image theory is only applicable in simple cases but can be useful in obtaining approximations when the distance between the recording electrode and the surface of discontinuity is small, thereby approximating an infinite surface (Durand *et al.*, 1995).

5.2.3. EFFECTS OF APPLIED ELECTRIC FIELDS ON TRANSMEMBRANE POTENTIALS

The interaction between the extracellular voltage and axons can be studied with a simple passive membrane model. The membrane resistance can be modeled as the rest conductance of the sodium, potassium, and leakage channels using compartmental analysis (Rall, 1979). Current flowing from the outside to the inside of the cell increases the voltage across the membrane and causes hyperpolarization. When current flows from the inside to the outside, the membrane is depolarized (Ranck, 1975) (Figure 5.4).

A quantitative analysis of the interaction between electric fields and neural tissue can be obtained by combining the passive membrane model with the extracellular voltage (V_e) generated by stimulating electrodes. The model makes several assumptions (McNeal, 1976). Assuming that the presence of the fiber does not affect the extracellular voltage, the extracellular voltage can be calculated using the equations previously derived (see Section 5.2 above). A passive electrical model of the axon can be built (Figure 5.4B) and the circuit can be solved using numerical methods for compartmental analysis (Koch and Segev, 1989). These methods have been analyzed using neuronal simulation packages such as Neuron (Hines, 1984). Each compartment in Figure 5.4B models a length Δx of axon. Applying Kirchoff's law at each node and taking the limit when Δx tends to zero, one obtains the following inhomogenous cable equation (Altman, 1988; Rattay, 1989):

$$\lambda^2 \frac{\partial^2 V_m}{\partial x^2} - \tau_m \frac{\partial V_m}{\partial t} - V_m = -\lambda^2 \frac{\partial^2 V e}{\partial x^2} \tag{5.19}$$

λ is the space constant of the fiber and is determined by the geometric and electric properties of the axon:

$$\lambda = \frac{1}{2} \sqrt{\frac{R_m^s d}{R_a^s}} \tag{5.20}$$

where R_m^s is the specific membrane resistance, R_a^s the axoplasmic specific resistance, and d the diameter of the axon. τ_m is the time constant of the axon and is given by

$$\tau_m = R_m C_m \tag{5.21}$$

The source term of the cable equation is negative and is the product of the square of the space constant with the second spatial derivative of the extracellular voltage. At the onset of a pulse ($t = 0$), the voltage on the cable is equal to zero and the change in voltage (dV/dt) is proportional to the equivalent voltage source:

$$V_{eq} = \lambda^2 \frac{\Delta^2 V_e}{\Delta x^2} = \lambda^2 \frac{d^2 V_e}{dx^2} \bigg|_{\Delta x \to 0} \tag{5.22}$$

The amplitude of this equivalent voltage sources is plotted in Figure 5.4C for a 10-μm axon stimulated by a 1-mA anodic current located 1 cm away from the axon. A positive value of V_{eq} indicates membrane depolarization while a negative value indicates hyperpolarization. The membrane polarization calculated from this equation can be predicted by simply examining the current flow pattern in and out of the axon. This analysis is valid only at the onset of the pulse, because during the pulse, currents will be distributed throughout the cable and will affect the transmembrane potential (Warman et al., 1992). However, for short

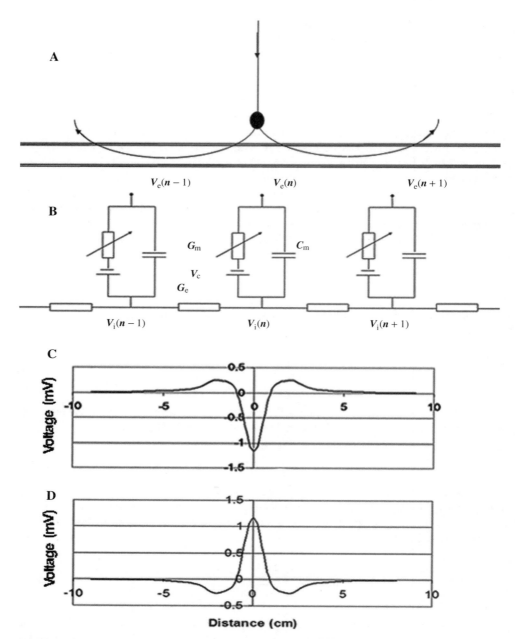

FIGURE 5.4. Membrane polarization by electrical stimulation. (A) An anode is placed close to a nerve. The current enters the membrane underneath the electrode and leaves the cell on both sides of the electrode. (B) Passive equivalent electrical model of the membrane. The intracellular resistance is modeled by an axoplasmic resistance G_a. The membrane is modeled by a capacitance C_m and a conductance G_m. The extracellular voltage is calculated from the knowledge of the current sources and applied to the model. (C) Equivalent voltage source (forcing function of the cable equation) for an anode (1 mA). Equivalent forcing function for a cathode (-1 mA). The arrows indicate the sites of maximum depolarization and the location of excitation.

pulses, these effects are small and the shape of the transmembrane voltage can be predicted using the equivalent voltage source. This analysis is valid for long axonal structures with uniform membrane. However, when cell bodies and dendrites and more realistic axons are simulated, the site of excitation is strongly influenced by the distribution of sodium channels and a more detailed analysis is required to predict the site of excitation (McIntyre and Grill, 1999).

5.2.3.1. Activation of Myelinated Axons

The cable equation derived above cannot describe the activation of myelinated fibers because the presence of myelin sheath around the axon forces the current to flow in and out of the membrane only at the nodes of Ranvier (see Figure 5.1). Therefore, the action potential also jumps from node to node (saltatory conduction). The interaction between myelinated fibers and applied fields can be described using the model shown in Figure 5.3 with R_m and C_m replaced by R_n and C_n. In this simple model the myelin sheath is assumed to have infinite resistance. Therefore, R_n and C_n represents the membrane resistance and capacitance at the node of Ranvier only.

$$R_n = \frac{R_n^s}{\pi \, dl} \qquad C_n = C_n^s \pi \, dl \qquad (5.23)$$

where R_n^s and C_n^s are the specific membrane capacitance and resistance at the node, d is the inner fiber diameter, l the width of the node, and L the internodal distance (Figure 5.1). The cable equation for a myelinated nerve can be derived using Kirchoff's law:

$$\frac{R_n}{R_a}\Delta^2 V_m - R_n C_n \frac{\partial V_m}{\partial t} - V_m = -\frac{R_n}{R_a}\Delta^2 V_e \qquad (5.24)$$

Δ^2 is the second difference operator, $\Delta^2 V = V_{n-1} - 2V_n + V_{n+1}$, and R_a represents the resistance between two nodes. The source term for this equation $(-(R_n/R_a)\Delta^2 V)$ is independent of the diameter because $L/D = 100$ and $d/D = 0.7$, where D is the outside diameter of the axon and l is constant. However, because the distance between the nodes increases with the diameter of the fiber, the second-order difference is a function of the diameter of the axons.

5.2.3.2. Effect of Polarity of Applied Stimulus

Activation of axons is determined by the amplitude of the membrane depolarization. The polarization of the membrane can be predicted directly from the sign of the equivalent voltage source V_{eq} (a positive sign indicates depolarization). This voltage is plotted in Figure 5.4 and it is clear from the figure that the membrane depolarization generated by the cathode is greater than that generated from the anode (Figure 5.4B and C). This result has been confirmed experimentally. Note also that the polarity of the stimulus affects the location of the excitation. Cathodic excitation produces excitation directly under the cathode, whereas anodic excitation produces excitation at two sites located away from the electrodes (see arrows in Figure 5.4). However, the location of excitation also depends on the largest density of sodium channels. For example, anodic stimulation of cortical pyramidal cells has a lower threshold than cathodic stimulation when applied to the surface of the brain.

5.2.3.3. *Effect of Space Constant of the Axon*

The equivalent voltage source of the cable equation is proportional to the square of the space constant λ. λ^2 is proportional to the diameter of the fiber (Eq. (20)) for unmyelinated fibers and to the square of the diameter for myelinated fibers (Durand, 1995). Therefore, in both cases, V_{eq} is higher for fibers with larger diameter and, therefore, large-diameter fibers have a lower threshold. Because the physiological recruitment order by the central nervous system is to first recruit the small fibers followed by large ones, electrical stimulation produces a reverse recruitment order. Techniques have been developed to recruit small fiber before large fibers by using a different stimulation waveform (Fang and Mortimer, 1991). Because λ^2 is also dependent on the electrical properties of the axons, it is possible to predict that fibers with a larger membrane resistance or lower axoplasmic resistance will also have lower thresholds.

5.2.3.4. *Effect of the Second-Order Spatial Derivative of the Electrical Field*

The equivalent source voltage is also proportional to the second spatial derivative of the voltage along the nerve. Therefore, a constant voltage applied along the nerve or even a constant field would not be capable of reaching threshold. A nonzero value of the second-order spatial derivative is required to generate excitation. The second-order term, also known as the activation function (Rattay, 1990), is given by the following equation for unmyelinated fibers:

$$f_{unmy} = \frac{d^2 V_e}{dx^2} \tag{5.25}$$

This function depends only on the extracellular potential and is independent of the properties of the fiber. For myelinated fibers, the extracellular voltage is evaluated at the nodes of Ranvier, and a different equation containing implicit information about the diameter is required.

$$f_{my} = \frac{\Delta^2 V_e}{\Delta x^2} \tag{5.26}$$

A discussion of the explicit dependence on the fiber diameter can be found in Durand (1995).

5.3. ELECTRODE–TISSUE INTERFACE

Electrical stimulation for restoration of function is typically delivered using metal electrodes implanted inside the body. Because current is carried by electrons in metals and by ions in the body, chemical reactions must take place at the interface between the metal electrode and the tissue. In general, this interface has a nonlinear impedance that is a function of the voltage across the interface. This interface impedance can affect the properties of stimulation, and electrochemical reactions at the electrode–tissue interface can lead to electrode dissolution and/or production of chemical species that may be damaging to tissue.

5.3.1. REGULATED VOLTAGE AND REGULATED CURRENT STIMULATION

The electronic circuit used to deliver the applied stimulus may be either a constant (regulated) voltage device or a constant (regulated) current device, and this will have a direct impact on the properties of excitation (Figure 5.5). In general, regulated current stimulators should be used, as this enables direct control of the extracellular electric field. However, consideration of the risk for tissue damage suggests that regulated voltage stimulators may be more appropriate for surface stimulation.

The effects of electrical stimulation on neurons is mediated by the extracellular electric field, and thus controlling neuronal excitation requires control of the electric field. Regulated current stimulators produce the same current flow through the tissue, and thus the same electric field, regardless of impedance of the electrode–tissue interface (Fig. 5.5). Therefore, the amplitude and time course of the stimulus can be controlled directly, even in the face of nonlinear or changing impedance of the electrode–tissue interface.

Conversely, when using a regulated voltage stimulator, a nonlinear or changing impedance of the electrode–tissue interface will lead to changes in the current flow through

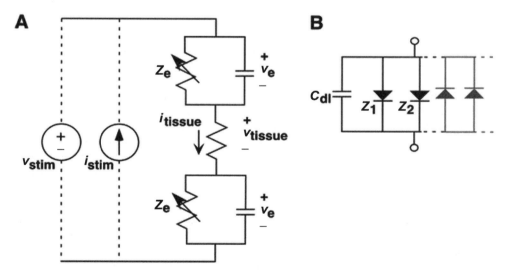

FIGURE 5.5. Equivalent circuit models of the electrode–tissue interface. (A) Equivalent circuit model of the stimulator, a pair of electrodes, and the tissue. v_{stim} and i_{stim} and the stimulator voltage and current for a regulated voltage or regulated current stimulator, respectively, Z_e is the impedance of the electrode–tissue interface, v_e is the voltage across the electrode–tissue interface, i_{tissue} is the current flowing through the tissue, and v_{tissue} is the voltage across the tissue. (B) Equivalent circuit models of a metal electrode in an ionic conducting medium. C_{dl} is the double layer capacitance, and Z_i represents potential-dependent electrochemical reactions that can occur at the electrode–tissue interface.

the tissue and thus changes in the excitation of the neurons. Because it is the voltage between the electrodes that is regulated, increases in the interface impedance will reduce the amount of current that flows in the tissue (Figure 5.5B) and decrease excitation, whereas decreases in the interface impedance will increase the current flow in the tissue and strengthen excitation.

Because constant-current stimulators maintain the same current strength, independent of the load impedance, they can create a risk for discomfort or tissue injury when using electrodes on the skin surface. If a surface electrode were to become partially detached, the interface impedance would increase, but a regulated current stimulator would deliver the same amount of current into a smaller area, resulting in an increase in current density that could cause pain or skin damage. With a constant-voltage stimulator, as the interface impedance increases because of electrode detachment, the amount of current flowing will decrease. Therefore, if a constant-current stimulator is used for surface stimulation the output voltage should be limited to prevent large current flows into high impedances (small areas).

5.3.2. TISSUE DAMAGE

Clinical implementations of neural prostheses have not encountered decreased prosthesis performance as a result of tissue damage. However, continued safe use of these devices requires an understanding of the mechanisms responsible for tissue damage. Most generally, tissue damage can be classified as either passive (resulting from the presence of the electrode) or active (resulting from the passage of stimulus current), although it is not always possible to differentiate between these two damage modes. Passive damage can result from surgical trauma and chemical and mechanical bio-incompatibility of the implanted device. Active damage results from electrochemical reaction products formed at the tissue–electrode interface and from physiological changes associated with neural excitation.

Chemically bio-incompatible materials and implants contaminated during manufacture or implantation lead to a chronic inflammatory response that may, by macrophage activity, destroy healthy tissue as well as the implant. Currently, noble metal conductors and medical-grade synthetic insulators, combined with exacting cleaning procedures ensure a minimal tissue response. These materials are also able to withstand the harsh environment within the body without degradation. Mechanical damage may result from surgical trauma such as nerve stretching or from the presence of the electrode(s). Furthermore, damage to peripheral nerves may result from compression during postsurgical edema, or from stretching if the electrode becomes anchored by scar tissue. Similarly, electrode arrays implanted on the surface of the brain have resulted in local mechanical damage including tissue compression, meningeal fibrosis, and molecular layer and fiber deformation. Insertion of penetrating microelectrodes may result in vascular as well as neural damage. By minimizing electrode diameter, and beveling rather than sharpening the electrode tip, mechanical trauma has been minimized (Yuen *et al.*, 1990).

Tissue damage due to the passage of current may arise from electrochemical reaction products formed at the electrode–tissue interface and/or from physiological changes in the neural and surrounding tissue that are associated with neural excitation. The finding that charge per phase and charge density are co-factors in determining the threshold for neural injury by stimulation in both the peripheral nervous system (Agnew *et al.*, 1989) and the central nervous system (McCreery *et al.*, 1990) suggests that both electrochemical and physiological mechanisms contribute to neural damage. The number and spatial extent of neurons activated are related to the charge injected in each stimulus pulse (see "The Strength–Duration and Charge–Duration Relationships", below), whereas the charge density (charge per unit area of the stimulating electrode) contributes to the type and rate of electrochemical reactions that occur at the electrode–tissue interface.

Tissue damage may result from the products of electrochemical reactions at the electrode–tissue interface where the charge carriers change from electrons in the metal to ions in the tissue (Robblee and Rose, 1990). Reactions that take place are dependent on the potential of the electrode as well as the time course of the stimulus pulse, and may be accelerated by increased current density. The electrode interface can be modeled by the parallel combination of a capacitor (C), representing the double-layer capacitance, and a series of diode-like elements, each representing an electrochemical reaction (Fig. 5.5B). The voltage developed across the electrode–tissue interface (V_e) is determined by the amount of charge in the stimulus pulse (Q), because $V = Q/C$. The voltage across the interface determines which chemical reactions will take place or, in the simple model, which diodes will turn on. The electrode capacitance is determined by the properties of the material and is proportional to the electrode area ($C \propto A$). Therefore, the potential developed across the interface is proportional to electrode area ($V_e \propto Q/A$). This relationship is the basis for the correlation between charge density and tissue damage and the assertion that the charge density is an indirect measure of the electrochemical contribution to tissue damage. If the interface voltage is kept within certain limits (i.e., between the diode threshold voltages) then chemical reactions can be avoided, and all charge transfer will occur by the charging and discharging of the double-layer capacitance (Brummer and Turner, 1977). However, in many instances, the electrode capacitance is not sufficient to store the charge necessary for the desired excitation without the electrode voltage reaching levels where reactions will occur. The principal approach to control the interface voltage has been the use of charge-balanced biphasic stimuli that have two phases that contain equal and opposite charge (Lilly *et al.*, 1955). However, even with charge-balanced pulses it is possible that the interface voltage may reach levels where electrochemical reactions can occur. If charge is lost through a reaction (diode turned on) in the first phase of the pulse, then the charge delivered in the second phase of the pulse will exceed the charge still on the capacitor and cause the electrode voltage to overshoot zero. This is the basis for imbalanced biphasic stimulation, which has been shown to reduce electrode corrosion (McHardy *et al.*, 1977) and enables greater charge densities without damaging muscle tissue (Scheiner *et al.*, 1990).

The correlation of charge per phase and tissue damage is thought to result from physiological changes associated with neural excitation. The number of excited neurons is dependent on the charge delivered in the stimulus pulse. Thus, physiological changes occurring because of synchronous activation of a population of neurons increase as the number of excited neurons increases. This relationship may create the correlation between charge per phase and neural damage. There is substantial support for this hypothesis in both the peripheral nervous system (PNS) and the central nervous system (CNS). In the PNS, peripheral nerve injury can still occur at very low charge densities (Agnew *et al.*, 1989), and anesthetic block of electrical activity occludes the damaging effect of stimulation (Agnew *et al.*, 1990). In the CNS, equivalent tissue damage was seen under platinum electrodes and tantalum pentoxide capacitor-type electrodes (which were presumed not to have electrochemical reactions occurring) (McCreery *et al.*, 1988). These studies, which have attempted to differentiate electrochemically induced and activity-induced tissue damage, support that tissue injury results from physiological changes in the neural environment resulting from synchronous activation of populations of neurons. However, there appears to be a synergism between damage due to electrochemical products and activity-induced tissue damage, and the presence of one should not eliminate concern for the other. There also exists a synergism between stimulus parameters over a wide range of values (McCreery *et al.*, 1995), making it difficult to obtain limits for individual parameters that ensure that tissue damage will not occur.

5.3.3. EFFECT OF WAVEFORM

In addition to influencing tissue damage by driving electrochemical reactions at the electrode–tissue interface, the stimulation waveform has a strong impact on the pattern of neuronal excitation that is generated.

Strength–Duration and Charge–Duration Relationships: The strength–duration curve describes the empirically observed relationship between the duration of a rectangular stimulus pulse and the threshold stimulus amplitude. The stimulus amplitude necessary for excitation, I_{th}, increases as the duration of the stimulus is decreased. The strength–duration relationship is given by the following equation, and an example of a strength–duration curve is shown in Figure 5.6A.

$$I_{th} = I_{rh}[1 + (T_{ch}/PW)] \tag{5.27}$$

The parameter I_{rh} is the rheobase current, and is defined as the minimum current amplitude necessary to excite the neuron with a pulse of infinite duration. The parameter T_{ch} is the chronaxie time and is defined as the pulse duration necessary to excite the neuron with a pulse amplitude equal to twice the rheobase current.

The amount of charge necessary for excitation can be determined directly from the strength–duration relationship. The resulting charge–duration relationship is given by the following equation and is shown in Figure 5.6B.

$$Q_{th} = PWI_{rh}[1 + (T_{ch}/PW)] \tag{5.28}$$

The charge required for excitation decreases as the duration of the pulses decreases (Figure 5.6B). Thus, although short pulses require higher currents for excitation, shorter pulses are more efficient at generating excitation than are longer pulses. Reducing the charge required for excitation reduces the probability of electrode corrosion or tissue damage and reduces stimulator power requirements. Shorter pulses also decrease the gain between the stimulus magnitude and the number of nerve fibers activated by increasing the threshold difference between different-diameter nerve fibers (Gorman and Mortimer, 1983). Similarly, shorter pulses increase the spatial selectivity of stimulation by increasing the threshold

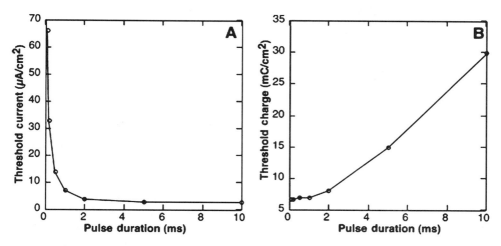

FIGURE 5.6. Strength–duration (A) and charge–duration (B) curves for excitation of neurons. The data were generated using intracellular stimulation in a space-clamped patch of neuronal membrane with Hodgkin–Huxley dynamics.

difference between nerve fibers lying at different distances from the electrode (Grill and Mortimer, 1996a).

5.3.3.1. Anodic vs. Cathodic Stimulation

The polarity of a monophasic stimulation pulse has a direct influence on the threshold for and pattern of stimulation. Cathodic stimuli applied extracellularly depolarize neuronal membrane in the vicinity of the electrode, and an action potential is generated. In contrast, anodic stimuli hyperpolarize the membrane immediately adjacent to the electrode (Figure 5.4C).

There are two ways in which anodic stimuli, which hyperpolarize the membrane in the vicinity of the electrode, may generate excitation: anode break excitation and virtual cathodes. Anode break excitation arises from deinactivation of the voltage-gated sodium channels that occurs during membrane hyperpolarization (Hodgkin and Huxley, 1954; Mortimer, 1990). At the termination of a long-duration hyperpolarization, the membrane is in a hyperexcitable state, and the inflow of sodium that occurs upon repolarization can lead to an action potential. Because the action potential is initiated at the end of the stimulus pulse, this is referred to as anode break excitation. Anode break excitation requires long-duration ($\geq$500-μs) pulses, as the time constant for inactivation–deinactivation is quite long.

Anodic stimuli can also generate excitation at the regions of depolarized membrane (called virtual cathodes) that form adjacent to the region of the neuron hyperpolarized by the stimulus. Action potential initiation will occur at the virtual cathode(s) if the current amplitude is large enough to bring the depolarized portion of the membrane to threshold. Typically, current thresholds for excitation with an anodal current, via virtual cathodes, are 5–8 times larger than the threshold current for direct cathodal stimulation.

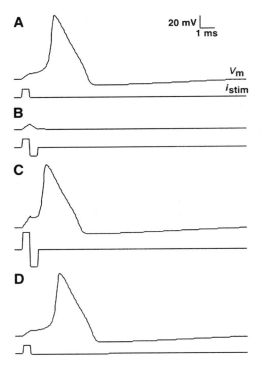

FIGURE 5.7. Effect of stimulation waveform on neuronal excitation. (A) Stimulation with a supra-threshold monophasic current pulse generates an action potential (pulse duration (PD) = 0.5 ms, pulse amplitude (PA) = 20 μA/cm^2). (B) The same stimulus, when followed by an equal amplitude pulse of opposite polarity (PD = 0.5 ms, PA = −20 μA/cm^2), fails to generate an action potential. (C) When the amplitude of both the first pulse (PD = 0.5 ms, PA = 40 μA/cm^2) and the second pulse (PD = 0.5 ms, PA = −40 μA/cm^2) is increased, the biphasic pulse generates an action potential. (D) If the duration of the second pulse is increased by a factor of 10 (PD = 5 ms) and its amplitude is decreased by a factor of 10 (PA = −2 μA/cm^2) then the original stimulus pulse generates an action potential. The data were generated using intracellular stimulation in a space-clamped patch of neuronal membrane with Hodgkin–Huxley dynamics.

5.3.3.2. *Monophasic vs. Biphasic Stimuli*

Under most conditions, biphasic stimulus pulses are used because they prevent damage to stimulating electrodes or the underlying tissue. The second phase of the stimulus removes the charge put onto the double layer by the first phase of the pulse, may reverse electrochemical reactions, and shifts the electrode potential in a direction opposite to that of the primary phase. The second phase of the stimulation waveform, primarily in place for charge recovery, also has effects on excitation (van den Honert and Mortimer, 1979; Gorman and Mortimer, 1983). Specifically, the second phase may arrest an action potential generated by the first pulse and will thus increase the threshold for excitation (Figure 5.7). This effect is dependent on the duration and amplitude of primary as well as secondary phase of the stimulus.

5.4. NEUROMUSCULAR PROSTHESES

Injury or disease of the human nervous system can impair the control of numerous motor functions, including limb movements and internal body functions such as respiration, micturition, and defecation. These impairments often severely limit the independence of individuals with neurological deficits, increasing their dependence on family members or other attendants for most activities of daily living. Neuroprostheses are devices that use electrical stimulation of the nervous system to substitute for the natural neural activation that has been lost, with the goal of restoring function. The movements restored by functional electrical stimulation (FES) can thus increase the independence of disabled individuals, potentially allowing users of these systems to participate more fully in social and occupational activities while also reducing the economic impact of the disability.

Most existing neuroprostheses restore function by stimulating motoneurons that innervate the muscles powering desired movements. Thus, neuroprostheses can currently be applied to any neurological condition where the motoneurons and muscles are still functional. Although a number of neurological conditions (e.g., multiple sclerosis and head injury) could potentially be addressed by neuroprostheses, most applications of FES to date have been in individuals with stroke and spinal cord injury (SCI), largely because individuals in both these groups tend to reach stable states where FES can produce reliable results. There are approximately 750,000 incidents of stroke in the United States each year (Broderick *et al.*, 1998). About 10% stroke survivors stabilize, with moderate to severe movement disorders, but attain a reasonable level of general health and cognitive ability to potentially benefit from neural prostheses. SCI affects 7500–10,000 individuals per year (Stover and Fine, 1986; Harvey *et al.*, 1990). The severity and extent of the resulting movement deficits depend on the location and extent of the injury. Paraplegia (paralysis of the legs and pelvis) results from injuries at thoracic (mid-back) or lower levels, whereas tetraplegia (paralysis of the legs, trunk, and arms) results from injuries at cervical levels. The extent of functional loss increases as the spinal cord injury level moves closer to the head because the motoneuron pools to more and more muscle groups become isolated from the brain. Depending on the level of injury, individuals with SCI could benefit from restoration of many different motor functions.

5.4.1. RECRUITMENT PROPERTIES

Neuroprostheses operate by passing current through an electrode to excite nearby neural tissue and thereby elicit muscular contractions that produce a function of interest. Effective restoration of function requires reliable control of desired force levels. Typically, however, the magnitude of muscle contraction produced by electrical stimulation depends in a complex manner on the type of electrode used, the shape and amplitude of the stimulation waveform, and the location of the electrode relative to the targeted neural structures. Specific electrode types will be discussed in the following section, and the effects of stimulus waveform were already discussed above. This section will therefore focus on the effects of stimulation intensity and the location of the electrode.

The force produced by a muscle is naturally modulated by two different mechanisms. Varying the frequency of activation of a given motoneuron ("rate modulation") causes the rather sluggish muscle force responses to adjacent stimulus pulses to begin to overlap and summate. Eventually, the responses completely merge into a "fused" response, and no greater force can be generated regardless of stimulation frequency. Because the high stimulation frequencies required to reach complete fusion are usually accompanied by significant fatigue (Bigland-Ritchie *et al.*, 1979), most neuroprostheses use a fixed stimulus frequency that is fairly low (12–20 Hz) (McNeal, 1973) and rely upon the second force modulation mechanism—"recruitment"—to grade force (Crago *et al.*, 1980). Recruitment

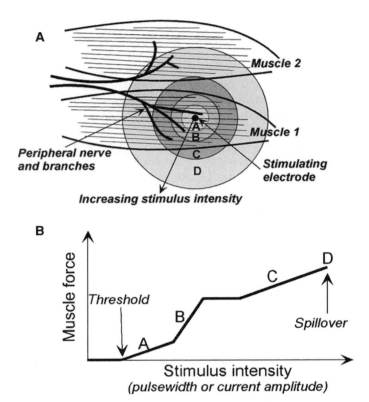

FIGURE 5.8. Hypothetical recruitment mechanism (part a) and resulting force recruitment curve (part b) for a muscle-based electrode. Part (a) illustrates two muscles and one electrode, with the concentric circles representing the territory activated by the electrode as the stimulating current increases. Part (b) represents a force recruitment curve that could result from simulation of the electrode. The force levels indicated by the labels A–D in part (b) correspond to the indicated territory in part (a).

is the mechanism by which progressive excitation of more and more motoneurons (and the muscle fibers they serve) is used to modulate muscle force. As discussed above, recruitment by electrical stimulation is significantly different from natural neural recruitment in that larger motoneurons tend to be activated at lower stimulus intensities than smaller motoneurons, the reverse of the natural mechanism.

However, electrical recruitment also depends significantly on the anatomy of the nerve within the muscle, in particular the branching pattern of the nerve within the muscle, and the location of the electrode relative to the various branches (Crago *et al.*, 1980; Grandjean and Mortimer, 1986). Figure 5.8A illustrates a hypothetical relationship of two muscles and their neural supply to a stimulating electrode in one of the muscles. The concentric circles surrounding the electrode indicate the increasing territory over which current passed through the electrode can excite neural tissue as the stimulation intensity is increased. A hypothetical relationship between the stimulation intensity and the resulting force, known as the "recruitment curve," is indicated in Figure 5.8B. For very low stimulus intensities, no motoneurons exceed threshold and no contraction occurs. This corresponds to the region to the left of the "threshold" in Figure 5.8B. For slightly higher stimulus intensities (labeled A in both figures), only a single nerve branch is activated in muscle #1. As the intensity is increased, more and more motoneurons from this single branch are recruited and the force increases smoothly and progressively. When the stimulus intensity reaches the level indicated by B, a second nerve branch in muscle #1 exceeds threshold, and the slope of the recruitment curve initially increases to reflect continued recruitment in both nerve branches. For somewhat higher stimulation intensities, all motoneurons in the two branches are recruited and no additional force is produced for a range of increasing intensities, resulting in a plateau region. However, a third branch is then recruited in muscle #1 and the force begins to rise again (region C). As the stimulation intensity is increased even further, the territory reached by the electrode becomes large enough to activate motoneurons in muscle #2 (region D). This "spillover" to other muscles can occur before full recruitment of the targeted muscle is achieved and is therefore undesirable unless the newly recruited muscle is synergistic with the targeted muscle. Although this hypothetical example of a recruitment curve illustrates the general factors that determine its shape, a real muscle can exhibit many different recruitment properties depending on the size of the muscle, the branching pattern of its nerve, and the type of electrode used. Thus, smooth recruitment can be achieved over the full range of stimulation intensities, one or more force plateaus can be seen, and/or spillover can occur at very low to very high stimulation levels (Kilgore *et al.*, 1990, 1993a,b). Muscle recruitment can also be complicated by changes in the spatial relationship between the electrodes and the nerve branches caused by contraction and subsequent movement, also known as "length-dependent recruitment" (Crago *et al.*, 2000). Muscle fatigue, a common consequence of FES attributable to the recruitment reversal produced by conventional electrodes and waveforms, changes the relationship between motoneuron activation and muscle force generation and thus also can affect recruitment properties.

Recruitment properties can have significant effects on the functional performance of neuroprostheses. The maximum achievable force level can be limited by spillover, whereas the ability to smoothly grade force levels may be hampered by nonlinearities in the recruitment properties. These nonlinear properties manifest as ranges of stimulation intensities where force either changes very rapidly ("high gain" regions) or not at all ("plateau" regions). To effective control muscle force output, all neuroprostheses must explicitly or implicitly invert these characteristics (Kilgore *et al.*, 1990, 1993a,b) so that a smooth mapping between a desired force and the actual force is achieved despite recruitment nonlinearities.

5.4.2. ELECTRODES FOR MUSCLE STIMULATION

Neuroprostheses operate by passing current through an electrode to excite nearby neural tissue. Neuroprostheses can activate paralyzed neurons at a number of levels of the nervous system, including the spinal cord (Grill *et al.*, 1999; Mushahwar *et al.*, 2000; Prochazka *et al.*, 2002;), spinal roots (Rushton *et al.*, 1997a,b), peripheral nerves (McNeal and Bowman, 1985; McNeal *et al.*, 1989; Grill and Mortimer, 1996a,b) and intramuscular nerve branches (Mortimer and Peckham, 1973; Crago *et al.*, 1980; Grandjean and Mortimer, 1986; Memberg *et al.*, 1993). Because most existing neuroprostheses have used muscle-based electrodes, we will focus on activation of intramuscular nerve branches in this review.

Although these electrodes are often called "muscle electrodes," it should be noted that with rare exceptions (Kern *et al.*, 1999; 2002) all such electrodes actually activate nerve branches within the muscle rather than the muscle fibers themselves owing to the much lower thresholds of neurons (Crago *et al.*, 2000).

A number of different muscle-based electrode types have been developed and used, with varying properties and degrees of invasiveness. Surface electrodes are simplest and least invasive, typically consisting of large conductive pads that are placed on the surface of the skin over the muscle to be stimulated. Such electrodes are most effective for large and superficial muscles. The high resistance of the skin requires that higher currents be passed from the electrode before the muscle is activated, with the side effect of stimulating the multitude of sensory afferents in the skin, often making the stimulation uncomfortable. Because of the large size of the electrodes, spillover is a significant problem. Currents sufficient to activate deep muscles will inevitably activate the superficial muscles first, so deeper muscles cannot be activated independently. Finally, the relative motion between a contracting muscle and the skin can be substantial, so length-dependent recruitment can be a significant problem (Crago *et al.*, 2000).

Implanted electrodes avoid many of the limitations of surface electrodes, although at the cost of invasiveness because they must be placed surgically. Figure 5.9A illustrates epimysial (Grandjean and Mortimer, 1986) and intramuscular (Memberg *et al.*, 1993; Scheiner *et al.*, 1994) electrodes. These are widely used in existing neuroprostheses because they are much more selective than surface electrodes, have fewer problems with movement-related recruitment, and avoid the sensory-laden skin. Typically, the lead wires from these implanted electrodes are routed subcutaneously back to an implanted stimulator device (Figure 5.9B),

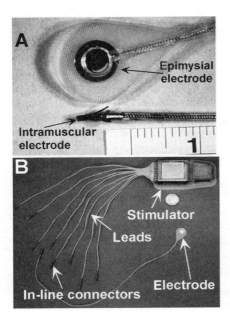

FIGURE 5.9. Typical muscle-based electrodes and implanted stimulator. Part (a) illustrates both an epimysial electrode and surgically implanted intramuscular electrode. Part (b) illustrates an eight-channel implantable stimulator, along with the lead wires, in-line connectors, and electrodes needed to complete an implanted neuroprosthesis.

so that the entire stimulation system is contained within the body. Epimysial electrodes are metallic disks that have a polymer backing for suturing to the muscle epimysium near the entry point of the nerve. They must be installed in an open surgical procedure, which is invasive but allows clear visualization of the spatial relationship of the electrode to the nerve. This electrode has been used in an implanted hand grasp neuroprosthesis (Grandjean and Mortimer, 1986; Smith *et al.*, 1987; Keith *et al.*, 1989; Kilgore *et al.*, 1997) and in lower extremity systems (Triolo *et al.*, 1996a,b). Intramuscular electrodes are typically formed from insulated multistrand wire that is wound into a helical coil, the tip of which is deinsulated to provide the electrode surface. Typically, the end of the wire is bent backwards to form a barb that secures the electrode in place. Intramuscular electrodes are inserted into a muscle using a hypodermic needle. Once the needle is properly located, it is withdrawn, and the electrode remains in place because of the barbed end. Intramuscular electrodes can be inserted through the skin into the muscle and have a percutaneous interface (Marsolais and Kobetic, 1986) or can be inserted directly into the muscle in an open surgical procedure. Percutaneous systems have been used for both temporary (Kirsch *et al.*, 1998; Yu *et al.*, 2001) and permanent applications (Handa *et al.*, 1985, 1992, 1998; Marsolais and Kobetic, 1986, 1988; Hoshimiya *et al.*, 1989; Kobetic *et al.*, 1997; Kameyama *et al.*, 1999), and for both functional and therapeutic (Daly *et al.*, 2001) applications. Surgically placed intramuscular electrodes (Memberg *et al.*, 1993) have been found to be useful for smaller muscles (Lauer *et al.*, 1999) and for muscles where the nerve entry point is difficult to visualize because of its deep location, for example the erector spinae (Triolo *et al.*, 2001; Uhlir *et al.*, 2001).

5.4.3. UPPER EXTREMITY APPLICATIONS

A number of different neuroprostheses have been developed for restoring hand grasp and release, using different electrodes. Two hand grasp systems using surface electrodes, the Handmaster (Ness Ltd., Israel) (Nathan, 1993; Triolo *et al.*, 1996a,b) and the Bionic Glove (Prochazka *et al.*, 1997), have been developed. The Handmaster device has been used primarily in individuals with hemiplegia resulting from stroke, but also in individuals with C4-C7. The Bionic Glove is primarily targeted to individuals with C6-C7 SCI because they retain voluntary wrist extension. As noted above, these systems have the advantage of being noninvasive, but they also exhibit several disadvantages. The surface electrodes must be placed accurately for each use. Because there is no surgery, complementary reconstructive surgery to compensate for denervated muscles or other soft tissue deformities (Keith and Lacey, 1991; Keith *et al.*, 1996) cannot be performed.

Intramuscular electrodes with percutaneous leads have also been used to implement hand grasp neuroprostheses (Handa *et al.*, 1985; Keith *et al.*, 1988; Peckham *et al.*, 1992; Triolo *et al.*, 1996a,b). The use of intramuscular electrodes has the benefits of avoiding open surgery, providing selective stimulation, allowing access to deeper muscles, and fixed electrode locations. However, reconstructive surgery is again not available and the skin exit sites of the electrode leads must be carefully maintained over long periods of time. Applications of this approach have been made in individuals with cervical SCI (C4-C6) (Hoshimiya *et al.*, 1989; Kirsch *et al.*, 1998; Yu *et al.*, 2001) and hemiplegia (Handa *et al.*, 1992, 1998), restoring motions of the hand, forearm, elbow, and shoulder.

Upper extremity neuroprostheses utilizing surgically implanted epimysial and intra-muscular electrodes, together with a fully implanted stimulator, have also been developed (Smith *et al.*, 1987; Peckham and Keith, 1992; Kilgore *et al.*, 1997). A commercial system based on this earlier work was commercially available until recently as the Freehand System (NeuroControl, Inc., Cleveland). This system (illustrated in Figure 5.10) uses a pacemaker-like stimulator implanted in the upper chest (Figure 5.9B) and seven to eight epimysial electrodes implanted in muscles of the hand and forearm to restore both palmar grasp and key grip in individuals with C5-C6 SCI. The stimulator receives power and stimulation commands via an electromagnetic link through the skin. The user controls the stimulation levels using a joystick-like device that measures voluntary movements of the contralateral shoulder (Johnson and Peckham, 1990) or ipsilateral wrist (Hart *et al.*, 1998). This im-planted system is initially more expensive because of the required surgical procedures, but the implantation is usually combined with other surgical procedures (e.g., muscle tendon transfers) that augment the actions of the neural prosthesis (Keith *et al.*, 1996). Furthermore, the internal components are protected and quite reliable, and the user must put on and take off only a few external components, simplifying use.

Ongoing research is extending the functionality of this basic system by enhancing wrist extension (Lemay and Crago, 1997), forearm pronation (Lemay *et al.,* 1996), elbow extension (Crago *et al.*, 1998; Grill and Peckham, 1998), and shoulder function (Kirsch *et al.*, 1998; Yu *et al.*, 2001). Improvement of hand grasp postures via stimulation of intrinsic hand muscles (Lauer *et al.*, 1999) and providing hand function to both hands have also been investigated (Scott *et al.*, 1996).

5.4.4. *LOWER EXTREMITY APPLICATIONS*

Neuroprostheses have also been developed for movement deficits associated with both SCI and stroke. Following a stroke or incomplete SCI, paralysis of ankle dorsiflexor muscles eliminates the ability to lift the toes during the swing phase of gait, leading to the foot being dragged along and potentially causing the individual to stumble. Several groups (Strojnik *et al.*, 1987; Kljajic *et al.*, 1992; Kralj *et al.*, 1993b; Burridge *et al.*, 1997; Taylor *et al.*, 1999) have developed footdrop neuroprostheses that initiate stimulation of the common peroneal nerve (through surface or implanted electrodes) when a contact switch in the shoe (or other sensor) detects that the foot has left the ground. Stimulation of the peroneal nerve directly activates the paralyzed dorsiflexor muscles to lift the toes, but it also evokes a flexor withdrawal reflex (owing to excitation of sensory fibers within the nerve), which causes the ankle, knee, and hip all to flex and further raise the foot off the ground. More than 5000 individuals have used a footdrop neural prosthesis. Currently, at least two commercial sys-tems (the Footlifter, Elmetec A/S, Denmark; and the Walkaide, Neuromotion, Inc., Canada) are available.

A similar approach has been extended to provide walking and standing in individuals with thoracic SCI or hemiplegia (Kralj and Bajd, 1989; Kralj *et al.*, 1993a; Graupe and Kohn, 1997). Surface stimulation of the quadriceps muscles is used to stand up, with assistance from the voluntarily controlled upper extremities. Walking is produced by stimulating the peroneal nerve of one leg to elicit a flexion withdrawal, which generates a crude stepping motion, and simultaneously stimulating the quadriceps on the opposite leg to provide weight support. This stimulation pattern is then reversed between the two legs to step with the other

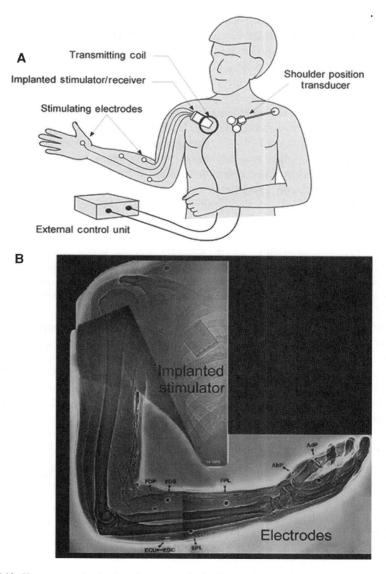

FIGURE 5.10. Upper extremity implanted neuroprosthesis. Part (a) is a schematic representation of a hand grasp neuroprosthesis that includes an implanted stimulator (illustrated in Figure RFK-2(b)), eight stimulating electrodes, a shoulder position transducer for user control of the stimulation, and an external controller device that implements the control algorithm and powers the implanted stimulator via an RF link. Part (b) is a composite radiograph of this system implanted in a user.

leg and produce a full gait cycle. Forward propulsion is provided only by the voluntary actions of the upper extremities, not through stimulation of lower extremity muscles. Crutches or a rolling walker are required for stability and support. The commands for eliciting a step are produced by hand or foot switches on the user. A commercial system based on these principles, the Parastep device (Sigmedics, Inc. Northfield, IL), has been developed (Graupe and Kohn, 1997). Surface-electrode-based lower extremity neuroprostheses suffer from the same limitations as surface-based upper extremity systems—deep muscles cannot be used and the user must accurately place the electrodes before each use. Furthermore, walking systems based on a flexor reflex are limited by accommodation of the reflex during the repeated activation required during gait.

Neuroprostheses that restore standing or walking via intramuscular or implanted electrodes have been developed (Marsolais and Kobetic, 1986, 1988; Triolo *et al.*, 1996a,b; Kobetic *et al.*, 1997; Davis *et al.*, 2001). This approach uses individual electrodes for each muscle to be stimulated rather than relying on the flexor reflex. Forward propulsion is assisted by stimulated contractions, although all users also employ a walker or crutches for stability and safety. Systems using percutaneous intramuscular electrodes are introduced in a minimally invasive manner via hypodermic needles (Marsolais and Kobetic, 1986). Percutaneous-based neural prostheses have been shown to restore standing and walking (Kobetic *et al.*, 1997) in individuals with paraplegia. More recent work has focused on the use of implanted stimulators, electrodes surgically implanted in or on a muscle, and the stimulation of peripheral nerves or spinal roots (Davis *et al.*, 1994, 2001; Triolo *et al.*, 1996a,b; Rushton *et al.*, 1997). Figure 5.11 is a schematic representation of the implanted standing and transfer neuroprosthesis developed by Triolo and colleagues (Triolo *et al.*, 1996a,b; Davis *et al.*, 2001).

Several groups (Andrews *et al.*, 1988; Solomonow *et al.*, 1997; Ferguson *et al.*, 1999; Marsolais *et al.*, 2000) have devised "hybrid" systems that use electrical stimulation of a few lower extremity muscles to provide forward propulsion while using extensive external bracing [e.g., a reciprocating gait orthosis (RGO)] to provide stability. By eliminating the need for stimulated muscle contractions to support body weight, the metabolic cost of restoring standing and locomotion can be significantly reduced. RGO systems, with or without electrical stimulation, are relatively inexpensive and have been found to be successful in allowing users to stand and walk with fairly low energy consumption. However, the external bracing is often cosmetically unacceptable to the users, they are difficult to put on and take off, and they work well only over flat, even surfaces.

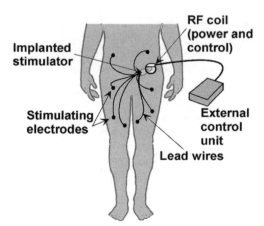

FIGURE 5.11. Schematic representation of a lower extremity implanted neuroprosthesis. The stimulator is the same used in the upper extremity neuroprosthesis of Figure 5.10, although lower extremity and trunk muscles are now targeted.

5.4.5. BLADDER PROSTHESES

The principal functions of the urinary bladder are accumulation and storage of urine (continence) and evacuation of urine at an appropriate time and place (micturition). Control of bladder function involves spinal, supra-spinal, and peripheral (ganglionic) elements (de Groat *et al.*, 1993, Chai and Steers, 1996). In able-bodied persons there is a synergic relationship between the bladder and the urethral sphincter such that during continence, pressure in the bladder is low but activity in the external urethral sphincter is high and, conversely, during micturition, pressure in the bladder is high and activity in the external urethral sphincter is low. Loss of voluntary control of bladder and bowel function occurs after disease and injury, including spinal cord injury (Blaivas, 1982; Watanabe *et al.*, 1996), stroke (Sakakibara *et al.*, 1996), and multiple sclerosis (Gallien *et al.*, 1998). Complications include loss of voluntary control of bladder function, development of bladder hyperreflexia (the bladder contracts reflexively at very low volumes), and bladder–sphincter dyssynergia (the sphincter contracts in synchrony with the bladder, leading to absent or incomplete bladder emptying). These problems can lead to significant medical complications, lead to severe loss in quality of life, and result in high medical costs.

A number of approaches to restore bladder function using electrical stimulation have been tried and are outlined in Figure 5.12 (Rijkhoff *et al.*, 1997a; Jezernik *et al.*, 2002). In many cases, direct or reflex activation of the urethral sphincter, which closes the outlet as the bladder is contracting, has prevented efficient bladder emptying. Large numbers of patients have achieved significant clinical benefit by electrical stimulation of the sacral nerves innervating the lower urinary tract (Brindley *et al.*, 1982; Brindley, 1994; Creasey *et al.*, 2001).

Direct stimulation of the bladder wall has met with limited success and has virtually been abandoned, although it may be useful in cases of denervation of the bladder. The failure of this approach was primarily due to the small region of the bladder activated by direct stimulation and the difficulty of creating stable and reliable electrical interfaces in contact with the bladder wall. The second approach is direct stimulation of the pelvic nerves. Although this would seem to be the most logical method to generate selective activation of the bladder, this approach has been hindered by the difficultly in accessing and interfacing with the pelvic nerve and the co-contraction of the urethral sphincter. The third location that has been attempted is direct stimulation of the spinal cord using penetrating electrodes. Pairs of electrodes implanted in the sacral spinal cord to stimulate the parasympathetic innervation of the bladder produced good results in 16 of 27 patients followed for as long as 10 years, but no further implants have been performed (Nashold *et al.*, 1982). Intraspinal microstimulation for control of bladder function is an active area of research and development (Carter *et al.*, 1995; Grill *et al.*, 1999). Electrical stimulation of a group of neurons around the central canal, previously identified to be active during reflex micturition (Grill *et al.*, 1998), produces micturition-like responses (Grill *et al.*, 1999). Another approach to activation of the spinal circuits for micturition is by stimulation of sensory nerve fibers (Grill, 2000). Electrical stimulation of urethral afferents in the pudendal nerve produces bladder contraction and relaxation of the urethral sphincter (Shefchyk and Buss, 1998; Grill *et al.*, 2001), whereas stimulation of other sensory components of the pudendal nerve produces bladder inhibition (Wheeler *et al.*, 1993; Grill *et al.*, 2001; Lee and Creasey, 2002).

The sacral spinal nerves are where electrical stimulation has produced the most widespread clinical success. The sacral roots contain the small-diameter parasympathetic

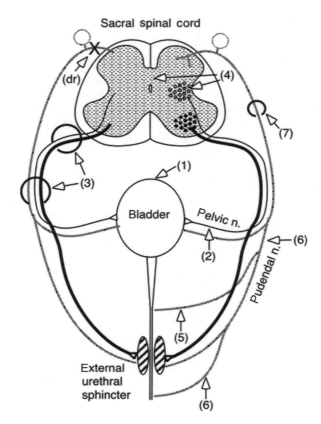

FIGURE 5.12. Sites of application of electrical stimulation for restoration of bladder function. Electrical stimulation has been applied directly to the bladder wall (1), to the pelvic nerve (2), the sacral spinal roots (3), within the intermediolateral and pericanicular regions of the sacral spinal cord (4), and to peripheral sensory nerves (5). Surgical transection of the sacral posterior roots (dorsal rhizotomy, dr) to abolish bladder hyperreflexia is usually conducted in parallel with stimulation of the sacral roots for bladder emptying. As an alternative, electrical stimulation for inhibition of bladder hyperreflexia has been applied to the sensory pudendal nerve (6) or the posterior sacral spinal roots (7).

fibers innervating the bladder via the pelvic nerve and the larger-diameter somatic fibers innervating the external urethral sphincter via the pudendal nerve (Figure 5.12). Because large-diameter fibers have lower excitation thresholds than small-diameter fibers, stimulation of the sacral root results in activation of the urethral sphincter at low amplitudes and coactivation of both bladder and the sphincter at high amplitudes.

Several methods have been developed to prevent coactivation of the sphincter with the bladder during sacral root stimulation, including surgical transection of the pudendal nerve, electrical block of pudendal nerve transmission, stimulation-induced fatigue of the sphincter, and intermittent stimulation. Intermittent stimulation has achieved substantial success in emptying the bladder by taking advantage of the difference in the speed of contraction and relaxation of the bladder and external urethral sphincter (Brindley *et al.*, 1982). The bladder consists of slowly contracting and relaxing smooth muscle, and the external urethral sphincter consists of rapidly contracting and relaxing striated muscle.

Intermittent bursts of stimulation (e.g., 3–6 s On, 9 s Off) generate sustained contraction of the bladder, but between bursts the sphincter relaxes and urine flows. Electrodes placed either intradurally on the ventral (motor) roots or extradurally on the combined (motor and sensory) roots to deliver intermitted stimulation combined with surgical interruption of the sacral dorsal (sensory) roots has restored bladder control in large numbers of individuals (Brindley, 1994). Selective stimulation of the small fibers innervating the bladder is another method to prevent coactivation of the bladder and external urethral sphincter (Rijkhoff *et al.*, 1997b). Activation of just the small fibers innervating the bladder can be accomplished by stimulating both large and small fibers and then arresting action potentials in large fibers (Fang and Mortimer, 1991a,b) or by increasing the threshold of large fibers above that of the small fibers using novel stimulus waveforms (Grill and Mortimer, 1997).

5.5. CONCLUSIONS

Electrical stimulation of nerves has been applied in many physiological systems to restore functions such as vision, hearing, gait, and hand grasp, with varying amount of success. The cochlear prosthesis has been particularly successful whereas several others such as the visual prosthesis have been very difficult to implement. Electrical stimulation can restore limited function of the hand, but few paralyzed patients have benefited from this technology. The gait prosthesis has not yet been successfully implemented as a practical method. Although scientists and engineers have a good understanding of the interaction between nerve and applied electric fields (Sections 5.1 and 5.2), as well as the physiology of the neural systems of interest (Section 5.3), the interface between the neural tissue and the electrode is a limiting factor that prevents rapid progress toward effective and practical neural prostheses. The size, biocompatibility, and mechanical difference between excitable biological neural tissue and the electrodes for stimulating and recording suggest that significant improvements are needed. It is hoped that some of this improvement can come from new technologies such as micromachining and nanotechnology. The success of the cochlear prosthesis has shown that neural prostheses can provide great benefits to patients and with the development of improved interfacing technology, other neural prostheses will meet with similar success.

ACKNOWLEDGMENTS

Preparation of this chapter was supported by NIH grants 2RO1 NS32856 and R21-NS-43450.

REFERENCES

Agnew, W. F., McCreery, D. B., Yuen, T. G., and Bullara, L. A., 1989, Histologic and physiologic evaluation of electrically stimulated peripheral nerve: Considerations for the selection of parameters, *Ann. Biomed. Eng.* **17**(1):39–60.

Agnew, W. F., McCreery, D. B., Yuen, T. G., and Bullara, L. A., 1990, Local anaesthetic block protects against electrically-induced damage in peripheral nerve, *J. Biomed. Eng.* **12**:301–309.

Altman, K. W., and Plonsey, R., 1988, Development of a model for point source electrical fibre bundle stimulation, *Med. Biol. Eng. Conmput.* **26:**466–475.

Andrews, B. J., Baxendale, R. H., Barnett, R., Phillips, G. F., Yamazaki, T., Paul, J. P., and Freeman, P. A., 1988, Hybrid FES orthosis incorporating closed loop control and sensory feedback, *J. Biomed. Eng.* **10:**189–195.

Bigland-Ritchie, B., Jones, D. A., and Woods, J. J., 1979, Excitation frequency and muscle fatigue: Electrical responses during human voluntary and stimulated contractions, *Exp. Neurol.* **64:**414–427.

Blaivas, J. G., 1982, The neurophysiology of micturition: A clinical study of 550 patients, *J. Urol.* **127:**958–963.

Brindley, G. S., 1994, The first 500 patients with sacral anterior root stimulator implants: General description, *Paraplegia* **32:**795–805.

Brindley, G. S., and Lewin, W. S., 1968, The sensations produced by electrical stimulation of the visual cortex, *J. Physiol.* **106:**479–493.

Brindley, G. S., Polkey, C. E., and Rushton, D. N., 1982, Sacral anterior root stimulators for bladder control in paraplegia, *Paraplegia* **20:**365–381.

Broderick, J., Brott, T., Kothari, R., Miller, R., Khoury, J., Pancioli, A., Gebel, J., Mills, D., Minneci, L., and Shukla, R., 1998, The Greater Cincinnati/Northern Kentucky Stroke Study: Preliminary first-ever and total incidence rates of strokes among blacks, *Stroke* **29:**415–412.

Brummer, S. B., and Turner, M. J., 1977, Electrochemical considerations for safe electrical stimulation of the nervous system with platinum electrodes, *IEEE Trans. Biomed. Eng.* **24:**59–63.

Burridge, J. H., Taylor, P. N., Hagan, S. A., Wood, D. E., and Swain, I. D., 1997, The effects of common peroneal stimulation on the effort and speed of walking: A randomized controlled trial with chronic hemiplegic patients, *Clin. Rehabil.* **11**(3):201–10.

Carter, R. R., McCreery, D. B., Woodford, B. J., Bullara, L. A., and Agnew, W. F., 1995, Micturition control by microstimulation of the sacral spinal cord of the cat: Acute studies, *IEEE Trans. Rehabil. Eng.* **3:**206–214.

Chai, T. C., and Steers, W. D., 1996, Neurophysiology of micturition and continence, *Urol. Clin. North Am.* **23:**221–236.

Clark, G. M., Tong, Y. C., and Patrick, J. F., 1990, *Cochlear Prostheses,* Churchill Linvingston, NY.

Crago, P. E., Kirsch, R. F., and Triolo, R. J., 2000, Movement synthesis and regulation in neuroprostheses, in: *Biomechanic and Neural Control of Movement* (J. M. Winters and P. E. Crago, eds.), pp. 573–589.

Crago, P. E., Memberg, W. D., Usey, M. K., Keith, M. W., Kirsch, R. F., Chapman, G. J., Katorgi, M. A., and Perreault, E. J., 1998, An elbow extension neuroprosthesis for individuals with tetraplegia, *IEEE Trans. Rehabil. Eng.* **6:**1–6.

Crago, P. E., Peckham, P. H., and Thrope, G. B., 1980, Modulation of muscle force by recruitment during intramuscular stimulation, *IEEE Trans. Biomed. Eng.* **27:**679–684.

Creasey, G. H., Grill, J. H., Korsten, M., U. HS, Betz, R., Anderson, R., and Walter, J., 2001, An implantable neuroprosthesis for restoring bladder and bowel control to patients with spinal cord injuries: A multicenter trial, *Arch. Phys. Med. Rehabil.* **82:**1512–1519.

Daly, J. J., Kollar, K., Debogorski, A. A., Strasshofer, B., Marsolais, E. B., Scheiner, A., Snyder, S., and Ruff, R. L., 2001, Performance of an intramuscular electrode during functional neuromuscular stimulation for gait training post stroke, *J. Rehabil. Res. Dev.* **38**(5):513–526.

Davis, J. A., Jr., Triolo, R. J., Uhlir, J. P., Bhadra, N., Lissy, D. A., Nandurkar, S., and Marsolais, E. B., 2001, Surgical technique for installing an eight-channel neuroprosthesis for standing, *Clin. Orthop.* (385):237–252.

Davis, R., MacFarland, W. C., and Emmons, S. E., 1994, Initial results of the nucleus FES-22-implanted system for limb movement in paraplegia, *Stereotact. Funct. Neurosurg.* **63:**192–197.

de Groat, W. C., Booth, A. M., and Yoshimura, N., 1993, Neurophysiology of micturition and its modification in animal models of human disease, in: *The Autonomic Nervous System*, Vol. 3 (C. A. Maggi, ed.), Harwood Academic Publishers, London, pp. 227–290.

Dostrovsky, J. O., and Lozano, A. M., 1992, Mechanisms of deep brain stimulation, *Mov. Disord.* **17**(Suppl 3):S63–S68. [Review]

Durand, D. M., 1995, Electrical stimulation of excitable tissue, *Handbook of Biomedical Engineering*, CRC Press, Boca Raton, pp. 229–251.

Durand, D., Ferguson, A. S. F., and Dalbasti, T., 1995, Effects of surface boundary on neuronal magnetic stimulation, *IEEE Trans. Biomed. Eng.* **37:**588–597.

Fang, Z. P., and Mortimer, J. T., 1991, Selective activation of small motor axons by quasitrapezoidal current pulses, *IEEE Trans. Biomed. Eng.* **38:**168–174.

Ferguson, K. A., Polando, G., Kobetic, R., Triolo, R. J., and Marsolais, E. B., 1999, Walking with a hybrid orthosis system, *Spinal Cord* **37**(11):800–804.

Gallien, P., Robineau, S., Nicolas, B., Le Bot, M. P., Brissot, R., and Verin, M., 1998, Vesicourethral dysfunction and urodynamic findings in multiple sclerosis: A study of 149 cases, *Arch. Phys. Med. Rehabil.* **79**:255–257.

George, M. S., Nahas, Z., Bohning, D. E., Kozel, F. A., Anderson, B., Chae, J. H., Lomarev, M., Denslow, S., Li, X., and Mu, C., 2002, Vagus nerve stimulation therapy: A research update, *Neurology* **59**(6 Suppl. 3):S56–S61.

Glenn, W. W., Hogen, J. F., Coke, J. S., Ciesieski, T. E., Phelps, M. L., Roweder, R., 1984, Ventilatory support by pacing the conditioned diaphragm in quadriplegia, *N. Engl. J. Med.* **3**(310):1550–1555.

Gorman, P. H., and Mortimer, J. T., 1983, The effect of stimulus parameters on the recruitment characteristics of direct nerve stimulation, *IEEE Trans. Biomed. Eng.* **30**:407–414.

Grandjean, P. A., and Mortimer, J. T., 1986, Recruitment properties of monopolar and bipolar epimysial electrodes, *Ann. Biomed. Eng.* **14**:53–66.

Graupe, D., and Kohn, K. H., 1997, Transcutaneous functional neuromuscular stimulation of certain traumatic complete thoracic paraplegics for independent short-distance ambulation, *Neurol. Res.* **19**:323–333.

Grill, J. H., and Peckham, P. H., 1998, Functional neuromuscular stimulation for combined control of elbow extension and hand grasp in C5 and C6 quadriplegics, *IEEE Trans. Rehabil. Eng.* **6**:190–199.

Grill, W. M., Bhadra, N., and Wang, B., 1999, Bladder and urethral pressures evoked by microstimulation of the sacral spinal cord in cats, *Brain Res.* **836**:19–30.

Grill, W. M., Craggs, M., Foreman, R., Ludlow, C., and Buller, J., 2001, Emerging clinical applications of electrical stimulation: Opportunities for restoration of function, *J. Rehabil. Res. Devel.* **38**(6):641–653.

Grill, W. M., and Kirsch, R. F., 1999, Neuroprosthetic applications of electrical stimulation, *IEEE Trans. Rehabil. Eng.* **7**(2):150–158.

Grill, W. M., and Kirsch, R. F., 2000, Neuroprosthetic applications of electrical stimulation, *Assist. Technol.* **12**:6–20.

Grill, W. M., and Mortimer, J. T., 1996a, Effect of stimulus pulse duration on selectivity of neural stimulation, *IEEE Trans. Biomed. Eng.* **43**:161–166.

Grill, W. M., and Mortimer, J. T., 1996b, Quantification of recruitment properties of multiple contact cuff electrodes, *IEEE Trans. Rehabil. Eng.* **4**:49–62.

Grill, W. M., and Mortimer, J. T., 1997, Inversion of the current distance relationship by transient depolarization, *IEEE Trans. Biomed. Eng.* **44**(1):1–9.

Hambrecht, F. T., 1979, Neural prostheses, *Annu. Rev. Biophys. Bioeng.* **8**:239–267.

Handa, Y., Handa, T., Ichie, M., Murakami, H., Hoshimiya, N., Ishikawa, S., and Ohkubo, K., 1992, Functional electrical stimulation (FES) systems for restoration of motor function of paralyzed muscles—versatile systems and a portable system, *Front. Med. Biol. Eng.* **4**:241–255.

Handa, Y., Yagi, R., and Hoshimiya, N., 1998, Application of functional electrical stimulation to the paralyzed extremities, *Neurol. Med. Chir. (Tokyo)* **38**(11):784–788.

Handa, Y., Handa, T., and Nakatsuchi, Y., 1985, A voice controlled functional electrical stimulation system for the paralyzed hand, *Jpn. J. Med. Electron. Biol. Eng.* **25**:292–298.

Hart, R. L., Kilgore, K. L., and Peckham, P. H., 1998, A comparison between control methods for implanted FES hand-grasp systems, *IEEE Trans. Rehabil. Eng.* **6**(2):208–218.

Harvey, C., Rothschild, R., Asmann, A., and Stripling, T., 1990, New estimates of traumatic SCI prevalence: A survey-based approach, *Paraplegia* **28**:537–544.

Hines, M., 1984, Efficient computation of branched nerve equations, *Int. J. Biol. Med. Comp.* **15**:69–76.

Hodgkin, A. L., and Huxley, A. F., 1954, The dual effect of membrane potential on sodium conductance in the giant axon of Loligo, *J. Physiol.* **116**:497–506.

Hoshimiya, N., Naito, A., Yajima, M., and Handa, Y., 1989, A multichannel FES system for the restoration of motor functions in high spinal cord injury patients: A respiration-controlled system for multijoint upper extremity, *IEEE Trans. Biomed. Eng.* **36**(7):754–760.

Jezernik, S., Craggs, M., Grill, W. M., Creasey, G. H., and Rijkhoff, N. J. M., 2002, Electrical stimulation for treatment of bladder dysfunction: Current status and future possibilities, *Neurol. Res.* **24**:413–430.

Johnson, M. W., and Peckham, P. H., 1990, Evaluation of shoulder movement as a command control source, *IEEE Trans. Biomed. Eng.* **37**(9):876–885.

Kameyama, J., Handa, Y., Hoshimiya, N., and Sakurai, M., 1999, Restoration of shoulder movement in quadriplegic and hemiplegic patients by functional electrical stimulation using percutaneous multiple electrodes, *Tohoku J. Exp. Med.* **187**(4):329–337.

Keith, M. W., Kilgore, K. L., Peckham, P. H., Wuolle, K. S., Creasey, G., and Lemay, M., 1996, Tendon transfers and functional electrical stimulation for restoration of hand function in spinal cord injury, *Hand. Surg. (Am.)* **21**:89–99.

Keith, M. W., and Lacey, S. H., 1991, Surgical rehabilitation of the tetraplegic upper extremity, *J. Neurol. Rehabil.*, 75–87.

Keith, M. W., Peckham, P. H., Thrope, G. B., Buckett, J. R., Stroh, K. C., and Menger, V., 1988, Functional neuro-muscular stimulation neuroprostheses for the tetraplegic hand, *Clin. Orthop.* **233**:25–33.

Keith, M. W., Peckham, P. H., Thrope, G. B., Stroh, K. C., Smith, B., Buckett, J. R., Kilgore, K. L., and Jatich, J. W., 1989, Implantable functional neuromuscular stimulation in the tetraplegic hand, *J. Hand Surg. (Am.)* **14**(3):524–530.

Kern, H., Hofer, C., Modlin, M., Forstner, C., Raschka-Hogler, D., Mayr, W., and Stohr, H., 2002, Denervated muscles in humans: Limitations and problems of currently used functional electrical stimulation training protocols, *Artif. Organs* **26**(3):216–218.

Kern, H., Hofer, C., Strohhofer, M., Mayr, W., Richter, W., and Stohr, H., 1999, Standing up with denervated muscles in humans using functional electrical stimulation, *Artif. Organs* **23**(5):447–452.

Kilgore, K. L., and Peckham, P. H., 1993a, Grasp synthesis for upper-extremity FNS. Part 2. Evaluation of the influence of electrode recruitment properties, *Med. Biol. Eng. Comput.* **31**(6):615–622.

Kilgore, K. L., and Peckham, P. H., 1993b, Grasp synthesis for upper-extremity FNS. Part 1. Automated method for synthesising the stimulus map, *Med. Biol. Eng. Comput.* **31**(6):607–614.

Kilgore, K. L., Peckham, P. H., and Keith, M. W., 1990, Electrode characterization for functional application to upper extremity FNS, *IEEE Trans. Biomed. Eng.* **37**:12–21.

Kilgore, K. L., Peckham, P. H., Keith, M. W., Thrope, G. B., Wuolle, K. S., Bryden, A. M., and Hart, R. L., 1997, An implanted upper-extremity neuroprosthesis. Follow-up of five patients, *J. Bone Joint Surg. Am.* **79**(4):533–541.

Kirsch, R. F., Acosta, A. M., Yu, D., and Keith, M. W., 1998, Feasibility of restoring shoulder and elbow function in high tetraplegia by functional neuromuscular stimulation, in: *20th Annual International Conference IEEE Engineering in Medicine and Biology Society*, October 1998.

Kljajic, M., Malezic, M., Acimovic, R., Vavken, E., Stanic, U., Pangrsic, B., and Rozman, J., 1992, Gait evaluation in hemiparetic patients using subcutaneous peroneal electrical stimulation, *Scand. J. Rehabil. Med.* **24**:121–126.

Kobetic, R., Triolo, R. J., and Marsolais, E. B., 1997, Muscle selection and walking performance of multichannel FES systems for ambulation in paraplegia, *IEEE Trans. Rehabil. Eng.* **5**:23–29.

Koch, C., and Segev, I., 1989, *Methods in Neural Modelling,* MIT Press.

Kralj, A., Acimovic, R., and Stanic, U., 1993a, Enhancement of hemiplegic patient rehabilitation by means of functional electrical stimulation, *Prosthet. Orthot. Int.* **17**:107–114.

Kralj, A., and Bajd, T., 1989, *Functional Electrical Stimulation: Standing and Walking After Spinal Cord Injury,* CRC Press, Boca Raton, FL.

Kralj, A. R., Bajd, T., Munih, M., and Turk, R., 1993b, FES gait restoration and balance control in spinal cord-injured patients, *Prog. Brain Res.* **97**:387–396.

Lauer, R. T., Kilgore, K. L., Peckham, P. H., Bhadra, N., and Keith, M. W., 1999, The function of the finger intrinsic muscles in response to electrical stimulation, *IEEE Trans. Rehabil. Eng.* **7**:19–26.

Lee, Y. H., and Creasey, G. H., 2002, Self-controlled dorsal penile nerve stimulation to inhibit bladder hyperreflexia in incomplete spinal cord injury: A case report, *Arch. Phys. Med. Rehabil.* **83**(2):273–277.

Lemay, M. A., and Crago, P. E., 1997, Closed-loop wrist stabilization in C4 and C5 tetraplegia, *IEEE Trans. Rehabil. Eng.* **5**:244–252.

Lemay, M. A., Crago, P. E., and Keith, M. W., 1996, Restoration of pronosupination control by FNS in tetraplegia—experimental and biomechanical evaluation of feasibility, *J. Biomech.* **29**:435–442.

Lilly, J. C., Hughes, J. R., Alvord, E. C., and Galkin, T. W., 1955, Brief noninjurious electric wave form for stimulation of the brain, *Science* **121**:468–469.

Margalit, E., Maia, M., Weiland, J., Greenberg, R., Fujii, G., Torres, G., Piyathaisere, D., O'Hearn, T., Liu, W., Lazzi, G., Dagnelie, G., Scribner, D., de Juan, E., and Humayun, M., 2002, Retinal prosthesis for the blind, *Urv Ophthalmol.* **47**(4):335.

Marsolais, E. B., and Kobetic, R., 1986, Implantation techniques and experience with percutaneous intramuscular electrodes in the lower extremities, *J. Rehabil. Res. Dev.* **23**:1–8.

Marsolais, E. B., and Kobetic, R., 1988, Development of a practical electrical stimulation system for restoring gait in the paralyzed patient, *Clin. Ortho. Rel. Res.* **233**:64–74.

Marsolais, E. B., Kobetic, R., Polando, G., Ferguson, K., Tashman, S., Gaudio, R., Nandurkar, S., and Lehneis, H. R., 2000, The Case Western Reserve University hybrid gait orthosis, *J. Spinal Cord Med.* **23**(2):100–108.

McCreery, D. B., Agnew, W. F., Yuen, T. G. H., and Bullara, L. A., 1988, Comparison of neural damage induced by electrical stimualtion with faradaic and capacitor electrodes, *Annu. Biomed. Eng.* **16**:463–481.

McCreery, D. B., Agnew, W. F., Yuen, T. G., and Bullara, L., 1990, Charge density and charge per phase as cofactors in neural injury induced by electrical stimulation, *IEEE Trans. Biomed. Eng.* **37**(10):996–1001.

McCreery, D. B, Agnew, W. F., Yuen, T. G., and Bullara, L. A., 1995, Relationship between stimulus amplitude, stimulus frequency and neural damage during electrical stimulation of sciatic nerve of cat, *Med. Biol. Eng. Comput.* **33**(3 Spec No):426–429.

McDonald, J. W., and Sadowsky, C., 2002, Spinal cord injury, *Lancet* **359**(9304):417–425.

McHardy, J., Geller, D., and Brummer, S. B., 1977, An approach to corrosion control during electrical stimulation, *Annu. Biomed. Eng.* **5**:144–149.

McIntyre, C. C., and Grill, W. M., 1999, Excitation of central nervous system neurons by non-uniform fields, *Biophys. J.* **76**:878–888.

McNeal, D., 1973, Peripheral nerve stimulation—superficial and implanted, in: *Neural Organization and its Relevance to Prosthetics* (W. S. Fields and L. A. Leavitt, eds.), Intercontinental Medical Book Corp, New York.

McNeal, D. R., 1976, Analysis of a model for excitation of myelinated nerve, *IEEE Trans. BME* **23**:329–337.

McNeal, D. R., Baker, L. L., and Symons, J. T., 1989, Recruitment data for nerve cuff electrodes: Implications for design of implantable stimulators, *IEEE Trans. Biomed. Eng.* **36**:301–308.

McNeal, D. R., and Bowman, B. R., 1985, Selective activation of muscles using peripheral nerve electrodes, *Med. Biol. Eng. Comp.* **23**:249–253.

Memberg, W., Peckham, P. H., Thrope, G., Keith, M., and Kicher, T., 1993, An analysis of the reliability of percutaneous intramuscular electrodes in upper extremity FNS applications, *IEEE Trans. Rehabil. Eng.* **1**(2):1–8.

Mortimer, J. T., 1990, Electrical excitation of nerve, in: *Neural Prostheses: Fundamental Studies* (W. F. Agnew and D. B. McCreery, eds.), Prentice-Hall, Englewood Cliffs, NJ, 1990, pp. 67–84.

Mortimer, J. T., and Peckham, P. H., 1973, Intramuscular electrical stimulation, in: *Neural Organization and Its Relevance to Prosthetics* (W. S. Fields and L. A. Leavitt, eds.), Intercontinental Medical Book Corp., New York, pp. 77–99.

Mushahwar, V. K., Collins, D. F., and Prochazka, A., 2000, Spinal cord microstimulation generates functional limb movements in chronically implanted cats, *Exp. Neurol.* **163**(2):422–429.

Nagarajan, S. S., Durand, D., and Warman, E. N., 1993, Effects of induced electric fields on finite neuronal structures: A simulation study, *IEEE Trans. Biomed. Eng.* **40**:1175–1188.

Nashold, B. S., Friedman, H., and Grimes, J., 1982, Electrical stimulation of the conus medullaris to control bladder emptying in paraplegia: A ten-year review, *Appl. Neurophysiol.* **45**:40–43.

Nathan, R. H., 1993, Control strategies in FNS systems for the upper extremities, *CRC Crit. Rev. Biomed. Eng.* **21**:485–568.

Nunez, P. L., 1981, *Electric Fields in the Brain, The Neurophysics of EEG*, Oxford University Press, New York.

Peckham, P. H., and Keith, M. W., 1992, Motor prostheses for restoration of upper extremity function, in: *Neural Prostheses: Replacing Motor Function After Disease or Disability* (R. B. Stein, P. H. Peckham, and D. B. Popovic, eds.), Oxford University Press, New York, pp. 162–187.

Peckham, P. H., Marsolais, E. B., and Mortimer, J. T., 1980, Restoration of key grip and release in the C6 tetraplegic patient through functional electrical stimulation, *J. Hand Surg. (Am)* **5**(5):462–469.

Plonsey, R., 1969, *Bioelectric Phenomena*, McGraw-Hill Series in Bioengineering.

Prochazka, A., Gauthier, M., Wieler, M., and Kenwell, Z., 1997, The bionic glove: An electrical stimulator garment that provides controlled grasp and hand opening in quadriplegia, *Arch. Phys. Med. Rehabil.* **78**:608–614.

Prochazka, A., Mushahwar, V., and Yakovenko, S., 2002, Activation and coordination of spinal motoneuron pools after spinal cord injury, *Prog. Brain Res.* **137**:109–124.

Rall, W., 1979, Core conductor theory and cable properties of neurons, in: *Handbook of Physiology—The Nervous System I*, Bethesda, MD, Chapt. 3, pp. 39–96.

Ranck, J. B., 1975, Which elements are excited in electrical stimulation of mammalian central nervous system: A review, *Brain Res.* **98**:417–440.

Rattay, F., 1989, Analysis of models for extracellular fiber stimulation, *IEEE Trans. Biomed. Eng.* **36**:676–681.

Rattay, F., 1990, *Electrical Nerve Stimulation, Theory, Experiments and Applications*, Springer-Verlag, Wien.

Rijkhoff, N. J., Wijkstra, H., van Kerrebroeck, P. E., and Debruyne, F. M., 1997a, Urinary bladder control by electrical stimulation: Review of electrical stimulation techniques in spinal cord injury, *Neurourol. Urodyn.* **16**:39–53.

Rijkhoff, N. J., Wijkstra, H., van Kerrebroeck, P. E., and Debruyne, F. M., 1997b, Selective detrusor activation by electrical sacral nerve root stimulation in spinal cord injury, *J. Urol.* **157**:1504–1508.

Robblee, L. S., and Rose, T. L., 1990, Electrochemical guidelines for selection of protocols and electrode materials for neural stimulation, in: *Neural Prostheses: Fundamental Studies* (W. F. Agnew and D. B. McCreery, eds.), Prentice-Hall, Englewood Cliffs, NJ, pp. 25–66.

Roth, B. J., and Basser, P. J., 1990, A model for stimulation of a nerve fiber by electromagnetic induction, *IEEE Trans. Biomed. Eng.* **37**:588–597.

Rushton, D. N., Donaldson, N. D., Barr, F. M., Harper, V. J., Perkins, T. A., Taylor, P. N., and Tromans, A. M., 1997, Lumbar root stimulation for restoring leg function: Results in paraplegia, *Artif. Organs* **21**:180–182.

Rutecki, P., 1990, Anatomical, physiological and theoretical basis for the antiepileptic effects of vagus nerve stimulation, *Epilepsia* **31**(Suppl. 2):S1–S6.

Sakakibara, R., Hattori, T., Yasuda, K, and Yamanishi, T., 1996, Micturitional disturbances and the pontine tegmental lesion: Urodynamic and MRI analyses of vascular cases, *J. Neurol. Sci.* **141**:105–110.

Scheiner, A., Mortimer, J. T., and Roessmann, U., 1990, Imbalanced biphasic electrical stimulation: Muscle tissue damage, *Annu. Biomed. Eng.* **18**(4):407–425.

Scheiner, A., Polando, G., and Marsolais, E. B., 1994, Design and clinical application of a double helix electrode for functional electrical stimulation, *IEEE Trans. Biomed. Eng.* **41**(5):425–431.

Schmit, B. D., and Mortimer, J. T., 1999, The effects of epimysial electrode location on phrenic nerve recruitment and the relation between tidal volume and interpulse interval.

Scott, T. R., Peckham, P. H., and Kilgore, K. L., 1996, Tri-state myoelectric control of bilateral upper extremity neuroprostheses for tetraplegic individuals, *IEEE Trans. Rehabil. Eng.* **4**:251–263.

Shefchyk, S. J., and Buss, R. R., 1998, Urethral pudendal afferent-evoked bladder and sphincter reflexes in decerebrate and acute spinal cats, *Neurosci. Lett.* **244**:137–140.

Smith, B., Peckham, P. H., Keith, M. W., and Roscoe, D. D., 1987, An externally powered, multichannel, implantable stimulator for versatile control of paralyzed muscle, *IEEE Trans. Biomed. Eng.* **34**:499–508.

Solomonow, M., Aguilar, E., Reisin, E., Baratta, R. V., Best, R., Coetzee, T., and D'Ambrosia, R., 1997, Reciprocating gait orthosis powered with electrical muscle stimulation (RGOII). Part I: Performance evaluation of 70 paraplegic patients, *Orthopedics* **20**:315–324.

Stover, S. L., and Fine, P. R., 1986, *Spinal Cord Injury: The Facts and Figures*, The University of Alabama at Birmingham, Birmingham.

Strojnik, P., Acimovic, R., Vavken, E., Simic, V., and Stanic, U., 1987, Treatment of drop foot using an implantable peroneal underknee stimulator, *Scand. J. Rehabil. Med.* **19**:37–43.

Taylor, P. N., Burridge, J. H., Dunkerley, A. L., Wood, D. E., Norton, J. A., Singleton, C., and Swain, I. D., 1999, Clinical use of the Odstock dropped foot stimulator: Its effect on the speed and effort of walking, *Arch. Phys. Med. Rehabil.* **80**(12):1577–1583.

Triolo, R. J., Bieri, C., Uhlir, J., Kobetic, R., Scheiner, A., and Marsolais, E. B., 1996a, Implanted functional neuromuscular stimulation systems for individuals with cervical spinal cord injuries: Clinical case reports, *Arch. Phys. Med. Rehabil.* **77**:1119–1128.

Triolo, R. J., Liu, M. Q., Kobetic, R., and Uhlir, J. P., 2001, Selectivity of intramuscular stimulating electrodes in the lower limbs, *J. Rehabil. Res. Dev.* **38**(5):533–544.

Triolo, R. J., Nathan, R., Handa, Y., Keith, M. W., Betz, R. R., Carroll, S., and Kantor, C., 1996b, Challenges to clinical deployment of upper limb neural prostheses, *J. Rehabil. Res. Dev.* **33**:11–122.

Uhlir, J. P., Triolo, R. J., and Davis, J. A., 2001, The effect of stimulated trunk extension on the upright body weight distribution while standing with functional neuromuscular stimulation, *J. Spinal Cord Med.* **24**:S7.

van den Honert, C. H., and Mortimer, J. T., 1979, The response of the myelinated nerve fiber to short duration biphasic stimulating currents, *Ann. Biomed. Eng.* **7**:117–125.

Warman, E. N., Grill, W. M., and Durand, D., 1992, Modeling the effects of electric fields on nerve fibers: Determination of excitation thresholds, *IEEE Trans. Biomed. Eng.* **39**(12):1244–1254.

Watanabe, T., Rivas, D. A., and Chancellor, M. B., 1996, Urodynamics of spinal cord injury, *Urol. Clin. North Am.* **23:**459–473.

Wheeler, J. S., Walter, J. S., and Cai, W., 1993, Electrical stimulation for urinary incontinence, *Crit. Rev. Phys. Rehabil. Med.* **5:**31–55.

Yu, D. T., Kirsch, R. F., Bryden, A. M., Memberg, W. D., and Acosta, A. M., 2001, A neuroprosthesis for high tetraplegia, *J. Spinal Cord Med.* **24:**109–113.

Yuen, T. G. H., Agnew, W. F., Bullara, L. A., and McCreery, D. B., 1990, Biocompatibility of electrodes and materials in the central nervous system, in: *Neural Prostheses: Fundamental Studies* (W. F. Agnew and D. B. McCreery, eds.), Prentice-Hall, Englewood Cliffs, NJ, pp. 197–224.

6

NEURAL SIGNAL PROCESSING

Donna L. Hudson[1]* and Maurice E. Cohen[1,2]

[1]University of California, San Francisco
[2]California State University, Fresno

6.1. OVERVIEW

The major thrust of this chapter is on neural signal processing in the central nervous system (CNS). In order to establish the framework for this discussion, it is instructive to look at the biological foundations, from single neurons to the peripheral nervous systems, because these are important building blocks and provide input and output signals for the complex neuronal structure of the CNS. Section 6.2 gives an overview of biological structures and historical discoveries. Section 6.3 examines signal processing in the single neuron and how it contributes to the complex network of signals. Examination of the function of the central and peripheral nervous systems depends to a large extent on time series analysis. Basic techniques are summarized in Section 6.4. Section 6.5 describes the peripheral nervous system as input and output media for the CNS. Section 6.6 describes methods for analyzing signals emanating from the CNS. Section 6.7 discusses the use of signal analysis in the diagnosis and treatment of neurological disease. Because of the complexity of signal analysis, its use in diagnosis of disease is dependent on higher-order decision models, which are described in Section 6.8. Finally, Section 6.9 describes current frontiers of signal analysis and prospects for the future.

6.2. BIOLOGICAL FOUNDATIONS AND HISTORY

6.2.1. BIOLOGICAL FOUNDATIONS

The neuron is a cell that has the features of an ordinary cell along with additional special properties. A cell is made up of a semipermeable membrane between 70 and 100 Å in thickness that surrounds it. In the interior, components include the nucleus, the mitochondria, and the Golgi bodies. The nucleus is composed of nuclear sap and a nucleoprotein-rich network from which chromosomes and nucleoli arise. A nucleolus contains DNA templates for RNA.

* Address for correspondence: University of California, San Francisco, 2615 E. Clinton Avenue, Fresno, California 93703; e-mail: hudson@ucsfresno.edu.

The mitochondria produce energy for the cell through cellular respiration. Golgi bodies are involved in the packaging of secretory proteins (Rogers and Kabrisky, 1991). A neuron is an extension of the simple cell with two types of appendages: multiple dendrites and a single axon, as shown in Figure 6.1. The dendrites receive input from other neurons, while the axon is an output channel to other neurons. Neurons have important basic characteristics. The cell membrane has an electrical resting potential of -70 mV, which is maintained by pumping positive ions out of the cell principally using the sodium (Na^+) pump. The main difference between a neuron and an ordinary cell is that the neuron is excitable. In response to inputs from the dendrites, the cell may become unable to maintain the -70-mV resting potential, resulting in the generation of an action potential producing a pulse that is transmitted down the axon. The action potential occurs when a certain threshold value has been exceeded, typically -50 mV. After releasing the pulse, the neuron returns to its resting potential. The action potential causes a release of biochemical agents known as neurotransmitters, the means by which messages are transmitted to the dendrites of nearby neurons. These neural transmitters may have either an excitatory or an inhibitory effect on neighboring neurons. A number of biochemical transmitters are known, including acetylcholine (usually excitatory), catecholamines, such as dopamine, norepinephrine, and epinephrine, and other amino acid derivatives such as histamine, serotonin, glycine, and γ-aminobutyric acid (GABA). GABA and glycine are two important inhibitory transmitters (Butter, 1968).

6.2.2. CENTRAL NERVOUS SYSTEM

Deciphering how individual neurons are organized into complex neuronal structures has been addressed over the years by many researchers (Hudson and Cohen, 1999). Santiago Ramón de Cajal was the first to discover the complex interconnection structure in the cerebral cortex (DeFelipe and Jones, 1988). Along with his associate Camillo Golgi (1886), they were able to produce photographs of the structures by applying dyes that were absorbed differently. They were jointly awarded the 1906 Nobel Prize in medicine for this work. Lorente de Nó was one of Cajal's students. He examined the types of neurons in the cerebral cortex in the 1930s, showing 32 to 34 different types based on shape classification, not on function (Asanuma and Wilson, 1979). In the 1940s, Hodgkin and Huxley (Hodgkin, 1964) began their well-known work on the giant squid. The squid was chosen because it has two very large neurons. They were awarded the Nobel Prize for their investigations into threshold, inhibition, and excitation in the giant squid axon. Hubel and Wiesel (1962) did extensive investigation into the cerebral cortex of the cat. They were able to map many complex structures and track the path from the optic nerve to the lateral geniculate body, and to the visual cortex. They distinguished between simple cells, complex cells, and hypercomplex cells in the visual cortex. Their work also emphasized the parallel nature of the visual-processing system.

6.3. ANALYSIS OF SIGNALS FROM SINGLE NEURONS

The work of Hodgkin and Huxley established the basic properties of single neurons, on which most neural network models have been based. These models have served a dual role, as models of the functioning of the nervous system and as the basis for computerized

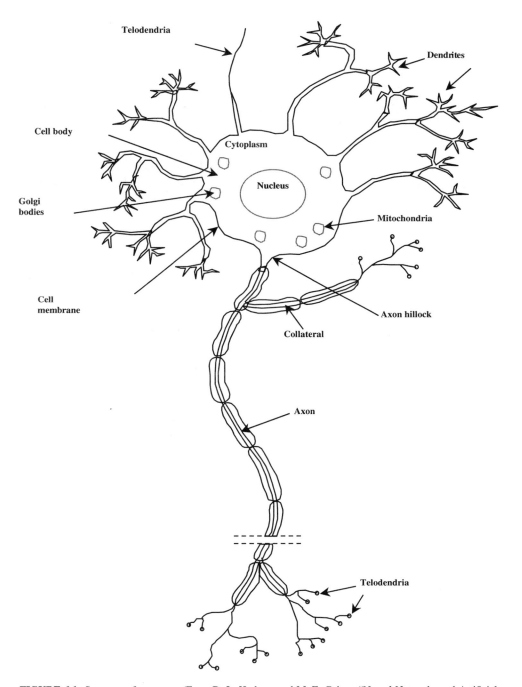

FIGURE 6.1. Structure of a neuron. (From D. L. Hudson, and M. E. Cohen, "Neural Networks and Artificial Intelligence for Biomedical Engineering," p. 15, © 1999, IEEE)

classification systems. This section describes basic work on modeling and analysis of a single neuron.

6.3.1. NEURON MODELS

Neural network research can be divided into two areas of investigation. The first is called the direct problem. The direct problem employs computer and engineering techniques to model the human brain (MacGregor, 1987; Aakerlund and Hemmingsen, 1998). This type of modeling is used extensively by cognitive scientists (Harley, 1998) and can be useful in a number of domains including neuropsychiatry (Rialle and Stip, 1994; Ruppin *et al.*, 1996) and neurophysiology (Saugstad, 1994). The direct problem approach has been useful in establishing models that mimic activities of the CNS. The second focus of neural network research is called the inverse problem. The inverse problem simulates biological structures with the objective of creating computer or engineering systems. The inverse problem is used extensively in building computer-assisted decision aids to be used in differential diagnosis, modeling of disease processes, and in building more complex biomedical models (Hudson and Cohen, 1999). These models are based on simulating the basic structure of the neuron, including the acceptance of multiple inputs and one output that results if a threshold is exceeded.

In the 1950s, Rosenblatt (1962) introduced models of the brain he called perceptrons. These represented artificial neurons based on the neuron models of McCulloch and Pitts. However, he made a departure form the McCulloch and Pitts model in that he based his model on probability theory rather than symbolic logic. The photoperceptron, defined by Rosenblatt, responded to optical patterns. It contained a sensory area, an association area, and a response area (Figure 6.2). The sensory area corresponds to the retinal structure. Each

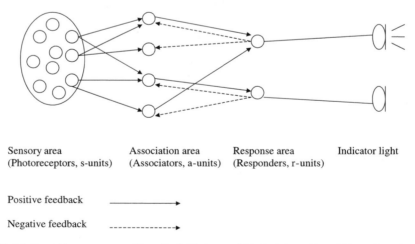

| Sensory area | Association area | Response area | Indicator light |
| (Photoreceptors, s-units) | (Associators, a-units) | (Responders, r-units) | |

Positive feedback ——————→

Negative feedback - - - - - - - - - →

FIGURE 6.2. Rosenblatt perceptron. (From D. L. Hudson and M. E. Cohen, "Neural Networks and Artificial Intelligence for Biomedical Engineering," p. 18, © 1999, IEEE)

point responds to light in an on/off manner. Input is then transmitted to the association area. The connections have three possible weights: 1 (excitatory), −1 (inhibitory), or 0. When a pattern is presented to the sensory area, a unit in the association area becomes active if its value exceeds a predetermined threshold θ. At time t, the output from the association area is defined as

$$y(t) = \text{sgn} \sum [x_i(t)w_i(t) - \theta] \qquad (6.1)$$

where sgn is either +1 (for positive argument) or −1 (for negative argument), $x_i(t)$ is the ith input signal, and $w_i(t)$ is the weight of the ith input to the node.

This digital model attempts to simulate the action of neurotransmitters through the use of weights, with positive weights representing excitatory transmitters and negative weights representing inhibitory influences. However, the action of the neuron is analog in nature, thus this representation is at best an approximation.

6.3.2. NEUROTRANSMITTERS

From a signal-analysis point of view, the action potential represents the electrical activity of the neuron. The action potential results in the release of neurotransmitters that then affect the electrical activity of adjacent cells. This electrical activity can be detected, but does not completely describe the action and interaction of the neurotransmitters. Many neurological diseases, as well as diseases that may be interpreted as psychological, are due to an improper balance of neurotransmitters. Signal analysis in itself is not sufficient to detect these problems.

6.3.3. ACTION POTENTIAL DETECTION

The analysis of extracellular signals requires the ability to detect action potentials, a problem that is complicated by the low signal-to-noise ratio. Statistical characteristics of background noise can be very similar and can result from potentials from neurons that are not coupled tightly to the electrode site. Nonlinear analyses based on wavelet analysis have been applied to improve threshold detection (Kim and Kim, 2003). This problem has been considered analogous to identification of the QRS complex in the electrocardiogram.

6.3.4. IMPLANTED ELECTRODES

Although most neural signal recordings are done using surface electrodes, in some cases, electrodes are implanted for deep brain stimulation to treat patients with diseases such as Parkinson's. These electrodes can then record electrical activity from human basal ganglia. In one study (Priori *et al.*, 2002) signals after voluntary movements were found to be in the high beta range in the subthalamic nucleus and human basal ganglia whereas the low beta range was found only in the human basal ganglia. Additionally, L-DOPA influenced power spectra changes. As such treatments become more common, additional opportunities may be presented for studying single-neuron output.

6.4. TIME SERIES ANALYSIS

Biological signals are a subset of time series. Basic techniques in time series are in general applicable to all type of biosignals. However, as we shall see in later sections, additional pre- and postprocessing are often necessary. This section summarizes basic methods.

6.4.1. PROPERTIES OF TIME SERIES

Time series data can be classified according to a number of basic properties. The basic division is between deterministic and random signals. In general, biomedical time series, including neural signals, fall into the deterministic category. In recent research, a subcategory of deterministic systems, denoted chaotic systems, has been identified. Chaotic systems result from complex interactions of multiple components. Most biomedical time series, including neural signals, fall into this category. Deterministic time series can also be classified as periodic, quasiperiodic, and transient nonperiodic. These parameters will determine to some degree the type of analysis that is appropriate.

Another property of time series is stationarity. This concept is important because some methods of analysis are applicable only to stationary time series. Let us define $x_k(t)$ as the kth sample function of the time series taken at point t and $p(x)$ as the probability density function associated with it. The mean value μ_x, and the expected value E are defined as

$$\mu_x = E(x_k(t)] = \int_{-\infty}^{\infty} xp(x)\,dx \tag{6.2}$$

For stationary processes, this value is independent of t.

6.4.2. CORRELATION AND COVARIANCE FUNCTIONS FOR STATIONARY PROCESSES

For two points in time, denoted t_1 and t_2, the covariance function for the two time series $x(t)$ and $y(t)$ is defined as

$$C_{xy}(t_1, t_2) = E[(x_k(t_1) - \mu_x(t_1))(y_k(t_2) - \mu_y(t_2))] \tag{6.3}$$

The covariance function is related to the correlation function R by

$$C_{xy}(t) = R_{xy}(t) - \mu_x\mu_y \tag{6.4}$$

For stationary processes, the results are a function of the difference t_1–t_2 rather than the actual values of t_1 and t_2.

6.4.3. CORRELATION AND COVARIANCE FUNCTIONS FOR NONSTATIONARY PROCESSES

For nonstationary processes, the results will vary depending on the actual values for t_1 and t_2. The correlation functions for nonstationary processes are defined in terms of expected values. The autocorrelation R_x is given by

$$R_x(t_1, t_2) = E[x(t_1)x(t_2)] \tag{6.5}$$

The cross correlation is

$$R_{xy}(t_1, t_2) = E[x(t_1)y(t_2)] \tag{6.6}$$

The covariance functions are defined by

$$C_x(t_1, t_2) = R_x(t_1, t_2) - \mu_x(t_1)\mu_x(t_2) \tag{6.7}$$
$$C_{xy}(t_1, t_2) = R_{xy}(t_1, t_2) - \mu_x(t_1)\mu_y(t_2) \tag{6.8}$$

For more details refer to Bendat and Piersol (1971).

6.4.4. FOURIER ANALYSIS

Basic approaches to signal analysis have relied on Fourier analysis, cross-correlation, autocorrelation, and other techniques to determine if the signal is stationary. The traditional approach to EEG analysis, Fourier analysis provides a quantitative tool to examine signal frequencies and their relative loads. An EEG falls into the category of transient nonperiodic data. An important characteristic of transient data is that a discrete spectral representation is not possible. Instead, a continuous spectral representation can be obtained using the Fourier integral:

$$X(f) = \int_{-\infty}^{\infty} x(t)e^{-j2ft}\, dt \tag{6.9}$$

In practice, this infinite interval is restricted to finite time

$$X(f, T) = \int_{0}^{T} x(t)e^{-j2ft}\, dt \tag{6.10}$$

For computational purposes a discrete version is constructed by assuming the $x(t)$ is sampled at N equally spaced points at a distance h apart. Then

$$x_n = x(nh) \quad n = 0, 1, 2, \ldots, N-1 \tag{6.11}$$

giving the following discrete version

$$X(f, T) = h \sum_{n=0}^{N-1} x_n \exp[-j2\pi f nh] \tag{6.12}$$

This expression can be simplified by including h with $X(f_k, T)$. Results are unique up to $k = N/2$, the Nyquist cutoff frequency. The final simplified computational form is

$$X(k) = \sum_{N=0}^{N-1} x(n)W(kn) \quad k = 0, 1, 2, \ldots, N-1 \tag{6.13}$$

where $W(u) = \exp[-j2\pi u/N]$, $X(k) = X_k$ and $x(n) = x_n$.

Figure 6.3 shows a portion of an EEG in the time domain. Figure 6.4 shows the corresponding spectrum in the frequency domain.

6.4.5. POWER SPECTRAL DENSITY FUNCTIONS

A spectral density function $S(f)$ shows the general frequency distribution of the data and can be defined in terms of Fourier transforms

$$S(f) = \int_{-\infty}^{\infty} R(\tau) \exp(-j2\pi f \tau) \, d\tau \tag{6.14}$$

6.4.6. WAVELET ANALYSIS

The wavelet acts as a mathematical microscope in which different parts of the signal can be observed by adjusting the focus. The wavelet is implemented in terms of a wavelet function convolved with a high-pass filter to produce the detailed signal that is subsequently convolved with a low-pass filter associated with the scaling function. This process is iterated until the desired wavelet scale is achieved. The wavelet approach has been used successfully in segmentation of EEGs (Inouye *et al.*, 1995).

The wavelet is defined as (Daubechies, 1988)

$$(W \cdot f)(a, b) = |a|^{-0.5} \int_{-\infty}^{\infty} f(t) \Psi((t - b)/a) \, dt \tag{6.15}$$

the inner product of f, a windowing function with translated and dilated versions of Ψ, the wavelet function. $(W \cdot f)$ is the result of applying a high pass filter to f. For practical purposes, a finite sum is needed so the algorithm can be implemented using finite impulse response (FIR) filters. A scaling function ϕ is associated with the wavelet Ψ that is considered as the impulse response to a filter analogous to W that is low-pass rather than high-pass. Wavelet analysis for an EEG is shown in Figure 6.5. Akay and Daubenspeck (1999) have used wavelets in the study of respiratory evoked potentials.

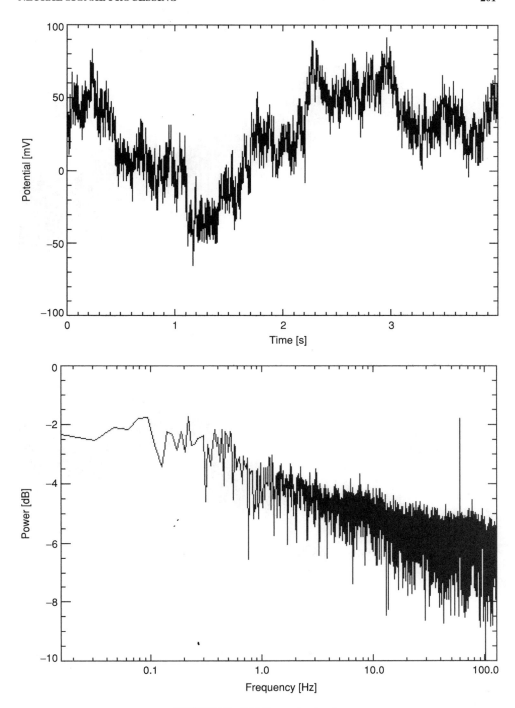

FIGURE 6.3. EEG time series signal.

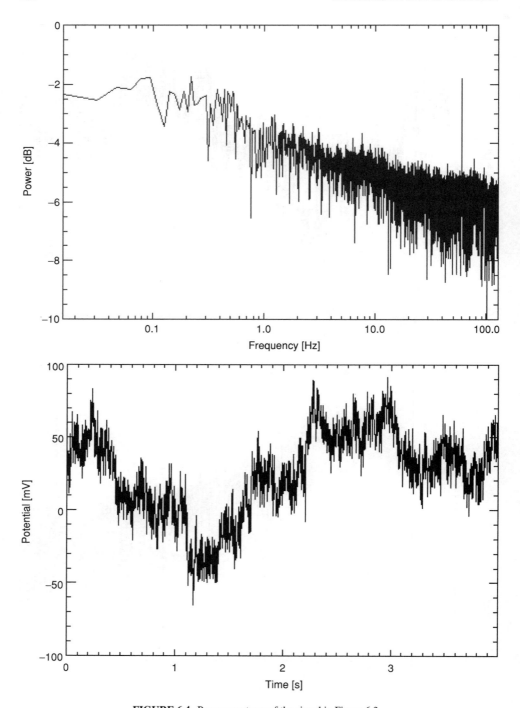

FIGURE 6.4. Power spectrum of the signal in Figure 6.3.

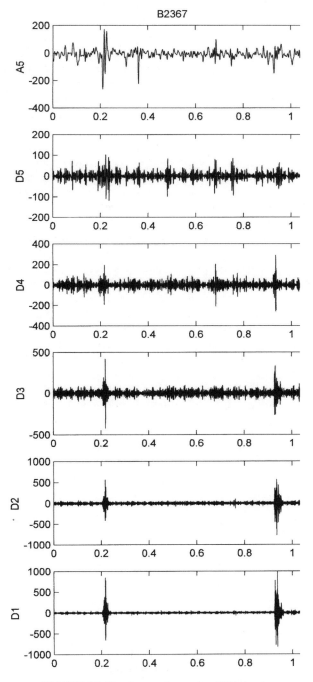

FIGURE 6.5. Wavelet transforms of an EEG signal.

6.4.7. CHAOTIC ANALYSIS

As a new area of research that has developed in the last 20 years, chaos theory has been shown to be especially promising in addressing problems in nonlinear dynamics. Most biomedical time series represent nonlinear dynamical processes. Chaos theory has been applied extensively in cardiology (Chialvo and Jalife, 1987; Goldberger, 1989) and to a lesser extent in neurology (Freeman, 1987). Chaotic analysis provides a new way of looking at nonlinear time series data that in general results in systems with intractable mathematical solutions.

From the point of view of decision-making systems, the contribution of chaos theory is a measure of either the presence or absence of chaos in a system or the degree to which chaos is present. There are two approaches to chaotic analysis: graphical and numerical. Graphical techniques include strange attractors, Poincaré plots, and second-order difference plots. Numerical techniques include the fractal dimension, the Lyapunov exponent, and central tendency measure (Cohen and Hudson, 1999).

An example of a chaotic method that uses both graphical and numerical representations based on a soft solution of the logistic equation has been successfully applied to the analysis of ECGs (Cohen et al., 1998), and more recently has been applied to EEG analysis (Cohen and Hudson, 2000b). The method is based on a second-order difference plot that is defined for T_n, the nth point in a time series. The plot consists of $T_{n+2}-T_{n+1}$ versus $T_{n+1}-T_n$. Figure 6.6 shows examples of theoretical second-order difference plots generated from the continuous solution of the logistic equation (Cohen et al., 1994). The more closely the points are clustered around the origin the less variability, or chaos, is present in the system. Figure 6.7 shows an experimental second-order difference plot for the EEG for delta and beta frequencies. Although this graphical measure is a good visual indicator of the degree of variability of time of the biomedical signal, some quantification is necessary in order to include this information in a higher-order decision model such as a neural network or

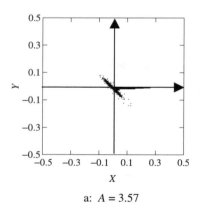
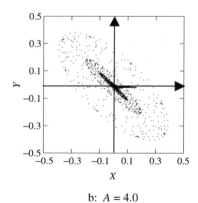

a: $A = 3.57$ b: $A = 4.0$

$$T_{n+2} - T_{n+1} \text{ versus } T_{n+1} - T_n$$
Values for A in logistic equation

FIGURE 6.6. Theoretical second-order difference plot. (From D. L. Hudson and M. E. Cohen, "Neural Networks and Artificial Intelligence for Biomedical Engineering," p. 282, © 1999, IEEE)

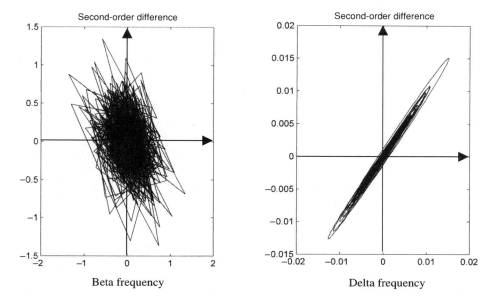

FIGURE 6.7. Second-order difference plots for various EEG frequencies.

a knowledge-based system. The central tendency measure (CTM) was developed for this purpose. The central tendency measure is computed by

$$\text{CTM} = \left[\sum_{i=1}^{t-2} \delta(d_i)] / (t-2) \right] \text{ where} \tag{6.16}$$

$$\delta(d_i) = \begin{array}{l} 1 \text{ if } [(T_{i+2} - T_{i+1})^2 + (T_{i+1} - T_i)2]^{1/2} < r \\ 0 \text{ otherwise} \end{array} \tag{6.17}$$

where r is a specified radius around the origin. This measure has produced excellent results in applications in cardiology (Cohen *et al.*, 1998) and is currently being tested in EEG analysis for the purpose of differentiating types of dementia.

6.4.8. LINEAR VERSUS NONLINEAR ANALYSIS

In general, analysis of biomedical time series relies on methods of nonlinear analysis. Although there is some controversy regarding nonlinear analysis for EEGs, new studies have shown that the complexity of the system requires the use of nonlinear methods, particularly in diseases with complex patterns. For a discussion of higher-level linear and nonlinear analysis, refer to Garrett *et al.* (2003).

6.4.9. BIOMEDICAL SIGNALS

Biomedical signals in general include any time series related to biological functioning, such as electrocardiograms (ECG), electroencephalograms (EEG), respiration patterns, and hemodynamic studies. These time series contain important diagnostic information and are used routinely in cardiac diagnosis and monitoring, neurology diagnosis, and ICU tracking. Specific aspects of the series that are useful depend on the application. For example, in ECG analysis, the repeated pattern associated with each heartbeat (QRS complex) is of major diagnostic value (Hudson *et al.*, 1998). On the other hand, EEGs lack specific repeated patterns, although the occurrence of certain repeated patterns, such as alpha, beta, and theta waves, are of known clinical significance (Leuchter *et al.*, 1993). In other cases, such as hemodynamic and respiration studies, it may be the change in the patterns over time that is of significance (Cohen *et al.*, 1992).

Analysis of these signals present many problems, including nonstationarity of signals, nonlinearity, noise, and very large data sets (the number of points in a biological time series often exceeds 100,000.) Basic approaches to signal analysis have relied on Fourier analysis, cross-correlation, autocorrelation, and other techniques to determine if the signal is stationary. Although these approaches have proved useful in many areas, analysis of some medical time series, such as EEGs, are still problematic. These problems are described in detail in Section 6.6. The traditional approach to EEG analysis, Fourier analysis, provides a quantitative tool to examine signal frequencies and their relative loads. It is almost certain that the conventional Fourier analysis cannot represent the entire spectrum of biological activities. In addition, some of the assumptions such as the stationarity of the signal are not valid. Signal averaging and analysis based on short intervals ranging from one to four sections are inadequate. These problems may bias the analysis (Blanco *et al.*, 1997).

Recent new theoretical developments have augmented traditional approaches to provide insight into the behavior of the time series. Two aspects of the analysis are of vital importance: the occurrence of patterns and the variability over time. Wavelet analysis is useful in detecting patterns on various scales (Akay, 1995), and chaos theory has given impressive results in quantifying the variability of patterns (Eberhart, 1989), particularly in cardiology (Cohen and Hudson, 2000a).

6.5. PERIPHERAL NEURAL SIGNALS

Although most neural signal analyses focus on the EEG, neural signals in the peripheral system are particularly important in the design of devices to return function to patients who have been injured or suffer from debilitating disease. In most cases, loss of function is due to loss of neural signals. The function of muscles is dependent on receipt of signals. Myoelectric activity always precedes muscle contractions. Electromyograms (EMG) measure this activity. Researchers attempting to restore muscle activity after an injury in which the signal transmission has been disrupted analyze signals from electromyograms in an attempt to use the information to stimulate muscles. Several methods are used including viewing EMG activity as an on–off system in which the presence activates a single-function prosthesis (Evans *et al.*, 1984), using a multistate control system by defining adaptive signal boundaries (Berube *et al.*, 1984), as well as considering systems with multiple electrodes to include activities in more than one muscle (Philipson, 1985). EMG processing has a number of applications, including functional neuromuscular stimulation (FNS) as described above. EMG signal processing (Hefftner *et al.*, 1988) uses pattern recognition and classification that assume the time signature of the signal is obtained from a single pair of surface electrodes. The temporal signature is dependent on which motor units in the vicinity of the electrode have been recruited. Analysis of the EMG signal uses standard time-series analysis methods such as autoregressive, moving-average, and mixed models.

Other applications use the EMG as a decision tool, often in conjunction with higher-order reasoning models, as illustrated in an example for staging sleep levels (Principe *et al.*, 1989). Often the EMG is combined with EEG results. Reasoning methodologies usually focus on classification and include neural networks, knowledge-based systems, fuzzy systems, Dempster–Shafer evidence theory, and hybrid systems. These methods are discussed in Section 6.8.

6.6. SIGNAL PROCESSING IN THE CNS

The major avenue for signal analysis in the CNS is the electroencephalogram (EEG), although in some cases, mainly during surgical procedures, implanted electrodes are used to collect more specific information. New methods in micro- and nanotechnology are increasing possibilities for individual signal analysis and are discussed in Section 6.9.

Conventional EEG evaluation methodologies are useful but limited and potentially problematic from both theoretical and practical standpoints. EEG signals are considered as the results of the combined dynamic activity of neuronal populations. Models including excitatory and inhibitory circuits with feedback loops have been adopted to explain the oscillation property of EEG activity (Lopes da Silva, 1993). The clinical correlations of the dominant signal frequencies and the visual detection of paroxysmal events such as spikes or sharp waves have been the mainstay of clinical neurological interpretation of EEG recording. We are far from understanding the exact mechanism of the generation of EEG signals. However, these limitations are partly due to the lack of appropriate theoretical models and appropriate measurements to adequately describe and dissect the EEG signals. The more comprehensive linear and nonlinear analyses of EEG signals not only have practical utility as outlined above but can also open new windows for studying the significance of the EEG signal in the understanding of the basic neurophysiological functioning of the human cerebral cortex.

6.6.1. EEG ANALYSIS

Analog EEG machines that plot the electrical signal directly have been replaced in most cases by digital EEG machines that provide the obvious advantage of facilitating analysis that was previously done by visual inspection only. The EEG is typically sampled at a rate of 200 samples per second and uses 18 or more electrodes to record signals from various lobes of the cerebral cortex. The major lobes include the frontal, temporal, parietal, and visual. Figure 6.8 shows typical electrode placements.

EEG recordings are done using 18 or more leads that are placed symmetrically on the scalp. A recording of 10 min produces approximately 75,000 points for each lead. The signal consists of spikes that are categorized according to frequency (f) in the following

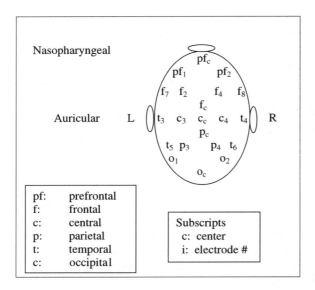

FIGURE 6.8. EEG lead placement.

groups that have clinical significance:

> Alpha (α): $8 \leq f \leq 13$ Hz: This is the principal resting rhythm of the brain that is common in awake, resting adults, especially in the occipital lobes. Alpha waves are suppressed when the eyes are open and visual stimulation is present.
> Beta (β): $f > 13$ Hz: High frequency beta waves appear as background activity in anxious subjects.
> Delta (δ): $0.5 \leq f \leq 4$ Hz: These appear at deep-sleep stages.
> Theta (θ): $4 \leq f \leq 8$ Hz: These appear at the beginning stages of sleep.

Depression or absence of expected rhythms in a specific state may indicate abnormality. Focal brain injuries produce abnormal slow waves in the affected regions. Bilateral asynchrony may indicate cortical pathway disturbance. Synchronized spikes and sharp waves are indicative of epileptic seizures. In alert subjects, the frequency of waves increases while the amplitude decreases.

It is almost certain that the conventional Fourier analysis cannot represent the entire spectrum of biological activities. In addition, some of the assumptions such as the stationarity of the signal are not valid. Signal averaging and analysis based on short intervals ranging from one to four sections are inadequate. These problems may bias the analysis (Blanco *et al.*, 1997). Traditional EEG analysis uses spectral representation of the data, as shown in Figure 6.4.

Clinical utility of the EEG is also limited by the frequent lack of specificity of the EEG abnormality. Generalized slowing during an EEG tracing unrelated to drowsiness can be an indication of generalized cerebral dysfunction due to metabolic derangement, neurodegenerative disorders, or infectious/inflammatory diseases.

The EEG is a difficult signal and in general requires preprocessing. The signal has a low signal-to-noise ratio, and poses a challenge in determining events. It does not have a periodic pattern such as the electrocardiogram, which produces a fixed shape for each heartbeat. The signal is also typically measured on the scalp rather than at the cortical level. A number of techniques are available for focusing on signals of various frequencies, the most common of which is wavelet analysis.

Although the history of EEG analysis appears to paint a bleak picture, new methods have been developed that have potential for the refinement of the EEG in three areas: preprocessing, analysis, and higher-order processing. New preprocessing techniques focus on methods for estimation of cortical signal potential that have the possibility of measuring the true potential on the cortical surface rather than the surface of the scalp and can also be used to localize the signal to a specific part of the brain. Chaotic methods that had previously been used for the analysis of 24-h ECG recordings have been applied to EEG data to give an overall pattern of the long-term behavior of the signal. Higher-order processing of the chaotic analysis is done using the overall decision model described in the later.

6.6.2. PREPROCESSING

Depending on the nature of the signal, some preprocessing may need to be done before starting the analysis. The preprocessor consists of four components.

Noise removal: Noise levels are usually removed using a thresholding approach. This process is particularly important and often difficult for signals such as the EEG, which have a low signal-to-noise ratio.

Event identification: For signals such as the ECG, the recurring QRS complex provides a fixed pattern for analysis, and thus no event identification is necessary. On the other hand, location of events in EEGs is necessary for some types of detailed analysis.

Windowing: Windowing is used to damp the Gibbs effect that results from truncation of an infinite series. A number of windowing functions can be used for this purpose.

Wavelet analysis: Wavelet analysis permits selection of desired frequencies in signals. The wavelet acts as a mathematical microscope in which different parts of the signal can be observed by adjusting the focus.

6.6.3. SIGNAL ANALYSIS

6.6.3.1. Cortical Potential

The electroencephalogram (EEG) provides neuroscientists with discrete temporal and spatial maps of the scalp surface potential. Although unsurpassed in temporal resolution, the scalp EEG suffers from limited spatial resolution owing to the large variation in conductivity between the cerebrospinal fluid (CSF) and skull. Neuroscientists have long sought methods for improving the spatial resolution of scalp EEG and for the noninvasive determination of the cortical surface potential. These techniques can be divided into two categories: techniques based on the surface Laplacian (SL) (Hjorth, 1975; Law *et al.*, 1993; Nunez, *et al.*, 1994; He *et al.*, 2001) and techniques based on cortical potential estimation (Kearfott *et al.*, 1991; Edlinger, 1998; He *et al.*, 2002).

The SL is a mathematical operator that can be applied to any sufficiently smooth function over any smooth, two-dimensional surface. Hjorth (1975) has shown that the divergence of the surface parallel component of the scalp current density is related to the SL of the scalp potential. The surface parallel divergence of the surface parallel scalp current density differs from zero in general; this reflects current injection to or from the underlying skull. The surface Laplacian of the scalp surface potential also provides a noninvasive, qualitative estimation of the cortical surface potential. On the basis of physical approximations and computer simulations, Nunez *et al.* (1994) and Law *et al.* (1993) have argued that the SL of the scalp surface potential is proportional to the cortical surface potential. Calculation of cortical surface potential will help to localize the source of the EEG in the spatial domain.

Recent work has focused on developing realistic geometry of the head including a Laplacian estimation technique (He *et al.*, 2001) and boundary element method based cortical potential techniques (He *et al.*, 2002). For more details refer to chapter 7 by He and Lian.

Summary Methods

As the EEG presents an enormous amount of data, summary measures are required, particularly if the EEG results are to be combined with other clinical parameters. The standard summary method has been the Fourier transform. More recent methods include chaotic modeling, which can provide an indication of the activity over time. These methods have been described above under Signal Analysis.

6.6.3.2. Evoked Potentials

An evoked potential is a specific method of EEG analysis in which a specific stimulus is presented to see if a standard response can be obtained from the EEG. The stimulus may be visual (e.g. displaying a specific color) or motor (e.g. the subject is asked to perform a certain movement). The method has been used both in attempts to achieve more specificity in the EEG analysis and in attempts to differentiate responses in subjects with different diseases. The method is used extensively in diagnostic EEG evaluation.

6.6.4. HIGHER-ORDER MODELING

The main objective of EEG analysis is to provide input for diagnosis of disease. In general, the EEG alone is not sufficient to clearly confirm the presence of a particular disease, except in specific instances such as epilepsy. However, if EEG results are used in conjunction with other clinical parameters, they can contribute significantly to disease classification. Classification systems depend on higher-order reasoning models that are discussed in Section 6.8.

6.7. NEURAL SIGNAL ANALYSIS AND DISEASE

6.7.1. EPILEPSY

The EEG has been most successful in confirming epilepsy, in which generalized synchronization occurs in multiple channels. The EEG plays a major role in evaluating epilepsy through the detection of interictal activity. The sleep EEG is particularly useful. Activation procedures are usually required and include hyperventilation or photic stimulation with separate trains of photoflashes of 10-s duration for each frequency, with frequencies ranging from 1 to 60 Hz (Flint *et al.*, 2002).

6.7.2. PARKINSON'S

There is growing evidence that diseases such as Parkinson's and Huntington's that are manifested in the form of movement disorders are linked to dysfunction of the basal ganglia. Seiss *et al.* (2003) used evoked potentials in an attempt to establish this connection. Both EEGs and EMGs were used for the study. The results showed that proprioception-related potentials reflect the bilateral activity of the postcentral sensory cortex, with an altered scalp distribution in Parkinson's disease and an ipsilateral hemisphere reduction in Huntington's disease.

6.7.3. HUNTINGTON'S

Huntington's disease is hereditary, with each offspring of an infected individual having a 50% chance of having the affliction, for which symptoms usually appear in the fourth decade of life. The disease causes progressive deterioration in functional capacity. Accurate and early assessment can lead to treatment before the disease manifests itself with severe motor symptoms. Typical methods include MRI images and cognitive testing. Neurophysiological abnormalities point to underlying functional changes. The EEG has been shown to exhibit amplitude reduction in Huntington patients. De Tomasso *et al.* (2003) used a neural network classifier with a 16-dimensional vector (output from 16 electrodes) to classify the level of functioning of Huntington subjects to evaluate the alpha, beta, theta, and delta ranges. Alpha was shown to be the most discriminating rhythm.

6.7.4. ALZHEIMER'S

Some success has been achieved in using quantitative EEG signals for the purpose of early diagnosis of dementia. Schreiter-Gasser *et al.* (1993) noted that the theta band was the best to differentiate early-onset Alzheimer's disease (AD) from cognitively normal controls. In contrast, the delta band is by far the best indicator for the degree of dementia (Schreiter-Gasser *et al.*, 1994). Most approaches rely on traditional methods such as Fourier transforms (Signorino *et al.*, 1995) and quantitative mapping of EEGs (Miyauchi *et al.*, 1994; Prichep *et al.*, 1994). New techniques such as wavelets (Kalayci and Ozdamar, 1995) have been applied to EEG analysis. A few researchers have used chaotic analysis in EEG analysis in conjunction with AD. Woyshville and Calabrese (1994) used fractal dimension for quantification, Pritchard *et al.* (1994) used a neural network model in nonlinear analysis, and Cohen and Hudson (2000b) used the CTM measure to quantify the level of variability. Several studies have found EEG analysis to be useful in the diagnosis of AD (Ihl *et al.*, 1992; Soininen and Riekkinen, 1992; Wszolek *et al.*, 1992). New methods have been developed that have the potential both for the refinement of the EEG signal and for differentiation between normal individuals, stroke patients, and AD patients (Cohen and Hudson, 2003a).

6.7.5. DIFFERENTIATION OF TYPES OF DEMENTIA

With improvement in population life expectancy, dementia becomes an ever more significant morbidity factor in geriatric medicine. A challenge in clinical medicine is the ability to diagnose specific type(s) of dementia and to find the most effective treatment modalities. Studies have shown that new methods for EEG analysis can contribute to differentiation of types of dementia, using chaos theory (Cohen and Hudson, 2001; Porcher and Thomas, 2001; Sarbadhikari and Chakrabarty, 2001), neural modeling (Petrosian *et al.*, 2001; Hudson and Cohen, 2002), and signal refinement techniques (Kim *et al.*, 2001) to study parameters that can help in early diagnosis of dementia related to Alzheimer's disease (Visser *et al.*, 2001; Storey *et al.*, 2002), Parkinson's disease, Lewy body disease (Suzuki *et al.*, 2002; Tiraboschi *et al.*, 2002), Lewy body variant of Alzheimer's disease (Santa Cruz *et al.*, 2002), vascular dementia (Varma *et al.*, 2002), and mixed types of dementia (Jellinger *et al.*, 2002; Kenny *et al.*, 2002).

A case study for dementia diagnosis using EEG data in a higher-order decision model is given in the next section.

6.8. HIGHER-ORDER DECISION MODELS

Signal analysis data alone can contribute important information for diagnosis and tracking of disease. Although electrocardiogram (ECG) results have made major contributions to cardiac diagnosis, the contribution is strengthened when combined with clinical parameters (Cohen *et al.*, 1998). The EEG has not been shown to be as useful in neurological diagnosis. In the case of the EEG it is even more important to include results in a comprehensive higher-order reasoning paradigm. Several possible methodologies exist, including knowledge-based approaches, data-based approaches, and hybrid systems; data-based approaches include neural networks, fuzzy systems, Bayesian methods, and Dempster–Shafer theory of evidence. Hybrid systems have been developed in order to take advantage of all information and multiple methodologies to produce a robust and comprehensive model. Variables from many sources and many data types can be used as input to the system without requirements of independence of variables. For many applications, signal analysis results are an important component. We have already seen one of these techniques—neural network modeling—that can be used to combine multiple signal outputs, such as the combination of EEG channels into a vector for neural network classification and the chaotic model in which various measures of variability were combined. In the latter application, an expanded neural network model also included other clinical parameters to reach a final conclusion. A case study illustrates a hybrid system that uses intelligent agents as a basis for dementia diagnosis.

6.8.1. CASE STUDY: DIAGNOSIS OF DEMENTIA

6.8.1.1. Intelligent Agents in Diagnosis

Intelligent agents had their origins in distributed artificial intelligence and have been used successfully in a number of business applications (Hofmann and Bodendorf, 2000). Each agent is an independent methodology with reasoning capabilities working on a prescribed task. The goal of the overall system is to provide a cooperative environment in which two or more agents can be combined to solve a problem through the use of a mediator or facilitator (Swigger and Duckworth, 2000) that provides a common means of communication. The intelligent agent approach is a natural extension of hybrid systems for combining various methodologies without altering the independent agents or algorithms. The use of intelligent agents in biomedical systems has been limited, with the major focus on health

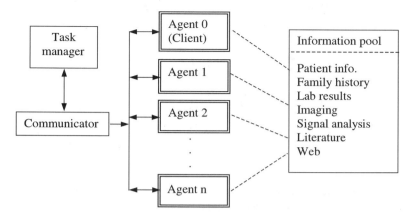

FIGURE 6.9. Intelligent agent system structure.

care delivery (Hsu and Goldberg, 1999; Mack *et al.*, 2000). Some complex medical decision making problems have been addressed (Silverman, 1998; Lanzola *et al.*, 1999). The goal of the intelligent agent system is to arrive at a decision. In the specific application described here, the objective is diagnosis of dementia.

Components

Figure 6.9 shows the components of the intelligent agent system. The agents are defined as follows:

> *Agent 0: Medical Professional (MP).* The human decision maker functions as agent 0 and interacts with the communicator using natural language. In the current design, agent 0 can also perform as one of the other agents, as discussed below.
>
> *Agent 1: Knowledge-Based System (Emerge).* A system previously developed by the authors, EMERGE, is used as the knowledge-based component (Hudson and Cohen, 1988). The original application was analysis of chest pain in the emergency room. The method has subsequently been used on numerous applications through the derivation of new knowledge bases. The inference engine uses approximate reasoning techniques to include weighted antecedents, partial presence of findings, and partial substantiation of rules. The general rule structure is shown in Table 6.1. The truth of proposition P, represented by this structure, is determined by assuming there exists some subset C of V such that (1) the number of elements in C satisfies Q; or (2) each element in C satisfies the property A. The degree S to which C satisfies P is

$$S = \max_{C \in A}\{V_P(c)\} \tag{6.18}$$

where

$$V_P(c) = \max\left[\left(Q\sum_{i=1}^{n} c_i^{\wedge} a_i\right)^{\wedge} \min_{i=1,\dots,n} (w_i^{c_i^{\wedge} a_i})\right] \tag{6.19}$$

TABLE 6.1. Rule Structure

	Antecedent	Weighting factor	Degree of substantiation
IF	1	w_1	a_1
	2	w_2	a_2
	.		
	.		
	.		
	n	w_n	a_n
THEN	Conclusion if S > Threshold		

where $\hat{}$ indicates minimum, w_i and a_i are the weighting factor and degree of substantiation, respectively, of the ith antecedent, $c_i \in \{0, 1\}$, and n is the number of antecedents. The sources of information that are relevant for rule base development for dementia evaluation are listed in Table 6.2.

Agent 2: Neural Network Model (Hypernet). The neural network model is based on the authors' Hypernet system (Hudson and Cohen, 1999). Hypernet uses an expanded potential function approach in a supervised learning algorithm resulting in a nonlinear network structure with three or more layers. Input parameters can be any type of ordered data. In this application, the neural network is used to assess treatment strategies. Output is in the form of a decision function:

$$D(x) = \sum_{i=1}^{n} w_i x_i + \sum_{i=1}^{n} \sum_{j=1\,(i \neq j)}^{n} w_{ij} x_i x_j \qquad (6.20)$$

Agent 3: Chaotic Analyzer (CATS). The EEG is evaluated in terms of both amplitude and frequency of wave occurrence using a method previously developed by the authors for ECG analysis (Hudson and Cohen, 2001). The method uses second-order difference plots and the CTM measure described above.

Agent 4: Image Analysis (IA). Two types of imaging are relevant to diagnosis of dementia: anatomical and functional. Anatomical can be done using CT scans or MRI scans while functional is done with MRI only. Functional imaging can supply information regarding levels of activity in each of the lobes. This information can be

TABLE 6.2. Clinical Parameters

Cognitive Factors
    Mini-Mental State Examination (MMSE)
    Clinician evaluation of level of function
    Caregiver evaluation of level of function
Family History
    Number of first- or second-degree relatives with dementia
    Number of first- or second-degree relatives with AD
    Number of first- or second-degree relatives with early-onset dementia
    Number of first- or second-degree relatives with early-onset AD
Genetic Testing
    Presence of E4 isoform of apoE

obtained from radiological interpretation (symbolic) or by automated image analysis programs that may supply information in either symbolic or numeric format.

Results from signal analysis provide vital information for clinical decision making for numerous diseases. In many cases, however, these results must be combined with other clinical and historical information to provide a comprehensive conclusion. Although physicians often perform this process manually, computer-assisted decision support systems have not completely addressed the use of multiple models in arriving at one overall decision. Although traditional hybrid systems have dealt with this problem, often modifications are needed in the individual algorithms to permit combination of results. The intelligent agent approach allows free interaction among agents by providing an external facilitator to deal with communications issues and with combination of results. This approach is especially useful for the incorporation of signal analysis data that is often supplied in the form of lists of abnormalities or as summary data that is not easily combined with traditional clinical parameters.

6.9. FRONTIERS OF NEURAL SIGNAL PROCESSING

Nanotechnology offers the promise of providing means to solve the problem of non-specificity in the EEG signal mentioned in the previous section. If specific signals can be obtained on the neuronal level, higher-level processing as described above can lead to significant advances in the diagnosis of neurological disorders at an early stage when treatment may be more efficacious. In addition, nanotechnology-based devices can be used to deliver treatment both through precise drug delivery and electrical stimulation, the latter of which has been demonstrated to be effective in the treatment of Parkinson's disease and epilepsy. Exciting possibilities exist for diagnosis, treatment, and monitoring of neurological disorders.

Recently developed user-programmable nanocontrollers permit the usage of mixed digital and analog systems (Frenger, 2002). These new features can be used to create a better artificial neuron with implications for both the direct and inverse problem described above. In the long term this technology may permit neurological disorders to be treated through the use of implantable artificial neurons. Analysis of signal-processing components will be crucial to accomplish the interface with biological neurons. Emerging methods are discussed in Akay (2001).

Technological advances in miniaturization, biosensors, and computer processing, coupled with an increased understanding of illnesses at the molecular level, will lay the basis for a new generation of monitors (Kohli-Seth and Oropello, 2000). Monitoring will become noninvasive and will include monitoring of changes in the intracellular environment and signal messaging. In addition to traditional signal monitoring, new sensors will provide information not currently available, with the potential of great impact in drug delivery methodology and tracking of disease (Cohen and Hudson, 2003b).

REFERENCES

Aakerlund, L., and Hemmingsen, R., 1998, Neural networks as models of psychopathology, *Biol. Psychiatry* **43**(7):471–482.
Akay, M., 1995, Wavelets in biomedical engineering, *Ann. Biomed. Eng.* **23**:531–542.

Akay, M., 2001, Merging engineering and neuroscience, *Proc. IEEE* **89**(7):991–992.

Akay, M., and Daubenspeck, J. A., 1999, Spatial mapping of respiratory related evoked responses using wavelet transform method, *EMBS/BMES Joint Proc.* **21**:962.

Asanuma, H., and Wilson, V. J. (eds.), 1979, *Integration in the Nervous System: A Symposium in Honor of David P. C. Lloyd and Rafael Lorente de No, The Rockefeller University*, Tokyo, New York.

Bendat, J. S., and Piersol, A. G., 1971, *Random Data: Analysis and Measurement Procedures*, Wiley-Interscience, New York.

Berube, J. L., Parker, P. A., Gander, R. E., and Dunfield, V. A., 1984, Digital myoelectric signal processor with adaptive decision boundaries, *Med. Biol. Eng. Comput.* **22**(4):349–352.

Blanco, S., Kochen, S., Rosso, O. A., and Salgado, P., 1997, Applying time-frequency analysis to seizure EEG activity, *IEEE EMBS Mag.* **16**(1):64–71.

Butter, C. M., 1968, *Neuropsychology: The Study of Brain and Behavior*, Brooks/Cole Publishing Co., Belmont, CA.

Chialvo, R., and Jalife, J., 1987, Non-linear dynamics of cardiac excitation and impulse propagation, *Nature* **330**:749–752.

Cohen, M. E., and Hudson, D. L., 1999, Chaos and time series analysis, in *Encyclopedia of Electrical & Electronics Engineering* (J. G. Webster, ed.), John Wiley & Sons, New York, pp. 218–226.

Cohen, M. E., and Hudson, D. L., 2000a, New chaotic methods for biomedical signal analysis, *IEEE EMBS Inform. Technol. Applicat. Biomed.* **2000**:117–122.

Cohen, M. E., and Hudson, D. L., 2000b, Extension of chaotic techniques to electroencephalogram analysis, ISCA Comput. Applicat. Ind. Eng. **13**:82–85.

Cohen, M. E., and Hudson, D. L., 2001, EEG analysis based on chaotic evaluation of variability, *IEEE Eng. Med. Biol.* **23**.

Cohen, M. E., and Hudson, D. L., 2003a, Knowledge-based and data-based analysis of biomedical signals, *ISCA Comput. Their Applicat.* **18**:67–70.

Cohen, M. E., and Hudson, D. L., 2003b, Nonlinear analysis using continuous chaotic modeling, *Biomint Seminar, World Academy of Biotechnology*, UNESCO.

Cohen, M. E., Hudson, D. L., and Anderson, M. F., 1992, The effect of vasoactive drugs on the chaotic nature of blood flow, *MEDINFO* **92**:931–936.

Cohen, M. E., Hudson, D. L., Anderson, M. F., and Deedwania, P. C., 1994, A conjecture to the solution of the continuous logistic equation, *Int. J. Uncert. Fuzz. Knowl.-Based Syst.* **2**(4):445–461.

Cohen, M. E., Hudson, D. L., and Deedwania, P. C., 1998, The use of continuous chaotic modeling in differentiation of categories of heart disease, *Inform. Process. Manage. Uncert. Knowl.-Based Syst.* **7**:548–554.

Daubechies, I., 1988, Orthonormal bases of compactly supported wavelets, *Commun. Pure Appl. Math.* **XLI**:909–996.

DeFelipe, J., and Jones, E. G., eds., 1988, *Cajal on the Cerebral Cortex: An Annotated Translation of the Complete Writings*, Oxford University Press, New York.

De Tomasso, M., De Carlo, F., Difruscolo, O., Massagra, R., Sciruicchio, V., and Bellotti, R., 2003, Detection of subclinical brain electrical activity changes in Huntington's disease using artificial neural networks, *Clin. Neurophysiol.* **114**:1237–1245.

Eberhart, R. C., 1989, Chaos theory for the biomedical engineer, *IEEE EMB Mag.* Sept., 41–45.

Edlinger, G., Wach, P., and Pfurtscheller, G., 1998, On the realization of an analytic high-resolution EEG, *IEEE Trans. Biomed. Eng.* **45**:736–745.

Evans, H. B., Pan, Z., Philip, A. P., and Scott, R. N., 1984, Signal processing for proportional myoelectric control, *IEEE Trans. Biomed. Eng.* **31**:207–211.

Flint, R., Pederson, B., Guekht, A. B., Malmgren, K., Michelucci, R., Neville, B., Pinto, F., Stephani, U., and Ozkara, C., 2002, Guidelines for the use of EEG methodology in the diagnosis of epilepsy, *Acta Neurol. Scand.* **106**:1–7.

Freeman, W. J., 1987, Simulation of chaotic EEG patterns with a dynamic model of the olfactory system, *Biol. Cybernet.* **56**:139–150.

Frenger, P., 2002, Nanocontroller update: Building a better artificial neuron, *Biomed. Sci. Instrum.* **38**:441–445.

Garrett, D., Peterson, D. A., Anderson, C. W., and Thaut, M. H., 2003, Comparison of linear, nonlinear and feature selection methods for EEG signal classification, *IEEE Trans. Neural Syst. Rehabil. Eng.* **11**:141–144.

Goldberger, A. L., 1989, Cardiac chaos, *Science* **243**(2987):1419.

Golgi, C., 1886, *Sulla fina anatomia degli organi centrali del sistema nervoso*, Hoepli, Milano.

Harley, T. A., 1998, Connectionist modeling of the recovery of language functions following brain damage, *Brain Lang.* **52**(1):7–24.

He, B., Lian, J., and Li, G., 2001, High-resolution EEG: A new realistic geometry spline Lapacian estimation technique, *Clin. Neurophysiol.* **112**:845–852.

He, B., Zhang, Z., Lian, J., Sasaki, H., Wu, S., and Towle, V. L., 2002, Boundary element method based on cortical potential imaging of somatosensory evoked potentials using subjects; magnetic resonance imaging, *Neuroimage* **16**:564–576.

Hefftner, G., Zucchini, W., and Jaros, G. G., 1988, The electromyogram (EMG) as a control signal for functional neuromuscular stimulation—Part I: Autoregressive modeling as a means of EMG signature discrimination, *IEEE Trans. Biomed. Eng.* **35**(4):230–237.

Hjorth, B., 1975, An on line transformation of EEG scalp potentials into orthogonal source derivations, *Electroencephalogr. Clin. Neurophysiol.* **39**:526–530.

Hodgkin, A. L., 1964, *The Conduction of the Nervous Impulse*, Liverpool University Press.

Hofmann, O., and Bodendorf, F., 2000, A framework for agent mediated electronic business, *ISCA Comput. Applicat. Med. Care* **15**:120–123.

Hsu, C., and Goldberg, H. S., 1999, Knowledge-mediated retrieval of laboratory observations, *Proc. AMIA* **1999**:809–813.

Hubel, D. H., and Wiesel, T. N., 1962, Receptive fields, binocular interaction, and functional architecture of the cat visual cortex, *J. Physiol.* **160**(1):106–154.

Hudson, D. L., and Cohen, M. E., 1988, An approach to management of uncertainty in an expert system, *Int. J. Intell. Syst.* **3**(1):45–58.

Hudson, D. L., and Cohen, M. E., 1999, *Neural Networks and Artificial Intelligence for Biomedical Engineering*, IEEE Press-Wiley.

Hudson, D. L., and Cohen, M. E., 2001, Use of intelligent agents to include signal analysis data, *IEEE Eng. Med. Biol.* **23**. [CD]

Hudson, D. L., and Cohen, M. E., 2002, Pattern identification in electroencephalograms, *ISCA Comput. Their Applicat.* **17**:315–318.

Hudson, D. L., Cohen, M. E., and Deedwania, P. C., 1998, Chaotic ECG analysis using combined models, *IEEE Eng. Med. Biol.* **20**:1553–1556.

Ihl, R., Dierks, T., Martin, E. M., Frolich, L., and Maurer, K., 1992, Importance of the EEG in early and differential diagnosis of dementia of the Alzheimer type, *Fortschritte der Neurologie-Psychiatrie* **60**(12):451–459.

Inouye, T., Toi, S., and Matsumoto, Y., 1995, A new segmentation method of electroencephalograms by use of Akaike's information criterion, *Brain Res.* **3**(1):33–40.

Jellinger, K. A., Seppi, K., Wenning, G. K., and Poewe, W., 2002, Impact of coexistent Alzheimer pathology on the natural history of Parkinson's disease, *J. Neural Transm.* **109**(3):329–339.

Kalayci, T., and Ozdamar, O., 1995, Wavelet preprocessing for automated neural network detection of EEG spikes, *IEEE EMBS Mag.* **14**(2):160–166.

Kearfott, R. B., Sidman, R. D., Major, D. A., and Hill, C. D., 1991, Numerical tests of a method for simulating electric potentials on the cortical surface, *IEEE Trans. Biomed. Eng.* **38**:294–299.

Kenny, R. A., Kalaria, R., and Ballard, C., 2002, Neurocardiovascular instability in cognitive impairment and dementia, *Ann. N.Y. Acad. Sci.* **977**:183–195.

Kim, H., Kim, S., Go, H., and Kim, D., 2001, Synergetic analysis of spatio-temporal EEG patterns: Alzheimer's disease, *Biol. Cybernet.* **85**:1–17.

Kim, K. H., and Kim, S. J., 2003, A wavelet-based method for action potential detection from extracellular neural signal recording with low signal-to-noise ratio, *IEEE Trans. Biomed. Eng.* **50**(8):999–1011.

Kohli-Seth, R., and Oropello, J. M., 2000, The future of bedside monitoring, *Crit. Care Clin.* **16**(4):557–578.

Lanzola, G., Gatti, L., Falasconi, S., and Stefanelli, M., 1999, A framework for building cooperative software agents in medical applications, *Artif. Intell. Med.* **16**:223–249.

Law, S. K., Nunez, P. L., and Wijesinghe, R. S., 1993, High-resolution EEG using spline generated surface laplacians on spherical and ellipsoidal surfaces, *IEEE Trans. Biomed. Eng.* **40**:145–153.

Leuchter, A. F., Cook, I. A., Newton, T. F., and Weiner, H., 1993, Regional differences in brain electrical activity in dementia: Use of spectral power and spectral ratio measures, *Electroencephalogr. Clin. Neurophysiol.* **87**(6):385–393.

Lopes da Silva, F., 1993, Dynamics of EEGs as signals of neuronal populations: models and theoretical considerations, in: *Electroencephalography: Basic Principles, Clinical Applications, and Related Fields*, 3rd edn. (E. Niedermeyer and F. Lopes da Silva, eds.), Williams and Williams, Baltimore, pp. 63–77.

MacGregor, R. J., 1987, *Neural and Brain Modeling*, Academic Press, San Diego.

Mack, S. J., Holstein, J., Kleber, K., and Grönemeyer, D. H., 2000, New aspects of image distribution and workflow in radiology, *J. Digit. Imag.* **13**(2):17–21.

Miyauchi, T., Hagimoto, H., Ishii, M., Endo, S., Tanaka, K., Kajiwara, S., Endo, K., Kajiwara, A., and Kosaka, K., 1994, Quantitative EEG in patients with presenile and senile dementia of the Alzheimer type, *Acta Neurol. Scand.* **89**(1):56–64.

Nunez, P. L., Silberstein, R. B., Cadusch, P. J., Wijesinghe, R. S., Westdorp, A. F., and Srinivasan, R., 1994, A theoretical and experimental study of high resolution EEG based on surface laplacians and cortical imaging, *Electroencephalogr. Clin. Neurophysiol.* **90**:40–57.

Petrosian, A. A., Prokhorov, D. V., Lahara-Nanson, W., and Schiffer, R. B., 2001, Recurrent neural network-based approach for early recognition of Alzheimer's disease in EEG, *Clin. Neurophysiol.* **112**(8):1378–1387.

Philipson, G., 1985, Adaptable myoelectric prosthetic control with functional visual feedback using microprocessor techniques, *Med. Biol. Eng. Comput.* **23**:8–14.

Porcher, R., and Thomas, G., 2001, Estimating Lyapunov exponents in biomedical time series, *Phys. Rev. E: Stat. Phys. Plasmas Fluids Relat. Interdiscip. Top.* **64**(1–1):010902.

Prichep, L. S., John, E. R., Ferris, S. H., Reisberg, B., Almas, M., Alper, K., and Cancro, R., 1994, Quantitative EEG correlates of cognitive deterioration in the elderly, *Neurobiol. Aging* **15**(1):85–90.

Principe, J., Gala, S. K., and Chang, T. G., 1989, Sleep staging automaton based on the theory of evidence, *IEEE Trans. Biomed. Eng.* **36**(5):503–509.

Priori, A., Foffani, G., Pesent, A., Bianchi, A., Chiesa, V., Baselli, G., Caputo, E, Tamma, F., Rampini, P., Egidi, M., Locatelli, M, Barbieri, S., and Scarlato, G., 2002, Movement-related modulation of neural activity in human basal ganglia and its L-DOPA dependency: recordings from deep brain stimulation in patients with Parkinson's disease, *Neurol. Sci.* **23**(Suppl. 2):S101–102.

Pritchard, W. S., Duke, D. W., Coburn, K. L., Moore, N. C., Tucker, K. A., Jann, W. S., and Hostetler, R. M., 1994, EEG-based, neural-net predictive classification of Alzheimer's disease versus control subjects is augmented by non-linear EEG measures, *Electroencephalogr. Clin. Neurophysiol.* **91**(2):118–130.

Rialle, V., and Stip, E., 1994, Cognitive models in psychiatry: from symbolic models to parallel and distributed models, *J. Psychiatry Neurosci.* **19**(3):178–192.

Rogers, S. K., and Kabrisky, M., 1991, *An Introduction to Biological and Artificial Neural Networks for Pattern Recognition*, SPIE Optical Engineering Press.

Rosenblatt, F., 1962, *Principles of Neurodynamics*, Spartan Books, New York.

Ruppin, E., Reggia, J. A., and Horn, D., 1996, Pathogensis of schizophrenic delusions and hallucinations: A neural network model, *Schizophren. Bull.* **22**(1):105–123.

Santa Cruz, K. S., Tasaki, C. S., Kim, R. C., and Cotman, C. W., 2002, Brainstem and cortical Lewy bodies in patients presenting with Alzheimer's disease, *J. Alzheim. Dis.* **4**(1):11–17.

Sarbadhikari, S. N., and Chakrabarty, K., 2001, Chaos in the brain: A short review alluding to epilepsy, depression, exercise and lateralization, *Med. Eng. Phys.* **23**(7):445–455.

Saugstad, L. F., 1994, Deviation in cerebral excitability: Possible clinical implications, *Int. J. Psychophysiol.* **18**(3):205–212.

Schreiter-Gasser, U., Gasser, T., and Ziegler, P., 1993, Quantitative EEG analysis in early onset Alzheimer's disease: A controlled study, *Electroencephalogr. Clin. Neurophysiol.* **86**(1):15–22.

Schreiter-Gasser, U., Gasser, T., and Ziegler, P., 1994, Quantitative EEG analysis in early onset Alzheimer's disease: Correlations with severity, clinical characteristics, visual EEG and CCT, *Electroencephalogr. Clin. Neurophysiol.* **90**(4):267–272.

Seiss, E., Praamstra, C., Hesse, C. W., and Rickards, H., 2003, Proprioceptive sensory function in Parkinson's disease and Huntington's disease: Evidence from proprioception-related EEG potentials, *Exp. Brain Res.* **148**:308–319.

Signorino, M., Pucci, E., Belardinelli, N., Nolfe, G., and Angeleri, F., 1995, EEG spectral analysis in vascular and Alzheimer dementia, *Electroencephalogr. Clin. Neurophysiol.* **94**(5):313–325.

Silverman, B. G., 1998, The role of Web agents in medical knowledge management, *MD Comput.* **15**(4):221–231.

Soininen, H., and Riekkinen, P. J., Sr., 1992, EEG in diagnostics and follow-up of Alzheimer's disease, *Acta Neurol. Scand.* **139**(Suppl.):36–39.

Storey, E., Slavin, M. J., and Kinsella, G. J., 2002, Patterns of cognitive impairment in Alzheimer's disease: Assessment and differential diagnosis, *Front. Biosci.* **7**:E155–E184.

Suzuki, M., Desmond, T. J., Albin, R. L., and Frey, K. A., 2002, Striatal monoamnergic terminals in Lewy body and Alzheimer's dementias, *Ann. Neurol.* **51**(6):767–771.

Swigger, K. M., and Ducksworth, L., 2000, Supporting computer-mediated collaboration through user-defined agents, *ISCA CAINE* **13**:43–46.

Tiraboschi, P., Hansen, L. A., Alford, M., *et al.*, 2002, Early and widespread cholinergic losses differentiate dementia with Lewy bodies form Alzheimer disease, *Arch. Gen. Psychiatry* **59**(10):946–951.

Varma, A. R., Laitt, R., Lloyd, J. J., Carson, K. J., Snowden, J. S., Neary, D., and Jackson, A., 2002, Diagnostic value of high signal abnormalities on T2 weighted MRI in the differentiation of Alzheimer's, frontotemporal and vascular dementias, *Acta Neurol. Scand.* **105**(5):355–364.

Visser, P. J., Verhey, F. R., Ponds, R. W., and Jolles, J., 2001, Diagnosis of preclinical Alzheimer's disease in a clinical setting, *Int. Psychogeriatr.* **13**(4):411–423.

Woyshville, M. J., and Calabrese, J. R., 1994, Quantification of occipital EEG changes in Alzheimer's disease utilizing a new metric: the fractal dimension, *Biol. Psychiatry* **35**(6):381–387.

Wszolek, Z. K., Herkes, G. K., Lagerlund, T. D., and Kokmen, E., 1992, Comparison of EEG back-ground frequency analysis, psychological test scores, short test of mental status, and quantitative SPECT in dementia, *J. Geriatr. Psychiatry Neurol.* **5**(1):22–30.

7

ELECTROPHYSIOLOGICAL NEUROIMAGING

Bin He[1]* and Jie Lian[2]

[1]Department of Biomedical Engineering, University of Minnesota
[2]Microsystems Engineering, Inc.

7.1. INTRODUCTION

7.1.1. THE GENERATION AND MEASUREMENT OF THE EEG

Although electrical activity recorded from the exposed cerebral cortex of a monkey was reported in 1875 (Caton, 1875), it was not until 1929 that Hans Berger, a psychiatrist in Jena, Germany, first recorded rhythmic electrical activity from the human head (Berger, 1929). Since then, the electroencephalogram (EEG) has become one of the most prominent methods for noninvasive examination of brain activity. Tremendous effort has been made in order to describe the phenomena of the EEG in normal individuals and in those with various diseases. In particular, the EEG has been demonstrated to be a valuable tool for both researchers and clinicians in the fields of sleep physiology and epilepsy, although other applications are also promising, such as in the fields of psychiatry and psychophysiology.

The EEG is generated mainly by inhibitory and excitatory postsynaptic potentials of cortical nerve cells. The discharge of a single neuron or single nerve fiber in the brain generates a very small potential field, and does not contribute significantly to scalp potential recordings. Instead, the recorded scalp EEG represents the summation of the far field potentials generated by many thousands or even millions of neurons or fibers when they fire synchronously. In other words, the intensity of the scalp EEG is determined mainly by the number of neurons and fibers that fire in synchrony with one another, not by the total level of electrical activity in the brain.

The intensities of the scalp EEG range from 0 to 200 μV, and their frequencies range from once every few seconds to 50 or more per second. The EEG recording involves the application of a set of electrodes to standard positions on the scalp. The most commonly used electrode placement montage is the International 10/20 system, which uses the distances

---

* Address for correspondence: Department of Biomedical Engineering, University of Minnesota, 7-105 BSBE, 312 Church Street, Minneapolis, Minnesota 55455; e-mail: binhe@umn.edu.

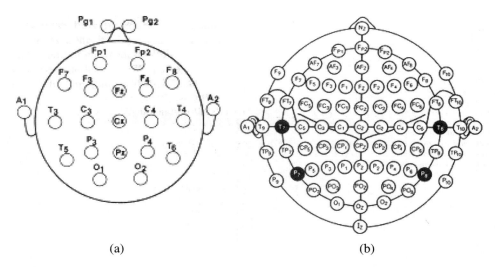

(a)   (b)

FIGURE 7.1. Electrode placement montage for EEG measurement: (a) standard International 10/20 system; (b) expanded 10/20 system. (From Fisch, Elsevier, 1991, with permission)

between bony landmarks of the head to generate a system of lines which run across the head and intersect at intervals of 10 or 20% of their total length (Figure 7.1a). Additional electrodes can also be introduced according to the expanded 10/20 system as proposed by the American EEG society (Figure 7.1b). Besides spontaneous EEG recording, some associated measurements have also been widely practiced. For example, the event-related potentials (ERPs) measure the brain responses that are time-locked to some external "event," whereas its subclass, the evoked potentials (EPs), are usually elicited in response to the sensory stimuli, such as the visual evoked potential (VEP), auditory evoked potential (AEP), and the somatosensory evoked potential (SEP).

7.1.2. SPATIAL AND TEMPORAL RESOLUTION OF THE EEG

Brain electrical activation is a spatiotemporal process, which means that its activity is three-dimensionally distributed and evolves with time. The most significant merit of EEG is its unsurpassed millisecond-scale temporal resolution, which is essential for resolving the rapid change of neurophysiological process. However, the conventional EEG has limited spatial resolution, mainly due to two factors. One factor is the limited spatial sampling. The standard 10/20 EEG recording montage results in interelectrode distances of about 6 cm (Nunez *et al.*, 1994). A remarkable development in the past decade is that high-resolution EEG systems with 64 to 256 electrodes have been commercially available. For example, with up to 124 scalp electrodes, the average interelectrode distance can be reduced to about 2.5 cm (Gevins *et al.*, 1994). The second factor is the head volume conduction effect. The electric potentials generated from the neural sources are attenuated, distorted, and blurred as they pass through the neural tissue, cerebrospinal fluid, meninges, and the low-conductivity skull, and scalp (Nunez, 1981, 1995). Therefore, advanced EEG imaging techniques are highly desired in order to compensate for the head volume conduction effect and enhance the spatial resolution of the EEG. The solutions of two separate but closely related problems, EEG forward problem and EEG inverse problem, are required for a high-resolution mapping of brain electric activity based on external EEG measurement.

7.1.3. EEG FORWARD PROBLEM AND INVERSE PROBLEM

Given the known information on the brain electric source distribution and the head volume conduction properties, the EEG forward problem determines the source-generated electric field. The EEG forward solution can be electric potentials, such as the cortical potential or the scalp potential, and can also be other metrics, for example, the current density distribution. The EEG forward problem is well defined and has a unique solution, governed by the quasi-static limit of Maxwell's equations (Plonsey, 1969; Nunez, 1981; Malmivuo and Plonsey, 1995; Gulrajani, 1998; He, 2004).

By solving the EEG forward problem, the relationship between the neuronal sources and the external sensor measurements can be established. In particular, under linear approximation, the EEG measurements and the underlying brain electric sources can be related by the so-called transfer matrix or lead field matrix, which is only dependent on the geometry and the conductive characteristics of the head volume conductor.

On the other hand, given the known electric field (e.g., the scalp EEG measurement) and the head volume conductor properties, the EEG inverse problem estimates the location and extent of the brain electric sources. Unlike the forward problem, the EEG inverse problem is fundamentally ill-posed in that there are an infinite number of source configurations that could explain a given data set of scalp potential measurement (Von Helmholtz, 1853). In order to obtain unique inverse solution, additional constraints have to be imposed, for example the anatomical constraints, the physiological constraints, spatiotemporal constraints, or the functional constraints, provided by other imaging modalities.

In terms of dimension of the solution space, the EEG inverse solution can be classified to isolated source model, such as the single or multiple dipole sources, and the distributed source model, such as the cortical surface sources, or volume sources distributed in the three dimensions of the brain. The choice of different source models depends on particular applications, although the primary goal of the EEG inverse problem remains the same, which is to find an equivalent representation of the brain electric sources that can account for the external EEG measurement. The electrophysiological neuroimaging, based on the EEG inverse solutions, will therefore provide a noninvasive and economic probe for high-resolution mapping of brain activity and function.

7.1.4. HEAD VOLUME CONDUCTOR MODELS AND SOURCE MODELS

In terms of geometry, the head volume conductor can be represented by simple spherical models, or by realistically shaped head models. According to the conductive properties of the tissues, the head volume conductor can also be characterized as homogeneous or inhomogeneous models. The most commonly used head volume conductor models include the one-sphere homogeneous model, the three- or four-sphere inhomogeneous model, the realistic geometry homogeneous model, the realistic geometry inhomogeneous model, and so on. The spherical models have been widely used in computer simulations because the forward analytic solutions are generally available and can be used for evaluation and validation study. On the other hand, the realistic geometry models can more accurately represent the head volume conductor, and are usually implemented by the numerical methods, such as the boundary element method (BEM), the finite element method (FEM), the finite volume method (FVM), the finite difference method (FDM), and so on.

Several source models have been proposed for equivalently representing brain electric sources. The primary bioelectric sources can be represented as an impressed current density $\vec{J}^i$, which is driven by the electrochemical process of excitable cells. In other words, it is a nonconservative current that arises from the bioelectric activity of nerve and muscle cells due to the conversion of energy from the chemical to the electrical form (Plonsey, 1969).

The simplest brain electric source model is a point current source or monopole source. For example, if the volume conductor is infinite and homogeneous with a conductivity of σ, the bioelectric potential obeys Possion's equation under quasi-static conditions (Plonsey, 1969):

$$\nabla^2 \Phi = \frac{\nabla \cdot \vec{J}^i}{\sigma} = -\frac{I_v}{\sigma} \tag{7.1}$$

Equation (7.1) is a partial differential equation satisfied by the electrical potential Φ in which I_v is the source function. The solution of Eq. (7.1) for the scalar function Φ for a region that is uniform and infinite in extent is (Stratton, 1941; Plonsey, 1969; Jackson, 1975)

$$\Phi = -\frac{1}{4\pi\sigma} \int_V \frac{\nabla \cdot \vec{J}^i}{r} \, dv \tag{7.2}$$

where r refers to the distance from the source to the observation point. Because the source element $\nabla \cdot \vec{J}^i dv$ in Eq. (7.2) behaves like a point source, in that it sets up a field that varies as $1/r$, the expression $I_v = -\nabla \cdot \vec{J}^i$ can be considered as an equivalent monopole source density (Plonsey, 1969; Malmivuo and Plonsey, 1995; Gulrajani, 1998; He et al., 2002a). Therefore, the brain electric activity can be equivalently represented by the source function I_v, which behaves as a fundamental driving force establishing the electrical potentials inside the brain and over the passive medium of the head volume conductor.

On the other hand, in a living system, one can never have a single isolated monopole current source because of electrical neutrality. However, collections of positive and negative monopole sources are physically realizable if the total sum of current is zero. The simplest collection of monopole sources is a dipole, which consists of two monopoles of opposite sign but equal strength separated by an infinitely small distance.

Still considering the infinite homogeneous volume conductor model, using the vector identity $\nabla \cdot (\vec{J}^i/r) = \nabla(1/r) \cdot \vec{J}^i + (1/r)\nabla \cdot \vec{J}^i$ and the divergence (or Gauss's) theorem, Eq. (7.2) can be transformed to (Stratton, 1941; Malmivuo and Plonsey, 1995; Gulrajani, 1998)

$$\Phi = \frac{1}{4\pi\sigma} \int_V \nabla\left(\frac{1}{r}\right) \cdot \vec{J}^i dv \tag{7.3}$$

Here, the source element $\vec{J}^i dv$ behaves like a dipole source, with a field that varies as $1/r^2$. Therefore, the impressed current density $\vec{J}^i$ may be interpreted as an equivalent dipole source density, which behaves as a fundamental driving force establishing the electrical potentials within the head volume conductor. Although higher-order equivalent source models such as the quadrupole have also been studied to represent the bioelectric sources (Geselowitz, 1960; Jerbi *et al.*, 2002), the dipole model has been so far the most commonly used brain electric source model.

7.1.5. ELECTRICAL POTENTIALS IN A CONCENTRIC THREE-SPHERE VOLUME CONDUCTOR MODEL

As illustrated in Figure 7.2, the concentric three-sphere inhomogeneous model (Rush and Driscoll, 1969) has been widely used to represent the head volume conductor. If the brain source model is represented by an isolated point current source (monopole) or dipole current source, the electrical potentials on the scalp and inside the brain can be analytically calculated. For distributed sources, the principle of superposition can be applied to calculate the potential distribution. As examples, below we present the formula for calculating the electrical potentials in a concentric three-sphere volume conductor model, generated by a point current source and a dipole current source.

The scalp potential produced by a point current source I located on the z-axis with $z = r_0$ can be represented by (He *et al.*, 2002a)

$$\Phi_{\text{scalp,monopole}} = \sum_{l=1}^{\infty} \frac{If^l}{4\pi\sigma_3 c} \frac{s(2l+1)^3}{d_l(l+1)l} P_l(\cos\theta) \qquad (7.4)$$

and the scalp potential produced by the z-component P_z of a dipole located on the z-axis is (He *et al.*, 2002a)

$$\Phi_{\text{scalp,dipole}} = \sum_{l=1}^{\infty} \frac{P_z l f^{l-1}}{4\pi\sigma_3 c^2} \frac{s(2l+1)^3}{d_l(l+1)l} P_l(\cos\theta) \qquad (7.5)$$

FIGURE 7.2. Schematic illustration of the concentric three-sphere head volume conductor model. The normalized radii of the brain, skull, and scalp spheres are 0.87, 0.92, and 1.0, respectively. The brain electric sources can be equivalently represented by a closed-surface dipole layer that is close to the cortical surface. (From He *et al.*, *Human Brain Mapping*, 2001a, with permission)

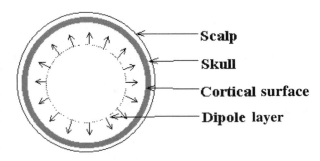

where

$$s = \frac{\sigma_2}{\sigma_1} = \frac{\sigma_2}{\sigma_3}$$

$$d_l = ((l+1)s + l)\left(\frac{ls}{l+1} + 1\right) + (1-s)((l+1)s + l)\left(f_1^{l_1} - f_2^{l_1}\right) - l(1-s)^2 \left(\frac{f_1}{f_2}\right)^{l_1}$$

$$l_1 = 2l + 1$$

$$f_1 = a/c$$

$$f_2 = b/c$$

$$f = r_0/c$$

Here, a, b, and c respectively represent the eccentricity of the cortical, skull, and scalp layers, while σ_1, σ_2, and σ_3 respectively represent their conductivity values (assume $\sigma_1 = \sigma_3$). $P_l(\cos\theta)$ is the associated Legendre function.

Similarly, the potential on the cortical surface produced by a point current source I is (He *et al.*, 2002a)

$$\Phi_{\text{cortical,monopole}} = \sum_{l=1}^{\infty} \frac{If^l}{4\pi\sigma_1 c} \frac{c_l}{d_l} P_l(\cos\theta) \tag{7.6}$$

and the potential on the cortical surface produced by the z-component P_z of a dipole located on the z-axis is (He *et al.*, 2002a)

$$\Phi_{\text{cortical,dipole}} = \sum_{l=1}^{\infty} \frac{P_z l f^{l-1}}{4\pi\sigma_1 c^2} \frac{c_l}{d_l} P_l(\cos\theta) \tag{7.7}$$

where

$$c_l = (2l+1)\left(f_1^l + \frac{\alpha}{\beta}f_1^{-(l+1)}\right)$$

$$d_l = l(1-s)f_1^{2l+1} + \frac{\alpha}{\beta}(l + (1+l)s)$$

$$\alpha = f_2^{2l+1}(1-s) - \left(1 + \frac{ls}{1+l}\right)$$

$$\beta = f_2^{-(2l+1)}(1-s) - \left(1 + \frac{l+1}{l}s\right)$$

7.2. DIPOLE SOURCE LOCALIZATION

7.2.1. EQUIVALENT CURRENT DIPOLE MODELS

The most commonly used brain electric source model is the equivalent current dipole model, which assumes that the scalp EEG is generated by one or a few focal sources. Each of the focal sources can be modeled by an equivalent current dipole with six parameters: three location parameters and three directional component parameters.

The simplest dipole model is the single fixed dipole, whose three location parameters are fixed, while its orientation and magnitude are variable. An extension of the single fixed dipole model is the single moving dipole, which has varying magnitude and orientation, as well as variable location; therefore, it has six independent variables. In the rotating dipole model, the location of the dipole is fixed over a selected latency range, whereas its three directional components are free to change independently for each time point within the period. The multiple-dipole model includes several dipoles, each representing a certain anatomical region of the brain. These dipoles have varying magnitude and varying orientation, whereas their locations could be either fixed or variable. If both the location and orientation are fixed, each dipole has only one independent variable, the magnitude. Then the number of independent variables is equal to the number of the dipoles.

7.2.2. EEG-BASED DIPOLE SOURCE LOCALIZATION

Given a specific dipole source model, the dipole source localization (DSL) solves the EEG inverse problem by using a nonlinear multidimensional minimization procedure, to estimate the dipole parameters that can best explain the observed scalp potential measurements in a least-square sense (Kavanagh *et al.*, 1978; Scherg and Von Cramon; 1985; He *et al.*, 1987; He and Musha, 1992; Homma *et al.*, 1994; Cuffin, 1995; Roth *et al.*, 1997; Musha and Okamoto, 1999; Gulrajani *et al.*, 2001). Further improvement of the DSL can be achieved by combining EEG with the magnetoencephalographic (MEG) data, which may increase information content and improve the overall signal-to-noise ratio (Diekmann *et al.*, 1998; Fuchs *et al.*, 1998).

7.2.2.1. Single Time-Slice Source Localization

Generally, there are two approaches for the EEG-based DSL. One approach is the single time-slice source localization, in which the dipole parameters are fitted at an instance in time, based on the single time "snapshots" of the measured scalp EEG (Kavanagh *et al.*, 1978; He *et al.*, 1987; He and Musha, 1992; Cuffin, 1995). For example, the scalp potentials at the single time-slice could be collected into a column vector $\vec{\phi}$, each row of which is the potential data recorded from one electrode. The problem then is to find a column vector $\vec{\psi}$, the collection of the potentials at the same electrode sites but generated by the assumed sources inside the brain. In practice, an initial starting point (also termed seed point) is estimated, then using an iterative procedure, the assumed dipole sources are moved around inside the brain in an attempt to produce the best match between $\vec{\phi}$ and $\vec{\psi}$. This involves solving the forward problem repetitively and calculating the difference between the measured and estimated potential vectors at each step. The most commonly used measure is the squared distance between the two vectors, which is given by

$$J = \left\| \vec{\phi} - \vec{\psi} \right\|^2 \tag{7.8}$$

where J is the objective function which is to be minimized. The inverse solution of the dipole sources is obtained after the process of moving the sources stops when this objective function is minimized. Different methods could be applied to solve this nonlinear optimization problem, whereas the Simplex method (Nelder and Mead, 1965; He *et al.*, 1987) is widely used because of its simplicity and relative robustness to local minima. In addition, reciprocal approaches have also been explored in solving the dipole source localization, in an attempt to improve the numerical accuracy of the transfer matrices (Fletcher *et al.*, 1995; Finke and Gulrajani, 2001; Gulrajani *et al.*, 2001).

7.2.2.2. Spatiotemporal Source Localization

Another approach is the multiple time-slice source localization, also termed spatiotemporal source localization, by incorporating both the spatial and temporal components of the EEG in model fitting (Scherg and Von Cramon, 1985). In this approach, multiple dipole sources are assumed to have fixed locations inside the brain during a certain time interval, and the variations in scalp potentials are due only to variations in the strengths of these sources. Under linear condition, the dipole sources $\vec{S}$ are coupled to the scalp potentials $\vec{\Phi}$ by the lead field matrix A, which is only dependent on the head volume conductor properties and the source-sensor configurations:

$$\vec{\Phi} = A\vec{S} \tag{7.9}$$

Here, $\vec{\Phi}$ is the N electrodes by T time-slices EEG data matrix, and $\vec{S}$ is the M dipoles by T time-slices source waveform matrix. The task of the spatiotemporal DSL is to determine the locations of multiple dipoles, whose parameters could best account for the spatial distribution as well as the temporal waveforms of the scalp EEG measurement. Similarly, an iteration procedure is needed to adjust the source parameters with the aim to minimize the following objective function:

$$J = \left\| \vec{\Phi} - A\vec{S} \right\|^2 = \left\| (I - AA^+)\vec{\Phi} \right\|^2 \tag{7.10}$$

where I is the identity matrix, and A^+ is the pseudo-inverse of matrix A. With the incorporation of the EEG temporal information in the model fitting, the spatiotemporal DSL is more robust against measurement noise and artifacts than the single time-slice DSL.

All equivalent dipole algorithms need an *a priori* knowledge of the number and class of the underlying dipole sources. If the number of dipoles is underestimated for a given model, then the DSL inverse solution is biased by the missing dipoles. On the other hand, if too many dipoles are specified, then spurious dipoles are introduced, which may be indiscernible from the true dipoles. Moreover, since the computation complexity of the least-squares estimation problem is highly dependent on the number of nonlinear parameters that must be estimated, then too many dipoles also adds needless computational burden.

In practice, the principal component analysis (PCA) has been used to approximately estimate the number of field patterns contained in the scalp EEG data (Soong and Koles, 1995), and multiple signal classification (MUSIC) algorithms (Mosher *et al.*, 1992) have been used to perform source localization by scanning the source region. For example, the MUSIC algorithm scans through the 3D brain volume (solution space) to identify sources that produce potential patterns that lie within the signal subspace of the EEG measurements (Mosher *et al.*, 1992). Assuming there are M independent dipole sources, the eigen decomposition of the measured-data covariance matrix (with a dimension of $N \times N$) will generate M eigenvalues arising from the signal sources and $N - M$ eigenvalues arising from the noise, and their corresponding eigenvectors span the signal subspace and noise subspace, respectively. Because the lead field vector at each source location should be orthogonal to the noise subspace, the locations of the dipole sources can be estimated by evaluating a scan metric at each possible spatial location. The scan metric measures the orthogonality between the lead field vector and the noise subspace, and the locations that produce a peak in the scan metric are chosen as probable source locations. Recently, a recursive method (RAP-MUSIC) has also been proposed in order to overcome the "multiple-peak picking" problem of the original MUSIC scan (Mosher and Leahy, 1998). A FINES algorithm has also been introduced to solve the EEG DSL problem (Xu *et al.*, 2004), in which projections are made onto a subspace by a small set of particular vectors in the estimated noise-only subspace.

7.2.3. CONSTRAINED DIPOLE SOURCE LOCALIZATION

When *a priori* knowledge of the source information is available, additional constraints can be incorporated in the DSL in order to improve source reconstruction. The most commonly used constraints include the anatomical constraint and the functional constraint.

7.2.3.1. Anatomically Constrained Dipole Source Localization

By limiting the solution space to a subset of the whole three-dimensional (3D) brain volume, the anatomically constrained DSL can greatly improve the computation efficiency of the optimization procedure owing to a decrease in the dimension of the search space. For example, unrealistic source locations such as the white matter could be excluded from the solution space. If the observed EEG signals are known to be mainly produced by cortical sources, then the solution space can be restricted to the cortex surface while excluding deep source locations such as the brainstem. In particular situations, when *a priori* information is known on the possible source region (such as the sensory EP data, or based on preliminary diagnosis of the epilepsy), the solution space can be restricted to only half of the brain or even more focused on a certain lobe.

An increasing interest in anatomical constrained DSL has arisen by searching dipole source locations in a realistic geometry head model constructed from the magnetic resonance images (MRIs), by means of BEM or FEM. Computer simulation and experimental studies have demonstrated that the inverse solution of the DSL is more accurate when using the realistic geometry head model than when using the simplified spherical head model (Cuffin, 1996; Waberski *et al.*, 1998; Crouzeix *et al.*, 1999; Ollikainen *et al.*, 1999; Benar and Gotman, 2002). Furthermore, by registration with the magnetic resonance (MR) images, the 3D coordinates of the estimated dipole sources can be visualized relative to the brain anatomy. Therefore, it has great potential to reveal the electrophysiologically active neural substrate underlying the scalp potential measurements, facilitate comparison with other functional imaging modalities (Snyder *et al.*, 1995; Martinez *et al.*, 1999; Northoff *et al.*, 2000; Mangun *et al.*, 2001), and has clinical significance in detecting the epileptic foci (Wong, 1991; Huppertz *et al.*, 2001), presurgical localization of the sensorimotor cortex (Cakmur *et al.*, 1997; Mine *et al.*, 1998), and some other applications.

For instance, compared with the structural lesions, Huppertz *et al.* (2001) demonstrated satisfactory accuracy of the electric source reconstruction in 14 patients with focal intracerebral lesions and epileptic seizures using DSL based on individual head anatomy as obtained from MRI. In this study, a 64-channel recording system was used for high-resolution scalp EEG measurement, and the DSL was based on available delta EEG activity and/or interictal epileptiform activity. As illustrated by an example shown in Figure 7.3, the patient's MRI showed a large resection hole comprising parts of the left postcentral and supramarginal gyrus and residual embolized angiomatous tissue in the adjacent supramarginal and superior temporal gyrus. Both the maximum dipole density (Vieth *et al.*, 1996) resulting from the source reconstruction of the delta activity and the DSL derived from the averaged interictal epileptiform activity were found within pathologically altered tissue near the resection hole, with distances to lesion margin of 6 and 2 mm, respectively.

7.2.3.2. *Functionally Constrained Dipole Source Localization*

The DSL with functional constraint derived from the hemodynamic imaging modalities has drawn great attention during the past few years. The underlying assumption of this approach is that regions in the brain that show increased metabolic activity are also on average more electrically active over time. In early investigations, the ERPs and positron emission tomography (PET) have been integrated to study the brain activity during visual spatial attention in humans (Heinze *et al.*, 1994; Woldorff *et al.*, 1997). The PET activation foci were used to *seed* the iterative optimization procedure of the DSL, in order to provide an objective initial guess of the dipole-source locations. Further effort has been made in fMRI-seeded dipole modeling to localize brain sources during visual processing and target-detection tasks (Menon *et al.*, 1997; Opitz *et al.*, 1999; Wang *et al.*, 1999). In addition, the multiple-dipole modeling based on MEG while using fMRI as a constraint has also been explored by several research groups(George *et al.*, 1995; Ahlfors *et al.*, 1999; Korvenoja *et al.*, 1999).

Figure 7.4 shows an example of the fMRI-constrained DSL when healthy subjects were presented with auditory oddball tasks (Opitz *et al.*, 1999). Unattended deviants elicited a mismatch negativity in the ERP with corresponding fMRI activation in the bilateral transverse/superior temporal gyri, and the fMRI-seeded DSL revealed significant activation close to Heschl's gyri underlying the mismatch negativity. On the other hand, attended deviants generated a mismatch negativity followed by an N200/P300 complex in the ERP with corresponding fMRI activation in both superior temporal gyri and the neostriatum, and the fMRI-seeded DSL revealed two stable dipole sources in the left and right superior temporal gyri between 320 and 380 ms.

7.3. *DISTRIBUTED SOURCE IMAGING*

Although the DSL has been demonstrated to be useful in locating a spatially restricted brain electric event, it has a major limitation in that its simplified source model may not adequately describe sources of significant extent (Snyder, 1991; Cuffin, 1998). Therefore, distributed source imaging has been aggressively studied in the past decade, particularly when studying higher-order brain functions.

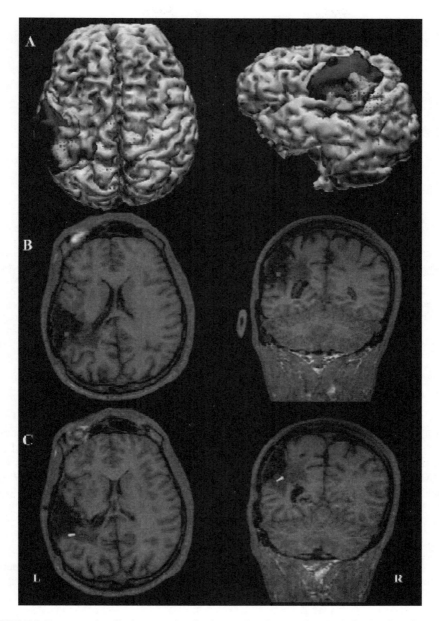

FIGURE 7.3. Representative dipole source localization results of one patient (embolized and partly resected vascular malformation): (A) 3D view of the segmented cortex from above and from the side of the lesion. The segmented resection hole is shown in black, and the pathologically altered brain tissue in brown. The dipole density resulting from the source reconstruction of delta EEG activity is encoded in squares of different sizes, which are projected onto the segmented cortex. (B) Maximum of the dipole density shown in axial and coronal slices of the patient's MRI. (C) Equivalent dipole source location corresponding to the averaged interictal epileptiform activity. (From Fig. 2 in Huppertz *et al.*, *NeuroImage*, 2001, with permission)

Left/right TTG   Left STG/SMG   Right STG   Neostriatum

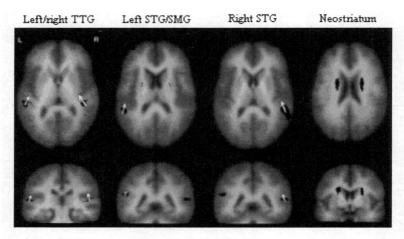

FIGURE 7.4. Brain areas that showed significant fMRI activation to deviant stimuli were superimposed on average structural MRI in Talairach space (averaged brain). The fMRI-seeded best-fitting dipoles are indicated by arrows. (a) In the unattended condition, the fMRI revealed two significant clusters of activity in the left and right transverse temporal gyri (TTG), with the size of activation being larger in the right TTG. The fMRI-seeded dipole solutions revealed two dipoles in the left and right TTG, with orientations almost perpendicular to the gray matter. (b)–(d) In the attended condition, the fMRI revealed four clusters of activation in the posterior part of the left and right superior temporal gyri (STG), adjacent to the supramarginal gyri (SMG) and the left and right neostriatum. The fMRI-seeded dipole solutions revealed only dipoles located at both STG. (Modified from Fig. 2 and Fig. 4 from Opitz *et al.*, *Psychophysiology*, 1999, with permission)

7.3.1. *DISTRIBUTED SOURCE MODELS*

Unlike the point dipole source models, the distributed source models do not make any ad hoc assumption on the number of brain electric sources. Instead, the equivalent sources (could be dipoles, monopoles, potentials, etc.) are distributed in a two-dimensional (2D) sheet such as the epicortical surface, or the 3D volume of the brain.

Assuming quasi-static condition and linear properties of the head volume conductor, the brain electric sources and the scalp EEG measurements could be mathematically described by the following linear matrix equation:

$$\vec{\phi} = A\vec{X} + \vec{n} \qquad (7.11)$$

where $\vec{\phi}$ is the vector of scalp potential measurement, $\vec{X}$ is the vector of source distribution, $\vec{n}$ is the vector of additive measurement noise, and A is the transfer matrix relating $\vec{\phi}$ and $\vec{X}$.

The aim of the distributed source imaging is to reconstruct source distributions (or topography) from the noninvasive scalp EEG measurements, or mathematically, is to design an inverse filter B which can project the measured data into the solution space:

$$\vec{X} = B\vec{\phi} \qquad (7.12)$$

This linear inverse approach, however, is intrinsically underdetermined, because the number of unknown distributed sources is much larger than the limited number of sensoring electrodes over the scalp. The inverse solution is nonunique, that is there are infinite number of source configurations that could explain a given data set of scalp potential measurement. Additional constraints have to be imposed in order to obtain unique linear inverse solutions.

7.3.2. LINEAR INVERSE FILTERS

7.3.2.1. General Inverse

The general inverse, also termed the minimum norm least-squares (MNLS) inverse, minimizes the least-square error of the estimated inverse solution $\vec{X}$ under the constraint $\vec{\phi} = A\vec{X}$ in the absence of noise. In mathematical terms, the MNLS inverse filter B_{MNLS} is determined when the following objective function is minimized:

$$J_{\text{MNLS}} = \left\| \vec{\phi} - A\vec{X} \right\|^2 \tag{7.13}$$

For an underdetermined system, if AA^T is nonsingular, we have

$$B_{\text{MNLS}} = A^+ = (A^T A)^- A^T \tag{7.14}$$

where $()^T$ and $()^-$ denote matrix transpose and matrix inversion, respectively. The general inverse solution is also a minimum norm solution among the infinite set of solutions, which satisfy the scalp potential measurements (Hamalainen and Ilmoniemi, 1984, 1994).

However, when the rank of A is less than the number of its rows, AA^T is singular and its inverse does not exist. In such a case, the general inverse can be sought by the method of singular value decomposition (SVD) (Biglieri and Yao, 1989). For an $m \times n$ matrix A, its SVD is given by

$$A = U\Sigma V^T \tag{7.15}$$

where $U = [u_1, u_2, \ldots, u_m]$, $V = [v_1, v_2, \ldots, v_n]$, $\Sigma = \text{diag}(\lambda_1, \lambda_2, \ldots, \lambda_p)$, $\lambda_1 > \lambda_2 > \cdots > \lambda_p$, and $p = \min(m, n)$. The vectors u_i and v_i are the orthonormal eigenvectors of AA^T and $A^T A$, respectively. The λ_i are the singular values of matrix A, and Σ is a diagonal matrix with the singular values on its main diagonal. On the basis of the SVD of matrix A, the general inverse of matrix A can be solved by

$$A^+ = V\Sigma^{-1}U^T = \sum_{i=1}^{p} \frac{1}{\lambda_i} v_i u_i^T \tag{7.16}$$

where $()^+$ is also known as the Moore–Penrose inverse, or the pseudo-inverse. For the linear system of Eq. (7.11), the inverse solution estimated by Eq. (7.16) is given by

$$\vec{X} = A^+\vec{\phi} = V\Sigma^{-1}U^T\vec{\phi} = \sum_{i=1}^{p} \frac{1}{\lambda_i} v_i (u_i^T \vec{\phi}) \tag{7.17}$$

Although the general inverse leads to a unique inverse solution with the smallest residual-error-giving constraint in Eq. (7.13), it is often impractical for real applications because of the ill-posed nature of the EEG inverse problem. In other words, the small measurement errors in $\vec{\phi}$ will be amplified by the small or near-zero singular values, leading to large perturbations in the inverse solution, as seen from Eq. (7.16).

7.3.2.2. Tikhonov Regularization

A common approach to overcome this numerical instability is the Tikhonov regularization (TIK), in which the inverse filter is designed to minimize an alternative objective function (Tikhonov and Arsenin, 1977):

$$J_{\text{TIK}} = \left\| \vec{\phi} - A\vec{X} \right\|^2 + \lambda \left\| G\vec{\phi} \right\|^2 \tag{7.18}$$

where λ is a small positive number known as the Tikhonov regularization parameter, G can be the identity, gradient, or Laplacian matrix, corresponding to the 0th-, 1st-, and 2nd-order Tikhonov regularizations, respectively. The underlying concept of this approach involves minimizing both the scalp potential residual and the inverse solution (distribution, gradient, or the curvature) together with the relative weighting parameter λ, in order to suppress unwanted oscillations in the inverse solution. The corresponding inverse filter is given by (Tikhonov and Arsenin, 1977)

$$B_{\text{TIK}} = A^{\text{T}}(AA^{\text{T}} + \lambda GG^{\text{T}})^+ \tag{7.19}$$

Large values of λ make the solution smoother because the second term in Eq. (7.18) dominates, whereas for a small value of λ the first term in Eq. (7.18) dominates, and the MNLS general inverse filter corresponding to the special case when $\lambda = 0$.

7.3.2.3. Truncated SVD

Another frequently used technique to overcome the ill-poseness of the inverse problem is the truncated SVD (TSVD), which is simply carried out by truncating at an index $k < p$ in the evaluation of A^+ given by Eq. (7.16), or mathematically (Shim and Cho, 1981)

$$B_{\text{TSVD}} = V\Sigma_k^{-1}U^{\text{T}} = \sum_{i=1}^{k} \frac{1}{\lambda_i} v_i u_i^{\text{T}} \tag{7.20}$$

The effects of measurement noise on the inverse solution is reduced, while the high-frequency spatial information contributed by the small singular values is also lost as a trade-off. The balance between the stability and accuracy of the inverse solution is controlled by the truncation parameter k.

7.3.2.4. Weiner Filter and Parametric Weiner Filter

When the statistical information of the signal and noise are available, the Weiner filter (WF) can be applied to the EEG inverse problem, by minimizing the expected residual error:

$$J_{\mathrm{WF}} = E\left\|\vec{\phi} - A\vec{X}\right\|^2 \tag{7.21}$$

where E denotes the expectation operator. Denoting the signal and noise covariance matrices derived from the expectation over the signal $\{\vec{X}\}$ and noise $\{\vec{n}\}$ ensembles, respectively, by R and Q, the Wiener filter is given by (Dale and Sereno, 1993; Sekihara and Scholz, 1995, 1996; Philips *et al.*, 1997a)

$$B_{\mathrm{WF}} = RA^{\mathrm{T}}(ARA^{\mathrm{T}} + Q)^{+} \tag{7.22}$$

Similarly, a regularization parameter λ can be introduced to weight the noise component, which would lead to the parametric Wiener filter (PWF):

$$B_{\mathrm{PWF}} = RA^{\mathrm{T}}(ARA^{\mathrm{T}} + \lambda Q)^{+} \tag{7.23}$$

If $R = Q = I$, then Eq. (7.23) is reduced to the zero-order Tikhonov regularization method, Eq. (7.19).

7.3.2.5. Projection Filter and Parametric Projection Filter

In many experimental and clinical situations, the statistical information of the signal may not be accurately estimated, whereas the noise covariance matrix can be estimated from data that are known to be source-free, such as the prestimulus data in EPs and ERPs. In such cases, the projection filter (PF) can be applied, which provides the orthogonal projection of the signal onto the range of the inverse filter that minimizes the expectation of the noise component in the inverse solution (Oja and Ogawa, 1986; Ogawa and Oja, 1987). In other words, the projection filter determines B by minimizing the following cost function:

$$J_{\mathrm{PF}} = E\left\|B\vec{n}\right\|^2 = \mathrm{tr}(BQB^{\mathrm{T}}) \tag{7.24}$$

where $\mathrm{tr}()$ is the trace operator. In the case of a nonsingular noise covariance Q, the concomitant solution is given by

$$B_{\mathrm{PF}} = (A^{\mathrm{T}}Q^{-}A)^{+}A^{\mathrm{T}}Q^{-} \tag{7.25}$$

In order to suppress the noise component in the inverse solution, a combined criterion was also introduced: find out the operator B, which minimizes

$$J_{\mathrm{PPF}} = E\left\|I - BA\right\|^2 + \lambda E\left\|B\vec{n}\right\|^2 = \mathrm{tr}(I - BA)(I - BA)^{\mathrm{T}} + \lambda\,\mathrm{tr}(BQB^{\mathrm{T}}) \tag{7.26}$$

Similarly, a regularization parameter λ in Eq. (7.26) controls the mutual weights of two error terms. The parametric projection filter (PPF), satisfying Eq. (7.26), is derived by (Oja and Ogawa, 1986; Ogawa and Oja, 1987; Hori and He, 2001)

$$B_{\mathrm{PPF}} = A^{\mathrm{T}}(AA^{\mathrm{T}} + \lambda Q)^{+} \tag{7.27}$$

Note, the PPF is a special case of the PWF, when $R = I$ in Eq. (7.23).

7.3.3. REGULARIZATION PARAMETERS

As noted above, in order to improve the stability of the inverse problem, a free regularization parameter λ in TIK (Eq. (7.19)), PWF (Eq. (7.23)), and PPF (Eq. (7.27)), or k in TSVD (Eq. (7.20)) should be determined. Proper selection of this parameter is critical for the inverse problem to balance the stability and accuracy of the inverse solution.

The optimal regularization parameter should be determined by minimizing the relative error (RE) or maximizing the correlation coefficient (CC) between the true source distribution $\vec{X}_{\text{true}}$ and the inversely estimated source distribution $\vec{X}_{\text{inv}}$:

$$RE = \frac{\left\| \vec{X}_{\text{true}} - \vec{X}_{\text{inv}} \right\|}{\left\| \vec{X}_{\text{true}} \right\|} \tag{7.28}$$

$$CC = \frac{\vec{X}_{\text{true}} \cdot \vec{X}_{\text{inv}}}{\left\| \vec{X}_{\text{true}} \right\| \cdot \left\| \vec{X}_{\text{inv}} \right\|} \tag{7.29}$$

Unfortunately, in real applications, the true source distribution is unknown, and alternative methods that do not depend on *a priori* knowledge of $\vec{X}_{\text{true}}$ should be used.

7.3.3.1. L-Curve Method

Hansen (1990, 1992) popularized the L-curve approach to determine a regularization parameter, which was first described by Miller (1970) and by Lawson and Hanson (1974). The L-curve approach involves a plot using a log–log scale of the norm of the solution, $\left\| \vec{X} \right\|$, on the ordinate against the norm of the residual, $\left\| \vec{\phi} - A\vec{X} \right\|$, on the abscissa, with λ or k as a parameter along the resulting curve. In most cases, the shape of this curve is in the form of an L, and the λ or k value at the corner of the L is taken as the result (Figure 7.5a). At the corner, clearly both $\left\| \vec{\phi} - A\vec{X} \right\|$ and $\left\| \vec{X} \right\|$ attain simultaneous individual minima that

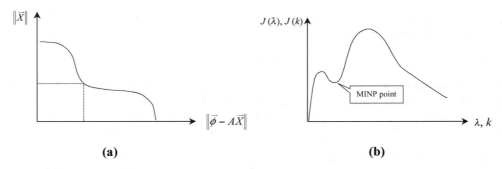

(a)   (b)

FIGURE 7.5. (a) Illustration of the L-curve. By plotting the norm of the inverse solution versus the norm of the residual as functions of regularization parameter (λ or k), an L-shaped curve occurs and the optimal parameter is placed near the "corner" of the curve. (b) Illustration of the product curve, by plotting the product of the norm of the inverse solution versus the norm of the residual, as functions of λ or k. The parameters are determined when the product reaches the relative minimal point. (From Lian and He, *Brain Topography*, 2001, with permission)

intuitively suggests an optimal solution. A numerical algorithm to compute the site of the L-curve corner, when it exists, has been given by Hansen and O'Leary (1993). The algorithm defines the corner as the point on the L-curve with maximum curvature.

7.3.3.2. Minimal Product Method

Recently, Lian et al. (1998) and Lian and He (2001) developed a minimal product (MINP) method to determine the regularization or truncation parameter. In this approach, the parameter λ or k is determined such that the following objective function reaches a relative minimum:

$$J_{\text{MINP}} = \left\| \vec{X} \right\|^{\alpha} \cdot \left\| \vec{\phi} - A\vec{X} \right\|^{\beta}$$
(7.30)

The MINP method is a trade-off between the upper bounds of the solution and residual, and aims to minimize both terms simultaneously. Here, α and β are used to determine the relative weight of the two components. Particularly when $\alpha = \beta = 1$, the solution and residual parts are taken as equal-weight. The relative minimal product of the solution and residual corresponds to the point on the L-curve where the enclosed area bounded by its x and y coordinates is relatively smallest. Under the ideal condition when the L-curve is apparent, the result of the MINP method corresponds to exactly the "corner" point of the L-curve (Figure 7.5b). Therefore, the underlying concept of the MINP method is consistent with the L-curve approach. More importantly, it is easy for implementation, and can be used in both constrained inverse methods (TIK, PWF, PPF) and TSVD.

7.3.3.3. CRESO Method

Colli Franzone et al. (1985) proposed an empirical approach for determining the TIK regularization parameter, denoted composite residual and smoothing operator (CRESO). In this approach, the λ in Eq. (7.19) is determined as the smallest positive value that results in a relative maximum of the objective function:

$$J_{\text{CRESO}} = \left\| \vec{X} \right\|^{2} + 2\lambda \left(\frac{d\left\| \vec{X} \right\|^{2}}{d\lambda} \right)$$
(7.30)

Note, however, since the parameter λ is explicitly used in the algorithm, that this approach can only be used for constrained inverse methods such as the TIK, but not applicable for the TSVD method.

7.3.3.4. Zero-Crossing Method

It was shown that the J_{CRESO} in Eq. (7.30) is the derivative of the function J_{ZC} with respect to λ (Johnston and Gulrajani, 2000), where

$$J_{\text{ZC}} = \lambda \left\| \vec{X} \right\|^{2} - \left\| \vec{\phi} - A\vec{X} \right\|^{2}$$
(7.31)

It follows that the λ determined by the CRESO method corresponds to the first point where J_{ZC} changes concavity. By choosing a λ value such that $J_{ZC} = 0$, Johnston and Gulrajani (1997) proposed a zero-crossing method for the TIK regularization parameter determination. In a recent report (Johnston and Gulrajani, 2000), they demonstrated that the MINP method is equivalent to the zero-crossing method in TIK, and both methods have better performance than the CRESO method in TIK.

7.3.3.5. Discrepancy Principle

The discrepancy principle is another widely used method for determination of the regularization parameter λ or k, usually attributed to Morozov (1984). In this approach, the parameter λ or k is selected so that the residual norm is equal to an *a priori* upper bound δ_e of the noise norm $\left\| \vec{n} \right\|$:

$$\left\| \vec{\phi} - A\vec{X} \right\| = \delta_e, \quad \text{where } \left\| \vec{n} \right\| \leq \delta_e \tag{7.32}$$

For practical applications, we could let $\left\| \vec{n} \right\| = \delta_e$ for the sake of simplicity.

7.3.3.6. Recursive Method

A recursive method for determining the optimal regularization parameter has been recently proposed (Hori and He, 2001). In this approach, a new objective function was introduced:

$$J = \left\| \vec{X} - BA\vec{X} \right\|^2 + \text{tr}(BQB^T) \tag{7.33}$$

The recursive procedure is as follows: (i) An initial inverse solution $\vec{X}_0$ is obtained using an initial regularization parameter λ, which should be selected to reduce the second term in Eq. (7.33), or in other words, generate an over-regularized initial inverse solution to suppress the noise effect. (ii) Substitute $\vec{X}$ with $\vec{X}_0$ in Eq. (7.33) and calculate the objective function. (iii) Obtain a new optimal regularization parameter λ' by minimizing the objective function in Eq. (7.33). (iv) Repeat (i)–(iii), replacing λ with the new λ' until $\left\| \lambda - \lambda' \right\| / \left\| \lambda \right\| < \varepsilon$, where ε is a preset small number representing the condition of convergence. If the correlation between signal and noise is not negligible, the objective function in Eq. (7.33) can be rewritten in a more general form (Hori *et al.*, 2002):

$$J = \left\| \vec{X} - BA\vec{X} \right\|^2 + \text{tr}(BQB^T) + 2 < B\vec{\phi} - BA\vec{X}, \vec{X} - BA\vec{X} > \tag{7.34}$$

where <> denotes the inner product.

7.3.3.7. Other Statistical Methods

Other methods based on statistical considerations have also been proposed for the regularization parameter determination. For example, if the expectations of the noise and measurement are both available, the truncation parameter of TSVD in Eq. (7.20) could be determined by (Shim and Cho, 1981; Jeffs *et al.*, 1987; Gencer *et al.*, 1996)

$$k = \max_i \left\{ i \,\middle|\, \frac{\lambda_i^2}{\lambda_1^2} \geq \frac{E\left(\|\vec{n}\|^2\right)}{E\left(\|\vec{\phi}\|^2\right)} \right\} \tag{7.35}$$

Another popular method for choosing the regularization parameter is the generalized cross-validation (GCV) method proposed by Golub *et al.* (1979). The GCV technique is based on the statistical considerations that a good value of the regularization parameter should predict missing data values, therefore no *a priori* knowledge about the error norms is required.

7.4. TWO-DIMENSIONAL CORTICAL IMAGING TECHNIQUE

7.4.1. CONCEPT OF CORTICAL IMAGING TECHNIQUE

The recent development of cortical imaging technique (CIT) employs a distributed source model, in which the equivalent sources are distributed in 2D cortical surface, and no ad hoc assumption on the number of source dipoles is needed. Using an explicit biophysical model of the passive conducting properties of a head, the CIT attempts to deconvolve a measured scalp potential distribution into a distribution of the electrical potential or current dipoles over the cortical surface. The schematic diagram is shown in Figure 7.6 to illustrate the concept of the CIT.

In the cortical current imaging, an equivalent cortical current dipole distribution is directly estimated from the scalp potentials (Dale and Sereno, 1993; Srebro, 1996; Srebro and Oguz, 1997; Babiloni *et al.*, 2000; Cincotti *et al.*, 2001; Hori and He, 2001; Hori *et al.*, 2004). The rationale for the cortical current imaging is based on the observation that the scalp EEG signals are mainly contributed by cortical pyramidal cells, whose primary current flow

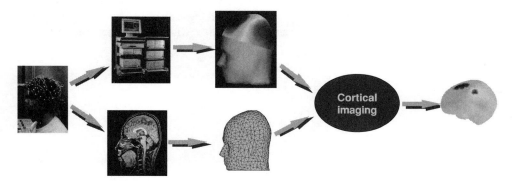

FIGURE 7.6. The schematic diagram of cortical imaging technique (CIT). The scalp potentials are recorded using a multichannel data acquisition system. The realistic geometry head volume conductor model can be constructed from the MR images of the subject. By solving the inverse problem, the CIT deconvolves the blurred scalp potential distribution into an equivalent cortical source distribution with higher spatial resolution.

axis is normal to the local cortical surface (Mitzdorf, 1985). Another approach is the cortical potential imaging, which estimates the epicortical potentials from the scalp EEG (Sidman *et al.*, 1990; Le and Gevins, 1993; Srebro *et al.*, 1993; Gevins *et al.*, 1994; Nunez *et al.*, 1994; He *et al.*, 1996, 1999, 2001a, 2002b; Babiloni *et al.*, 1997; Edlinger *et al.*, 1998; He, 1998, 1999; Wang and He, 1998; Ollikainen *et al.*, 2001; Zhang *et al.*, 2003). Because the cortical potential distribution can be experimentally measured (Gevins *et al.*, 1994; Towle *et al.*, 1995; He *et al.*, 2002b; Zhang *et al.*, 2003) and compared to the inverse imaging results, the cortical potential imaging approach is of clinical importance. Essentially not affected by the insulating skull layer, both cortical current imaging and cortical potential imaging offer much enhanced spatial resolution in assessing the underlying brain activity as compared to the blurred scalp potentials.

7.4.2. CORTICAL CURRENT IMAGING

The cortical current imaging employs an equivalent cortical dipole layer (DL) source model, in which a layer of current dipoles is normally oriented with respect to the local cortical surface. Therefore the cortical current imaging is also termed the cortical dipole-layer imaging (CDI) (He *et al.*, 2002c).

7.4.2.1. Forward Theory of the Equivalent Cortical Dipole Layer Model

Although the scalp EEG is primarily generated by cortical sources, brain sources may also be located in subcortical regions. Therefore, it is important to provide a theoretical justification on the use of equivalent cortical dipole layer (DL) to account for the neural electric activity within the entire brain. An equivalent DL theory based on electromagnetics has been recently introduced (He *et al.*, 2002c), which gives insight on the equivalence of representing brain electric sources by means of a closed-current DL surrounding the cortex.

Considering an arbitrary-shaped head model as illustrated in Figure 7.7, where S' represents the scalp surface and S represents the DL. Denote V_0 as the volume bounded within S, and V the volume bounded by S and S'. If all brain electric sources are in V_0 and there is no active source in V, then for any given point x within V, the potential function

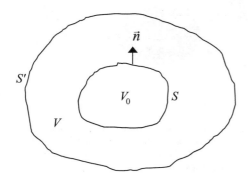

FIGURE 7.7. Illustration of the equivalent dipole layer in an arbitrary-shaped head model.

$\varphi(x)$ can be expressed as (Jackson, 1975)

$$\varphi(x) = \oint_S \left[G(x, x') \left(\sigma \frac{\partial \varphi(x')}{\partial n'} \right) - \sigma \varphi(x') \frac{\partial G(x, x')}{\partial n'} \right] ds', \quad \text{for } x' \text{ on } S \quad (7.36)$$

where $\partial/\partial n'$ is the normal derivative at surface S, σ is the conductivity in V_0, $G(x,x')$ is the Green function, which is the potential solution of a point current source inside a volume conductor model.

Different from the conventional Green function method (Jackson, 1975), Eq. (7.36) can be solved by keeping G unchanged, while modifying function φ by decomposing it into two potential components produced by the sources in volumes V and V_0, respectively. Because the sources in volume V make no contribution to the surface integral in Eq. (7.36), we can replace $\varphi(x)$ with a proper $\varphi'(x)$ so that

$$\varphi'(x') = \varphi'(x')_V + \varphi(x')_{V_0}, \quad \text{for } x' \text{ on } S \quad (7.37a)$$

and

$$\frac{\partial \varphi'(x')}{\partial n'} = \frac{\partial \varphi'(x')_V}{\partial n'} + \frac{\partial \varphi(x')_{V_0}}{\partial n'} = 0, \quad \text{for } x' \text{ on } S \quad (7.37b)$$

Then Eq. (7.36) can be simplified to

$$\varphi(x) = - \oint_S \sigma \varphi'(x') \frac{\partial G(x, x')}{\partial n'} ds' = \oint_S \sigma \varphi'(x') G_d(x, x') ds', \quad \text{for } x' \text{ on } S \quad (7.38)$$

where $G_d = -\partial G/\partial n'$ is the Green function of a dipole on surface S oriented normally outward. Therefore, Eq. (7.38) can be considered as an equivalent DL model, with an equivalent dipole source density of $f_d = \sigma \varphi'(x')$, and with Green function of G_d. Equation (7.38) is valid for an arbitrary geometric model and numerical methods can be applied to calculate G_d. Particularly, when the equivalent DL is considered to be a spherical surface, the closed solution of f_d can be obtained, with $\varphi'(x')$ being the surface potential produced by a dipole inside a homogeneous conducting sphere with boundary condition satisfying Eqs. (7.37a,b). This implies that f_d is proportional to the potential over the same spherical surface when the exterior space of the DL is replaced by air. In other words, the equivalent DL source density distribution may reflect the potential distribution over this layer, had the upper medium been removed, such as during open-skull surgery (He *et al.*, 2002c).

7.4.2.2. Inverse Problem of Cortical Dipole-Layer Imaging

On the basis of the above forward theory, the cortical current imaging or CDI solves the EEG inverse problem by reconstructing the equivalent dipole source distribution over the DL, which is essentially not affected by the low-conductivity skull layer, thus has higher spatial resolution as compared to the scalp potential map.

Discretizing Eq. (7.38) as

$$\varphi(x) = \sum_i (f_d(i)\Delta s_i)G_d(x, x'), \quad \text{for } x' \text{ on } S \tag{7.39}$$

Denote the source density weighed by discrete grid area $f_d(i)\Delta s_i$ as the equivalent DL source strength. Then Eq. (7.39) linearly relates the potential φ for x in V, with the equivalent DL source strength, by the discrete Green function G_d. Particularly, when φ is the measured scalp potential $\tilde{\phi}$, and assuming the equivalent DL source strength to be the unknown vector $\vec{X}$, then a linear relationship as stated in Eq. (7.11) could be established (in the presence of noise $\vec{n}$), where the transfer matrix A is calculated from the discrete Green function G_d.

Different linear inverse filters and regularization techniques can be applied to solve the inverse problem of the CDI. Using the TSVD inverse filter and the MINP regularization method, Figure 7.8 shows two typical examples of cortical current imaging of simulated

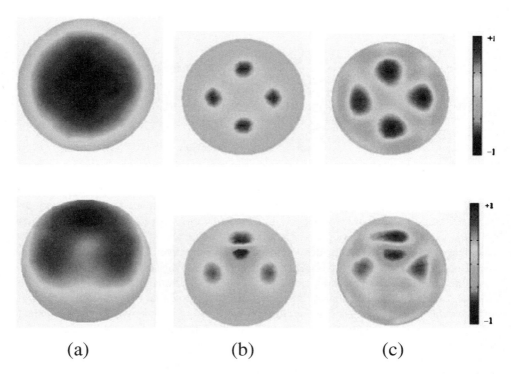

(a)   (b)   (c)

FIGURE 7.8. Two simulation examples of cortical dipole imaging using a three-sphere inhomogeneous head model. The top panels correspond to four radial dipoles positioned at $0.7 \cdot (\pm \sin(\pi/6), 0, \cos(\pi/6))$ and $0.7 \cdot (0, \pm \sin(\pi/6), \cos(\pi/6))$, and the bottom panels correspond to the configuration of three dipole sources, with two radial dipoles positioned at $0.6 \cdot (\pm \sin(\pi/6), 0, \cos(\pi/6))$ and one tangential dipole positioned at $0.7 \cdot (0, \sin(\pi/6), \cos(\pi/6))$: (a) scalp potential maps contaminated with 5% Gaussion white noise; (b) forward solution of the cortical equivalent DL maps; and (c) inversely estimated strength maps of cortical equivalent DL. (From Fig. 3 in He *et al.*, *Clinical Neurophysiology*, 2002c, with permission)

brain electric sources in a three-sphere inhomogeneous head model (He et al., 2002c). In both examples, (a) displays the 5% Gaussion white noise (GWN) contaminated scalp potential map, (b) shows the forward solution of the cortical equivalent DL, and (c) shows the inversely estimated strength map of cortical equivalent DL. Notably, the scalp potential maps were blurred and distorted by the head volume conductor and additive noise. The forward-calculated strength maps of cortical equivalent DL clearly indicate the well-localized brain electric activities corresponding to the primary dipole sources in both examples. The inverse maps are highly consistent with the forward solution of the cortical equivalent DL, and show strong capability of the CDI in correcting the blurring effect caused by the head volume conductor.

7.4.3. CORTICAL POTENTIAL IMAGING

The cortical potential imaging is also referred as downward continuation (Le and Gevins, 1993; Gevins et al., 1994), in which the electrical potentials over the epicortical surface are reconstructed from the electrical potentials over the scalp surface.

7.4.3.1. Indirect Approach

An early attempt to reconstruct the cortical potentials used an intermediate hemisphere equivalent DL to generate an inward harmonic potential function in a homogeneous sphere head volume conductor model (Sidman et al., 1990). The inverse procedure estimated the equivalent DL strength distribution from the scalp EEG, then the cortical potentials were reconstructed by solving the forward problem, from the estimated equivalent DL to the cortical potentials.

Recently, several approaches have been reported to further the effort along this line. He and co-workers used a concentric three-sphere head model to include the significant conductivity inhomogeneity, the skull, in the head volume conductor, and a closed-spherical DL with higher density to improve the numerical accuracy of reconstructing the cortical potentials corresponding to superficial sources (He et al., 1996; He, 1998, 1999; Wang and He, 1998). Babiloni et al. (1997) further extended the approach to include both the skull inhomogeneity and realistic geometry of the head by means of the BEM, based on the isolated problem approach (Hamalainen and Sarvas, 1989; Meijs et al., 1989). The dipole layer was also extended to a surface with similar geometric shape to that of the dura mater compartment.

Figure 7.9 shows a representative example of the indirect cortical potential imaging of the P300 and Novelty P3 components using the three-sphere head model (He et al., 2001a). The ERP data were acquired using 129 electrodes from 15 healthy subjects by running the auditory oddball paradigm (Spencer et al., 1999). The rare events elicited the classical P300 component characterized by the diffused centroparietal scalp potential distribution (Figure 7.9a). On the other hand, the novel events first elicited a maximal positivity at the frontal electrodes, succeeded by a more parietal positivity during the later segments of the P300 epoch (Figure 7.9b). In comparison to the scalp potential maps, both frontal and parietal activities can be well separated from the reconstructed cortical potential maps corresponding to either rare or novel events (Figure 7.9c,d). The cortical potential maps in response to rare and novel events differ mainly in the relative amplitude of the components

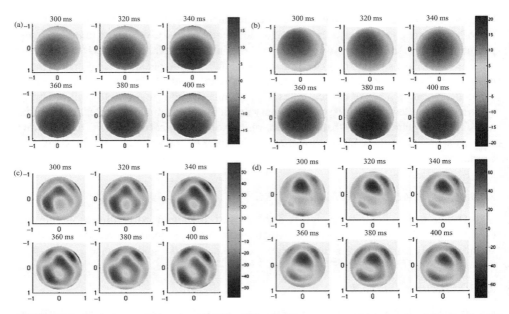

FIGURE 7.9. Cortical potential imaging of P300 and Novelty P3 components: (a) scalp potential maps elicited by the rare event; (b) scalp potential maps elicited by the novel event; (c) cortical potential maps in response to the rare event; and (d) cortical potential maps in response to the novel event. Maps are shown at six different time instances. Colorbar unit: microvolts. (From He *et al.*, *Human Brain Mapping*, 2001a, with permission)

they elicit. For the rare events, the parietal component is strong all along the P300 interval. For the novel events, the frontal component is more dominant at the early stage of P300 interval, and the parietal activity progressively increases thereafter. The separation of frontal and parietal activities in the reconstructed cortical potential maps provides clear evidence that the Novelty P3 and P300 are two independent ERP components elicited by deviant events.

7.4.3.2. Direct Approach

Gevins and co-workers (Le and Gevins, 1993; Gevins *et al.*, 1994) developed the "Deblurring" approach to estimate directly the cortical potentials from the scalp EEG recordings using the FEM. In this approach, each subject's finite element head model was constructed from the MR images. Poisson's equation was applied to the conducting volume between the scalp and the cortical surface, and the FEM was applied to handle the complex geometry and varying conductivity of the head. An initial empirical validation (without quantitative comparison) of their approach was conducted by comparing estimated cortical potentials with those measured with subdural grid recordings from two neurosurgical patients, and promising results were reported in their experimental studies and dramatic improvement of spatial resolution was achieved in the cases shown.

An alternative approach has been explored by Srebro *et al.* (1993), who directly linked the evoked potential field on the scalp with the cortical potential field by means of the

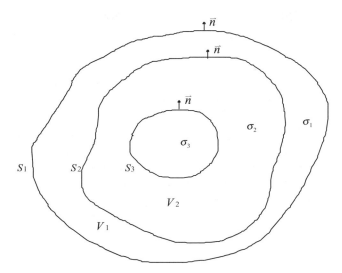

FIGURE 7.10. Schematic illustration of the three-shell head volume conductor model.

BEM (Barr *et al.*, 1977). The volume conductivity between the surfaces was assumed to be homogeneous, and detailed anatomical information for each subject was obtained from MR images. Regularized inversion was applied to obtain the cortical surface potential estimation. Although their physical and human visual evoked potential experiments demonstrated the more localized nature of the estimated cortical potentials than their scalp field counterparts, the effect of significant conductivity inhomogeneity—the skull—was not considered in their head model.

Recently a new cortical potential imaging algorithm has been reported (He *et al.*, 1999), in which both the realistic geometry and the inhomogeneity of the head can be taken into consideration using the BEM. In this approach, the head is modeled by a three-shell volume conductor as shown in Figure 7.10. The three shells represent the scalp, the skull, and the brain, respectively, and each shell is homogeneous but has different electrical conductivities. Because brain electric sources exist only inside the brain, there are no active sources existing in the scalp V_1 and skull V_2. So Green's second identity can be applied to V_1 and V_2 separately. Applying Green's second identity to the volume V_1, discretizing the scalp surface S_1 and the skull surface S_2 into triangular elements, and taking the limit of observation point approaching the surface element on S_1 and S_2, respectively, from the inside of V_1, the following equations can be obtained:

$$P_{11}\vec{U}_1 + P_{12}\vec{U}_2 + G_{12}\vec{\Gamma}_2 = 0 \qquad (7.40)$$

$$P_{21}\vec{U}_1 + P_{22}\vec{U}_2 + G_{22}\vec{\Gamma}_2 = 0 \qquad (7.41)$$

where $\vec{U}_k$ is the vector consisting of the electrical potentials at every surface element on S_k, and $\vec{\Gamma}_k$ is the vector consisting of the normal derivatives of the electrical potentials at every triangle element on S_k but just inside of V_1. P_{11}, P_{12}, P_{21}, P_{22}, G_{12}, and G_{22} are coefficient matrices (Barr *et al.*, 1977).

Similarly, applying Green's second identity to the volume V_2 between the skull surface S_2 and cortical surface S_3, other two linear equations can be obtained (He et al., 1999):

$$P'_{22}\vec{U}'_2 + P_{23}\vec{U}_3 - G_{22}\vec{\Gamma}'_2 + G_{23}\vec{\Gamma}_3 = 0 \tag{7.42}$$

$$P_{32}\vec{U}'_2 + P_{33}\vec{U}_3 + G_{32}\vec{\Gamma}'_2 + G_{32}\vec{\Gamma}_3 = 0 \tag{7.43}$$

where $\vec{U}'_2$ is the vector consisting of the electrical potentials at every surface element on S_2, and $\vec{\Gamma}'_2$ is the vector consisting of the normal derivatives of the electrical potentials at every triangle element on S_2 but just inside of V_2. Combining Eqs. (7.40)–(7.43), and using the boundary conditions on S_2 that the electrical potential and the normal component of current must be continuous, the cortical potential $\vec{U}_3$ can be related to the scalp potential $\vec{U}_1$ by (He et al., 1999)

$$\vec{U}_1 = T_{13}\vec{U}_3 \tag{7.44}$$

where T_{13} is the transfer matrix from the cortical potential to the scalp potentials. In practice, the vector of the measured scalp potentials $\vec{\phi}$ is a subset of the potential vector $\vec{U}_1$, and the vector $\vec{U}_3$ is the unknown source distribution $\vec{X}$. Therefore, in the presence of noise $\vec{n}$, $\vec{\phi}$ can be connected with the cortical potentials by the linear equation (7.11), where A is the submatrix of T_{13}. To account for the low-conductivity skull layer, an adaptive approach has been developed to achieve high numerical accuracy in the transfer matrix A(He et al., 1999).

This BEM-based algorithm offers unique features of connecting directly and efficiently the cortical potentials to the scalp potentials via a transfer matrix with inclusion of the low-conductivity skull layer. The excellent performance of this new approach has been systematically evaluated by computer simulations (He et al., 1999, 2002b). In addition, SEP experiments were conducted in three patients to validate the algorithm by a quantitative comparison of the estimated cortical potentials with the direct potential recordings from a subdural grid over the somatosensory cortex (He et al., 2002b). As a typical example, Figure 7.11 shows (a) the measured scalp potential maps, (b) the direct recorded grid potentials, and (c) the estimated grid potentials for one subject, at eight time instants about 30 ms after the onset of right median nerve stimuli. The scalp potential maps show dipolar pattern of N/P30, with frontal negativity and parietal positivity over the left scalp. The smearing effect of the scalp potential maps was greatly reduced in the inversely estimated cortical potential maps, which show much more localized areas of positivity and negativity in the posterior edge of the electrode grid. Note the estimated and recorded grid potentials have similar distribution patterns, with an averaged CC value of 0.84 ± 0.01. Moreover, the central sulcus was clearly demarcated in both the estimated and recorded grid potential maps, by the separation of negative and positive potential extrema.

7.4.4. MULTIMODAL INTEGRATION

In an effort to further improve the localization accuracy and spatial resolution of the CIT inverse solution, other imaging modalities could be incorporated with EEG measurement to provide complementary information.

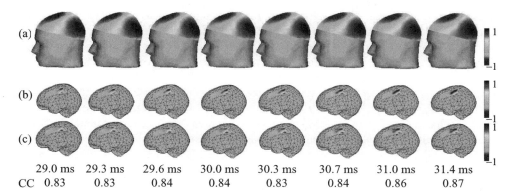

29.0 ms	29.3 ms	29.6 ms	30.0 ms	30.3 ms	30.7 ms	31.0 ms	31.4 ms
CC 0.83	0.83	0.84	0.84	0.83	0.84	0.86	0.87

FIGURE 7.11. At eight time instants about 30 ms after the onset of right median nerve stimuli for one patient: (a) the recorded scalp potential maps; (b) the direct recorded subdural grid potentials; and (c) the estimated subdural grid potentials. All the maps are normalized and the colorbars are shown on the right. The CC values between the estimated and recorded subdural grid potentials for each time instant are listed at the bottom. (From He *et al.*, *NeuroImage*, 2002b, with permission)

7.4.4.1. *Integration of EEG and MEG*

One representative approach is the EEG–MEG integration (Dale and Sereno, 1993; Phillips *et al.*, 1997b; Gencer and Williamson, 1998; Baillet *et al.*, 1999; Babiloni *et al.*, 2001). EEG has contribution from both radial and tangential components of the primary current. On the other hand, MEG measurements are insensitive to the volume conductor effects, and are more sensitive to tangential superficial sources. Therefore, the complementary aspects of the bioelectric and biomagnetic measurements suggest that the combination of EEG and MEG may provide enhanced reliability and accuracy of the CIT.

Figure 7.12 shows an example of the cortical imaging of the movement-related brain activity based on EEG, MEG, and combined EEG–MEG data, when the subject was performing a self-paced right middle finger extension task (Babiloni *et al.*, 2001). After onset

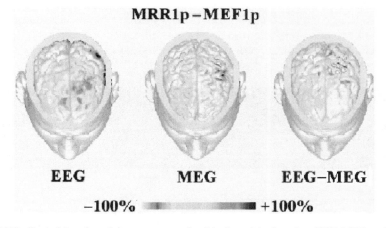

FIGURE 7.12. Cortical imaging of the movement-related brain activity based on EEG, MEG, and combined EEG–MEG data. (From Babiloni *et al.*, *Human Brain Mapping*, 2001, with permission)

of the movement, the MEG recorded a peak in the movement-evoked field at about 105 ms (MEF1p), whereas the EEG recorded a peak in the movement-related response at about 90 ms (MRR1p). The cortical imaging of the MRR1p (EEG) revealed a bilateral large frontal negativity and a slight centroparietal positivity across the central sulcus of the contralateral side. The cortical imaging of the MEF1p (MEG) showed a more restricted and contralaterally preponderant negativity over the primary motor-sensory areas (M1-S1). The cortical imaging of the MRR1p–MEF1p (EEG–MEG) also showed multiple restricted negative foci located mainly in the supplementary motor area (SMA) and contralateral M1-S1. Meanwhile, this estimate showed moderate positive activity across the whole sensorimotor cortex, in relation to those obtained by using EEG or MEG data alone. On the basis of the spatiotemporal analysis of cortical activation maps, it was suggested that the performance of cortical imaging was improved by the integration of EEG and MEG.

7.4.4.2. Integration of EEG and fMRI

During the past several years, the functional MRI (fMRI) constrained CIT has drawn great attention (Dale and Sereno, 1993; Liu *et al.*, 1998; Babiloni *et al.*, 2000; Dale *et al.*, 2000; Cincotti *et al.*, 2001; Bonmassar *et al.*, 2001; Babiloni *et al.*, in press).

On the basis of the Wiener inverse filter, Dale and Sereno (1993) proposed a framework to integrate EEG, MEG, and MRI (anatomical constraint from MRI and activation constraint from fMRI) to improve the source localization performance of the CIT. Note that in Eq. (7.22) the implementation of the Wiener filter requires an estimation of the noise covariance matrix Q and the signal covariance matrix R. In practice, Q can be estimated from the measurement data that is known to be source-free, such as the pre-stimulus EP data (Sekihara *et al.*, 1997; Van Veen *et al.*, 1997), whereas R can be estimated from the observed sensor covariance matrix by means of a linear approach (Sekihara and Scholz, 1995, 1996), or by means of a subspace formulation (Dale and Sereno, 1993). On the other hand, the signal covariance matrix R can also be estimated by using available spatial information on the hemodynamic responses obtained in fMRI, under the assumption that the brain regions that show increased metabolic activities are also the ones that are on average more electrically active over time (Dale and Sereno, 1993).

Using the hemodynamic response of fMRI as constrained with proper weighting (Liu *et al.*, 1998), Bonmassar *et al.* (2001) conducted a spatiotemporal cortical imaging of the visual evoked potential (VEP) in response to the full-field checkerboard pattern reversal visual stimuli. As an example, Figure 7.13 shows seven snapshots of the inversely estimated

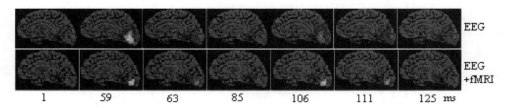

FIGURE 7.13. Spatiotemporal cortical imaging of the VEP for one subject. Millisecond-to-millisecond activations of the occipital region of the neocortex during the flashing checkerboard task are demonstrated. The upper row shows localizations for EEG alone. The bottom row shows localizations for fMRI-constrainted EEG. (Modified from Fig. 8 of Bonmassar *et al.*, *NeuroImage*, 2001, with permission)

cortical activity movie elicited by the visual stimulation, without and with fMRI constraint, for one subject. The spatiotemporal maps obtained using both methods reveal two peak activities corresponding to the N75 and the P100 components located in the occipital visual cortex. Nonetheless, the spatial extent of the fMRI-constrained source localization is more focal than the results based on EEG measurement alone. Therefore, the authors suggested that the combined EEG and fMRI analysis can estimate the cortical sources with higher spatiotemporal resolution than either approach alone.

7.4.5. SURFACE LAPLACIAN

In parallel to the development of the CIT, noteworthy is another high-resolution brain electric source imaging technique—surface Laplacian (SL). The SL does not need to solve the linear inverse problem; instead, it applies a spatial Laplacian filter to compensate for the head volume conduction effect and achieve a high-resolution source mapping directly over the scalp surface.

The SL has been considered an estimate of the local current density flowing perpendicular to the skull into the scalp, thus it has also been termed current source density or scalp current density (Perrin *et al.*, 1987; Nunez *et al.*, 1994). The SL has also been considered as an equivalent surface charge density corresponding to the surface potential (He and Cohen, 1992). In addition, the relationship between the SL and the cortical potentials has also been explored (Nunez *et al.*, 1994). Compared to the EEG inverse approaches, the SL approach does not require an exact knowledge of the conductivity distribution inside the head and has unique advantage of reference-independence.

Since Hjorth's early exploration on scalp Laplacian EEG (Hjorth, 1975), tremendous effort has been made to develop reliable and easy-to-use SL techniques. Noteworthy is the development of spherical spline SL (Perrin *et al.*, 1987), ellipsoidal spline SL (Law *et al.*, 1993), and the realistic geometry spline SL (Babiloni *et al.*, 1996, 1998; He *et al.*, 2001b). For a detailed review on the SL, see ref (He and Lian, 2004).

7.5. THREE-DIMENSIONAL BRAIN ELECTRIC SOURCE IMAGING

7.5.1. CHALLENGES OF 3D NEUROIMAGING

Tremendous progress has been made during the past several years in 3D neuroimaging, in which the brain electric sources are distributed in the 3D brain volume. Similar to the CIT inverse problem, the 3D neuroimaging approach is also based on a distributed source model, and is implemented by solving the linear inverse problem as detailed in Section 7.3. On the other hand, the 3D neuroimaging approach faces greater technical challenges: by extending the solution space from 2D cortical surface to 3D brain volume, the number of unknown sources increases dramatically. As a result, the inverse problem is even more underdetermined and the inverse solution is usually smeared because of regularization procedure. In addition, it becomes more important to retrieve depth information of the sources in 3D neuroimaging. Although the cortex can be modeled as a folded surface in the cortical imaging approach so that sources in sulci and gyri have different eccentricities, deeper sources probably exist below the cortical layer, such as in the amygdala and hippocampal formation.

7.5.2. INVERSE PROBLEM OF THE 3D NEUROIMAGING

As in the cortical imaging approach, the most popular linear inverse solution is the minimum-norm solution, which estimates the 3D brain source distribution with the smallest L2-norm solution vector that would match the measured data (Hamalainen and Ilmoniemi, 1984, 1994). Different regularization techniques as detailed in Section 7.3 can be used to suppress the effects of noise.

However, the standard minimum-norm solution has intrinsic bias that favors superficial sources because the weak sources close to the sensors can produce scalp EEG with similar strength as strong sources at deep locations. To compensate for the undesired depth dependency of the original minimum-norm solution, different weighting methods were introduced. The representative approaches include the normalized weighted minimum norm (WMN) solution (Jeffs *et al.*, 1987; Gorodnitsky *et al.*, 1995) and the Laplacian weighted minimum norm (LWMN) solution, also termed LORETA (Pascual-Marqui *et al.*, 1994; Pascual-Marqui and Michel, 1994; Lantz *et al.*, 1997; Mulert *et al.*, 2001).

The WMN compensates for the lower gains of deeper sources by using lead field normalization. In the absence of noise, Eq. (7.11) can be rewritten as

$$\vec{\phi} = A W^{-} W \vec{X} \tag{7.45}$$

The concomitant WMN inverse solution is given by (Jeffs *et al.*, 1987; Gorodnitsky *et al.*, 1995)

$$\vec{X}_{\text{WMN}} = W W^{\text{T}} A^{\text{T}} (A W W^{\text{T}} A^{\text{T}})^{-} \vec{\phi} \tag{7.46}$$

where W is the weighting matrix acting on the solution space. Most commonly, W is constructed as a diagonal matrix (Jeffs *et al.*, 1987; Gorodnitsky *et al.*, 1995):

$$W = \text{diag}(\|a_1\|, \|a_2\|, \ldots, \|a_n\|) \tag{7.47}$$

where $A = (a_1, a_2, \ldots, a_n)$. Thus, by using the norm of each column of the transfer matrix as the weighting factor for the corresponding position in the solution space, the contributions of the entries of the transfer matrix to a solution are normalized.

The LWMN approach defines a combined weighting operator LW, where L is a Laplacian operator acting on the solution space and W is same as in Eq. (7.47). For a 3D solution space, it is a 3D discrete Laplacian operator, and for a 2D solution space, for example on the cortical surface, it is a 2D Laplacian operator. The corresponding LWMN inverse solution is (Pascual-Marqui *et al.*, 1994; Pascual-Marqui and Michel, 1994)

$$\vec{X}_{\text{LWMN}} = (W L^{\text{T}} L W)^{-} A^{\text{T}} (A (W L^{\text{T}} L W)^{-} A^{\text{T}})^{-} \vec{\phi} \tag{7.48}$$

This approach combines the lead field normalization with the spatial Laplacian operator, and thus gives the depth-compensated inverse solution under the constraint of smoothly distributed sources.

Other variants of the minimum-norm solution were also proposed, by incorporating *a priori* information as constraint in a Bayesian formulation (Baillet and Garnero, 1997;

Phillips *et al.*, 1997a), or by estimating the source-current covariance matrix from the measured data in a Wiener formulation (Sekihara and Scholz, 1995, 1996). In addition, a "weighted resolution optimization" (WROP) method has been proposed (Grave de Peralta-Menendez *et al.*, 1997) in an effort to optimize the resolution matrix (Backus and Gilbert, 1968; Menke, 1984), which can be used to evaluate the goodness of the inverse filter in terms of several figures of merit, such as source identifiability, source visibility, and so on (Grave de Peralta-Menendez *et al.*, 1996; Grave de Peralta-Menendez and Gonzalez-Andino, 1998a,b).

In addition, beam-former techniques have also been used for 3D brain electric source localization. To localize distributed brain electric sources, a linearly constrained minimum-variance beam-former approach has been developed for EEG source localization, by designing a bank of narrow-band spatial filters where each filter passes signals originating from a specified location within the brain while attenuating signals from other locations (Van Veen *et al.*, 1997). Recently, an adaptive beam-former technique has also been developed for solving the MEG inverse problem, and the numerical experiments demonstrated that this technique performed significantly better than the previous minimum-variance-based beam-former technique, with respect to the spatial resolution and the output signal-to-noise ratio (Sekihara *et al.*, 2001).

7.5.3. 3D BRAIN ELECTRIC SOURCE MODELS

As stated in Section 7.1.4, brain electric sources can be estimated by using different equivalent source models. Therefore, alternative approaches have been developed to solve the 3D inverse problem by changing the commonly used equivalent dipole source (EDS) model.

One of the approaches, termed ELECTRA (Grave de Peralta-Menendez *et al.*, 2000), reformulated the inverse problem to solve some different physical magnitudes, for example, the 3D electric potential distribution over the brain volume. Another approach developed by He and colleagues (He *et al.*, 2002a) employed an equivalent current source (ECS) model, based on the observation that the equivalent volume current source (monopole) can be used to equivalently represent the bioelectric sources originating from neuronal membrane excitation (Plonsey, 1969; Malmivuo and Plonsey, 1995; Gulrajani, 1998). The major advantage of these approaches is that the number of unknowns in the new source models is reduced to one third of that in the conventional equivalent dipole source model, because each dipole has three directional components whereas the potential distribution and the ECS distribution are both scalar fields. The reduction in the dimension of the solution space not only can improve the computational efficiency, but also can reduce the underdetermination of the 3D inverse problem and improve the stability of the inverse solution.

As an example, Figure 7.14 shows the results of 3D source imaging of P100 activity in a pattern reversal VEP experiment based on ECS and EDS models (He *et al.*, 2002a). The LWMN approach was used to solve the linear inverse problem. Two subjects were respectively given left and right visual field stimuli, with expectation of visual cortex activation on the contralateral hemisphere of the brain. But paradoxically, the half visual field stimuli elicited stronger positive potential distribution over the midline or ipsilateral side of the scalp during the P100. Nonetheless, both the ECS and the EDS estimates clearly indicate that the contralateral visual cortex was activated, thus effectively eliminating the misleading

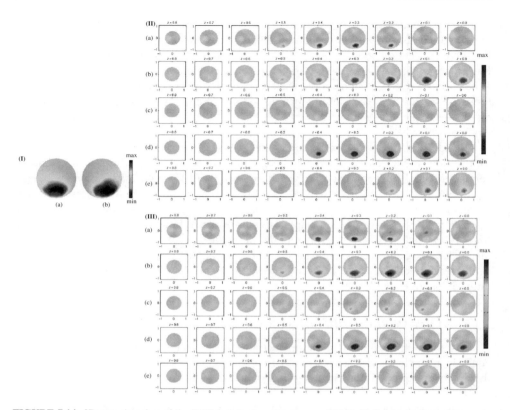

FIGURE 7.14. 3D neuroimaging of the P100 activity in pattern reversal VEP. (I) Top-back view of the scalp potential maps at P100 of VEP experiment: (a) Subject A was presented with the left visual field stimuli, and (b) Subject B was presented with the right visual field stimuli. (II) The estimated (a) ECS, (b) EDS moment, (c) EDS x-component, (d) EDS y-component, and (e) EDS z-component distributions corresponding to the scalp potential map shown in (I-a). (III) The estimated (a) ECS, (b) EDS moment, (c) EDS x-component, (d) EDS y-component, and (e) EDS z-component distributions corresponding to the scalp potential map shown in (I-b). (From He *et al.*, 2002a, with permission, © IEEE)

far field observed in the scalp potential. In addition, the source/sink distribution estimated by the ECS approach suggests a current flow pathway consistent with the EDS imaging results. Note that although the EDS imaging can also reveal the direction information of the dipole sources, by simultaneously displaying its x-, y-, and z-component distributions, the ECS imaging approach reduces the solution space to one third of that of the EDS imaging approach, thus substantially improving the computational efficiency.

7.5.4. NONLINEAR INVERSE PROBLEM

Because the 3D EEG inverse problem is highly underdetermined, the solutions obtained by the minimum-norm inverse and its variants are usually associated with relatively low spatial resolution. To overcome this problem, several nonlinear inverse approaches have been introduced to achieve more localized imaging results.

One method is to solve the inverse problem based on the L1-norm instead of the commonly used L2-norm (Matsuura and Okabe, 1995, 1997; Uutela et al., 1999). The use of the L1-norm requires solving a nonlinear system of equations for the same number of unknowns as the L2-norm inverse approach, thereby requiring much more computational effort. However, the L1-norm approach usually provides much more focal solutions and a more robust behavior against outliers in the measured data (Fuchs et al., 1999). Different nonlinear optimization approaches have been suggested, including the iteratively reweighted least squares method and the linear programming techniques (Fuchs et al., 1999).

Srebro (1996) proposed an iterative inverse method, termed shrinking ellipsoid method, and applied it to the spatiotemporal cortical imaging of the human VEP activities (Srebro and Oguz, 1997). In this approach, a minimum-norm solution is used to define a region of interest, an ellipsoid, which spans a size-reduced solution space. Then the minimum-norm approach is used iteratively, and the ellipsoid shrinks until finally there are too few dipole sources to predict the observed scalp potentials within a reasonable error bound. Gavit et al. (2001) proposed to build several ellipsoids instead of a unique area around all the activity centers and define the new source space at their intersection with the cortical surface. An adaptive regularization technique was also introduced to achieve multiresolution source imaging.

Through a different approach, a nonparametric algorithm for finding localized 3D inverse solutions, termed focal underdetermined system solution (FOCUSS), was proposed by Gorodnitsky et al. (1995). This algorithm has two integral parts: a low-resolution initial estimate of the inverse solution, such as the minimum-norm inverse solution, and the iteration process that refines the initial estimate to the final localized source solution. The iterations are based on weighted norm minimization of the dependent variable, with the weights being a function of the preceding iterative solutions. In other words, the weighting matrix at the kth iteration is given by (Gorodnitsky et al., 1995; Gorodnitsky and Rao, 1997)

$$W_k = \vec{X}_{k-1}^{-} \text{diag}(\|a_1\|, \|a_2\|, \ldots, \|a_n\|) \tag{7.49}$$

where $\vec{X}_{k-1}$ is the inverse solution obtained from the $(k-1)$th iteration, and $\vec{X}_0$ is the WMN solution given by Eqs. (7.46)–(7.47). By taking W_k to be updated from previous step solution, the stronger components in the solution space will be enhanced, and the process will continue until most source elements are reduced to zero.

Recently, the integration of FOCUSS with LWMN (or LORETA) has been proposed to enhance the spatial resolution of LWMN (LORETA) (Yao and He, 2001). Furthermore, a self-coherence enhancement algorithm (SCEA) has also been proposed to enhance the spatial resolution of the 3D inverse estimate (Yao and He, 2001). This algorithm provides a noniterative self-coherence solution, which is a function of the high-order self-coherence estimate of an unbiased smooth estimate of the underdetermined 3D inverse solution:

$$\vec{X} = \alpha \vec{X}' = \alpha \left(x_1^T f(\|x_1\|), x_2^T f(\|x_2\|), \ldots, x_n^T f(\|x_n\|) \right) \tag{7.50}$$

where $\vec{X}_0 = (x_1, x_2, \ldots, x_n)$ is the initial estimate obtained by the LWMN procedure. The parameter α is used to rectify the amplitude bias of the initial estimate compared to the actual source strength, and can be estimated by

$$\alpha \approx \sqrt{\frac{\vec{\phi}^T \vec{\phi} - \vec{n}^T \vec{n}}{(A\vec{X}')^T (A\vec{X}')}} \tag{7.51}$$

The function f in Eq. (7.50) is a high-order self-coherence function, which is defined as

$$f(\|x_i\|) = \|x_i\|^K / \max \left\{ \|x_1\|^K, \|x_2\|^K, \ldots, \|x_n\|^K \right\} \tag{7.52}$$

where the order K is used to rectify the blurring level of the initial smooth estimate, and could be determined by the blurring level of the actual source distribution as represented by a normalized blurring index (Yao and He, 2001).

7.6. DISCUSSION

The ultimate goal of electrophysiological neuroimaging is to image brain electric activity with a high resolution in both time and space domains based on noninvasive EEG recordings. Such noninvasive and high-resolution brain mapping techniques would bring significant advancement in the fields of clinical neurosurgery, neural pathophysiology, cognitive neuroscience, and neurophysiology. For example, it will facilitate presurgical planning and noninvasive localization; delineate the epileptic zone in seizure patients; characterize the brain dysfunction in schizophrenia, depression, alcoholism, and Alzheimer's; localize and image cortical regions contributing to cognitive tasks; and help understand how the "Mind" is implemented in the brain.

During the past decades, numerous techniques have been developed for brain electric source imaging by solving the EEG inverse problem. Dipole source localization is particularly useful for localizing isolated focal brain electric sources, whereas the distributed source imaging has the capability of imaging spatially distributed sources, such as 2D cortical imaging and 3D brain tomographic imaging. The choice of inverse approach depends on the particular application, because each inverse algorithm has its own advantage and limitation.

The major limitation of the dipole source localization is that it requires *a priori* knowledge of the number of dipole sources, which is usually small considering the computation efficiency, and its estimation is usually difficult in practice. The distributed source imaging, on the one hand, makes no assumption of the number of neural sources and, on the other hand, it has to deal with the high underdetermination of the inverse problem. Cortical imaging technique has a great potential to compensate for the head volume conduction effect and achieve a high-resolution mapping of cortical activities, whereas the 3D neuroimaging approach has the capability of retrieving the depth information of the distributed brain electric sources. A recent trend in the 3D distributed source imaging is by using the realistic geometry volume conductor model constructed from the MR or CT images, through which the anatomical constraints become feasible and more clinically meaningful result interpretations may be achieved. Another major trend in 3D neuroimaging is the development of novel techniques that aim to overcome the smoothing effect of the inverse solution, either by reducing the underdetermination of the inverse problem (Grave de Peralta-Menendez *et al.*, 2000; He *et al.*, 2002a) or by some nonlinear inverse approaches (Gorodnitsky *et al.*, 1995; Srebro, 1996; Fuchs *et al.*, 1999; Yao and He, 2001).

The performance of the distributed source imaging is closely associated with the linear inverse filter and regularization technique being selected. The regularization technique is critical in suppressing noise and obtaining a stable inverse solution. Although many regularization techniques have been proposed (Golub *et al.*, 1979; Shim and Cho, 1981; Morozov, 1984; Colli Franzone, 1985; Hansen, 1990; Johnston and Gulrajani, 1997; Hori and He, 2001; Lian and He, 2001), none of them have been demonstrated to be universal, and different methods should be selected with respect to different cases. On the other hand, different inverse filters have been developed to target different applications based on various assumptions, such as the presence or absence of noise, the availability of statistical information on signal and noise, and so on. Not surprisingly, more robust and accurate inverse solutions can be obtained by incorporating the *a priori* information as a constraint, for example, the anatomical constraint, the temporal constraint, and the functional constraint. The anatomical constraint can be easily implemented by coregistration the EEG inverse solution with the structural brain images obtained from MRI (Dale and Sereno, 1993; George *et al.*, 1995; Baillet and Garnero, 1997). The temporal constraint can be achieved by selecting an epoch of EEG data as input to the inverse procedure under the assumption that the underlying bioelectric sources remain relatively invariant (Scherg and Von Cramon, 1985; Baillet and Garnero, 1997). The functional constraint has shown great promise during the past several years, by combining the electromagnetic and hemodynamic imaging modalities that were recorded using the same paradigm in the same subjects (Babiloni *et al.*, 2000, in press; Dale *et al.*, 2000; Bonmassar *et al.*, 2001; Cincotti *et al.*, 2001). The rationale for this multimodal integration is that neural activity generating EEG potentials increases glucose and oxygen demands (Magistretti *et al.*, 1999). A growing body of evidence suggests that there is close spatial coupling between elecrophysiologic signals and hemodynamic responses (Hess *et al.*, 2000; Ogawa *et al.*, 2000). However, many technical challenges still exist and care should be taken in order to make unbiased physiological interpretations based on coregistration studies (Nunez and Silberstein, 2000).

In conclusion, electrophysiological neuroimaging by solving the EEG inverse problem has a great potential for the noninvasive mapping of brain activation and function with a high spatiotemporal resolution. Despite many challenges, with the integrated effort of algorithm development, computer simulation, experimental exploration, clinical validation, and the availability of more powerful computing resources, it can be foreseen that electrophysiological neuroimaging will become an important probe for imaging neural abnormalities and understanding the human mind.

7.7. ACKNOWLEDGMENTS

This work was supported in part by grants NIH R01EB00178, NSF BES-0218736, NSF BES-0411898, NSF CAREER Award BES-9875344, and a grant from the IRIB Program.

REFERENCES

Ahlfors, S. P., Simpson, G. V., Dale, A. M., Belliveau, J. W., Liu, A. K., Korvenoja, A., Virtanen, J., Huotilainen, M., Tootell, R. B., Aronen, H. J., and Ilmoniemi, R. J., 1999, Spatiotemporal activity of a cortical network for processing visual motion revealed by MEG and fMRI, *J. Neurophysiol.* **82:**2545–2555.

Babiloni, F., Babiloni, C., Carducci, F., Fattorini, L., Onorati, P., and Urbano. A., 1996, Spline Laplacian estimate of EEG potentials over a realistic magnetic resonance-constructed scalp surface model, *Electroencephalogr. Clin. Neurophysiol.* **98:**363–373.

Babiloni, F., Babiloni, C., Carducci, F., Fattorini, L., Anello, C., Onorati, P., and Urbano, A., 1997, High resolution EEG: A new model-dependent spatial deblurring method using a realistically-shaped MR-constructed subject's head model, *Electroencephalogr. Clin. Neurophysiol.* **102**:69–80.

Babiloni, F., Carducci, F., Babiloni. C., and Urbano, A., 1998, Improved realistic Laplacian estimate of highly-sampled EEG potentials by regularization techniques, *Electroencephalogr. Clin. Neurophysiol.* **106**:336–343.

Babiloni, F., Carducci, F., Cincotti, F., Del Gratta, C., Roberti, G. M., Romani, G. L., Rossini, P. M., and Babiloni, C., 2000, Integration of high resolution EEG and functional magnetic resonance in the study of human movement-related potentials, *Methods Inf. Med.* **39**:179–182.

Babiloni, F., Carducci, F., Cincotti, F., Gratta, C. D., Pizzella, V., Romani, G. L., Rossini, P. M., Tecchio, F., and Babiloni, C., 2001, Linear inverse source estimate of combined EEG and MEG data related to voluntary movements, *Hum. Brain Mapp.* **14**:197–209.

Babiloni, F., Babiloni, C., Carducci, F., Cincotti, F., Astolfi, L., Basilisco, A., Rossini, P. M., Ding, L., Ni, Y., Cheng, J., Christine, K., Sweeney, J., and He, B., in press, Assessing time-varying cortical functional connectivity with the multimodel integration of high resolution EEG and fMRI data by Directed Transfer Function, NeuroImage.

Backus, G. E., and Gilbert, J. F., 1968, The resolving power of gross earth data, *Geophys. J. R. Astronom. Soc.* **16**:169–205.

Baillet, S., and Garnero, L., 1997, A Bayesian approach to introducing anatomo-functional priors in the EEG/MEG inverse problem, *IEEE Trans. Biomed. Eng.* **44**:374–385.

Baillet, S., Garnero, L., Marin, C., and Hugonin, J. P., 1999, Combined MEG and EEG source imaging by minimization of mutual information, *IEEE Trans. Biomed. Eng.* **46**:522–534.

Barr, R. C., Ramsey, M., III, and Spach, M. S., 1977, Relating epicardial to body surface potential distributions by means of transfer coefficients based on geometry measurements, *IEEE Trans. Biomed. Eng.* **24**:1–11.

Benar, C. G., and Gotman, J., 2002, Modeling of post-surgical brain and skull defects in the EEG inverse problem with the boundary element method, *Clin. Neurophysiol.* **113**:48–56.

Berger, H., 1929, Über das Elektrenkephalogramm des Menschen I, *Arch. Psychiatr.* **87**:527–570.

Biglieri, E., and Yao, K., 1989, Some properties of SVD and their application to digital signal processing, *Signal Process.* **18**:227–289.

Bonmassar, G., Schwartz, D. P., Liu, A. K., Kwong, K. K., Dale, A. M., and Belliveau, J. W., 2001, Spatiotemporal brain imaging of visual-evoked activity using interleaved EEG and fMRI recordings, *NeuroImage* **13**:1035–1043.

Cakmur, R., Towle, V. L., Mullan, J. F., Suarez, D., and Spire, J. P., 1997, Intra-operative localization of sensorimotor cortex by cortical somatosensory evoked potentials: From analysis of waveforms to dipole source modeling, *Acta Neurochir. (Wien)* **139**:1117–1124.

Caton, R., 1875, The electrical currents of the brain, *Br. Med. J.* **2**:278.

Cincotti, F., Babiloni, C., Carducci, F., Rossini, P. M., Del Gratta, C., Romani, G. L., Angelone, L., and Babiloni, F., 2001, fMRI priors for the linear inverse estimation of EEG cortical sources, *Electromagnetics* **21**:579–592.

Colli Franzone, P., Guerri, L., Taccardi, B., and Viganotti, C., 1985, Finite element approximation of regularized solutions of the inverse potential problem of electrocardiography and applications to experimental data, *Calcolo* **XXII**:I, 91–186.

Crouzeix, A., Yvert, B., Bertrand, O., and Pernier, J., 1999, An evaluation of dipole reconstruction accuracy with spherical and realistic head models in MEG, *Clin. Neurophysiol.* **110**:2176–2188.

Cuffin, B. N., 1995, A method for localizing EEG sources in realistic head models, *IEEE Trans. Biomed. Eng.* **42**:68–71.

Cuffin, B. N., 1996, EEG localization accuracy improvements using realistically shaped head models, *IEEE Trans. Biomed. Eng.* **43**:299–303.

Cuffin, B. N., 1998, EEG dipole source localization, *IEEE Eng. Med. Biol. Mag.* **17**:118–122.

Dale, A. M., and Sereno, M. I., 1993, Improved localization of cortical activity by combining EEG and MEG with MRI cortical surface reconstruction: A linear approach, *J. Cog. Neurosci.* **5**:162–176.

Dale, A. M., Liu, A. K., Fischl, B. R., Buckner, R. L., Belliveau, J. W., Lewine, J. D., and Halgren, E., 2000, Dynamic statistical parametric mapping: Combining fMRI and MEG for high-resolution imaging of cortical activity, *Neuron* **26**:55–67.

Diekmann, V., Becker, W., Jurgens, R., Grozinger, B., Kleiser, B., Richter, H. P., and Wollinsky, K. H., 1998, Localisation of epileptic foci with electric, magnetic, and combined electromagnetic models, *Electroencephalogr. Clin. Neurophysiol.* **106**:297–313.

Edlinger, G., Wach, P., and Pfurtscheller, G., 1998, On the realization of an analytic high-resolution EEG, *IEEE Trans. Biomed. Eng.* **45:**736–745.

Finke, S., and Gulrajani, R. M., 2001, Conventional and reciprocal approaches to the forward problem of electroencephalography, *Electromagnetics* **21:**513–530.

Fisch, B. J., 1991, *Spehlmann's EEG Primer*, 2nd ed., Elsevier, New York.

Fletcher, D. J., Amir, A., Jewett, D. L., and Fein, G., 1995, Improved method for computation of potentials in a realistic head shape model, *IEEE Trans. Biomed. Eng.* **42:**1094–1104.

Fuchs, M., Wagner, M., Kohler, T., and Wischmann, H. A., 1999, Linear and nonlinear current density reconstructions, *J. Clin. Neurophysiol.* **16:**267–295.

Fuchs, M., Wagner, M., Wischmann, H. A., Kohler, T., Theiben, A., Drenckhahn, R., and Buchner, H., 1998, Improving source reconstructions by combining bioelectric and biomagnetic data, *Electroencephalogr. Clin. Neurophysiol.* **107:**93–111.

Gavit, L., Baillet, S., Mangin, J. F., Pescatore, J., and Garnero, L., 2001, A multiresolution framework to MEG/EEG source imaging, *IEEE Trans. Biomed. Eng.* **48:**1080–1087.

Gencer, N. G., Ider, Y. Z., and Williamson, S. J., 1996, Electrical impedance tomography: Induced-current imaging achieved with a multiple coil system, *IEEE Trans. Biomed. Eng.* **43:**139–149.

Gencer, N. G., and Williamson, S. J., 1998, Differential characterization of neural sources with the bimodal truncated SVD pseudo-inverse for EEG and MEG measurements, *IEEE Trans. Biomed. Eng.* **45:**827–837.

George, J. S., Aine, C. J., Mosher, J. C., Schmidt, D. M., Ranken, D. M., Schlitt, H. A., Wood, C. C., Lewine, J. D., Sanders, J. A., and Belliveau, J. W., 1995, Mapping function in the human brain with magnetoencephalography, anatomical magnetic resonance imaging, and functional magnetic resonance imaging, *J. Clin. Neurophysiol.* **12:**406–431.

Geselowitz, D. B., 1960, Multiple representation for an equivalent cardiac generator, *Proc. IRE* **48:**75–79.

Gevins, A., Le, J., Martin, N. K., Brickett, P., Desmond, J., and Reutter, B., 1994, High resolution EEG: 124-channel recording, spatial deblurring and MRI integration methods, *Electroencephalogr. Clin. Neurophysiol.* **90:**337–358.

Golub, G. H., Heath, M., and Wahba, G., 1979, Generalized cross-validation as a method for choosing a good ridge parameter, *Technometrics* **21:**215–223.

Gorodnitsky, I. F., George, J. S., and Rao, B. D., 1995, Neuromagnetic source imaging with FOCUSS: A recursive weighted minimum norm algorithm, *Electroencephalogr. Clin. Neurophysiol.* **95:**231–251.

Gorodnitsky, I. F., and Rao, B. D., 1997, Sparse signal reconstruction from limited data using FOCUSS: A re-weighted minimum norm algorithm, *IEEE Trans. Signal Process.* **45:**600–616.

Grave de Peralta-Menendez, R., and Gonzalez-Andino, S., 1998a, A critical analysis of linear inverse solutions to the neuroelectromagnetic inverse problem, *IEEE Trans. Biomed. Eng.* **45:**440–448.

Grave de Peralta Menendez, R., and Gonzalez-Andino, S., 1998b, Distributed source models: Standard solutions and new developments. In: *Analysis of Neurophysiological Brain Functioning* (C. Uhl, ed.), Springer Verlag, pp. 176–201.

Grave de Peralta-Menendez, R., Gonzalez-Andino, S., and Lutkenhonner, B., 1996, Figures of merit to compare linear distributed inverse solutions, *Brain Topogr.* **9:**117–124.

Grave de Peralta-Menendez, R., Gonzalez-Andino, S., Morand, S., Michel, C. M., and Landis, T., 2000, Imaging the electrical activity of the brain: ELECTRA, *Hum. Brain Mapp.* **9:**1–12.

Grave de Peralta-Menendez, R., Hauk, O., Gonzalez-Andino, S., Vogt, H., and Michel, C., 1997, Linear inverse solutions with optimal resolution kernels applied to electromagnetic tomography, *Hum. Brain Mapp.* **5:**454–467.

Gulrajani, R. M., 1998, *Bioelectricity and Biomagnetism*, John Wiley & Sons, New York.

Gulrajani, R. M., Finke, S., and Gotman, J., 2001, Reciprocal transfer-coefficient matrices and the inverse problem of electroencephalography, *Biomedizinische Technik* **46**(Suppl. 2)**:**13–15.

Hamalainen, M., and Ilmoniemi, R. J., 1984, Interpreting measured magnetic fields of the brain: Estimates of current distributions, Technical Report TKF-F-A559, Helsinki University of Technology.

Hamalainen, M. S., and Ilmoniemi, R. J., 1994, Interpreting magnetic fields of the brain: Minimum norm estimates, *Med. Biol. Eng. Comput.* **32:**35–42.

Hamalainen, M., and Sarvas, J., 1989, Realistic conductor geometry model of the human head for interpretation of neuromagnetic data, *IEEE Trans. Biomed. Eng.* **36:**165–171.

Hansen, P. C., 1990, Truncated singular value decomposition solutions to discrete ill-posed problems with ill-determined numerical rank, *SIAM J. Sci. Stat. Comput.* **11:**503–518.

Hansen, P. C., 1992, Analysis of discrete ill-posed problems by means of the L-curve, *SIAM Rev.* **34:**561–580.

Hansen, P. C., and O'leary, D. P., 1993, The use of the L-curve in the regularization of discrete ill-posed problems, *SIAM J. Sci. Comput.* **14**:1487–1503.

He, B., 1998, High resolution source imaging of brain electrical activity, *IEEE Eng. Med. Biol. Mag.* **17**:123–129.

He, B., 1999, Brain electric source imaging: Scalp Laplacian mapping and cortical imaging, *Crit. Rev. Biomed. Eng.* **27**:149–188.

He, B. (ed.), 2004, *Modeling and Imaging of Bioelectrical Activity—Principles and Applications*, Kluwer Academic Publishers.

He, B., and Cohen, R. J., 1992, Body surface Laplacian ECG mapping, *IEEE Trans. Biomed. Eng.* **39**:1179–1191.

He, B., and Lian, J., 2004, Body surface Laplacian mapping of bioelectric sources, in: *Modeling and Imaging of Bioelectrical Activity—Principles and Applications* (B. He, ed.), Kluwer Academic Publishers, pp. 183–212.

He, B., Lian, J., and Li, G., 2001, High-resolution EEG: a new realistic geometry spline Laplacian estimation technique, *Clin. Neurophysiol.* **112**:845–852.

He, B., Lian, J., Spencer, K. M., Dien, J., and Donchin, E., 2001a, A cortical potential imaging analysis of the P300 and novelty P3 components, *Hum. Brain Mapp.* **12**:120–130.

He, B., and Musha, T., 1992, Equivalent dipole estimation of spontaneous EEG alpha activity: Two-moving dipole approach, *Med. Biol. Eng. Comput.* **30**:324–332.

He, B., Musha, T., Okamoto, Y., Homma, S., Nakajima, Y., and Sato, T., 1987, Electrical dipole tracing in the brain by means of the boundary element method and its accuracy, *IEEE Trans. Biomed. Eng.* **34**:406–414.

He, B., Wang, Y., Pak, S., and Ling, Y., 1996, Cortical source imaging from scalp electroencephalograms, *Med. Biol. Eng. Comput.* **34**:257–258.

He, B., Wang, Y., and Wu, D., 1999, Estimating cortical potentials from scalp EEG's in a realistically shaped inhomogeneous head model, *IEEE Trans. Biomed. Eng.* **46**:1264–1268.

He, B., Yao, D., Lian, J., and Wu, D., 2002a, An equivalent current source model and Laplacian weighted minimum norm current estimates of brain electrical activity, *IEEE Trans. Biomed. Eng.* **49**:277–288.

He, B., Zhang, X., Lian, J., Sasaki, H., Wu, D., and Towle, V. L., 2002b, Boundary element method based cortical potential imaging of somatosensory evoked potentials using subjects' magnetic resonance images, *NeuroImage* **16**:564–576.

He, B., Yao, D., and Lian, J., 2002c, High resolution EEG: On the cortical equivalent dipole layer imaging, *Clin. Neurophysiol.* **113**:227–235.

Heinze, H. J., Mangun, G. R., Burchert, W., Hinrichs, H., Scholz, M., Munte, T. F., Gos, A., Scherg, M., Johannes, S., Hundeshagen, H., Gazzaniga, M. S., and Hillyard, S. A., 1994, Combined spatial and temporal imaging of brain activity during visual selective attention in humans, *Nature* **372**:543–546.

Hess, A., Stiller, D., Kaulisch, T., Heil, P., and Scheich, H., 2000, New insights into the hemodynamic blood oxygenation level–dependent response through combination of functional magnetic resonance imaging and optical recording in gerbil barrel cortex, *J. Neurosci.* **20**:3328–3338.

Homma, S., Musha, T., Nakajima, Y., Okamoto, Y., Blom, S., Flink, R., Hagbarth, K. E., and Mostrom, U., 1994, Location of electric current sources in the human brain estimated by the dipole tracing method of the scalp–skull–brain (SSB) head model, *Electroencephalogr. Clin. Neurophysiol.* **91**:374–382.

Hori, J., Aiba, M., and He, B., 2004, Spatio-temporal cortical source imaging of brain electrical activity by means of time-varying parametric projection filter, *IEEE Trans. Biomed. Eng.* **51**:768–777.

Hori, J., and He, B., 2001, Equivalent dipole source imaging of brain electric activity by means of parametric projection filter, *Ann. Biomed. Eng.* **29**:436–445.

Hori, J., Lian, J., and He, B., 2004, Cortical potential imaging of brain electrical activity by means of parametric projection filter, *Methods Info. Med.* **43**(1):66–69.

Hjorth, B., 1975, An on-line transformation of EEG scalp potentials into orthogonal source derivations, *Electroencephalogr. Clin. Neurophysiol.* **39**:526–530.

Huppertz, H. J., Hof, E., Klisch, J., Wagner, M., Lucking, C. H., and Kristeva-Feige, R., 2001, Localization of interictal delta and epileptiform EEG activity associated with focal epileptogenic brain lesions, *NeuroImage* **13**:15–28.

Jackson, J. D., 1975, *Classical Electrodynamics*, 2nd ed., John Wiley & Sons, New York.

Jeffs, B., Leahy, R., and Singh, M., 1987, An evaluation of methods for neuromagnetic image reconstruction, *IEEE Trans. Biomed. Eng.* **34**:713–723.

Jerbi, K., Mosher, J. C., Baillet, S., and Leahy, R. M., 2002, On MEG forward modelling using multipolar expansions, *Phys. Med. Biol.* **47**(4):523–555.

Johnston, P. R., and Gulrajani, R. M., 1997, A new method for regularization parameter determination in the inverse problem of electrocardiography, *IEEE Trans. Biomed. Eng.* **44**:19–39.

Johnston, P. R., and Gulrajani, R. M., 2000, Selecting the corner in the L-curve approach to Tikhonov regularization, *IEEE Trans. Biomed. Eng.* **47**(9):1293–1296.

Kavanagh, R. N., Darcey, T. M., Lehmann, D., and Fender, D. H., 1978, Evaluation of methods for three-dimensional localization of electrical sources in the human brain, *IEEE Trans. Biomed. Eng.* **25**:421–429.

Korvenoja, A., Huttunen, J., Salli, E., Pohjonen, H., Martinkauppi, S., Palva, J. M., Lauronen, L., Virtanen, J., Ilmoniemi, R. J., and Aronen, H. J., 1999, Activation of multiple cortical areas in response to somatosensory stimulation: Combined magnetoencephalographic and functional magnetic resonance imaging, *Hum. Brain Mapp.* **8**:13–27.

Lantz, G., Michel, C. M., Pascual-Marqui, R. D., Spinelli, L., Seeck, M., Seri, S., Landis, T., and Rosen, I., 1997, Extracranial localization of intracranial interictal epileptiform activity using LORETA (low resolution electromagnetic tomography), *Electroencephalogr. Clin. Neurophysiol.* **102**:414–422.

Law, S. K., Nunez, P. L., and Wijesinghe, R. S., 1993, High-resolution EEG using spline generated surface on spherical and ellipsoidal surfaces, *IEEE Trans. Biomed. Eng.* **40**:145–153.

Lawson, C. L., and Hanson, R. J., 1974, *Solving Least Squares Problems*, Prentice-Hall, Englewood Cliffs, NJ.

Le, J., and Gevins, A., 1993, Method to reduce blur distortion from EEG's using a realistic head model, *IEEE Trans. Biomed. Eng.* **40**:517–528.

Lian, J., and He, B., 2001, A minimal product method and its application to cortical imaging, *Brain Topogr.* **13**:209–217.

Lian, J., Yao, D., and He, B., 1998, A new method for implementation of regularization in cortical potential imaging, *Proc. Annu. Int. Conf. IEEE Eng. Med. Biol. Soc.* 2155–2158.

Liu, A. K., Belliveau, J. W., and Dale, A. M., 1998, Spatiotemporal imaging of human brain activity using fMRI constrained MEG data: Monte Carlo simulations, *Proc. Natl. Acad. Sci. USA* **95**:8945–8950.

Magistretti, P. J., Pellerin, L., Rothman, D. L., and Shulman, R. G., 1999, Energy on demand, *Science* **283**:496–497.

Malmivuo, J., and Plonsey, R., 1995, *Bioelectromagnetism*, Oxford University Press, New York.

Mangun, G. R., Hinrichs, H., Scholz, M., Mueller-Gaertner, H. W., Herzog, H., Krause, B. J., Tellman, L., Kemna, L., and Heinze, H. J., 2001, Integrating electrophysiology and neuroimaging of spatial selective attention to simple isolated visual stimuli, *Vision Res.* **41**:1423–1435.

Martinez, A., Anllo-Vento, L., Sereno, M. I., Frank, L. R., Buxton, R. B., Dubowitz, D. J., Wong, E. C., Hinrichs, H., Heinze, H. J., and Hillyard, S. A., 1999, Involvement of striate and extrastriate visual cortical areas in spatial attention, *Nat. Neurosci.* **2**:364–369.

Matsuura, K., and Okabe, Y., 1995, Selective minimum-norm solution of the biomagnetic inverse problem, *IEEE Trans. Biomed. Eng.* **42**:608–615.

Matsuura, K., and Okabe, Y., 1997, A robust reconstruction of sparse biomagnetic sources, *IEEE Trans. Biomed. Eng.* **44**:720–726.

Meijs, J. W. H., Weier, O. W., Peters, M. J., and Van Oosterom, A., 1989, On the numerical accuracy of the boundary element method, *IEEE Trans. Biomed. Eng.* **36**:1038–1049.

Menke, W., 1984, *Geophysical Data Analysis: Discrete Inverse Theory*, Academic Press, New York.

Menon, V., Ford, J. M., Lim, K. O., Glover, G. H., and Pfefferbaum, A., 1997, Combined event-related fMRI and EEG evidence for temporal–parietal cortex activation during target detection, *Neuroreport* **8**:3029–3037.

Miller, K., 1970, Least squares methods for ill-posed problems with a prescribed bound, *SIAM J. Math. Anal.* **1**:52–74.

Mine, S., Oka, N., Yamaura, A., and Nakajima, Y., 1998, Presurgical functional localization of primary somatosensory cortex by dipole tracing method of scalp–skull–brain head model applied to somatosory evoked potential, *Electroencephalogr. Clin. Neurophysiol.* **108**:226–233.

Mitzdorf, U., 1985, Current source-density method and application in cat cerebral cortex: investigation of evoked potentials and EEG phenomena, *Physiol. Rev.* **65**:37–100.

Morozov, V. A., 1984, *Methods for Solving Incorrectly Posed Problems*, Springer-Verlag, Berlin.

Mosher, J. C., and Leahy, R. M., 1998, Recursive MUSIC: A framework for EEG and MEG source localization, *IEEE Trans. Biomed. Eng.* **45**:1342–1354.

Mosher, J. C., Lewis, P. S., and Leahy, R. M., 1992, Multiple dipole modeling and localization from spatio-temporal MEG data, *IEEE Trans. Biomed. Eng.* **39**:541–557.

Mulert, C., Gallinat, J., Pascual-Marqui, R., Dorn, H., Frick, K., Schlattmann, P., Mientus, S., Herrmann, W. M., and Winterer, G., 2001, Reduced event-related current density in the anterior cingulate cortex in schizophrenia, *NeuroImage* **13**:589–600.

Musha, T., and Okamoto, Y., 1999, Forward and inverse problems of EEG dipole localization, *Crit. Rev. Biomed. Eng.* **27**:189–239.

Nelder, J. A., and Mead, R., 1965, A Simplex method for function minimization, *Comput. J.* **7**:308–313.

Northoff, G., Richter, A., Gessner, M., Schlagenhauf, F., Fell, J., Baumgart, F., Kaulisch, T., Kotter, R., Stephan, K. E., Leschinger, A., Hagner, T., Bargel, B., Witzel, T., Hinrichs, H., Bogerts, B., Scheich, H., and Heinze, H. J., 2000, Functional dissociation between medial and lateral prefrontal cortical spatiotemporal activation in negative and positive emotions: A combined fMRI/MEG study, *Cereb. Cortex* **10**:93–107.

Nunez, P. L., 1981, *Electric Field of the Brain*, Oxford University Press, London.

Nunez, P. L., 1995, *Neocortical Dynamics and Human EEG Rhythms*, Oxford University Press, New York.

Nunez, P. L., and Silberstein, R. B., 2000, On the relationship of synaptic activity to macroscopic measurements: Does co-registration of EEG with fMRI make sense? *Brain Topogr.* **13**:79–96.

Nunez, P. L., Silberstein, R. B., Cdush, P. J., Wijesinghe, R. S., Westdrop, A. F., and Srinivasan, R., 1994, A theoretical and experimental study of high resolution EEG based on surface Laplacian and cortical imaging, *Electroencephalogr. Clin. Neurophysiol.* **90**:40–57.

Ogawa, S., Lee, T. M., Stepnoski, R., Chen, W., Zhu, X. H., and Ugurbil, K., 2000, An approach to probe some neural systems interaction by functional MRI at neural time scale down to milliseconds, *Proc. Natl. Acad. Sci. USA* **97**:11026–11031.

Ogawa, H., and Oja, E., 1987, Projection filter, Weiner filter, and Kahunen–Loeve subspaces in digital image processing, *J. Math. Anal. Appl.* **114**:37–51.

Oja, E., and Ogawa, H., 1986, Parametric projection filter for image and signal restoration, *IEEE Trans. Acoust. Speech, Signal Process.* **34**:1643–1653.

Ollikainen, J. O., Vauhkonen, M., Karjalainen, P. A., and Kaipio, J. P., 1999, Effects of local skull inhomogeneties on EEG source estimation, *Med. Eng. Phys.* **21**:143–154.

Ollikainen, J. O., Vauhkonen, M., Karjalainen, P. A., and Kaipio, J. P., 2001, A new computational approach for cortical imaging, *IEEE Trans. Med. Imag.* **20**:325–332.

Opitz, B., Mecklinger, A., Von Cramon, D. Y., and Kruggel, F., 1999, Combining electrophysiological and hemo-dynamic measures of the auditory oddball, *Psychophysiology* **36**:142–147.

Pascual-Marqui, R. D., and Michel, C. M., 1994, LORETA (low resolution brain electromagnetic tomography): New authentic 3D functional images of the brain, *ISBET Newslett.* **5**:4–8.

Pascual-Marqui, R. D., Michel, C. M., and Lehmann, D., 1994, Low resolution electromagnetic tomography: A new method for localizing electrical activity in the brain, *Int. J. Psychophysiol.* **18**:49–65.

Perrin, F., Bertrand, O., and Pernier, J., 1987, Scalp current density mapping: Value and estimation from potential data, *IEEE Trans. Biomed. Eng.* **34**:283–288.

Phillips, J. W., Leahy, R., and Mosher, J. C., 1997b, Imaging neural activity using MEG and EEG, *IEEE Eng. Med. Biol. Mag.* **16**:34–41.

Phillips, J. W., Leahy, R. M., Mosher, J. C., and Timsari, B., 1997a, Imaging neural electrical activity from MEG and EEG, *IEEE Trans. Med. Imag.* **16**:338–348.

Plonsey, R., 1969, *Bioelectric Phenomena*, McGraw-Hill, New York.

Roth, B. J., Ko, D., von Albertini-Carletti, I. R., Scaffidi, D., and Sato, S., 1997, Dipole localization in patients with epilepsy using the realistically shaped head model, *Electroencephalogr. Clin. Neurophysiol.* **102**:159–166.

Rush, S., and Driscoll, D. A., 1969, EEG electrode sensitivity – an application of reciprocity, *IEEE Trans. Biomed. Eng.* **16**:15–22.

Scherg, M., and Von Cramon, D., 1985, Two bilateral sources of the AEP as identified by a spatio-temporal dipole model, *Electroencephalogr. Clin. Neurophysiol.* **62**:32–44.

Sekihara, K., Nagarajan, S. S., Poeppel, D., Marantz, A., and Miyashita, Y., 2001, Reconstructing spatio-temporal activities of neural sources using an MEG vector beamformer technique, *IEEE Trans. Biomed. Eng.* **48**:760–771.

Sekihara, K., Poeppel, D., Marantz, A., Koizumi, H., and Miyashita, Y., 1997, Noise covariance incorporated MEG-MUSIC algorithm: A method for multiple-dipole estimation tolerant of the influence of background brain activity, *IEEE Trans. Biomed. Eng.* **44**:839–847.

Sekihara, K., and Scholz, B., 1995, Average-intensity reconstruction and Wiener reconstruction of bioelectric current distribution based on its estimated covariance matrix, *IEEE Trans. Biomed. Eng.* **42**:149–157.

Sekihara, K., and Scholz, B., 1996, Generalized Wiener estimation of three-dimensional current distribution from magnetic measurements, *IEEE Trans. Biomed. Eng.* **43**:281–291.

Shim, Y. S., and Cho, Z. H., 1981, SVD pseudoinversion image reconstruction, *IEEE Trans. Acoust. Speech. Process.* **29**:904–909.

Sidman, R., Ford, M., Ramsey, G., and Schlichting, C., 1990, Age-related features of the resting and P300 auditory evoked responses using the dipole localization method and cortical imaging technique, *J. Neurosci. Methods* **33:**22–32.

Snyder, A. Z., 1991, Dipole source localization in the study of EP generators: A critique, *Electroencephalogr. Clin. Neurophysiol.* **80:**321–325.

Snyder, A. Z., Abdullaev, Y., Posner, M. I., and Raichle, M. E., 1995, Scalp electrical potentials reflect regional cerebral blood flow responses during processing of written words, *Proc. Natl. Acad. Sci. USA* **92:**1689–1693.

Soong, A. C., and Koles, Z. J., 1995, Principal-component localization of the sources of the background EEG, *IEEE Trans. Biomed. Eng.* **42:**59–67.

Spencer, K. M., Dien, J., and Donchin, E., 1999, A componential analysis of the ERP elicited by novel events using a dense electrode array, *Psychophysiology* **36:**409–414.

Srebro, R., 1996, Iterative refinement of the minimum norm solution of the bioelectric inverse problem, *IEEE Trans. Biomed. Eng.* **43:**547–552.

Srebro, R., and Oguz, R. M., 1997, Estimating cortical activity from VEPs with the shrinking ellipsoid inverse, *Electroencephalogr. Clin. Neurophysiol.* **102:**343–355.

Srebro, R., Oguz, R. M., Hughlett, K., and Purdy, P. D., 1993, Estimating regional brain activity from evoked potential field on the scalp, *IEEE Trans. Biomed. Eng.* **40:**509–516.

Stratton, J. A., 1941, *Electromagnetic Theory*, McGraw-Hill, New York.

Tikhonov, A. N., and Arsenin, V. Y., 1977, *Solutions of Ill-Posed Problems*, Wiley, New York.

Towle, V. L., Cohen, S., Alperin, N., Hoffmann, K., Cogen, P., Milton, J., Grzeszczuk, R., Pelizzari, C., Syed, I., and Spire, J. P., 1995, Displaying electrocorticographic findings on gyral anatomy, *Electroencephalogr. Clin. Neurophysiol.* **94:**221–228.

Uutela, K., Hamalainen, M., and Somersalo, E., 1999, Visualization of magnetoencephalographic data using minimum current estimates, *NeuroImage* **10**(2):173–180.

Van Veen, B. D., van Drongelen, W., Yuchtman, M., and Suzuki, A., 1997, Localization of brain electrical activity via linearly constrained minimum variance spatial filtering, *IEEE Trans. Biomed. Eng.* **44:**867–880.

Vieth, J. B., Kober, H., and Grummich, P., 1996, Sources of spontaneous slow waves associated with brain lesions, localized by using the MEG, *Brain Topogr.* **8:**215–221.

Von Helmholtz, H., 1853, Uber einige Gesetzeder Verbeitung elektrischer Strome in Koperlichen Leitern mit Anwendung auf die theorischelektrischen Versuche, *Ann. Physik. U. Chem.* **89:**211–233, 353–377.

Waberski, T. D., Buchner, H., Lehnertz, K., Hufnagel, A., Fuchs, M., Beckmann, R., and Rienacker, A., 1998, Properties of advanced headmodelling and source reconstruction for the localization of epileptiform activity, *Brain Topogr.* **10:**283–290.

Wang, Y., and He, B., 1998, A computer simulation study of cortical imaging from scalp potentials, *IEEE Trans. Biomed. Eng.* **45:**724–735.

Wang, J., Zhou, T., Qiu, M., Du, A., Cai, K., Wang, Z., Zhou, C., Meng, M., Zhuo, Y., Fan, S., and Chen, L., 1999, Relationship between ventral stream for object vision and dorsal stream for spatial vision: An fMRI + ERP study, *Hum. Brain Mapp.* **8:**170–181.

Woldorff, M. G., Fox, P. T., Matzke, M., Lancaster, J. L., Veeraswamy, S., Zamarripa, F., Seabolt, M., Glass, T., Gao, J. H., Martin, C. C., and Jerabek, P., 1997, Retinotopic organization of early visual spatial attention effects as revealed by PET and ERPs, *Hum. Brain Mapp.* **5:**280–286.

Wong, P. K., 1991, Source modeling of the rolandic focus, *Brain Topogr.* **4:**105–112.

Xu, X. L., Xu, B., and He, B., 2004, An alternative subspace approach to EEG dipole source localization, *Phys. Med. Biol.* **49**(2):327–343.

Yao, D., and He, B., 2001, A self-coherence enhancement algorithm and its application to enhancing three-dimensional source estimation from EEGs, *Ann. Biomed. Eng.* **29:**1019–1027.

Zhang, X., van Drongelen, W., Hecox, K., Towle, V. L., Frim, D. M., McGee, A., and He, B., 2003, High resolution EEG: Cortical potential mapping of interictal spikes, *Clin. Neurophysiol.* **114:**1963–1973.

8

MECHANISMS OF CORTICAL COMPUTATION

Leif H. Finkel[1]* and Diego Contreras[2]

[1]Department of Bioengineering, University of Pennsylvania, Philadelphia, Pennsylvania
[2]Department of Neuroscience, University of Pennsylvania, Philadelphia, Pennsylvania

8.1. INTRODUCTION

The purpose of this chapter is to explore the computational principles underlying cortical function. We will consider ideas proposed in a large number of recent theoretical models that present a range of interesting, and sometimes conflicting, mechanisms. We will try to tie these theoretical principles to the underlying biology, and will spend most of our time considering the link between the intrinsic properties of neurons and the information-processing abilities of cortical circuits. We will consider computations carried out across different cortical areas, associated with processes ranging from sensory detection in vision, audition, and olfaction, to recognition, memory, and categorization.

Understanding the function of a biological system has historically been tied to de-lineating its structure. Harvey's discovery of venous valves uncovered the function of the circulatory system; deciphering the structure of the immunoglobulins gave birth to modern immunology. The nervous system, however, presents a structure of unparalleled complexity, containing the largest number of cell types, and manifesting the most intricate and varied interactions between cells. This overwhelming complexity has stymied most attempts to understand functional principles, even in those cases where the structural principles are most clear. As an example, in the 1960s and 1970s a concerted international effort was organized to trace out the anatomical connectivity of the cerebellum, which owing to its crystalline-like repetitive circuit structure presents an ideal target for analysis. These studies yielded much useful information but disappointingly little progress toward understanding what the cerebellum actually does. Even at the supposedly simpler invertebrate level, despite several decades of brilliant and illuminating investigations, one would still be hard pressed to fully explain the function of the 30-neuron stomatogastric ganglion of the lobster, or to create, *in silico*, a functional Aplysia.

---

* Address for correspondence: Department of Bioengineering, University of Pennsylvania, Philadelphia, Pennsylvania.

Part of the problem is that neuroscience itself has become overwhelmingly complex, with studies on many different systems at many different levels. Yet, despite exponential growth in the volume of neuroscientific data, gains in understanding are growing at best linearly. One cause of this Malthusian dilemma is the difficulty in bridging insights across domains. If understanding achieved in one domain could be automatically translated into others, e.g., from molecular to systems, from Aplysia to hippocampus, exponential advances would emerge. However, in practice, it has proven problematic to relate findings from one cortical area to another, one cell type to another, one channel type to another, from *in vitro* to *in vivo*, from 23° to 37°.

Bridging across levels is supposed to be the strong point of computational modeling. Yet to date, theoretical approaches have also usually failed to bridge domains. With several notable exceptions, models of the brain have historically fallen into two strategic camps: models of information processing that totally neglect the intrinsic properties of neurons, versus detailed models of neuronal intrinsic properties that have no computational function. Technological limitations on computing speed have been partly to blame, and it has only recently been feasible to simulate large networks of biophysically complex neurons—but the real stumbling block has been a fundamental confusion over the roles of individual cell properties and network connectivity in cortical function. Most models have taken the first approach of drastically simplifying neuronal properties. This allows circuits to be constructed with excitatory and inhibitory interactions that produce interesting dynamics and generate behaviors that mimic a range of perceptual or motor processes. This reductionist approach is founded on the premise that the differences in function across the cortex are mainly at the level of circuit layout. For example, consider two processes carried out by the visual system: stereopsis (the reconstruction of 3D visual information by integrating the slightly different views seen by left and right eyes) and shape-from-shading (determining information about an object's shape from the pattern of shading across its surface). Is there really a fundamental difference between these processes in terms of their use of biophysical-level neuronal properties? Or are both carried out using a toolbox of similar neuronal elements?

More to the point, there would appear to be principles of neural function that can be understood at the circuit level without consideration of intrinsic neuronal properties. Take for example the classic model of visual texture discrimination of Jitendra Malik and Pietro Perona (Malik and Perona, 1990). Texture discrimination is the ability to distinguish surface regions that differ, statistically, in their features (contrasts, orientations, specularities, etc.; see Figure 8.1). Malik and Perona showed, by correlating the performance of their model with human psychophysics, that cells with odd-symmetric receptive fields are not necessary for texture discrimination. Such a result would appear to be independent of any biophysical-level mechanism. Another example is the recent study of Steven Grossberg and colleagues, in which they have proposed a model of how feedback information from higher cortical areas is integrated with bottom-up feedforward and horizontal information (Raizada and Grossberg, 2003). Their model suggests that feedback from higher cortical areas reenters via cortical layer VI, and by balanced activation of excitatory and inhibitory cells in layer IV it modulates feedfoward processing. It would appear that the predictions of such a model can be verified and the mechanisms tested purely at the circuit level.

What then are the reasons for incorporating neuronal intrinsic properties into circuit level models? One motivation is to study the actual mechanisms used by the nervous system

FIGURE 8.1. Texture discrimination. The texture of the gravel in the central region of this figure differs from that of the surrounding stones. The visual system is able to discriminate textures based on statistical differences in visual features, such as color, brightness, orientations, line endings, curvatures, specularities, and a host of other properties.

(Llinas, 1988). For example, Figure 8.2 shows four of the common firing patterns found in cortical cells—does the occurrence of one of these patterns signify something, or are they each useful for certain types of computations? Intrinsic properties must convey a significant benefit to justify the cost of producing dozens of channels, receptors, second messengers, and maintaining all the different morphologies, neurochemistries, and synaptic properties. Complex dendritic morphologies and biophysical mechanisms must endow a cell with greater computational power. But once these cellular tricks are understood, might it be possible to model them using simplified neurons? Can an intelligent machine be made of simplified elements (as Turing proposed), or is there something in neural processing that intrinsically requires a significant level of complexity at the cellular level?

Coming at the problem from the other side one might ask, is it possible that the key insights into cortical information processing might arise from studies of simplified circuits of complex neurons? Suppose we could construct a detailed biophysical-level model that replicates with total accuracy the electrophysiological properties of a hippocampal CA1 pyramidal cell. The hippocampus is an evolutionarily old part of cortex that plays a critical role in consolidating memory. As shown in Figure 8.3, the hippocampus is anatomically divided into several areas (dentate gyrus, CA3, and CA1) that differ in their cell types, and intrinsic and extrinsic connectivity. The hippocampus is an intensely studied structure,

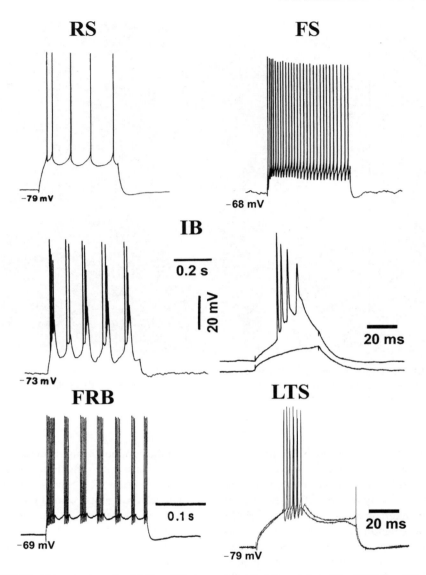

FIGURE 8.2. Firing patterns recorded from different cortical cell types: regular spiking (RS), fast spiking (FS), intrinsically bursting (IB), fast rhythmic bursting (FRB), and low-threshold spike bursting (LTS). Is there a computational role for each of these different firing patterns?

owing to its importance in memory and learning, its role in diseases such as Alzheimer's, epilepsy, and schizophrenia and the relatively stereotyped pattern of connectivity in each subarea. Suppose detailed models of several CA1 cell types were on hand (pyramidal cells and interneurons), and they were connected together with total fidelity to *in vivo* connectivity. Such a system, which is nearly accomplished in several laboratories, would provide a powerful tool to test hypotheses. However, such a model would not provide

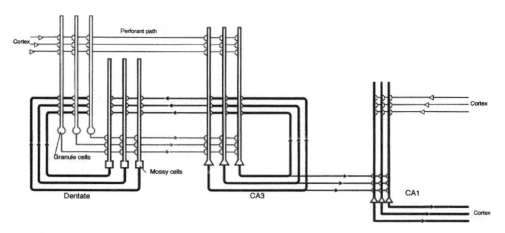

FIGURE 8.3. Connectivity diagram of the hippocampus. The hippocampus consists of several interconnected regions, the dentate gyrus, area CA3, and area CA1 (among other regions not shown here). The cortex sends inputs to the hippocampus via the perforant path, which synapses onto granule cells in the dentate gyrus, and onto pyramidal cells in areas CA3 and CA1. Pyramidal cells in CA3 are densely interconnected, and they also send axon collaterals to area CA1 and back to mossy cells in the dentate. The so-called trisynaptic loop consists of the cortical connections to dentate, from dentate to CA3, from CA3 to CA1, and back to cortex. However, there are also recurrent loops from CA3 back to dentate, and within CA3 itself. (Modified from Lisman and Otmakhova, 2001)

the answer to what CA1 does, any more than an *in vitro* hippocampal slice provides that answer.

Thus, to understand cortical mechanisms we need to consider both intrinisic properties and circuit properties, but perhaps it is sufficient to study these in separate models and synthesize the insights conceptually. Are the intrinsic properties and information processing separable?

These questions form the objectives of this chapter. Our goal is to investigate, by reviewing a number of recent models, how intrinsic properties can be incorporated into circuit models to the benefit of information processing. The discussion of these models will also take the reader on a tour of some of the most interesting current work in computational neuroscience. The processes considered include synaptic mechanisms of learning, precession of spike times in hippocampal-placed cells, mechanisms of visual hyperacuity, speech, olfactory, and biological motion recognition. Although our main focus is on cortical computation, we will discuss ideas that originate from other brain areas as well. The models reviewed are motivated by a variety of questions, but when considered together, they point to a possible new way of viewing cortical computation.

8.2. *LEARNING AND SYNAPTIC PLASTICITY*

The brain is at its core a learning machine. Learning may be to brain function what evolution is to physiology. Namely, while learning is most commonly viewed as a means of tuning physiological properties, it may also be the case that learning is itself the ultimate purpose of the brain. Learning is not the same thing as information processing because it is dynamic, adaptive, and requires interacting with the world. One might argue that learning is the most indispensable component of intelligent behavior.

In the last decade, the study of learning has seen advances on two fronts. In the domain of machine learning, a new generation of algorithms has achieved results that far outstrip previous accomplishments. These insights, which clearly are independent of neuronal properties, include sophisticated training procedures like Expectation–Maximization, routines for data mining, and applications of Bayesian networks and game theory to learning. We will return to these approaches below.

There has also been great progress made on understanding the role of intrinsic neuronal properties in learning. Since the classic work of Hebb (1949), and actually since the discovery of the synapse (in the 1890s), it has been postulated that the key to learning was the correlation of activity across the synapse. For example, Hebb's famous rule states that the strength of a synapse changes based upon the correlation of presynaptic and postsynaptic firing (see Churchland and Sejnowski, 1992 for an elegant review). Recent studies by Sakmann, Markram and colleagues (Markram *et al.*, 1997) have discovered that the mechanism has strict temporal constraints. First, note that a cell can rarely, if ever, be activated by a single synaptic input. Markram and colleagues found that if a presynaptic input arrives in a narrow time window ($\sim$10 ms) before the postsynaptic cell fires, the synapse will be facilitated (i.e., subsequent firings of the presynaptic cell will evoke larger activations of the postsynaptic cell), whereas if the presynaptic input arrives in a narrow time window after postsynaptic activation, the synapse is weakened. If presynaptic and postsynaptic activations occur outside these time windows, little or no change in synaptic efficacy occurs.

The Markram synaptic mechanism can be formulated and applied as an algorithm; however, it is important to note that intrinsic neuronal properties are the key to its function. When the postsynaptic cell fires, its action potential (AP) originates in the soma or axon, and backpropagates into the dendritic tree. A time delay of $\sim$10 ms is required for excitatory postsynaptic potentials (EPSPs) arriving at the dendrites to travel to the soma, initiate the AP, and for the AP to backpropagate back to the synapse in the dendritic tree. The time delay inherent in the synaptic modification mechanism thus enforces causality—namely, that the presynaptic input contributed to postsynaptic firing. If the presynaptic input did not contribute to postsynaptic firing (which the cell can only assess by the relative timing of the two events) then there is no reason to strengthen the association by strengthening the synapse. The biophysics also allows the mechanism to be adaptive. For example, modulation of dendritic ion channels (e.g., a fast, transiently active potassium channel known as the K_A channel) can alter the amplitude and timing of backpropagating APs. Release of a neuromodulator, such as acetylcholine, can decrease the K_A current, allowing backpropagation to reach more synapses located more distally on the dendritic tree. Thus, the intrinsic properties of the neuron along with the spatiotemporal pattern of synaptic input determine which synapses will be facilitated, suppressed, or unmodulated.

On the basis of these mechanisms, neural events occurring within an $\sim$10-ms temporal window can be linked by synaptic changes. However, in terms of behavior in the real world, associations can be learned between events occurring seconds, minutes, or hours apart (e.g., bait avoidance, where a rat learns not to eat food that makes him sick hours later; similarly, you might avoid potato salad for years if you happen to get sick hours after eating it). One solution to this "time-scale" problem is to store event-related activity in short-term memory until associations can be made. How might this occur?

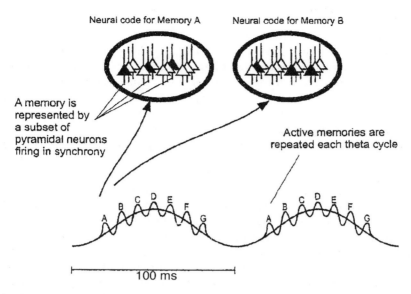

FIGURE 8.4. Storage of event stimuli in hippocampal short-term memory. Mechanism proposed by Lisman and colleagues for a short-term memory in the hippocampus. Firing of hippocampal cells is modulated by a slow (8–10 Hz) theta oscillation. Subpopulations of cells fire synchronously at a particular phase of the theta wave. The pattern of cells (e.g., pattern A) firing at a particular phase corresponds to the stimulus, and serves as a short-term memory, as it is repeated on each subsequent theta cycle. Separate memories can be stored at nearby phases of the theta oscillation (e.g., pattern B). Pattern A will repeat for minutes or hours after the occurrence of the stimulus, and pattern B can be entered any time during this period. Despite the fact that the two stimulus events occurred minutes or hours apart, their neuronal representations occur with a separation on the order of ∼10 ms, allowing the Markram-type synaptic mechanism to link the corresponding cell populations. (From Lisman and Otmakhova, 2001)

John Lisman (Lisman and Otmakhova, 2001) has suggested a mechanism, based on his studies of the hippocampus, in which the memory of an event is stored as repetitive activity of a subpopulation of neurons (an idea first proposed, without biological detail, by Hebb). Recordings from cells in the hippocampus show prominent oscillations in the global firing pattern; while animals are exploring, sniffing, and interacting with the environment, there is a global 6–10-Hz modulation of the activity of large populations of cells, known as the *theta rhythm*. Lisman proposes that memory of an event is represented by a subpopulation of cells that all fire synchronously at a fixed phase of the ongoing theta oscillation. A short-term reverberatory memory would be formed from repetition of the pattern at precisely the same phase of each subsequent theta cycle. If two patterns, corresponding to two different real world events, were stored at nearby theta phases (10–20 ms apart), an association between different events could be learned, even if the events themselves occurred some time apart (see Figure 8.4). The neuronal properties provide a "memory register" into which events can be read over real times and then synaptically linked. Use of the theta rhythm as a short-term memory has drawbacks—it is difficult to maintain the same pattern of activity over extended repetitions, and theta oscillations disappear entirely when the animal switches behavioral state (Buzsaki, 2002). The learned associations have to be transferred from the hippocampus back to the neocortex for long-term storage; however, there is experimental support for such a transfer (Wilson and McNaughton, 1994).

Similar temporal considerations arise in motor learning where sequences of actions must be learned. Consider the sequence of finger motions involved in playing a passage of a piano sonata—these require learning a sequence of spatiotemporal activations of motor neurons. Motor learning involves the basal ganglia and cerebellum, and the temporal delay learning process has been extensively modeled in these structures (Sutton and Barto, 1998). Recently, Rao and Sejnowski have proposed that the Sakmann-type backpropagation mechanism can mediate temporal delay learning (Rao and Sejnowski, 2001). Cells activated in a temporal sequence would strengthen their connections on the basis of timing: earlier cells in the sequence would promote the firing of later cells.

One particularly interesting application of this synaptic mechanism is as a learning mechanism for feedback-based predictive coding, as proposed by Rao and Ballard (Rao and Ballard, 1997). The predictive coding model puts forward the idea that cortical feedback serves to predict subsequent input stimuli based on learned patterns of previous inputs. Functioning like a neural Kalman filter, the feedback generated at time t represents an error signal that attempts to precisely inhibit the next feedforward pattern, occurring at time $t + 1$. If a series of inputs matches a previously learned sequence, appropriate cells can be activated in time to generate the necessary feedback signal to match (and cancel) the temporally evolving input. Failure to cancel the feedforward information signals an error in recognition.

The Rao and Ballard model was originally formulated without consideration of intrinsic properties; however, the incorporation of the Sakmann-type learning mechanism connects the information processing mechanisms with biophysical detail (Rao and Sejnowski, 2001). Distal dendritic inputs are identified with feedback inputs from higher cortical areas, whereas the more proximal inputs are supplied by feedforward pathways. Horizontal connections link cells in sequence. The backpropgating AP mechanism assumes a central, information-processing role—it integrates the top-down and bottom-up information based on relative timing of inputs from learned sources.

8.3. SPIKE-BASED COMPUTATION

One fundamental advantage of endowing a neuron with intrinsic properties is the ability to precisely control spike timing. The quantity, type, and anatomical distribution of channels determine a neuron's firing characteristics and the precision with which it can respond to stimulus input. Fast-spiking cortical interneurons, for example, express particular species of potassium channels, the Kv3.1–3.2 channels, which have extremely fast kinetics (Erisir et al., 1999). Following an AP, the Kv3.1 channel rapidly opens and the outward current rapidly brings the membrane potential back toward resting levels; as the voltage drops, the channel rapidly deactivates, allowing excitatory inputs to generate the next AP. Thanks to the Kv3 channel, fast-spiking cells along with some pyramidal cells in the auditory system are able to fire at close to 1 kHz. Rapid firing is not necessarily the same as precisely timed firing, but as phase locking in the auditory system suggests, the properties are closely related.

Intrinsic properties allow computations to make use of precise spike timing; however, the question remains of whether spike-based computation offers any advantages. Several recent studies suggest interesting answers.

Consider the problem of visual motion detection—how do we detect that an object has moved? There are many modeling approaches to this problem—most follow from the original proposals by Barlow (Barlow and Levick, 1965) and by Reichardt (Poggio and Reichardt, 1973; van Santen and Sperling, 1985) which involve (with some differences) sampling inputs from two nearby spatial locations, introducing a temporal delay to one of the inputs, and then combining them. If the temporal delay introduced matches the temporal delay required for the object to travel between the locations, an enhanced input is generated. Physiological studies (Saul and Humphrey, 1990) have shown that cells in the thalamus may introduce the temporal delay: so-called lagged cells in the LGN respond to retinal ganglion cell inputs with an initial hyperpolarization and delayed depolarization, whereas nonlagged cells depolarize immediately. Although the source of inhibition and the precise circuitry underlying lagged-cell behavior is not completely resolved, motion detection models can and have been constructed on the basis of this mechanism. Most of these models, however, suffer from a problem stemming from the fixed nature of the temporal delay. This can result in confusion between the velocity of a stimulus and the magnitude of its contrast. This problem is not encountered by motion models such as that of Watson and Ahumada (Watson and Ahumada, 1985; Zanker et al., 1999) that use more sophisticated filtering mechanisms, such as Hilbert transforms, to adjust the temporal delay depending upon velocity. Thus, a computational solution is available, but it is not clear how the retinocortical circuitry implements such a mechanism.

The insight is provided by a spike-based model recently developed by Kwabena Boahen. The model is constructed as a neuromorphic model of the retina implemented in a c-MOS VLSI chip (Boahen, 2002). Boahen's model features four classes of retinal ganglion cells, corresponding to those found physiologically: ON-center cells fire in response to light regions on darker backgrounds, OFF-center cells respond to dark regions on lighter backgrounds, Transient cells fire in response to changes in contrast, Sustained cells fire in response to static stimuli. Thus, ON and OFF Sustained cells spike in response to static contrast differences; ON and OFF Transient cells fire in response to increases or decreases in contrast, respectively. Retinal ganglion cells, the output cells of the retina, receive an input from bipolar cells that represents a spatial differention of the input image. As shown in Figure 8.5, the four types of ganglion cells fire in a specific temporal order, depending on whether the stimulus is moving leftwards or rightwards, and depending upon the contrast of the stimulus. For example, at the leading edge of a leftwards-moving white bar the firing sequence is, Decreasing, Increasing, Decreasing; whereas at the trailing edge the sequence is, Increasing, Decreasing, Increasing. ON cells are activated when the white bar is over them, OFF cells when the dark region is passing. Switching the direction of motion or the contrast of the bar results in a change in the order of cell firing. Thus, complete information is conveyed in the timing of the spikes—and the sequence order distinguishes both the stimulus (contrast) and the direction of motion.

In Boahen's retinomorphic model spike timing must be precise enough to allow the four ganglion cell classes to fire at clear distinguishable times—requiring a level of precision dictated by the velocity of the stimulus. However it is instructive to ask more generally, to what extent is the robust information processing performance of this model due to the use of spike timing, and to what extent could it be achieved with simpler neurons? In the absence of timing information, a model could still accurately determine the stimulus direction by comparing changes at the leading and trailing edges of the bar. But this is a significantly

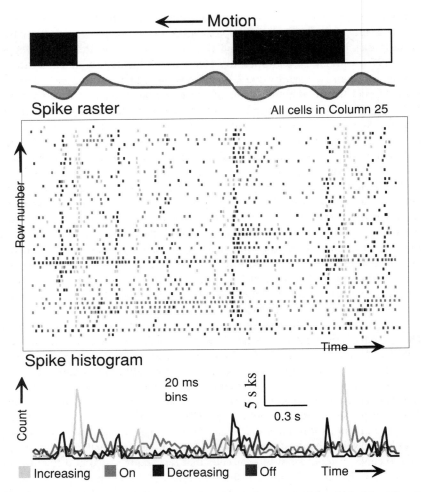

FIGURE 8.5. Spike-based model of motion detection. Boahen's neuromorphic model is able to detect motion independently of contrast. A moving stimulus (above) induces a temporal pattern of activity in bipolar cells (purple waveform) that reflects a spatial differentiation of the signal. The raster plot shows spike times of the four types of modeled ganglion cells (color code indicated at bottom). The summed activity of each cell type (plots at bottom) show that the direction of motion of each edge is uniquely reflected in the temporal order of firing of the different ganglion cell types.

more complex computation, requiring a logical-type operation on inputs from different spatial positions. Because the width of the bar cannot be known in advance, the circuit must allow for inputs from a range of spatial positions, which decreases the signal-to-noise ratio, since most inputs are irrelevant to the computation. In addition, such a circuit-based mechanism requires solving a correspondence-type problem, since there may be multiple edges in the scene, of varying contrasts, moving in different directions. The spike-based mechanism carries out its computation on cells with identical receptive fields, reducing wiring demands and increasing computational speed.

8.4. SPATIOTEMPORAL PATTERN RECOGNITION

All of the sensory recognition systems—visual, auditory, olfactory, somatosensory—have the ability to recognize spatiotemporal patterns of incoming stimuli, and to distinguish learned patterns from distractors and background noise. For example, in olfaction, each receptor is broadly tuned to a wide range of molecules. Depending on the intensity of the odor (distance to and strength of the source), the relative degree of activation of a given receptor varies considerably. Because the population of receptors activated by a stimulus odor and distractors can overlap substantially, it is a difficult computational problem to recognize an odor in the presence of multiple other odors. Over the last decade, John Hopfield has explored a set of models that use spike-based computation to solve this difficult recognition problem, both in the olfactory domain as well as in speech recognition (Hopfield, 1995; Hopfield and Brody, 2000, 2001).

Recognizing a spoken word involves categorizing a complex auditory spatiotemporal pattern. Hopfield and Brody (Hopfield and Brody, 2001) propose a spike-based mechanism for speech recognition that is based on two well-known neuronal properties: firing rate adaptation and spike synchronization. The speech signal is first analyzed by a set of neurons, each tuned to a simple auditory feature (onset or offset of sound energy in a particular frequency band). Activation of a feature neuron initiates spiking in a particular set of cells. The firing rate of each cell in the set adapts at a different rate, some rapidly decrease their firing rate to zero, other cells decrease more slowly. The next auditory feature detected initiates firing in another set of cells, specific for that feature, and each of these cells also begin to adapt, each with a different time constant. Given the brief time interval between the features (e.g., phonemes in a spoken word), each subsequent set of cells begins its adaptation at a slightly later time. As shown in Figure 8.6, at some later point in time, given the different rates of adaptation, some cells from each of the sets will be firing at the same rate. These cells, firing at the same rate, synchronize their firing. Because the firing rates continue to adapt at different rates there will only be a transient interval over which the synchronization can be maintained. However, the occurrence of synchronization is a signal that the network has recognized a particular word. This mechanism has two powerful computational properties: first, it is extremely insensitive to noise, and because the recognition process involves the convergence of the firing rates of a large number of cells, adding or subtracting a few cells has little effect. Second, the mechanism has the property of time invariance—it can generalize to words spoken at different rates (faster or

FIGURE 8.6. Schematic of the Hopfield–Brody mechanism. Detection of a feature by a group of neurons initiates spiking, but because of adaptation, the rate declines with time. At some later time, the firing rate of all three neurons will coincide, given an appropriate choice of the adaptatation decay rate. The mechanism makes use of many cells, each with a different adaptation rate, so that some subpopulations will decay appropriately. Cells with similar firing rates synchronize their firing, and synchronized firing is the signature of detection of a temporal sequence of events. (Modified from Abbott, 2001)

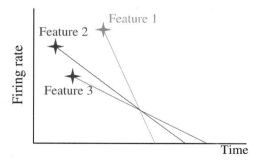

slower). Provided the relative time differences between events (phonemes) is maintained, the overall speaking rate can be increased or decreased, and there will still be a set of neurons whose adaptation rates allow convergence of the cell firing rates.

Hopfield and Brody cite the results of Barry Connors (Beierlein *et al.*, 2000), who has shown that such synchronization occurs in cortical networks. Cortical interneurons connected by both gap junctions and GABAergic connections rapidly synchronize when stimulated with similar injection currents. A number of theoretical studies have also shown that rapid synchronization is easily achieved provided cells, particularly interneurons, are firing at similar rates (Whittington *et al.*, 2000).

Thus, powerful recognition processes can arise from spike-timing mechanisms in cortical-based architectures. Clearly, there are commercially available speech recognition algorithms that perform impressively and bear no resemblance to neurons. The material point here, however, is how powerful a computational system can be generated from relatively simple cortical mechanisms requiring only a minute fraction of the biophysical machinery available. Other neuronal mechanisms may be more powerful or biologically plausible, and Hopfield and Brody themselves have proposed alternatives. But all such mechanisms require a minimum level of sophistication of the neurons. Although synchronization, adaptation and other processes can be modeled abstractly without specifying the biophysical mechanisms, incorporating the biophysics allows the mechanism to interact with other processes. For example, attention might improve recognition by speeding up synchronization.

The most interesting aspect of the Hopfield–Brody model is the identification of a new, neural-based computational principle. Analysis of the model led the authors to realize that the core process of the mechanism was to identify populations of cells exhibiting similar firing rates over a brief time window. Given a large population of cells, it is statistically unexpected for a sizable subpopulation to all fire with the same rate. Such an event signifies something unusual and may be an example of the kind of thing the cortex is well suited to compute. Rather than thinking of the network as "processing speech information," the cortex may be carrying out a generic computation of detecting an unusual firing pattern. Many different problems could be solved using the same computation, all that is required is to find the appropriate mapping from the stimulus domain.

Perhaps the strongest experimental support for spike-based computation comes from studies of the hippocampus and the phenomenon known as *theta phase precession*. Studies, mainly in rats, show that pyramidal cells in the CA1 region selectively fire when the animal is at a particular spatial location in a familiar environment. These so-called place cells are tuned to the combination of sensory cues—sights, sounds, smells—unique to a location. As the rat moves, the distances and bearings to these landmarks shift. Place cell fields are broadly tuned, so as a rat navigating a maze enters the "place field" for a particular cell it will begin to fire, and its firing rate will increase as the rat nears the center of the place field. As the rat continues to travel out of the place field the firing rate declines to baseline. A typical place cell might respond over a 10–cm-diameter region (roughly the body length of the rat), and it's firing rate might increase from 20 to 80 Hz and back to 20 Hz as it traverses the place field (Best *et al.*, 2001). For rats, the typical running speed is 20 cm/s, so the place cell might be active for 500 ms or approximately 5 theta cycles.

However, as the rat moves through its environment, its position is coded in a second manner as well, based on spike *timing* of the place cells. As discussed above, the hippocampus, and CA1 in particular, often shows a global rhythmic oscillation of activity at theta frequencies. In fact, theta oscillations are most prominent when the rat is behaviorally engaged in activities such as maze navigation. As O'Keefe and Reece first observed, as the rat approaches, reaches, and passes the center of the place field, the place cell spikes at successively earlier phases of the global theta oscillation. In other words, the spike time precesses (advances) with respect to the phase of the theta oscillation. The place cell first spikes (as the animal enters the place field) at the peak of the theta wave, and the final spike (as the animal leaves the place field) can occur at a theta phase of up to 180° earlier. Studies have shown that the spatial position of the animal can be reconstructed more accurately on the basis of spike timing than on changes in firing rate. It has also been pointed out that whereas firing rate increases and decreases, spike phase moves in a unilateral direction, thus provides a more explicit representation of position.

There are additional advantages to this spike-based coding. At any given time, a number of place cells will be active—those whose centers have been recently visited, those whose centers are at the current location of the animal, and those whose centers will soon be arrived at. Thus, on a single theta cycle it is possible to identify both the current location and past trajectory, and to generate a prediction about upcoming positions and trajectories. In the same manner as with the ganglion cell model for motion detection above, the temporal order of spiking identifies the path taken through the environment. In addition, the temporal dispersion of the spikes relative to theta allows synaptic learning to take place according to Sakmann's mechanism discussed above. Namely, during each theta cycle, a sequence of spikes ordered roughly 10–20 ms apart, will represent, like pearls on a string, the handful of adjacent place cell locations currently active. Because each cell spikes roughly 10–20 ms after the previous cell, a strong set of synaptic connections can be formed. These connections then form a memory of the navigated path, and allow the route to be predicted and followed in the future. Finally, the phase precession mechanism allows the representation to be independent of running speed—an effect analogous to the time invariance demonstrated by Hopfield's model.

Numerous models of place-cell-based navigation have been proposed in which phase precession is not used and the model neurons have minimal if any intrinsic properties. These models can generate attractor-type memories, learn trajectories, and replicate physiological findings. In addition to the advantages listed above, is there any more fundamental superiority of the spike-based scheme? Is the situation perhaps analogous to that in the auditory system, where the computation of binaural localization is primarily based on differences in sound intensity for frequencies above 3 kHz and primarily based on differences in time of arrival below 3 kHz? For example, if the place cell is only active for ~500 ms, and fires only ~5 spikes, it is difficult for the firing rate to change from 20 to 80 Hz and back, because the rate would have to change significantly over the time span of the interspike interval. Thus, a spike-time representation may be inherently more accurate. Additional insight comes from several recent studies that reveal the mechanism of the theta phase precession.

Gyorgi Buzsaki and colleagues (Kamondi *et al.*, 1998; Harris *et al.*, 2002) and Jeff Magee (Magee, 2001) have found that precession can be generated by delivering a stream of excitatory synaptic input to CA1 pyramidal cell dendrites and a theta frequency inhibitory drive to the soma. As the strength of the excitation increases, the phase of firing precesses with respect to the theta. Interneurons are known to be key generators of theta; they readily synchronize and provide an (inhibitory) oscillating input to pyramidal cell somata.

Excitatory inputs normally arrive at the distal dendrites from extrinsic sources, such as the perforant path (which provides cortical input to the hippocampus) and from intrinsic hippocampal sources, namely the axon collaterals of CA3 pyramidal cells, which synapse at more proximal locations on the CA1 dendritic tree. Weak excitatory inputs, as would occur when the place field is first entered are only able to fire the cell when somatic inhibition is at its nadir. As the excitation becomes stronger (nearing the center of the place field), spiking can occur at earlier theta phases, when the inhibition is slightly stronger. This mechanism requires that the excitatory inputs be relatively synchronized and roughly 180° out of phase with the inhibition, as is indeed found experimentally.

Modulation of neuronal intrinsic properties can alter the computation being carried out. Activation of another potassium channel, K_H, for example, has been shown to increase the gain of spike advancement, i.e., a larger phase advance occurs for the same excitatory input. Increased gain could have several benefits, including greater accuracy in positional reconstruction, restriction of synaptic connectivity, and ability to function at higher locomotor speeds. Most important, if precession is used as a means of predicting the path ahead, increased gain might translate into faster computation. The theta precession mechanism may be used more generally than for just navigation. Perhaps in processing sequences of memories, actions, or thoughts a similar precession occurs. Then increasing the gain might correspond to speeding up the process—playing the piano passage more rapidly, sifting through the memory more efficiently, or jumping to the conclusion of the thought process.

Hippocampal-based navigation is a paradigmatic problem, because it ties into the core of spatiotemporal pattern recognition. The spike-based phase precession mechanism computes movement with respect to identified landmarks. The ancient Greeks discovered a mnemonic tool—if you want to memorize a list of items, it helps to visualize placing them at landmarks successively encountered as you walk along a well-known path. Each item becomes associated in memory with a known landmark, and imagining yourself walking along the path then helps recall each novel item in turn. Perhaps the key to this trick is to activate the spike-based representations of each object at the appropriate time to allow plasticity between them, thus forming an attractor sequence of the new items.

Spatiotemporal recognition thus appears to make a strong case for spike-based computation. Consider two final examples. *Point-light walkers* are produced by dressing a person in black, marking his joints with small lights, and watching (or filming) him walking against a dark background (Ahlstrom *et al.*, 1997). All one sees is the motion of the dozen small lights, and if a snapshot is taken, it is difficult to determine what the shape is. But when the sequence of motions is viewed, it is immediately perceived as someone walking. (Any action is easily recognized: dancing, jumping jacks, etc., and observers can sometimes discriminate the gender of the actor, or even recognize a friend by their gait.) The point-light walker can be recognized against noisy backgrounds of randomly moving lights. FMRI studies have identified cortical regions selective for these types of biological motion stimuli (Grossman and Blake, 2002).

There is no evidence suggesting that recognition of point-light walkers requires spike-based mechanisms (there is no physiological evidence on this phenomenon at all). However, conceptually, recognizing biological motion appears to be a similar problem to the navigation problem. Each dot configuration might be recognized by a high-level visual cell, acting like a place cell—lets call it a "shape cell." As the dots change configurations, an ordered sequence of "shape cells" is activated. Theta activity is prominent in the visual cortex during active vision, and so it is conceivable that a similar kind of spike-timing precession occurs. Ostensibly, the synaptic connections between shape cells would be learned from viewing real live people walking. The point-light walkers, by following the same ordered sequence, activate the "walking" circuit, in the same manner that the Greek mnemonic makes use of the previous memory. Although totally speculative, this scheme suggests that recognition, in general, makes use of learned sequences of spatiotemporal patterns of activity, and that intrinsic neuronal properties are key to the mechanism (Das and Finkel, 2003).

The last example to be briefly considered reinforces this point. Studies of olfactory recognition by locusts, pioneered by Gilles Laurent, have shown that each odor (apple, quinine, etc.) produces a spatiotemporal pattern of activity beginning in the antennal lobe (Laurent *et al.*, 1996) (see Figure 8.7). Spikes are synchronized and occur at the peaks of a fast oscillation (~40 Hz). Over the course of roughly 100 ms, an ordered sequence of spatial patterns plays out across the nucleus. Each odor has its own characteristic spatiotemporal pattern, and if the synchrony is disrupted, odor recognition is behaviorally disrupted. It may be that the odor is temporally varying, particularly as the insect waves its antennae (containing the olfactory receptors) about. But these varied examples: olfactory recognition, point-light walkers, speech recognition, and rat navigation, are all difficult information-processing challenges and appear to demand similar neural mechanisms.

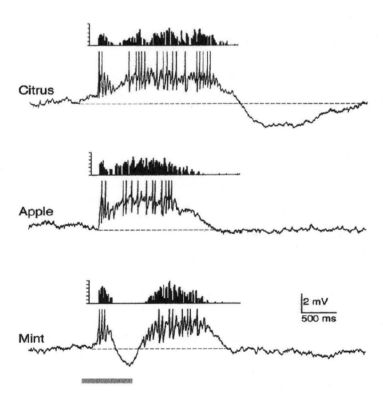

FIGURE 8.7. Spike responses of a single projection neurons in the locust antennal lobe to three odors. Distinct odors produce characteristic temporal patterns of spikes. (From Laurent *et al.*, 2001)

8.5. NEURONAL FIRING CHARACTERISTICS

In the introduction, the question was raised of why the cortex needs so many different types of cells. Cells differ in morphology, neurochemistry, firing patterns, and synaptic properties (whether they facilitate, depress, or show some combination of changes). Recent anatomical studies have shown that each cell type makes specific connections to other cells based on their cell type and location in particular cortical layers. Layer III pyramidal cells in the visual cortex have long axons that travel several millimeters horizontally and synapse with other pyramidal cells with related receptive field properties. Layer III cells in higher cortical areas have larger numbers of spines, allowing them to integrate larger numbers of inputs. Layer V pyramidal cells have dendrites that reach up to layer I, and thus receive inputs from cells in granular and supragranular layers. Thus each cell's morphological structure is crafted to a computational role—but what is that role?

Is there a basic computation carried out by the cortex, perhaps varied and elaborated in different regions, but stemming from an ur-computation reflecting the basic cortical architecture? Many authors have pointed to basic similarities shared by all cortical areas: the six horizontal layers, vertical columnar organization, similar total cell density (except for striate cortex), similar ratios of excitatory to inhibitory cells, preponderance of local corticocortical connections, etc., to argue that the major difference between information processing in different cortical areas is due to the nature of the inputs received. Experiments have shown that if during development visual inputs are rerouted to the auditory cortex (and the normal auditory inputs transected), then the auditory cortex develops columnar structure and receptive fields similar to those found in the normal visual cortex. On the other hand, close analysis reveals significant anatomical differences between cortical areas. The cell densities in various layers and sublayers differ between areas. For example, the primary somatosensory cortex, which receives tactile information from the entire body, has a large and dense layer IV, whereas the primary motor cortex has few cells in layer IV. It is difficult to see how these two areas could be carrying out the same computation.

In originally proposing the idea of a *canonical cortical computation*, Douglas, Martin, and colleagues attempted to abstract the key elements of the cortical circuit (Douglas and Martin, 1991). Their idea focuses on the extensive local corticocortical connectivity, which they propose amplifies activity produced by thalamic input. Amplification is due to mutual excitation of pyramidal cells in a cortical column, initiated by the thalamic input, which is then quickly quenched by cortical inhibition. The inhibition is slower to ramp up, but more powerful once generated.

The folded feedback model of Grossberg and colleagues, discussed above, while totally different in concept, is pitched at the same level as the Douglas and Martin model. Grossberg's model identifies an anatomical circuit for cortical feedback, and ties it to functional properties, namely, that feedback should modulate feedforward activity but should not induce activity in the absence of external stimuli (as that would correspond to hallucinations). This is achieved by balancing the drive to excitatory and inhibitory cells—in a manner different from the Douglas and Martin mechanism, but for the same purpose of avoiding network instability.

Either or both of these mechanisms may be correct, but it is informative to contrast them with a third mechanism recently proposed by Steven Zucker and colleagues (Miller and Zucker, 1999). The distinguishing feature of the Zucker model is that it ties a circuit mechanism to neuronal intrinsic properties. The model considers a local, densely connected "clique" of cells in supragranular (layers II and III) primary visual cortex. Cells in a clique have highly overlapped receptive fields, but each cell differs slightly in its exact receptive field center and orientation preference. A particular stimulus, say a short oriented bar, would activate a population of such cells, some more strongly than others, depending upon individual receptive field preferences. The dense connectivity among cells in the clique together with each cell's intrinsic properties make the local circuit act nonlinearly and cooperatively. If more than half of the cells are activated by the stimulus, then because of mutual excitation, all of the cells become activated. If less than half the cells receive afferent input, then firing quickly dies out. Thus, as in Douglas and Martin's model, the afferent input acts like a match igniting a conflagration. Here however, the cooperative behavior arises from a nonlinearity in the cell's response to input current. Studies (Douglas and Martin, 1991) have shown that as the injection current into a cell is increased, the firing rate changes nonlinearly: at injection currents below $\sim$0.6 nA the cell fires at a low rate; as the injection current increases above 0.6 nA, the firing rate increases dramatically. In response to a thalamic input (which provides a maximum of $\sim$0.6 nA), a cell generates at the most one spike; however, if multiple cells in the clique are simultaneously activated, each cell will receive a barrage of spikes from its neighbors, and this extra input current produces a dramatic rise in the firing rate of the cell. Increased firing by all the cells in the clique rapidly leads to all cells firing at the maximal level. After $\sim$25 ms, the barrage is terminated by local inhibition.

Thus, activation of a clique is signaled by rapid spiking of all the member cells. This coordinated, rapid firing among a group of cells would be a statistically significant event, and would be physiologically powerful, capable of transmitting this piece of information (e.g., excitatory clique #854 has just been activated) to other cortical and subcortical areas.

The nonlinearity of the neuronal intrinsic properties give the clique its "all or none" behavior. Activation of a clique requires that the stimulus matches the *intersection* of all the receptive field properties of the cells in the clique. The clique can thus be thought of as having its own receptive field, with far greater selectivity than any of its member cells. In fact, this mechanism is proposed by Zucker to be the basis of visual hyperacuity. Although each cell can only localize stimulus position to $\sim$12 arc-min^2 (the average receptive field size), the clique can localize stimulus position to 6 s of arc—which is less than the spacing of cones in the retina (1 min of arc). Through a rigorous analysis based on human hyperacuity performance, Zucker calculates that the number of cortical cells in an excitatory clique should be around 30. This number can be used to generate an incredibly accurate prediction for the density of cells in the striate cortex.

The Zucker model thus ties intrinsic properties to information processing more precisely than any model to date. In subsequent work, Zucker has examined the idea that the clique mechanism can be thought of in terms of game theory. Each cell and each clique may be viewed as trying to optimize its gain–loss ratio as defined in terms of activation. For example, if a cell can be activated with a given clique it would be a winner, but if it tries and fails to rouse a majority, it would be a loser—with attendant rewards and penalties. Perhaps the cortex is constructed so as to achieve superior play in this "game." These ideas add a new dimension to previous models of neuronal group selection (Edelman, 1989).

8.6. TIME CONSTRAINTS ON CORTICAL COMPUTATION

One issue that must still be addressed stems from recent psychophysical measurements of the speed of visual recognition. In the RSVP paradigm (rapid serial visual presentation), a sequence of ~100 images are successively flashed on a screen, each image persisting for as little as 30 ms (Keysers *et al.*, 2001). Even at these speeds, humans are remarkably good at detecting whether one of the flashed images contains a target, e.g., Marilyn Monroe. EEG studies show that it takes roughly ~100 ms for the information to reach the temporal cortex, but as soon as the information arrives (or within 30 ms) recognition is achieved. If, as these studies suggest, the computations associated with visual recognition require less than 30 ms, then the entire computation must occur using only one or two spikes (since even firing at 100 Hz, a cell would only generate 2–3 spikes in 30 ms). This recalls a famous prediction of David Marr's, that the hallmark of true cortical computation would be algorithms that generate the answer as soon as all the necessary information arrives, i.e., without the need for any additional iterations to converge to the solution [see Weiss (1997) for a discussion of this point].

Simon Thorpe and colleagues have recently proposed an interesting spike-timing model based on these RSVP findings (VanRullen and Thorpe, 2002). In Thorpe's model, the time required for a cell to generate a spike depends upon the salience of the stimulus received. More salient stimuli evoke earlier spiking. For the spikes arriving at layer IV of area V1 from the LGN, salience is taken to correspond to the contrast of the stimulus. However, in higher cortical areas, salience may depend upon other attributes. Thorpe shows that the initial few spikes generated contain the majority of information about the scene. For example, over 50% of the scene information is conveyed by the first 1% of spikes. Thus for rapid recognition, or other cortical processes, particularly when dealing with salient stimuli, the bulk of the computation could occur even before most spikes have arrived at the appropriate area.

An alternative approach to this problem, the *space rate code*, has been put forward by Wolfgang Maas and Thomas Natschliger (Maas, 1999). In this mechanism, information is represented by the fraction of a population of cells that spike within a narrow time window (e.g., 5 ms). Say, for example, that a particular orientation column consists of 100 cells; a bar positioned at the preferred orientation might elicit 95 spikes whereas one that is 40° off might elicit only 50 spikes, within the temporal window. Space rate coding has the advantage of accuracy (the larger the population, the more resolution), and the state of the population can vary rapidly (e.g., every 5 ms). The disadvantage is that large numbers of cells are required, and it is not immediately clear how to regulate the fraction of cells that respond and on what basis these cells differ. Unlike the Zucker model, the population does not carry out any computation separate from representing the input stimulus.

Either mechanism (spike time or space rate) could be used to rapidly transfer information in a cascade of feedforward networks. Ad Aertsen (Aertsen *et al.*, 1996) has shown that spike information can be reliably passed along such a chain of networks. A volley of synchronized spikes originating across a population of cells in one network can be passed from network to network, and the degree of synchronization can actually increase with each processing stage, thus assuring that despite the divergence and convergence between areas, volleys of spikes can be reliably transmitted. The idea that cortical processing occurs across chains of feedforward networks was championed by Moshe Abeles, based on his observations of cell populations in the frontal cortex (Vaadia *et al.*, 1989). Abeles also put forward the idea that cells may act more as coincidence detectors rather than spike counters. Using multielectrode recordings, Abeles found that some cells in the frontal cortex fired in specific temporal relationships much more commonly than would be expected statistically (i.e., A then B then C where the AB and BC intervals are maintained). He thus concluded that the cells must be located at different stages of a feedforward chain that was activated under some stimulus or behavioral condition.

Jean Bullier (Bullier, 2001) has recently catalogued the temporal delays between activation of all the visual cortical areas following a peripheral stimulus. A novel aspect of his findings is that some higher areas associated with the dorsal visual stream to the parietal cortex (areas generally concerned with spatial relationships) are activated well before more peripheral areas of the ventral stream (areas concerned with object properties). For example, area MT (an area specialized for motion analysis) is activated as early as 40 ms after retinal stimulation, whereas area V1 may not be activated till 60 ms. Thus, as Bullier importantly points out, what constitutes feedforward versus feedback depends upon the timing of activation as well as anatomical connectivity. The same pathway, e.g., from area IT to area V4 delivers "feedback" information (if V4 fires before IT) or "feedforward" (if V4 fires first). In addition, Bullier points out that feedforward and feedback connections between cortical areas are carried by fast-conducting myelinated fibers. In contrast, horizontal connections within areas are unmyelinated, and have surprisingly slow conduction velocities, on the order of 0.1–0.3 mm/ms. Thus, if processing is limited to a narrow temporal window, integration of horizontal information is limited to cells located within several millimeters.

If cortical computation can proceed only on the basis of the initial volley of spikes arriving at each area, does it imply that simplified cells are sufficient, as there is little time for horizontal or feedback information to be integrated? Given that the majority of cortical connections are horizontal or feedback connections, they most certainly are used for something. One possibility might be priming: feedback or horizontal inputs that precede bottom-up inputs might allow a particular cell to fire more quickly. We will consider other possibilities below.

8.7. PUTTING IT ALL TOGETHER

V.S. Ramachandran once called vision a "bag of tricks," namely, a set of special-purpose, individualized processes that together provide our perceptual armamentarium (Ramachandran, 1985). Visual cortex must carry out computations related to shape, color, brightness, texture, motion, and depth, with each process comprising multiple subprocesses (depth from stereo, depth from perspective, depth from focus, etc.). From an evolutionary point of view, it seems reasonable to assume that a primitive visual cortical circuit might have repeatedly diverged to generate these specialized mechanisms. During the first year of a child's life, various perceptual abilities successively come "online" in a maturational sequence, even the ability to recognize point-light walkers. Even if these developmental stages reflect myelination, rather than synaptic development, they still suggest the presence of independent circuits for each function. Specific functional losses associated with strokes and brain trauma also demonstrate a segregation of function (Zeki, 2003).

Is it conceivable that there are "general" principles of cortical function shared by all of these processes, not to mention the computations in auditory, motor, frontal, and other cortical areas? Does feedback from higher cortical areas work the same way in all cortical areas, do horizontal connections in the visual cortex carry out a similar process as horizontal connections in the frontal cortex? Can the principles of neural computation be abstracted, or are there merely specialized solutions for specialized problems? Is there an overarching principle, like the theory of natural selection, that explains brain function, or is neuroscience more like molecular biology, mechanism built upon mechanism?

Some insight into these questions may come from recent work on so-called Bayesian networks and related machine learning algorithms. Bayesian networks, pioneered by Judea Pearl (Pearl, 1988), are defined mathematically as directed graphs consisting of a set of nodes, each of which represents an unknown variable (e.g., the temperature in Philadelphia, annual profits of Microsoft) and connections between nodes representing conditional probability distributions. The graph structure represents the conditional dependencies between the data, and allows the probability distribution of any variable to be computed. The advantages of Bayesian networks over neural networks are that they (1) allow an exact inference to be computed, (2) can integrate any and all kinds of disparate information, and (3) deal extremely well with new or missing data. As such, they are the best current theoretical model of integration and inference, and it is tempting to seek ways in which cortical computation might make use of a similar process. However, Bayesian networks have certain clear differences from neurophysiology—for example, they become less manageable in the presence of feedback.

One of the most insightful attempts to bridge Bayesian networks to neural computation comes from Yair Weiss (Weiss, 1997), who applied this formalism to the thorny problem of determining figure–ground relations. Figure 8.8 shows a hand against a dark background, the bounding contour of the hand is extracted and the objective is then to determine which of the two surfaces (inside or outside) is the "figure." The figure is the surface that "owns" the contour, i.e., the contour is the edge of one of the surfaces, and whichever surface owns the contour is the figure (i.e., if the outside surface owned the contour, then the "hand" would correspond to a hole cut in a piece of paper). Imagine a set of orientation-selective neurons with receptive fields at each point along the contour. These cells are assumed to be sensitive to the "direction of figure," i.e., they can be thought of as pointing in the direction (orthogonal to the local contour orientation) in which the figure lies. Just as a directionally tuned cell responds to an oriented line moving in a particular direction, these cells respond to oriented lines bounding a surface located to one side. Rudiger von der Heydt (Zhou et al., 2000) has found cells in area V2 that display such figural selectivity, and theoretical models have investigated how such cells define surfaces (Sajda and Finkel, 1985).

In Weiss' model, the direction of figure is initially taken to be the direction of curvature, i.e., all curves are considered convex. Because the hand contains both convexities (fingertips) and concavities (interstices between fingers), this assumption introduces a global inconsistency in choosing which surface is the inside. The point of the model is to see how a Bayesian network is able to arrive at a globally consistent solution. Note that in order to arrive at a global solution, each neuron must receive and integrate information from many or all of the other neurons. However, this is precisely the kind of problem that frustrates neural networks—that they require multiple iterations to relax to a solution. Weiss applied several neural network algorithms to this image, including mean field Hopfield networks,

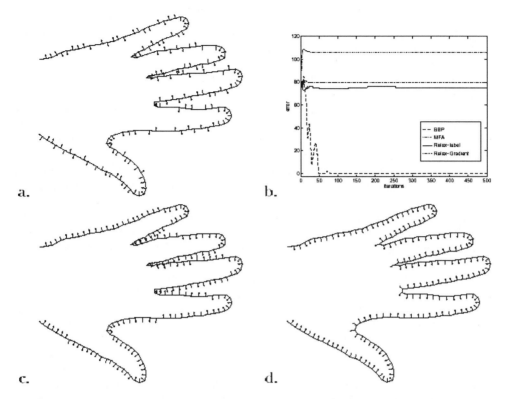

FIGURE 8.8. Bayesian belief propagation network of Yair Weiss. Cells responding to contour orientation must determine in which direction the surface lies (direction of figure). Initial condition is set so each arrow points in the direction of convex curvature. Local propagation of conditional probabilities leads to correct global solution (d). Neural networks fail to find global solution (c) and take far longer to converge. (From Weiss, 1997)

and gradient descent algorithms, and found that even after 500 iterations, none of these methods had converged to the correct solution (see Figure 8.8). In every case these algorithms get stuck in local minima with small domains of neurons stuck pointing in the wrong direction.

However, the Bayesian network finds the correct solution, and it does it in the minimum possible time. In other words, if there are 50 neurons sampling the contour it takes 50 iterations for information to be passed from end to end (only nearest-neighbor connectivity), then the solution will be arrived at with the 50th iteration. Moreover, at every iteration before the 50th, the solution is optimal given the information thus far received by each neuron.

Thus, the Bayesian network, in stark contrast to neural network algorithms, satisfies Marr's principle of not requiring additional processing beyond the arrival of the relevant information. The mechanism also fits with the single-spike neural models of Thorpe and Aersten. Most important, although Weiss' implementation is at present purely mathematical, the mechanism requires some sophisticated neuronal properties. The key to the Bayesian model is in what kind of information is passed and how it is passed. The information passed is a probability distribution—in this case, the probability of each surface being the figure. At each time step, each neuron makes a decision (which way to point) by combining its own local information (direction of curvature) with the current probabilities transmitted by its neighbors.

Josh Gold and Michael Shadlen have proposed that this kind of probabilistic decision making is the basis of neuronal function in the frontal cortex (Gold and Shadlen, 2001). They record firings from cells in monkey prefrontal cortex that make the decision whether to make a particular arm movement in a go/no-go behavioral paradigm. Their data suggests that evidence for opposing decisions (move LEFT vs. move RIGHT) is accumulated by competing cells, until one of the cells reaches a firing threshold and that action is initiated. A top-down bias, or a backchannel source of information, could easily predispose the decision in a particular direction. Gold and Shadlen propose that neurons transmit log-likelihoods rather than complete probability distributions. The likelihood is the ratio of the two probabilities (in a two-alternative decision), and taking the logarithm allows the inputs from different sources to be added rather than multiplied. Thus, each neuron represents a hypothesis, and a competition among hypotheses continues until one accumulates a threshold level of supporting evidence.

The power of the Bayesian approach is its generality—it can, in principle, be applied to all cortical processes. As a final example, Josh Tenebaum has recently developed a Bayesian model of the core cognitive process of categorization and generalization. Tenenbaum considers the problem of how we generalize from a few positive examples: shown one or two examples of a bicycle, a child quickly learns to recognize other bicycles. It is not necessary to point to multiple objects and teach the child "that's NOT a bicycle." Debates have raged over whether the underlying mechanisms of categorization involve similarity to prototypes or are rule-based. Tenenbaum shows that depending on the problem, a Bayesian model can appear to be either similarity-based or rule-based.

The essence of Tenenbaum's approach is to determine the probability of various *hypotheses* for the definition of the category. For example, suppose you're trying to figure out the normal range of concentrations of a hormone, and are only told that the following values are all within the normal range: 56, 54, 53, and 57. People have an intuitive notion of similarity that would lead most people to say that a value of 52 would likely be in the normal range, but a value of 75 would probably be outside the range. One might intuitively calculate that the value of 52 falls within two standard deviations of the mean of the mean of the values provided (55 ± 1.6), whereas 75 is over 10 standard deviations away. In fact, studies show that similarity tends to fall exponentially with differences in features. Tenenbaum makes two assumptions to derive the probabilities. First, he assumes that the samples provided (56, 54, 53, and 57) are drawn randomly from the category. And second, each putative hypothesis is normalized by the *size* of the hypothesis. In other words, the hypothesis "the normal range is 50–60" (which has a size of 10) is more likely than the hypothesis "the normal range is 40–70" (which has a size of 30). In other words, like Occam's razor, more specific hypotheses are rated more probable. Many hypotheses, H_i, can then be evaluated, and those rated most probable (i.e., the highest probability given the data, D), $P(H_i|D)$, are chosen. The conditional probability is calculated using Bayes rule and the above assumptions.

The remarkable thing about Tenenbaum's model is how well it explains a range of human judgments about similarity and generalization. It appears to approach a universal law of generalization, as envisioned by Roger Shephard. From our point of view, the model is doubly intriguing as it dovetails with many of the mechanisms we have considered. The alternative hypotheses could be computed in parallel by "decision" neurons in the frontal cortex, as postulated by Gold and Shadlen. The conditional probabilities might be processed, as in Weiss' model, based perhaps on spike timing relative to a global oscillation, or by means of a space rate code. The size principle may be tied to Hopfield's mechanism of detecting populations with similar firing rates, and/or to known cortical mechanisms of normalization and gain control. Might each neuron, or local clique of neurons, be viewed as evaluating the evidence in support of a particular hypothesis—and might neuromodulation serve to alter the hypothesis along certain directions? Might the sustained activity observed in frontal cortex neurons during working memory tasks represent not just reverberatory memory but a Bayesian-based recognition, generalization, and decision process?

The answers to these questions remain to be discovered. But the power of Bayesian-based computation, together with the stochastic nature of neuronal function, suggests that probabilistic models of neural computation may prove to be the key to cortical computation.

8.8. CONCLUSIONS

Having navigated our way through the landscape of recent cortical models, are we any closer to glimpsing the secrets of brain function? Are there any conclusions we can draw from looking at the elephant from so many sides? Have we learned anything about the role of neuronal intrinsic properties in computation?

One cortical operation touched on by many of the models is recognition. We have seen several interpretations of what recognition entails: detection, inference, categorization, decision, and prediction. Detection could occur by arrival of a single spike (Thorpe), activation of a clique of cells (Zucker), or development of unusually strong synchronization (Hopfield). Prediction might involve precession of spike times relative to a population oscillation, or synaptic sequencing and feedback cancellation (Rao and Ballard). Inference might be computed using probability distributions represented by space rate codes, or by summing log likelihoods. Categorization may correspond to calculating the most likely hypothesis. The cortex may be primarily designed as a learning machine, a spike-timing device, or the world's most sophisticated game engine. Whatever the final result, when the design is eventually understood, we will no doubt marvel at how beautifully neural function follows from its structure.

One additional consideration that might be of importance is the role of different neuronal firing modes. The Zucker model raises the idea that a shift in neuronal firing might be used to signal the outcomes of a computation. However, neurons exhibit distinctly different firing modes (regular spiking, intrinsic bursting, fast repetitive bursting; in addition to adaptation, potentiation, and other characteristics). Each firing mode may be best suited for a particular process—coincidence detection, synchronization, and plasticity. Alternatively, emergence of a particular mode (e.g., fast repeating bursts) may signal the outcome of a process—a detection or decision. Emergence of a new firing mode may require collective interactions in local circuits, or even the receipt of appropriate horizontal or feedback input.

In this regard, the role of horizontal connections remains unclear. Given their relatively slow conduction velocities, it may be that their dominant function is as intrinsic connectivity within local cliques. Longer range horizontal connections may mediate sequential interactions between cliques, as would occur in tracking a moving stimulus. Alternatively, horizontal connections could play a role in regulating spike timing; their conduction speeds are appropriately tuned to delay spikes in neighboring hypercolumns by $\sim$10 ms. Slow conduction along horizontal connections may allow contextual information to reach its target concurrently with feedback from higher cortical areas (Ito and Gilbert, 1999).

Many of the computational processes we have reviewed involve temporally varying stimuli—motion detection, speech recognition, and spatial navigation. However, it is important to note that the use of spike timing applies equally well to processes that are not intrinsically temporal in nature. We discussed the example of olfactory recognition (both the Gilles Laurent work and the Hopfield model) that uses spatiotemporal spike patterns to represent a spatial pattern of receptor activation. A related example comes from tactile recognition by the rat whisker barrel system. Diamond and colleagues (Petersen *et al.*, 2002) have shown that the relative timing of spikes in different barrels contains additional information to the firing rate of these cells. The cortex appears to use spatiotemporal spike patterns as a general computational process.

Indeed, all of the information processing problems we have considered can be thought of as involving directed spatiotemporal patterns of activity. The local cortical circuit, with specific roles for each cell type, integrates bottom-up, horizontal, and top-down inputs that together determine the pattern of activity. Neuromodulators, acting via changes in intrinsic properties, shape and drive the activity. Activity in one area can quickly initiate activity in other areas, and the whole notion of feedforward versus feedback may require elaboration. Through the use of temporal delay synaptic rules, learned sequences of activity can be used as predictors of future events. Activity patterns can be linked to conceptual hypotheses, and tested against incoming evidence until a decision is reached.

We asked the question whether there was something intrinsic in neural computation that requires intrinsic properties. Perhaps a partial answer lies in the integration of memory with perception and action. Intracellular signaling cascades, through their delayed modulation of neuronal properties, serve as a temporary repository and integrator of neuronal experience. They allow the cell to assimilate information over a time period, independently of the ongoing information-processing tasks. Then at the appropriate time, modifications to the neuron's function can be effected. This information must be stored at the cellular and synaptic level. Neuromodulators can titrate the effect of intracellular modifications—regulating the degree to which ongoing information is read into the intracellular domain, the kinds of computations allowed to proceed intracellularly, and the degree to which the results are read out as modifications in neuronal properties. Thus, the balance between online processing and adaptive change can be adjusted according to the behavioral state.

Perhaps the main lesson to be learned from all these considerations is patience and humility. Neuroscience is just a century old, and has only come of age recently. There are goals that still elude physics; we know next to nothing about development, intracellular signaling, transcriptional control, and most problems in biology. Understanding the neural bases of computation will be one of the great achievements of our species. Given the dynamic growth of experimental investigation and the continued exploration of new, exciting theoretical ideas, the solution will not long elude us.

REFERENCES

Abbott, L. F., 2001, The timing game, *Nat. Neurosci.* **4**:115–116.

Aertsen, A., Diesmann, M., and Gewaltig, M. O., 1996, Propagation of synchronous spiking activity in feedforward neural networks, *J. Physiol. Paris* **90**:243–247.

Ahlstrom, V., Blake, R., and Ahlstrom, U., 1997, Perception of biological motion, *Perception* **26**:1539–1548.

Barlow, H. B., and Levick, W. R., 1965, The mechanism of directionally selective units in rabbit's retina, *J. Physiol.* **178**:477–504.

Beierlein, M., Gibson, J. R., and Connors, B. W., 2000, A network of electrically coupled interneurons drives synchronized inhibition in neocortex, *Nat. Neurosci.* **3**:904–910.

Best, P. J., White, A. M., and Minai, A., 2001, Spatial processing in the brain: The activity of hippocampal place cells, *Annu. Rev. Neurosci.* **24**:459–486.

Boahen, K. A., 2002, A retinomorphic chip with parallel pathways: Encoding ON, OFF, Increasing, and Decreasing visual signals, *J. Analog Integr. Circ. Signal Process.* **30**:121–135.

Bullier, J., 2001, Integrated model of visual processing, *Brain Res. Brain Res. Rev.* **36**:96–107.

Buzsaki, G., 2002, Theta oscillations in the hippocampus, *Neuron* **33**:325–340.

Churchland, P. S., and Sejnowski, T. J., 1992, *The Computational Brain*, MIT Press, Cambridge.

Das, S., and Finkel, L. H., 2003, Cortical integration of bottom-up, top-down and horizontal information in biological motion recognition, in: *Proceedings of 1st International IEEE EMBS Neuroengineering Conference*, Capri.

Douglas, R. J., and Martin, K. A., 1991, A functional microcircuit for cat visual cortex, *J. Physiol.* **440**:735–769.

Edelman, G., 1989, *Neural Darwinism*, Basic Books, New York.

Erisir, A., Lau, D., Rudy, B., and Leonard, C. S., 1999, Function of specific K(+) channels in sustained high-frequency firing of fast-spiking neocortical interneurons, *J. Neurophysiol.* **82**:2476–2489.

Gold, J. I., and Shadlen, M. N., 2001, Neural computations that underlie decisions about sensory stimuli, *Trends Cogn. Sci.* **5**:10–16.

Grossman, E. D., and Blake, R., 2002, Brain areas active during visual perception of biological motion, *Neuron* **35**:1167–1175.

Harris, K. D., Henze, D. A., Hirase, H., Leinekugel, X., Dragoi, G., Czurko, A., and Buzsaki, G., 2002. Spike train dynamics predicts theta-related phase precession in hippocampal pyramidal cells, *Nature* **417**:738–741.

Hebb, D. O., 1949, *The Organization of Behavior: A Neuropsychological Theory*. Wiley, New York.

Hopfield, J. J., 1995, Pattern recognition computation using action potential timing for stimulus representation, *Nature* **376**:33–36.

Hopfield, J. J., and Brody, C. D., 2000, What is a moment? "Cortical" sensory integration over a brief interval, *Proc. Natl. Acad. Sci. USA* **97**:13919–13924.

Hopfield, J. J., and Brody, C. D., 2001, What is a moment? Transient synchrony as a collective mechanism for spatiotemporal integration, *Proc. Natl. Acad. Sci. USA* **98**:1282–1287.

Ito, M., and Gilbert, C. D., 1999, Attention modulates contextual influences in the primary visual cortex of alert monkeys, *Neuron* **22**:593–604.

Kamondi, A., Acsady, L., Wang, X. J., and Buzsaki, G., 1998, Theta oscillations in somata and dendrites of hippocampal pyramidal cells in vivo: Activity-dependent phase-precession of action potentials, *Hippocampus* **8**:244–261.

Keysers, C., Xiao, D. K., Foldiak, P., and Perrett, D. I., 2001, The speed of sight, *J. Cogn. Neurosci.* **13**:90–101.

Laurent, G., Wehr, M., and Davidowitz, H., 1996, Temporal representations of odors in an olfactory network, *J. Neurosci.* **16**:3837–3847.

Laurent, G., Stopfer, M., Friedrich, R. W., Rabinovich, M. I., Volkovskii, A., and Arbarbanel, H. D. I., 2001. Odor encoding as an active dynamical process: Experiments, computation, and theory, *Annnu. Rev. Neurosci.* **24**:263–297.

Lisman, J. E., and Otmakhova, N. A., 2001, Storage, recall, and novelty detection of sequences by the hippocampus: Elaborating on the SOCRATIC model to account for normal and aberrant effects of dopamine, *Hippocampus* **11**:551–568.

Llinas, R. R., 1988, The intrinsic electrophysiological properties of mammalian neurons: Insights into central nervous system function, *Science* **242**:1654–1664.

Maas, W., ed. 1999, *Computation with Spiking Neurons*, MIT Press, Cambridge.

Magee, J. C., 2001, Dendritic mechanisms of phase precession in hippocampal CA1 pyramidal neurons, *J. Neurophysiol.* **86**:528–532.

Malik, J., and Perona, P., 1990, Preattentive texture discrimination with early vision mechanisms, *J. Opt. Soc. Am. A* **7**:923–932.

Markram, H., Lubke, J., Frotscher, M., and Sakmann, B., 1997, Regulation of synaptic efficacy by coincidence of postsynaptic APs and EPSPs, *Science* **275**:213–215.

Miller, D. A., and Zucker, S. W., 1999, Computing with self-excitatory cliques: A model and an application to hyperacuity-scale computation in visual cortex, *Neural. Comput.* **11**:21–66.

Petersen, R. S., Panzeri, S., and Diamond, M. E., 2002, The role of individual spikes and spike patterns in population coding of stimulus location in rat somatosensory cortex, *Biosystems* **67**:187–193.

Pearl, J., 1988, *Probabilistic Reasoning in Intelligent Systems: Networks of Plausible Inference*, Morgan Kaufmann, San Mateo, CA.

Poggio, T., and Reichardt, W., 1973, Considerations on models of movement detection, *Kybernetik* **13**:223–227.

Raizada, R. D., and Grossberg, S., 2003, Towards a theory of the laminar architecture of cerebral cortex: Computational clues from the visual system, *Cereb. Cortex* **13**:100–113.

Ramachandran, V. S., 1985, The neurobiology of perception, *Perception* **14**:97–103.

Rao, R. P., and Ballard, D. H., 1997, Dynamic model of visual recognition predicts neural response properties in the visual cortex, *Neural Comput.* **9**:721–763.

Rao, R. P., and Sejnowski, T. J., 2001, Spike-timing-dependent Hebbian plasticity as temporal difference learning, *Neural Comput.* **13**:2221–2237.

Sajda, P., and Finkel, L. H., 1995, Intermediate-level visual representations and the construction of surface perception, *J. Cogn. Neurosci.* **7**:267–291.

Saul, A. B., and Humphrey, A. L., 1990, Spatial and temporal response properties of lagged and nonlagged cells in cat lateral geniculate nucleus, *J. Neurophysiol.* **64**:206–224.

Sutton, R. S., and Barto, A. B., 1998, *Reinforcement Learning: An Introduction*, MIT Press, Cambridge, MA.

Vaadia, E., Bergman, H., and Abeles, M., 1989, Neuronal activities related to higher brain functions–theoretical and experimental implications, *IEEE Trans. Biomed. Eng.* **36**:25–35.

van Santen, J. P., and Sperling, G., 1985, Elaborated Reichardt detectors, *J. Opt. Soc. Am. A* **2**:300–321.

VanRullen, R., and Thorpe, S. J., 2002, Surfing a spike wave down the ventral stream, *Vision Res.* **42**:2593–2615.

Watson, A. B., and Ahumada, A. J., Jr., 1985. Model of human visual-motion sensing, *J. Opt. Soc. Am. A* **2**:322–341.

Weiss, Y., 1997, Interpreting images by propagating Bayesian beliefs, *Adv. Neural Inform. Process. Syst.* **9**:908–915.

Whittington, M. A., Traub, R. D., Kopell, N., Ermentrout, B., and Buhl, E. H., 2000, Inhibition-based rhythms: Experimental and mathematical observations on network dynamics, *Int. J. Psychophysiol.* **38**:315–336.

Wilson, M. A., and McNaughton, B. L., 1994, Reactivation of hippocampal ensemble memories during sleep, *Science* **265**:676–679.

Zanker, J. M., Srinivasan, M. V., and Egelhaaf, M., 1999, Speed tuning in elementary motion detectors of the correlation type, *Biol. Cybern.* **80**:109–116.

Zeki, S., 2003, Improbable areas in the visual brain, *Trends Neurosci.* **26**:23–26.

Zhou, H., Friedman, H. S., and von der Heydt, R., 2000, Coding of border ownership in monkey visual cortex, *J. Neurosci.* **20**:6594–6611.

9

COMPUTATIONAL NEURAL NETWORKS

Dongming Xu,[1,*] Bryan Davis,[1] Mustafa Ozturk,[1] Liping Deng,[1]
Mark Skowronski,[1] John G. Harris,[1] Walter J. Freeman,[2]
and Jose C. Principe[1]

[1]Department of Electrical and Computer Engineering, University of Florida [2]Department of
Molecular and Cell Biology, University of California, Berkeley, California

9.1. INTRODUCTION

Brain function remains one of the most elusive and fascinating phenomena challenging modern science (Churchland, 1986). Although a lot is already known about the neuron and its functional characteristics, when we address the information-processing capabilities of a neural assembly, called here the mesoscopic description (Freeman, 1975), more often than not we are unable to quantify function and abstract the computation. The reasons can be found in the distributed nature of the system architecture, the lack of appropriate tools and metaphors to describe the communication among neurons (the neural code) (Rao *et al.*, 2002), and also very often due to the absence of a detailed knowledge of the function being implemented (Nicolelis, 2001). Hence, conducting studies in computational neuroscience requires a carefully planned methodology and experimental design.

In this chapter we will present neural nonlinear dynamics through a particular example by reporting on our current analysis and implementation of Freeman's model of the olfactory cortex (Freeman, 1975). Freeman's model quantifies the function of one of the oldest sensory cortices, where there is an established causal relation between stimulus and response. It also presents the function as an association between stimulus and stored information, in line with the content addressable memory (CAM) framework studied in neural networks (Llinas, 2001). Freeman utilizes the language of dynamics to model neural assemblies, which seems a natural solution owing to the known spatiotemporal characteristics of brain function (Kelso, 1995). The advantage of a dynamical framework to quantify mesoscopic interactions is related to the possibility of creating analog VLSI circuits that implement similar dynamics.

---

* Address for correspondence: Department for Electrical and Computer Engineering, University of Florida, Florida.

In this respect the dynamics is also independent of the hardware, mimicking the well-known hardware independence characteristics of formal systems. However, the dynamical approach to information processing is much less developed than the statistical reasoning used in pattern recognition.

Only recently did nonlinear dynamics start being used to describe computation (Grossberg, 1972) and there remains a long way to achieve a nonlinear dynamical theory of information processing. Hence we are at the same time developing the science and understanding the tool capabilities, which is far from the ideal situation. The challenge is particularly important in the case of Freeman's model, where the distributed system is locally stable but globally unstable, creating nonconvergent (eventually chaotic) dynamics. Nonconvergent dynamics are very different from the simple dynamical systems with point attractors studied by Hopfield (Hopfield, 1984), because they have positive Lyapunov exponents (Kaplan and Glass, 1995). Their properties for information processing are not yet well understood. Throughout this chapter, we will see a reflection of the issues addressed above, with an emphasis on dynamics, information-processing simulations, and analog VLSI implementations. Ultimately, we plan to use Freeman's model as a signal-to-symbol translator, quantify its performance and implement it in analog VLSI circuits for low-power real-time processing in intelligent sensory-processing applications.

9.2. REVIEW OF DYNAMICAL SYSTEMS ANALYSIS

In this section, we study the basic concepts of dynamical systems as well as the methodologies to analyze them. We start with the linear system, which is well studied and is a basic model for analyzing complex nonlinear systems. In Section 9.2.2, nonlinear systems and how their qualitative behavior is related to linear systems are studied. Sections 9.2.3 and 9.2.4 present two concepts that are important for our analysis through this chapter.

9.2.1. LINEAR TIME-INVARIANT SYSTEMS AND THEIR QUALITATIVE BEHAVIOR

A linear time-invariant dynamical system has the form

$$\frac{dx(t)}{dt} = Ax(t) \tag{9.1}$$

with $x \in R^N$ and a constant matrix $A \in R^{N \times N}$. The solution of Eq. (9.1) is $x(t) = x(0)\, e^{At}$.

The qualitative behavior of a linear system is determined by the eigenvalues of the constant matrix A and can be analyzed completely. Basically, the solution of the system defined by Eq. (9.1) has the following two possibilities:

a) Solutions of Eq. (9.1) are all stable if all the eigenvalues of the constant matrix A have a negative real part.
b) Solutions of Eq. (9.1) are all unstable if at least one of the eigenvalues of the constant matrix A has a positive real part.

In the case that some or all of the eigenvalues of A are purely imaginary, the system presents more complicated behaviors. Detailed discussions on this subject can be found in Arnold (1973) and Braun (1993).

9.2.2. NONLINEAR SYSTEMS

A general nonlinear dynamical system has the form

$$\frac{\mathrm{d}x(t)}{\mathrm{d}t} = f(x + (t), t) \tag{9.2}$$

with $x \in R^N$ and a nonlinear function $f(x)$. There are no existing methods to completely solve the general system defined by Eq. (9.2). However, in most cases, the behavior of a nonlinear system can be described according to the qualitative behavior around some equilibrium points (Hoppensteadt, 2000). Researchers linearize a nonlinear system to achieve an explicit solution and qualitative analysis in the neighborhood of equilibrium points. Here, we will only consider the case of an autonomous system defined as

$$\dot{x} = f(x) \tag{9.3}$$

where $x \in R^N$. The Hartman–Grobman theorem is the fundamental theorem that describes the condition when the phase portrait of Eq. (9.3) is qualitatively the same as that of a linearized system

$$\dot{x} = A(x_0)x \tag{9.4}$$

where x_0 is an equilibrium of Eq. (9.3) and A is the Jacobian matrix of $f(x)$ at x_0.

Hartman–Grobman theorem (Kuznetsov *et al.*, 1998). A system defined by Eq. (9.3) is locally topologically equivalent (preserving the parameterization) to its linearization as defined in Eq. (9.4) (i.e., there is a homomorphism in a neighborhood of the equilibrium that maps orbits of the nonlinear to the linear flows), if the linearization $A(x_0)$ has no purely imaginary eigenvalues. In this case, Eq. (9.3) is a locally hyperbolic dynamical system.◊

Thus, in many cases, study of the nonlinear system is based on the analysis of its linearized model.

9.2.3. BIFURCATIONS

Consider an autonomous dynamical system with its parameter set as

$$\dot{x} = f(x, \alpha) \tag{9.5}$$

with $x \in R^N$ and a parameter set $\alpha \in R^m$.

The parameter-dependent system defined by Eq. (9.5) may present different behaviors in phase space when the parameter passes through a certain point called a bifurcation point. Bifurcation occurs when a system is structurally different with respect to the variation of its parameter set. Although bifurcation analysis is important in order to understand complex systems, we will use it in this chapter to guarantee that the system behavior remains basically unchanged in the neighborhood of the operating point. This is important because when we implement a parametric nonlinear system in analog VLSI, there is unavoidable imprecision in the values of the components. However, we still would like to work in the same dynamic regime of the simulations. This is synonymous with choosing operating points far away from bifurcations.

9.2.4. *POINCARÉ–BENDIXON THEOREM*

Consider a second-order autonomous dynamical system in the form of

$$\begin{cases} \dot{x} = f(x, y) \\ \dot{y} = g(x, y) \end{cases} \tag{9.6}$$

The theorem describes how the system evolves with time in a compact region.

Poincaré–Bendixon Theorem (see Jordan and Smith, 1987; Hoppensteadt, 2000). Let D be a closed and bounded region in R^2, containing a finite number of equilibrium points of Eq. (9.6). A solution (i.e., $x_0(t)$ and $y_0(t)$) of Eq. (9.6) remains entirely in D. Then the solution has three possible behaviors: (1) it approaches an equilibrium point; (2) it approaches a closed path; (3) it approaches an orbit that joins a series of equilibrium points.◇

Poincaré–Bendixon theorem provides a clear view of possible behaviors in a second-order system, especially for the existence of limit cycles. It simplifies the qualitative analysis of higher-order systems if they can be reduced to a second-dimensional space. This theorem does not apply to systems of dimension higher than two, in which cases we should expect much more complex dynamics (Haken, 1996).

9.3. *THE OLFACTORY SYSTEM AS A DISTRIBUTED NEURAL NETWORK OF COUPLED OSCILLATORS*

We will now be studying a model of brain function developed by Walter Freeman. An extensive study of the model can be found in Freeman's 1975 book (Freeman, 1975) and in subsequent papers. The model is locally stable but globally chaotic in a very high dimensional space. The complexity of the whole system is somewhat simplified by a hierarchical embedding of similar, yet simpler, structures. Four different levels are included and are called K0, KI, KII, and KIII, due to the seminal work of Katchalsky (Freeman, 1975). Section 9.3.1 describes these levels. Section 9.3.2 studies the dynamical behavior of elements in the KII level.

9.3.1. *THE HIERARCHY STRUCTURE OF FREEMAN'S MODEL*

9.3.1.1. *The K0 Set and Network*

The K0 set is the most basic building block in the hierarchy. It includes three stages as illustrated in Figure 9.1. Spatial inputs to a K0 set are weighted and summed, and then the summation is passed through a linear time-invariant system with second-order dynamics. The output of the linear system is shaped by the asymmetric nonlinear function as defined by Eq. (9.8). Two categories of K0 sets (excitatory and inhibitory) are defined by the

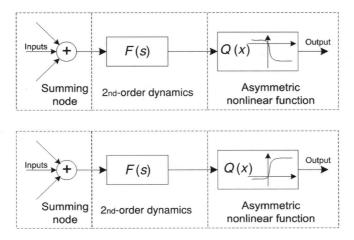

FIGURE 9.1. Diagram of a K0 set.

sign of the nonlinear function. There is no coupling among K0 sets when forming a K0 network.

9.3.1.2. The KI Network

K0 sets with a common sign (either excitatory or inhibitory) are connected through forward lateral feedback to construct a KI network. No autofeedback is allowed in the network. Figure 9.2 shows a KI network that has all excitatory K0 sets.

9.3.1.3. The KII Set and KII Network

A KII set in the model is a coupled oscillator that consists of two KI sets (or four K0 sets). Each set has fixed coupling coefficients obtained from biological experiments. A KII set is the basic computational element in the olfactory system. The measured output from any of the nonlinear functions has two stable states that are controlled by the external stimulus. The resting state occurs when external input is absent, whereas an oscillation occurs when the external input is present. Therefore a KII set is an oscillator controlled by the input.

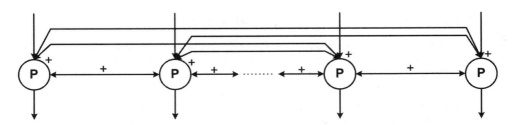

FIGURE 9.2. KI network that consists of interconnected K0 sets (periglomular cells).

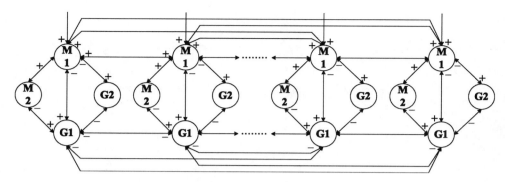

FIGURE 9.3. KII network. M: mitral cells; G: granular cells.

Interconnected KII sets form a KII network as shown in Figure 9.3. In addition to the internal connections of excitatory (M1) cells and inhibitory (G1) cells within a KII set, excitatory and inhibitory interconnections are built through M1 and G1 cells from different KII sets respectively. This interconnected structure represents a key stage of learning and memorizing in the olfactory system. Input patterns through M1 cells are mapped into spatially distributed outputs. Excitatory and inhibitory interconnections enable cooperative and competitive behaviors, respectively, in this network. The KII network functions as an encoder of input signals or as an autoassociative memory (Freeman *et al.*, 1988; Principe *et al.*, 2001).

9.3.1.4. The KIII Network as a Model for the Olfactory System

The KIII network is the computational model of the olfactory system. Figure 9.4 shows a KIII network with different layers representing real regions of a mammalian brain. In a KIII network, basic KII sets and a KII network are tightly coupled through dispersive connections, mimicking the different lengths and thicknesses of axons. Since the intrinsic oscillating frequencies of each one of the KII sets in different layers are incommensurate among themselves, this network of coupled oscillators will present chaotic behavior.

Although we believe that the full dynamical description of the KIII network is beyond our present analytical ability, we may still be able to understand the dynamics of the KII network from first principles.

9.3.2. DYNAMIC ANALYSIS OF A REDUCED KII SET

Instead of studying a full KII set with four K0s, we analyze the dynamics of a reduced KII set (Figure 9.5), which is a simplified version of a full KII set. m_2 and g_2 cells become redundant when the number of KII sets in a KII network increases. With only m_1 and g_1, we have a reduced KII set. Two neural masses are coupled, where $m(t)$ and $g(t)$ denote the neural masses of the mitral and granule cells respectively. K_{mg} and K_{gm} are the coupling coefficients between mitral and granule cells. $P(t)$ is a time-independent input that is given to the mitral cell population.

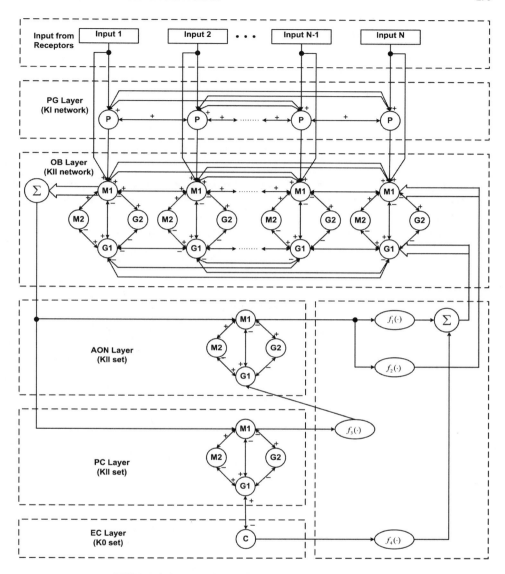

FIGURE 9.4. KIII network as a model for the olfactory system.

A reduced KII set described by two coupled second-order ordinary differential equations (ODEs) is as follows:

$$\begin{cases} \dfrac{1}{ab}\left(\dfrac{\mathrm{d}^2 m(t)}{\mathrm{d}t^2} + (a+b)\dfrac{\mathrm{d}m(t)}{\mathrm{d}t} + abm(t) \right) = K_{\mathrm{gm}} Q(g(t)) + P(t), & K_{\mathrm{gm}} < 0 \\[2mm] \dfrac{1}{ab}\left(\dfrac{\mathrm{d}^2 g(t)}{\mathrm{d}t^2} + (a+b)\dfrac{\mathrm{d}g(t)}{\mathrm{d}t} + abg(t) \right) = K_{\mathrm{mg}} Q(m(t)), & K_{\mathrm{mg}} > 0 \end{cases} \tag{9.7}$$

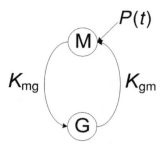

FIGURE 9.5. Diagram of a reduced KII set.

$Q(x)$ is the nonlinear function shown int Figure 9.6 that models the spatiotemporal integration of spikes into mesoscopic waves measured in the cortex (Freeman, 1975) and is defined by Eq. (9.8)

$$Q(x, Q_m) = \begin{cases} Q_m(1 - e^{-(e^x - 1)/Q_m}), & x > x_0 \\ -1, & \text{else} \end{cases}, \quad x_0 = \ln(1 - Q_m \ln(1 + 1/Q_m))$$

$$(9.8)$$

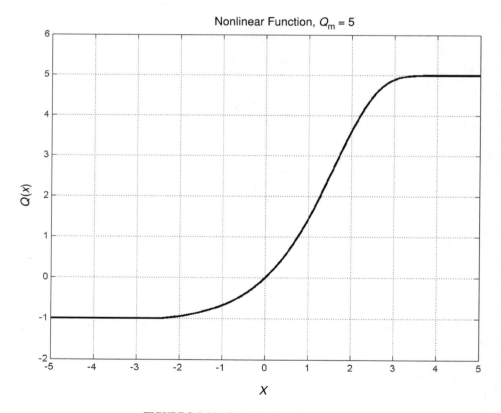

FIGURE 9.6. Nonlinear function with $Q_m = 5$.

Q_m is adjustable and is used to fit the function by experimental data. The values of a and b in this system are given by Freeman *et al.* (1988) as $a = 220s^{-1}$ and $b = 720s^{-1}$.

In this section, we will analyze the dynamical behavior of a reduced KII set, from which we expect to get further understanding of the whole KIII network and at the same time help with the circuit design of the olfactory system in analog VLSI. Because of the stable local behavior, the main goal is to find and then avoid bifurcation regions in parameter space, and ensure that the system does not present uncontrollable behavior.

9.3.2.1. Bifurcation Analysis of a Reduced KII Set

Controlled by the two possible states of the external input (i.e., $P(t) = 0$ and $P(t) > 0$), the desired dynamical behavior of a reduced KII set should have one of the two possibilities: a stable fixed point ($P(t) = 0$) or a stable limit cycle ($P(t) > 0$). In the following analysis, we will try to determine the conditions on K_{mg} and K_{gm} under which the desired system behavior is achieved. The analysis will help us understand the locally stable structure in Freeman's model. Also, it will facilitate the hardware design of a silicon cortex. For a second-order bounded nonlinear system, the Poincaré–Bendixon theorem characterizes the three possible system states by examining the qualitative behavior around equilibrium. Basically, a reduced KII set is a fourth-order bounded nonlinear system. We don't have similar theory to guarantee a simple approach but we put forward the following hypothesis.

Hypothesis: If a reduced KII set is not at a bifurcation point, its dynamical behavior is determined by the qualitative behavior at its equilibrium points. More specifically, if the equilibrium point is stable, the solution of a reduced KII set will approach a fixed point; if the equilibrium is unstable, the solution of a reduced KII set will approach a limit cycle.◇

Although we do not have a proof of this hypothesis yet, experimental results support the conclusions. Thus, the basic approach to understand the behavior of a reduced KII set is to determine the stability of its equilibrium point.

Equation (9.7) can also be expressed as a system of coupled first-order differential equations

$$\begin{cases} \dfrac{dm(t)}{dt} = m'(t) \\[2mm] \dfrac{dm(t)'}{dt} = -(a+b)m'(t) - abm(t) + ab\left(K_{gm}Q(g(t)) + P(t)\right) \\[2mm] \dfrac{dg(t)}{dt} = g'(t) \\[2mm] \dfrac{dg'(t)}{dt} = -(a+b)g'(t) - abg(t) + abK_{mg}Q(m(t)) \end{cases} \qquad (9.9)$$

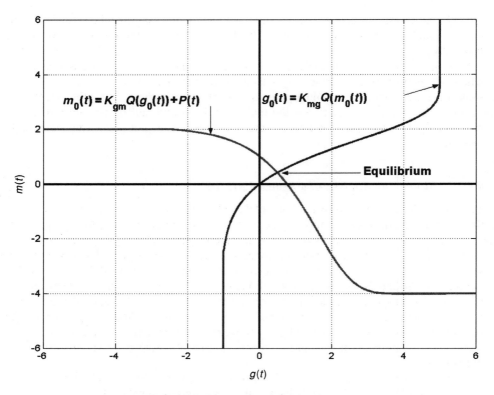

FIGURE 9.7. Equilibrium in a reduced KII set.

To find the fixed-point solution, we have

$$
\begin{cases}
\dfrac{dm(t)}{dt} = m'(t) = 0 \\[2mm]
\dfrac{dm(t)'}{dt} = -(a+b)m'(t) - abm(t) + ab\left(K_{gm}Q(g(t)) + P(t)\right) = 0 \\[2mm]
\dfrac{dg(t)}{dt} = g'(t) = 0 \\[2mm]
\dfrac{dg'(t)}{dt} = -(a+b)g'(t) - abg(t) + abK_{mg}Q(m(t)) = 0
\end{cases}
$$

Only one intersection between nullclines exists as shown in Figure 9.7, so we have only one set of equilibrium.

The Jacobian matrix of Eq. (9.9) is

$$
L = \begin{bmatrix}
0 & 1 & 0 & 0 \\
-ab & -(a+b) & abK_{gm}Q'(g_0(t)) & 0 \\
0 & 0 & 0 & 1 \\
abK_{mg}Q'(m_0(t)) & 0 & -ab & -(a+b)
\end{bmatrix} \tag{9.10}
$$

where $m_0(t)$ and $g_0(t)$ are fixed-point solutions.

By solving for the eigenvalues of Eq. (9.10), based on the analysis in Xu and Principe (2003), we obtained the condition, under which there is a pair of eigenvalues with zero real part (the system has bifurcations), as

$$\left| K_{mg} K_{gm} \right| = \frac{(a+b)^2}{ab} \frac{1}{R(K_{mg}, K_{gm})} \quad \text{and} \quad K_{mg} > 0, \quad K_{gm} < 0 \qquad (9.11)$$

According to Eq. (9.11),

$$\left| K_{mg} K_{gm} \right| > \frac{(a+b)^2}{ab} \frac{1}{R(K_{mg}, K_{gm})}$$

is the condition for hyperbolic and unstable equilibrium points, whereas

$$\left| K_{mg} K_{gm} \right| < \frac{(a+b)^2}{ab} \frac{1}{R(K_{mg}, K_{gm})}$$

leads to stable equilibrium points. Remember that we are more interested in determining K_{mg} and K_{gm} to make the system either have a stable state or go to a stable limit cycle. We want to avoid regions where the system behavior becomes complicated. So, in the following sections, we consider the cases where there is only hyperbolic equilibrium.

9.3.2.2. Analysis of Dynamical Behavior without External Stimulus

When there is no external input ($P(t) = 0$), the system defined by Eq. (9.7) has a stable equilibrium at the origin, i.e., $m_0(t) = 0$ and $g_0(t) = 0$. When $P(t) = 0$, $R(K_{mg}, K_{gm}) = Q'(m_0(t))Q'(g_0(t)) = 1$. Thus, we obtain

$$\left| K_{mg} K_{gm} \right| < \frac{(a+b)^2}{ab} \qquad (9.12)$$

Equation (9.12) gives the condition for the system to stay in a rest state when $P(t)$ is zero. Note that this upper bound is only dependent on the values of a and b. Using $a = 220 \text{ s}^{-1}$ and $b = 720 \text{ s}^{-1}$, we obtain $\left| K_{mg} K_{gm} \right| < 5.58$.

To verify the above analysis, we use two different sets of parameters to simulate the system (in both cases, $P(t) = 0$ and $Q_m = 5$):

a) In Figure 9.8, $\left| K_{mg} K_{gm} \right| = 5.5$ where $K_{mg} = 1$ and $K_{gm} = -5.5$
b) In Figure 9.9, $\left| K_{mg} K_{gm} \right| = 5.6$ where $K_{mg} = 1$ and $K_{gm} = -5.6$

Note that, in both cases, the actual values of K_{mg} and K_{gm} do not affect the stability of this system as long as their product remains constant.

From Figures 9.8 and 9.9 we see that Eq. (9.12) well estimates the dynamical behavior of the system defined by Eq. (9.7) given $P(t) = 0$. When $\left| K_{mg} K_{gm} \right| < 5.58$, the system is stable although the transient time is very long. When $\left| K_{mg} K_{gm} \right| > 5.58$, the system oscillates.

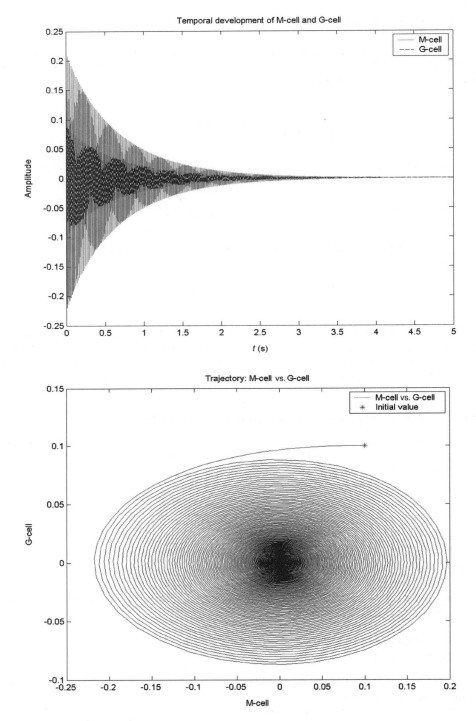

FIGURE 9.8. With $\left|K_{mg}K_{gm}\right| = 5.5$, $P(t) = 0$, and initial conditions $m_0(t) = 0.1$, and $g_0(t) = 0.1$. (a) Temporal development of M-cell $(m(t))$ and G-cell $(g(t))$; (b) Trajectory.

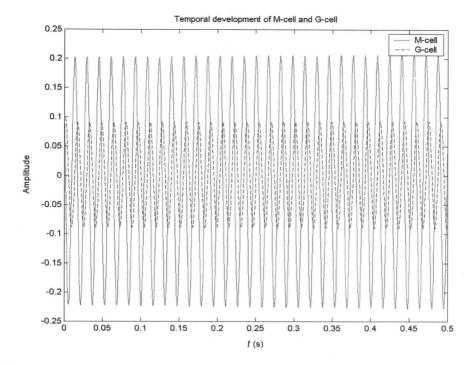

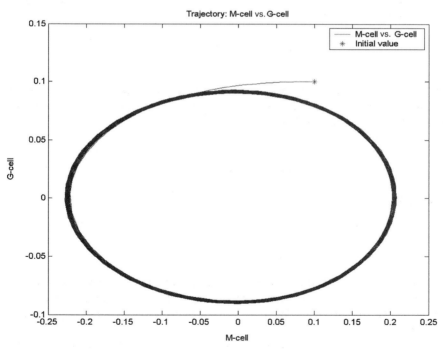

FIGURE 9.9. With $\left|K_{mg}K_{gm}\right| = 5.6$, $P(t) = 0$, and initial conditions $m_0(t) = 0.1$ and $g_0(t) = 0.1$. (a) Temporal development of M-cell ($m(t)$) and G-cell ($g(t)$); (b) Trajectory.

9.3.2.3. *Analysis of Dynamical Behavior with External Stimulus*

When $P(t)$ is present, i.e., $P(t) > 0$, we would like that the system has an unstable equilibrium point so that it will oscillate. Thus,

$$|K_{mg} K_{gm}| > \frac{(a+b)^2}{ab} \frac{1}{R(K_{mg}, K_{gm})}$$

sets the lower bound for the product of the two parameters. Generally, we may not be able to compute an analytical solution for the lower bound. However, we could adjust K_{mg} and K_{gm} to satisfy it through qualitative analysis and simulations. We give without proof the following facts (Xu and Principe, 2003):

Fact 1: Given a particular input $P(t) > 0$, we have one set of fixed points that are always positive, i.e., $0 \le m_0(t) < P(t)$ and $0 \le g_0(t) \le K_{mg} Q_m$.

Fact 2: We have the following properties of $R(K_{mg}, K_{gm})$: Recall that

$$R(K_{mg}, K_{gm}) = Q'(m_0(t))Q'(g_0(t))$$

where $Q'(x) = e^x e^{-(e^x - 1)/Q_m}$, $-1 \le x \le Q_m$, is the derivative of the nonlinear function $Q(x)$. Figure 9.10 shows the plot of $Q'(x)$ when $Q_m = 5$.

$Q'(x)$ has a maximum of $Q_m e^{-(Q_m - 1)/Q_m}$ when $x = \ln(Q_m)$. So we have a loose estimation of the lower bound as

$$|K_{mg} K_{gm}| > \frac{(a+b)^2}{ab} \frac{1}{R(K_{mg}, K_{gm})} \ge \frac{(a+b)^2}{ab} \frac{1}{(\ln Q_m)^2}$$

Clearly, to satisfy both the upper bound and lower bound as shown in Eq. (9.13), a necessary condition is to adjust K_{mg} and K_{gm} so that the fixed-point solutions are in the region where $Q'(m_0(t))Q'(g_0(t))$ is greater than 1.

$$\frac{(a+b)^2}{ab} \frac{1}{R(K_{mg}, K_{gm})} < |K_{mg} K_{gm}| < \frac{(a+b)^2}{ab} \tag{9.13}$$

Fact 3:

$$\begin{cases} m'_{K_{mg}} < 0, \, g'_{K_{mg}} > 0 \\ m'_{K_{|gm|}} < 0, \, g'_{K_{|gm|}} < 0 \end{cases}$$

Thus, the fixed-point solution $m_0(t)$ is a decreasing function with respect to K_{mg}, and $g_0(t)$ is an increasing function with respect to K_{mg}. Both $m_0(t)$ and $g_0(t)$ are decreasing functions with respect to $|K_{gm}|$. With this fact, we know how to move the equilibrium point to a desired value.

Assume that $Q_m = 5$, $a = 220 \text{ s}^{-1}$ and $b = 720 \text{ s}^{-1}$. From the above facts, we give the following steps to adjust parameters K_{mg} and K_{gm} so that both the upper and lower bounds given by Eq. (9.13) are satisfied. We will only consider how the product of the two coupling coefficients, rather than their individual values, determines the behavior of the system.

Steps to adjust K_{mg} and K_{gm}

a) Set $|K_{mg}K_{gm}| < 5.58$ so that the upper bound is satisfied and the system will have a stable state when $P(t) = 0$.

b) If we slightly reduce $|K_{mg}K_{gm}|$ from 5.58, Eq. (9.12) should always be satisfied. However, the response time when $P(t)$ goes to zero will be very long as shown in Figure 9.8 Therefore, we want to make $|K_{mg}K_{gm}|$ small enough to let the system die out quickly when $P(t) = 0$.

c) When $|K_{mg}K_{gm}|$ is much smaller than the upper bound, we cannot guarantee that the lower bound will also be satisfied because of the nonlinear relation between $|K_{mg}K_{gm}|$ and the lower bound. However, according to Fact 3, we know that we could adjust the values of K_{mg} and K_{gm} to move the equilibrium points within their possible regions without changing $|K_{mg}K_{gm}|$. So, after setting K_{mg} and K_{gm}, if the system does not oscillate when $P(t) > 0$, the first thing we want to know is the positions of the equilibrium points, which could be obtained by solving Eq. (9.14) or by checking the intersection of nullclines in Figure 9.7.

$$\begin{cases} -m_0(t) + K_{gm}Q(g_0(t) + P(t)) = 0 \\ -g_0(t) + K_{mg}Q(m_0(t)) = 0 \end{cases}, \quad K_{mg} > 0, \quad K_{gm} < 0 \qquad (9.14)$$

Knowing the positions, we could move the equilibria according to Fact 3 and keep $|K_{mg}K_{gm}|$ unchanged. The fixed-point solution may or may not reside in the region where $R(K_{mg}, K_{gm}) > 1$. If not, we need to decrease the values of $m_0(t)$ and $g_0(t)$(see Facts 1 and 2). Thus, according to Fact 3, we should mainly increase $|K_{gm}|$, and at the same time increase or decrease K_{mg} accordingly, until the system oscillates. Using a plot as shown in Figure 9.10 could be helpful for moving equilibrium to the right positions. However, if the system still cannot oscillate after both $m_0(t)$ and $g_0(t)$ pass through $\ln(Q_m)$, the system will never oscillate given current settings of Q_m and $P(t)$. If the system does not oscillate while both $m_0(t)$ and $g_0(t)$ are in the region where $Q'(x) > 1$, then $R(K_{mg}, K_{gm})$ is not large enough to satisfy the lower bound. In this case, we should either reduce or increase the values of $m_0(t)$ and $g_0(t)$ to make them as close to $\ln(Q_m)$ as possible.

As an example, we use the same system as in Section 9.3.2.2 but with an input $P(t) = 1$ and initial conditions being $m_0(t) = 0.1$ and $g_0(t) = 0.1$. First, we set $|K_{mg}K_{gm}| = 4$, $K_{mg} = 1$ and $K_{gm} = -4$. From Figure 9.11, we see that the system does not oscillate even when we provide an external stimulus. Figure 9.11(a) also gives an approximated value of the equilibrium at

$$\begin{cases} m_0(t) = 0.178 \\ g_0(t) = 0.191 \end{cases}$$

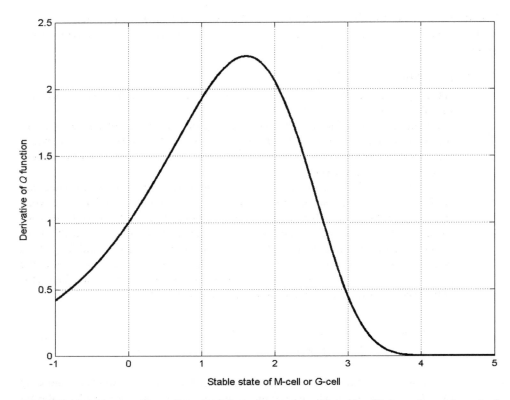

FIGURE 9.10. Derivative of the nonlinear function. (as an example, positions of equilibrium points are determined with $K_{mg} = 1$, $K_{gm} = -1$).

Thus,

$$R(K_{mg}, K_{gm}) = Q'(0.178)Q'(0.191) = 1.334$$

and

$$\frac{(a+b)^2}{ab} \frac{1}{R(K_{mg}, K_{gm})} = 4.182 > \left| K_{mg} K_{gm} \right| = 4$$

Apparently, $\left| K_{mg} K_{gm} \right|$ does not satisfy the lower bound required to make the system oscillate. However, we now know the positions of the equilibrium points.

In this case, we need to change the values of $m_0(t)$ and $g_0(t)$ to increase $R(K_{mg}, K_{gm})$. We set the parameters to $K_{mg} = 4$ and $K_{gm} = -1$ while keeping $\left| K_{mg} K_{gm} \right| = 4$ unchanged. We have

$$\begin{cases} m_0(t) = 0.154 \\ g_0(t) = 0.656 \end{cases}$$

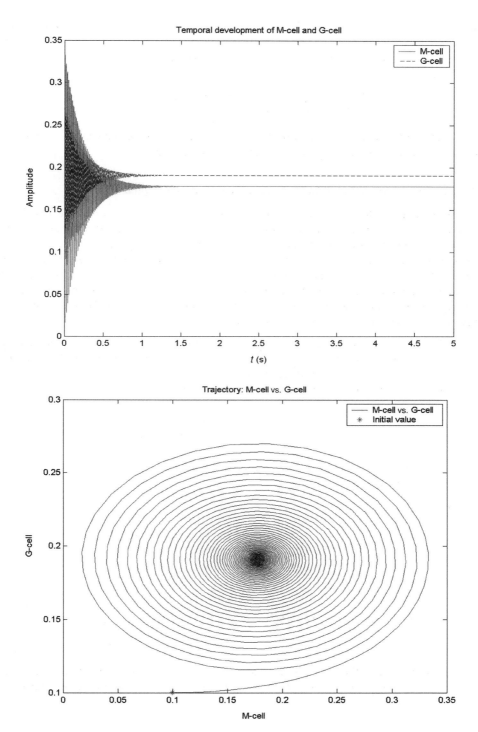

FIGURE 9.11. With $K_{\mathrm{mg}} = 1$, $K_{\mathrm{gm}} = -4$, $P(t) = 1$, and initial conditions $m_0(t) = 0.1$ and $g_0(t) = 0.1$. (a) Temporal development of M-cell ($m(t)$) and G-cell ($g(t)$); (b) Trajectory.

So we obtain $R(K_{mg}, K_{gm}) = Q'(0.154)Q'(g_0(0.656)) = 1.806$ and

$$\frac{(a+b)^2}{ab} \frac{1}{R(K_{mg}, K_{gm})} = 3.089 < |K_{mg}K_{gm}| = 4.$$

Figure 9.12 shows the simulation results after changing the parameters. We see that at increased values of $R(K_{mg}, K_{gm})$, the system starts oscillating.

9.3.2.4. System Sensitivity with Respect to the Coupling Coefficients

Knowing how to control the system by setting the coupling coefficients is helpful in both understanding the system and adjusting the weights when designing an analog circuit. However, we hope that we have a rather large valid region of K_{mg} and K_{gm} so that, once built, the system could bear large variations of the parameter set without a change in qualitative behavior. For example, in analog circuit design, there are offsets and noise due to circuit structure and fabrication that bring variations to each part of the circuit. In this section, we will try to analyze the system sensitivity w.r.t. the coupling coefficients. We will be showing that the valid sets of K_{mg} and K_{gm} are bounded.

(i) Regions containing K_{mg} and K_{gm} that satisfy $|K_{mg}K_{gm}| < (a+b)^2/ab$ are easy to find and are shown in Figure 9.13. Clearly, the requirement for the coupling coefficients is not very strict in this case.

(ii) We will be showing that if K_{mg} and K_{gm} satisfy

$$|K_{mg}K_{gm}| > \frac{(a+b)^2}{ab} \frac{1}{R(K_{mg}, K_{gm})}$$

then even in a bounded neighborhood, the condition is guaranteed to be satisfied everywhere.

From

$$Q'(x) = e^x \left(1 - \frac{Q(x)}{Q_m}\right)$$

we have

$$|K_{mg}K_{gm}| R(K_{mg}, K_{gm}) = |K_{mg}K_{gm}| e^{m_0+g_0} \left[1 - \frac{Q(m_0)}{Q_m}\right]\left[1 - \frac{Q(g_0)}{Q_m}\right] > \frac{(a+b)^2}{ab} \tag{9.15}$$

a) Now, assume that K_{mg}^* and K_{gm}^* satisfy the above condition. We will change the value of K_{gm} to see how far it can go, while keeping the conditions satisfied at the same time. Obviously,

$$0 < |K_{gm}| < \frac{(a+b)^2}{ab} \frac{1}{K_{mg}^*} \tag{9.16}$$

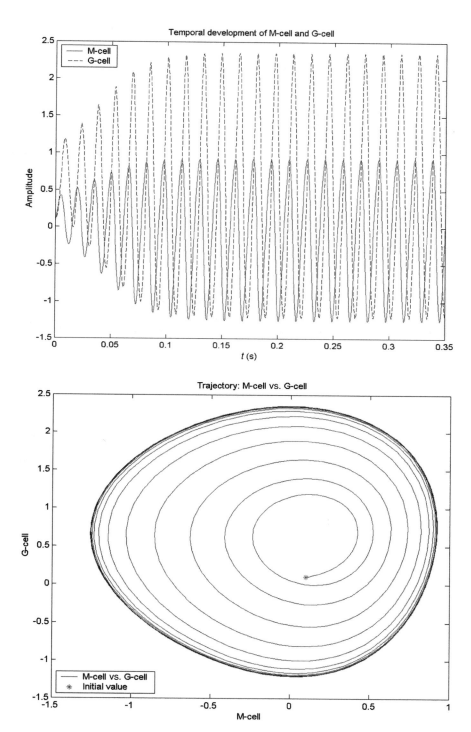

FIGURE 9.12. With $K_{mg} = 4$, $K_{gm} = -1$, $P(t) = 1$, and initial conditions $m_0(t) = 0.1$ and $g_0(t) = 0.1$. (a) Temporal development of M-cell ($m(t)$) and G-cell ($g(t)$); (b) Trajectory.

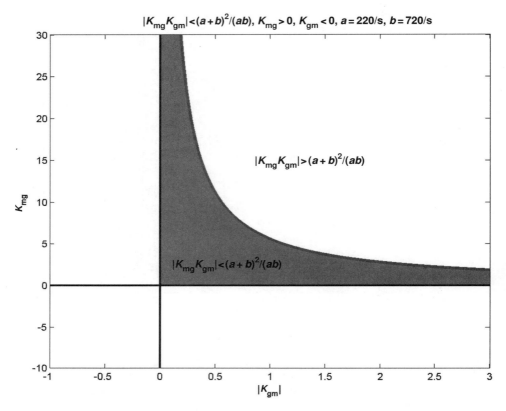

FIGURE 9.13. Valid region of the coupling coefficients that satisfy the upper bound.

When $|K_{gm}|$ increases, both g_0 and m_0 decrease; therefore

$$\left[1 - \frac{Q(m_0)}{Q_m}\right]\left[1 - \frac{Q(g_0)}{Q_m}\right]$$

increases and $e^{m_0+g_0}$ decreases. The value of $|K_{gm}|$ will either violate Eq. (9.15) or reach the upper bound in Eq. (9.16).
When $|K_{gm}|$ decreases, both g_0 and m_0 increase; therefore

$$\left[1 - \frac{Q(m_0)}{Q_m}\right]\left[1 - \frac{Q(g_0)}{Q_m}\right]$$

decreases and $e^{m_0+g_0}$ increases. The value of $|K_{gm}|$ will either violate Eq. (9.15) or reach zero. Hence, we see that $|K_{gm}|$ has both an upper and lower bound when K_{mg} is fixed.

b) Now, assume that we have K_{mg}^* and K_{gm}^* satisfying the above conditions. We will change the value of K_{mg} to see how far it can go while keeping the conditions satisfied at the same time. Apparently,

$$0 < K_{\mathrm{mg}} < \frac{(a+b)^2}{ab} \cdot \frac{1}{\left| K_{\mathrm{gm}}^* \right|} \tag{9.17}$$

When K_{mg} decreases and approaches zero, g_0 decreases and approaches zero while m_0 increases and is limited by P. So the left-hand side of the inequality in Eq. (9.15) also approaches zero. Thus, K_{mg} may have a lower bound that is much greater than zero.

When K_{mg} increases, g_0 increases and m_0 decreases. $[1 - (Q(g_0)/Q_{\mathrm{m}})]$ approaches zero. So, the value of K_{mg} will be bounded by either the upper bound in Eq. (9.17) or a smaller value.

9.4. DIGITAL SIMULATION OF THE FREEMAN MODEL

One rather important intermediate medium between theory and analog implementation is simulation. In particular when one is dealing with a distributed nonlinear dynamical system for which our analytical ability is still rather limited, simulations play a central role because they allow us to do the following:

- Verify the theoretical analysis.
- Simulate complex dynamical regimes.
- Experiment with the model for information-processing applications.
- Check the behavior of the analog VLSI chips we build.

In this section, we will discuss our approach to create flexible and efficient simulations of Freeman's model. The conventional way to evaluate solutions of dynamical systems defined by ODEs is by using Runge–Kutta integration (Lambert, 1991). Hence, it is just natural that we compare our approach with this widely used technique. For the purposes of this chapter, the focus of numerical simulations will be the reduced KII set (Figure 9.5).

9.4.1. DIGITAL IMPLEMENTATION APPROACHES

Consider the reduced KII set defined by Eq. (9.7) that is copied below for convenience

$$\begin{cases} \dfrac{1}{ab}\left(\dfrac{\mathrm{d}^2 m(t)}{\mathrm{d}t^2} + (a+b)\dfrac{\mathrm{d}m(t)}{\mathrm{d}t} + abm(t)\right) = K_{\mathrm{gm}} Q(g(t)) + P(t), \quad K_{\mathrm{gm}} < 0 \\[3mm] \dfrac{1}{ab}\left(\dfrac{\mathrm{d}^2 g(t)}{\mathrm{d}t^2} + (a+b)\dfrac{\mathrm{d}g(t)}{\mathrm{d}t} + abg(t)\right) = K_{\mathrm{mg}} Q(m(t)), \quad\quad\quad K_{\mathrm{mg}} > 0 \end{cases}$$

It is a coupled nonlinear system defined by continuous-time ODEs. In order to simulate continuous-time equations such as Eq. (9.7) in computer environment, discretization is employed. The most traditional method is the Runge–Kutta integration technique. However, it is not feasible to use Runge–Kutta in the Freeman model for our purposes for the following reasons.

(1) Although it is a well-accepted approximation to continuous time system dynamics, it requires heavy computation for a high-dimensional system such as Freeman's KIII, which translates into long simulation time. Therefore, it is not very efficient.
(2) It is not suitable for real-time implementations, because it is a mathematical construction, which is time consuming and not amenable to simple modifications in the topology.
(3) It is fairly difficult to incorporate an online learning algorithm in Runge–Kutta, which increases the computation load by requiring access to the output every time the update is computed.

Therefore, we will develop an equivalent digital model of the olfactory dynamics using a digital signal processing approach. Consider Eq. (9.7), both equations are ordinary linear differential equations of the form

$$\frac{1}{ab}\ddot{y}(t) + \frac{a+b}{ab}\dot{y}(t) + y(t) = u(t) \tag{9.18}$$

which is the linear dynamic part of a K0 set. In other words, we are interested in a special nonlinear system built from the interconnection of K0 models that possesses linear dynamics followed by a static nonlinearity. At least away from bifurcation points (Hartman-Grobman's theorem), we can use the well-known DSP techniques to transform the linear dynamics of the system into difference equations. However, at our stage of model development the Runge–Kutta solution of the Freeman model must be utilized as a benchmark to determine the performance of our digital system approximation.

9.4.1.1. System Identification with Adaptive Gamma Filter

In prior work, system identification techniques based on the gamma filter approximation architecture were used to achieve discretized approximations to the olfactory system (Tavares, 2001). The gamma filter implementation requires the decomposition of the desired impulse response using gamma basis functions (de Vries, 1991; Principe *et al.*, 1993). This filter structure is based on the first-order generalized delay unit, whose dynamics is defined as (de Vries, 1991; Principe *et al.*, 1993)

$$G(z) = \frac{\mu}{z - (1 - \mu)}$$

In a system identification framework, the optimal weights for a second-order gamma filter that best approximates the continuous-time system impulse response are $w_0 = -2.9 \times 10^{-3}$, $w_1 = 1.245 \times 10^{-1}$, $w_2 = 0.9016$ (Principe *et al.*, 2001). The feedback parameter, which determines the average memory depth of the filter, was found to be $\mu = 0.237 \times 10^{-1}$ (Principe *et al.*, 2001).

Because the gamma filter design is an approximation to the desired transfer function, it does not necessarily preserve the impulse response of the continuous-time system. Consequently, the step response is also not accurately approximated. The step-response accuracy is particularly important in the discretization of the Freeman model, because the inputs to the system are frequently constant voltage levels. There are at least three reasons for the significance of this type of input pattern. First, biological stimuli are represented as a level of an odor. Second, piecewise constant inputs lead to dynamics that are easier to analyze and understand. In addition, since the system exhibits a tightly coupled oscillatory behavior, any error will be amplified in time.

9.4.1.2. Direct Discretization with the Impulse Invariance Transformation

Alternatively, it is possible to design a discrete-time approximation to the olfactory system using the direct discretization approach. This approach eliminates the need to perform system identification and adaptive filter optimization tasks, reducing design complexity. An improvement in the performance of the approximation of the original continuous-time system by its discrete-time equivalent is expected, because the exact transfer function is used to generate the discretized approximation instead of learning the approximation with an adaptive gamma filter of a finite degree of freedom.

Considering the linear dynamical part of the K0 set defined by Eq. (9.18), the input–output transfer function is simply

$$H(s) = \frac{ab}{(s+a)(s+b)} = ab \left(\frac{s+a}{b-a} + \frac{s+b}{a-b} \right)$$

which leads to the time-domain signal

$$h(t) = ab \left(\frac{e^{-at}}{b-a} + \frac{e^{-bt}}{a-b} \right)$$

Using the impulse invariant transformation technique with sampling period T_s, the discrete approximation filter has the impulse response

$$h[n] = T_s h(nT_s) = T_s ab \left(\frac{e^{-anT_s}}{b-a} + \frac{e^{-bnT_s}}{a-b} \right)$$

Letting $\alpha_1 = e^{-aT_s}$, $\alpha_2 = e^{-bT_s}$, $c_1 = T_s ab/(b-a)$, and $c_2 = T_s ab/(a-b)$, the transfer function of the equivalent digital system is finally determined to be

$$H(z) = \frac{c_1}{1 - \alpha_1 z^{-1}} + \frac{c_2}{1 - \alpha_2 z^{-1}}. \tag{9.19}$$

In general, the discretization of continuous-time signals using the impulse-invariant transformation is susceptible to aliasing. However, in the K0 case, the continuous-time system exhibits a low-pass filter behavior, because the poles a and b are positive real numbers. Therefore, aliasing error can be controlled by making T_s small enough to guarantee a given error.

Implementation of the transfer function in Eq. (9.19) can be achieved using, for example, the parallel realization, which is shown in Figure 9.14. This architecture is supported by many standard commercial DSP software or hardware products, and can be efficiently implemented to operate accurately in real time.

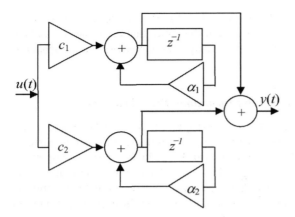

FIGURE 9.14. Parallel implementation of LTI dynamics.

Another DSP technique that can be used to discretize the Freeman model is the well-known bilinear z-transform (Oppenheim and Schafer, 1989). However, although the bilinear z-transform tackles the problem of aliasing by mapping all points in the s-plane onto a unique point on the z-plane, it distorts the frequency response, which will change the continuous time system behavior.

9.4.1.3. Comparison of Impulse Responses

In this section, we present a comparison of the impulse and step responses of three systems: continuous-time original, discrete-time gamma filter approximation, and discrete-time impulse-invariant approximation. The impulse responses are shown in Figure 9.15 and

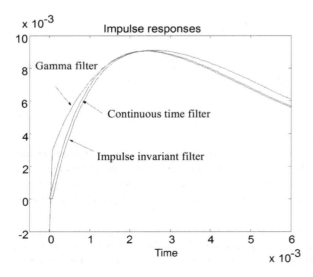

FIGURE 9.15. Impulse responses of the three filters.

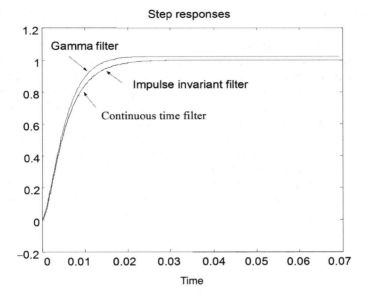

FIGURE 9.16. Responses of the three filters.

the step responses in Figure 9.16. As a quantitative measure of similarity, we employ the mean-square error (MSE) between the impulse responses of the discrete-time approximations and the original filter

$$\text{MSE} = \sum_n (h_d[n] - h_c(nT_s))^2$$

where h_d and h_c denote the impulse responses of the discrete and continuous-time filters. The MSE of the impulse response for the gamma filter and the impulse-invariant filter approximations are found as 4.73×10^{-5} and 4.72×10^{-6}, respectively. The noticeable order of magnitude of improvement is mainly due to the poor approximation incurred by the single pole of the gamma filter coupled with the low order of the approximation.

9.4.2. QUANTITATIVE PERFORMANCE ANALYSIS

The overall Freeman model is a highly complex nonlinear dynamical system with coupled oscillators. Hence, any discrepancy between the linear dynamics of the basic K0 sets will be propagated to the global behavior. In order to investigate the effect of the discrete-time approximation errors on the nonlinear behavior, we investigate the performance of the approximations on the reduced KII set, which might be considered as the smallest nonlinear subsystem representative of the global KIII dynamics.

In particular, the reduced KII set, with specific choices of the connection weights, possesses a fixed point and a limit cycle behavior controlled by the input. Therefore, there is a bifurcation which is controlled by only two parameters: K_{mg} and K_{gm}. In the following numerical examples, the output of the reduced KII set is defined to be the output of the mitral K0 set (see Figure 9.5).

For comparison, the continuous-time system is simulated using the ODE45 built-in Runge–Kutta integration technique in Matlab®. The digital network simulations are performed using the commercial software package Neurosolutions®, which is specialized for neural network simulations. Neurosolutions® has all the basic building blocks necessary to implement the discretized Freeman model. The software has a graphical user interface, which allows building an icon-based reduced KII model.

Ideally, with a positive input, if the connection weights are appropriate, the reduced KII model will exhibit sustained oscillations. Therefore, the digital simulators are expected to behave exactly the same way as the original system. The performance of the digital systems is measured by the distance between the points in the parameter space at which bifurcation occurs $[K_{mg}, K_{gm}]$. The following measure is defined to provide a percentage parameter error (PPE) to quantify the approximation quality, where

$$\text{PPE} = \frac{K_c - K_d}{K_c} \times 100 \tag{9.20}$$

In Eq. (9.20), K_c is the $|K_{gm}|$ (or $|K_{mg}|$) value at which the continuous-time system switches from nonoscillatory to oscillatory behavior while K_{gm} (or K_{mg}) is kept fixed. K_d is similarly the parameter value for the discrete-time approximation under consideration (either the gamma filter or the impulse-invariant filter).

In the following simulations, the input is assumed to be positive continuously and the sampling frequency for the impulse-invariant filter is 20 times the frequency of the faster pole in the continuous-time linear dynamics for alias-free approximation. Figure 9.17 shows the output of the continuous-time reduced KII sets where K_{mg} is kept constant and K_{gm} is varied until bifurcation is observed. According to the PPE measure, the approximation errors of the gamma and impulse-invariant filter approximations are determined as

$$\text{PPE(gamma)} = \frac{4.27 - 3.22}{4.27} \times 100 = 24$$
and
$$\text{PPE(imp - inv)} = \frac{4.27 - 3.95}{4.27} \times 100 = 7.49$$

The PPE for both approaches will decrease as the sampling frequency of the impulse response is increased. For every sampling frequency the gamma filter method requires solving a linear system identification problem with a nonlinear parameter. The impulse-invariant filter method, however, is easily modified for the new sampling frequency by reevaluating the values for the discrete transfer function parameters in Eq. (9.19). Therefore, only the performance of the impulse-invariant method is investigated versus increasing the sampling frequency. The performance (PPE) of the impulse-invariant method versus $1/T_s$ is shown in Figure 9.18. As expected, PPE decreases with increasing sampling frequency, but increasing $1/T_s$ will increase the computation requirements of the discretized systems. This creates a trade-off between modeling accuracy and computation speed, which is controlled by the sampling period.

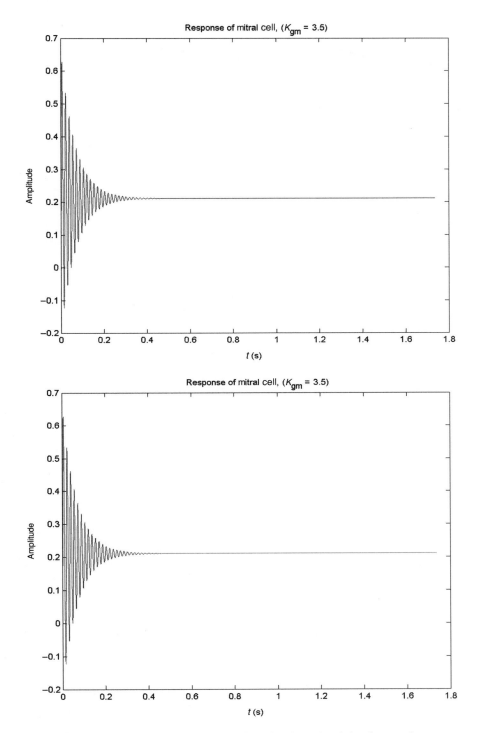

FIGURE 9.17. Behavior of continuous-time reduced KII set for $K_{mg} = 1$ switches from a point attractor to a limit cycle as $K_{gm} = -3, -3.5, -4.25, 4.5$. Bifurcation occurs at $K_{gm} = -4.27$.

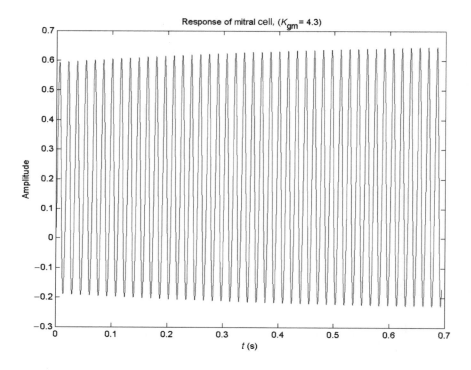

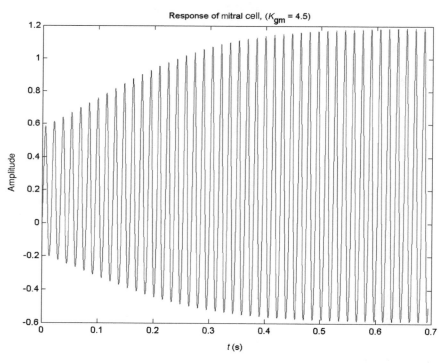

FIGURE 9.17. (*Continued*)

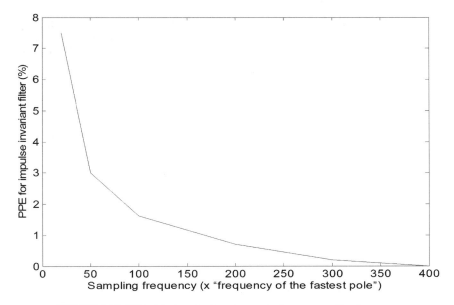

FIGURE 9.18. PPE of the impulse-invariant method versus sampling frequency.

A comparison between the behaviors of the reduced KII sets for the continuous-time system, gamma filter approximation, and the impulse-invariant approximation are given in Figure 9.19. The curves represent the boundaries between oscillatory and fixed-point dynamics in the parameter space of the reduced KII set for each system. As seen from the graph, the impulse-invariant filter provides a closer match to the boundary of the continuous-time system, when compared with the gamma filter.

9.5. THE REDUCED KII NETWORK AS AN ASSOCIATIVE MEMORY

An auto-associative memory is a content-addressable memory that stores a set of patterns during learning and recalls the learned pattern that is most similar to the present input. The utility of an associative memory can then be measured in terms of its ability to reconstruct the correct learned pattern given a corrupted version of that pattern, as well as the number of patterns that it can store simultaneously. There are basically two types of associative memories: static feedforward structures and the recurrent dynamic structures proposed by Hopfield (1982). Hopfield networks have a stable point attractor. Here we will be showing that the reduced KII network (and also the KIII) can also be used as dynamic associative memories, but they differ from the Hopfield type because they have nonconvergent dynamics (limit cycles or chaos). We are only considering multidimensional static binary input patterns (i.e., patterns that do not change in time, and have an arbitrary number of dimensions that can take on one of two values). An example is a memory that could store and recall the patterns 000111 and 010101.

Freeman's KII network is able to perform as a binary auto-associative memory because of three main properties: (1) Each KII set has two modes of operation separated by a

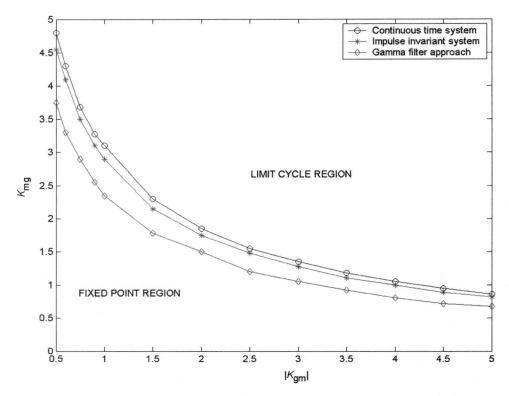

FIGURE 9.19. Boundaries between oscillatory and convergent dynamics in the parameter space of the reduced KII set.

bifurcation—this allows the KII to directly represent the state of a single binary output using the mode of one KII set. (2) Each KII set is fully connected to all other KII sets—this makes the network symmetrical about all of the individual KII sets, and each KII set can have a reinforcing or inhibiting effect on any other KII set. (3) The individual KII sets behave as low-pass filters—this allows them to average the input over time, effectively removing transient noise in the patterns.

Once an appropriate topology is chosen to create the memory store, the next step is to address learning. In order to get the KII network to perform as an auto-associative memory, the interconnection weights of the KII network should be modified by a learning rule according to the patterns that we would like to store. Kozma and Freeman (2001) describe an off-line learning rule resembling Hebb's rule: if both inputs to an excitatory KII network weight are high, this weight will be pulled high, whereas if one input is high and the other is low, the weight should be kept at a low value. In their work, the high and low values were empirically determined based on the dynamics of the KII set. In this chapter, we use a learning rule that works directly with the time signals available at the excitatory connections of the KII network. Finally, we show some results that illustrate the recall and noise rejection pattern completion capabilities of the network.

9.5.1. DESIGNING THE EXCITATORY INTERCONNECTIONS OF THE KII NETWORK

The interconnections between the individual KII sets in a KII network have inhibitory (negative) connections between one of the inhibitory K0 nodes among all of the KII sets. Likewise, there are excitatory connections between excitatory nodes in the KII set. These interconnections are fully connected—each KII set connects to every other KII set (excluding itself) as shown in Figure 9.3. According to Freeman's model, the inhibitory interconnection strengths are fixed, but the excitatory interconnection strengths are adaptable. If a group of KII sets are connected by strong excitatory interconnections, then this group will tend to oscillate together: when one KII is stimulated by a strong external input, it induces a strong response in the other nodes that are connected to it. Thus, even if only one member of a set is given a strong input, the entire group will oscillate together and in synchrony. If, on the other hand, two nodes are connected by very weak excitatory connections, then an oscillation in one set will inhibit oscillations in the other set because the inhibitory interconnections will dominate.

Using these tendencies, we can create an auto-associative memory by using a KII network. For example, suppose we have an eight-channel KII network (each channel represents one KII set), and we want the network to learn the two eight-bit patterns 11110000 and 00011110. In this case, the excitatory interconnection weight matrix would be as shown in Table 9.1.

These two patterns are representative of patterns obtained from real-world events, i.e., with overlapped bits (bit 4 is 1 in both patterns). It is known that bit overlaps among patterns limit the accuracy of the recall in associative memories, so they should be included in any testing.

9.5.2. TRAINING THE KII INTERCONNECTION WEIGHTS WITH OJA'S RULE

In order to create an associative memory, we must train the weights connecting individual KII sets. We fix the inhibitory weights at 0.6, and train the excitatory weights. The learning rule will be exploiting correlation, where the best example is Hebbian learning: increase the weight connecting two channels proportional to the product of the excitation

TABLE 9.1. Excitatory Interconnection Weight Matrix

		1	2	3	4	5	6	7	8
	1		S	S	M	W	W	W	W
	2	S		S	M	W	W	W	W
	3	S	S		M	W	W	W	W
$(i + 1)$th Channel	4	M	M	M		M	M	M	W
	5	W	W	W	M		S	S	W
	6	W	W	W	M	S		S	W
	7	W	W	W	M	S	S		W
	8	W	W	W	W	W	W	W	

Note. W indicates a weak excitatory connection and *S* indicates a strong excitatory connection. *M* indicates a medium connection. The diagonal terms are absent because a KII network can have no external connection directly to itself.

of the two channels. Hebbian learning can be expressed as

$$w_{ij} \leftarrow w_{ij} + \eta x_i x_j$$

where x_i is the ith channel activation, w_{ij} is the connection weight between the ith and jth channels, and η is the learning rate. If we start with weak excitatory interconnections, and apply the pattern 11100000 for a length of time, the first three KII sets will become strongly connected, while all others will remain weakly connected. If we then apply the other pattern 00011100, then we will obtain the matrix in Table 9.1. Unfortunately, Hebbian learning is unbounded, which means that if we apply the patterns long enough, the weights will eventually grow so large that the outputs will saturate. To mitigate this unbounded growth, we use Oja's rule (Oja, 1982)

$$w_{ij} \leftarrow w_{ij}(1 - \frac{\eta}{A}x_j^2) + \eta x_i x_j$$

where A is a normalization term such that $\sum_i w_{ij} = A$, $\forall j$. This is similar to the multiplicative forgetting factor on the weights called *habituation* used in Kozma and Freeman (2001) to limit the growth of the interconnection weights, except that here the forgetting factor is adaptive. The advantage of using Oja's rule versus exponential weight decay lies in the fact that it forces the sum of the weights connecting to a particular input to be a constant (habituation does not require this—the value of the weight is a weighted sum of past excitation levels). This is an advantage when designing the system, because it is less likely that the system will go out of its operating range when the interconnecting weights become too large or too small.

9.5.3. EXAMPLES

A simple application of an associative memory is that of character recognition. For this example, we created images of two characters (0 and 1) in an 8×8 display with pixel values of 1 (black) and 0 (white). The goal is to train the system so that if a corrupted version of one of these images is presented to the system, the system will respond with the uncorrupted image. To do this, we create a system composed of 64 reduced KII sets—one KII set for each pixel. We trained the system by alternately presenting the characters 0 and 1 for 10,000 time ticks each, and repeated the sequence 10 times. Notice that this form of presentation of information is different from the conventional pattern presentation in static memories, where each pattern can be presented in succession (this is due to the dynamics of this system). It is also different from the Hopfield network, where the input is initially presented and then removed.

Figure 9.20 shows the architecture of the reduced KII network as diagramed in the neural network simulation program Neurosolutions®. In the input box on the left, the 64-dimensional input retrieved from a file is displayed on an imaging grid. This input is connected to the excitatory nodes in the reduced KII network. The bottom icon in the KII network box is an array of KII sets with fixed connections, and the top icon in the box is the learned Hebbian connections. The output of the reduced KII network is oscillating and it contains biases that are not related to the stored information. We need to remove the

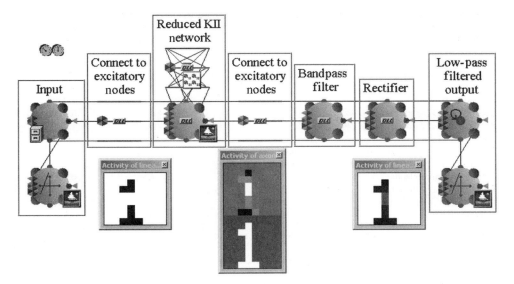

FIGURE 9.20. Neurosolutions® architecture for the reduced KII network. Information flows from left to right.

biases and display the amplitude of the primary oscillations to retrieve the desired output. To do this we first high-pass filter the output of the excitatory KII nodes to remove the DC component of the signal. Next, we full-wave rectify this result to show the amplitude of oscillation at each node. Finally, we low-pass filter the rectified output for smoothing, and normalize it between 0 and 1. On the bottom of Figure 9.20, the first image box displays the binary input. The second image box displays the excitatory (top half of image) and inhibitory (bottom half of half) activity of the reduced KII network. The right-most box displays the conditioned output. The initial high-pass filter is first-order with a 3dB cutoff frequency at 13 Hz, and the final low-pass filter is first-order with a 3dB cutoff frequency at 4 Hz. The other parameters of the network are shown in Table 9.2.

After 10 presentations of each pattern, the system learns the patterns quite well, as evidenced by the expected values of the interconnection weights. The utility of an associative memory lies in its ability to restore corrupted or incomplete patterns; therefore, we created a set of corrupted patterns that have missing pixels in the pattern and extra pixels in the background (Figure 9.21). As we can observe in this figure, the output of the KII network cleans the imperfections in the input quite well, very much like a Hopfield network. Although this is a very simple problem, it demonstrates the ability of a KII network to work as an auto-associative memory. During our experimentation we realized that the learning method used requires sparse inputs (inputs with most channels inactive) to learn properly. Also, when we presented all the 10 digits (0–9), the performance was not acceptable for this size of network. We suspect that this is largely due to the high degree of overlapping between the patterns. A preprocessing stage can take care of the orthogonalization of patterns, but we suspect that if we use a KIII network instead of a KII, the chaotic dynamics of the KIII will in essence uncorrelate the patterns over time and will provide improved performance.

TABLE 9.2. Parameters of the KII Network Used in the Digit Recognition Example

Parameter	Symbol	Value
Number of KII sets		64
Upper saturation limit of the K0 non-linearity	Q_m	5
Sampling rate	f_s	14.4 kHz
Second-order filter parameter	a	220 Hz
Second-order filter parameter	b	720 Hz
Active pixel input value		1.5
Inactive pixel input value		0
Internal connection weight from excitatory to inhibitory node	K_{mg}	2
Internal connection weight from inhibitory to excitatory node	K_{gm}	-1
Inhibitory interconnection weights	K_{ii}	0.01
Fixed excitatory interconnection weights	K_{ee}	0.003
Learning rate for Oja's rule	η	0.001
Normalization value	A	0.5
Transfer function of linear high-pass filter (applied before rectification)		$\dfrac{1 - z^{-1}}{1 - 0.995z^{-1}}$
Transfer function of linear low-pass filter (applied after rectification)		$\dfrac{1}{1 - 0.9995z^{-1}}$
Number of samples for which each pattern is presented		10,000
Number of times patterns are alternately applied		10

9.6. HARDWARE IMPLEMENTATION IN ANALOG VLSI

As indicated in Principe *et al.* (2001), analog implementation has the following advantages:

- It is a natural coupling to the analog real world.
- Computation is intrinsically parallel.
- The computation time is practically instantaneous (only set by the delay imposed by each component).
- There are no round-off problems, i.e., the amplitude resolution is dictated by device physics.
- It generally renders smaller circuits than digital devices.
- It is one order of magnitude more power efficient than digital implementations.

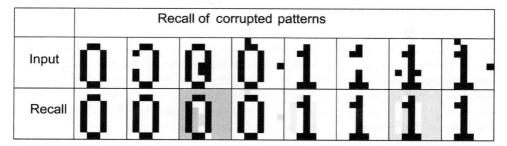

FIGURE 9.21. Recalling corrupted patterns using a trained 64-channel KII network. Amplitude for each channel is represented by the darkness of the image pixel. The image is normalized so that black and white represent the highest and lowest amplitudes in the image, respectively.

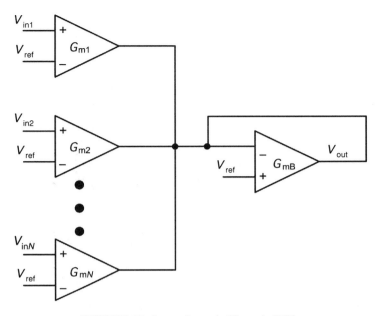

FIGURE 9.22. A summing node (Xu *et al.*, 2003).

Because of the hierarchical architecture of Freeman's model, where K0 is the most basic structure, our task is to design each part of the K0 set: input stage, nonlinear function, and second-order dynamics. All parts have been designed and fabricated using AMI 0.5-μm technology, and work in subthreshold region to achieve low power consumption.

9.6.1. INPUT STAGE

This part implements the weighted summation of the input stage in a K0 set. Although using discrete op-amps outside the chip could easily provide all the weights and summations needed, this solution becomes impractical when we build a large KII network. Therefore, we integrated the summing node on chip using transconductance amplifiers working in the subthreshold region. The design of wide-linear-range transconductance amplifiers (Furth and Ommani, 1997) is utilized here. Figure 9.22 shows the schematic of a summing node. We see that a current adder followed by a current-to-voltage converter is used to realize weights that are greater than 1. The output is a weighted summation of the inputs as

$$V_{\text{out}} = \sum_{i=1}^{N} \frac{G_{\text{mi}}}{G_{\text{mB}}} V_{\text{ini}}$$

9.6.2. NONLINEAR FUNCTION

Two considerations in the design of a nonlinear function are asymmetric and sigmoidal shapes. A transconductance amplifier working in the subthreshold region automatically

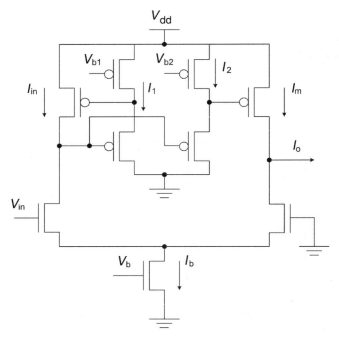

FIGURE 9.23. Asymmetric nonlinear function (Xu *et al.*, 2003).

gives a hyperbolic tangent function between the differential input voltage and the output current (Mead, 1989). It is not exactly the nonlinear function given by Freeman but provides a reasonable approximation (Principe *et al.*, 2001; Tavares, 2001). The asymmetric shape is implemented by an externally controlled current mirror. The ratio of positive saturation value over negative saturation value is set by the bias voltage provided outside the chip. Figure 9.23 shows the circuit that generates an asymmetric sigmoidal nonlinear function. Equation (9.21) describes how the current mirror generates the asymmetric

$$I_{\mathrm{m}} = \frac{I_1}{I_2} I_{\mathrm{in}} \tag{9.21}$$

I_1 and I_2 are controlled by V_{b1} and V_{b2}, respectively. The shape of the circuit is defined by

$$Q\,(V_{\mathrm{in}}) = I_{\mathrm{b}}\,\frac{\dfrac{I_1}{I_2}\mathrm{e}^{(V_{\mathrm{in}}/VT)} - 1}{\mathrm{e}^{(V_{\mathrm{in}}/VT)} + 1} \tag{9.22}$$

Note that there is a systematic offset in Eq. (9.22) whereas in Eq. (9.8) there is not. To cancel this offset, the circuit shown in Figure 9.24 is used (Tavares, 2001). Non1 and Non2 are two nonlinear functions that have exactly the same settings. Thus, we assume that Non1 and Non2 have the same systematic offset. The output of Non1 cancels the offset of Non2.

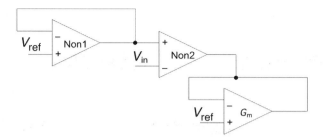

FIGURE 9.24. Offset cancellation scheme (Tavares, 2001; Xu *et al.*, 2003).

The output current is converted to voltage at the final stage by a transconductance amplifier Gm with negative feedback.

9.6.3. SECOND-ORDER DYNAMICS

The second-order linear time-invariant system is implemented using the Filter and Hold (F&H) method proposed in (Tavares, 2001). Figure 9.25 is a diagram of this implementation. Analog output of each stage in the second-order system is held for a certain period to increase the actual time constant. A duty cycle of $k\%$ gives an effective capacitance $100/k$ times larger than that without F&H. This provides a convenient way to design on-chip capacitance without using a large chip area. Readers can refer to Principe *et al.* (2001) and Tavares (2001) for detailed design issues.

9.6.4. CHIP MEASUREMENT RESULTS

In this section, we will show some measurement results from an analog VLSI chip. This chip was designed using AMI 0.5-μm technology and includes a reduced KII set. Each K0 set has an area of 600 μm × 100 μm and a power consumption of 26 μW. Figure 9.26 shows the response from the excitatory cell and the inhibitory cell. A square wave is connected to the input of the excitatory cell. As we expected, the qualitative behaviors of the KII set follows the theoretical conclusions, and exhibits two states of dynamical behavior controlled by an external stimulus.

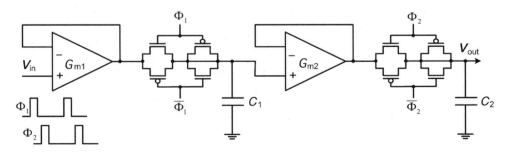

FIGURE 9.25. F&H implementation of second-order dynamics (Tavares, 2001; Xu *et al.*, 2003).

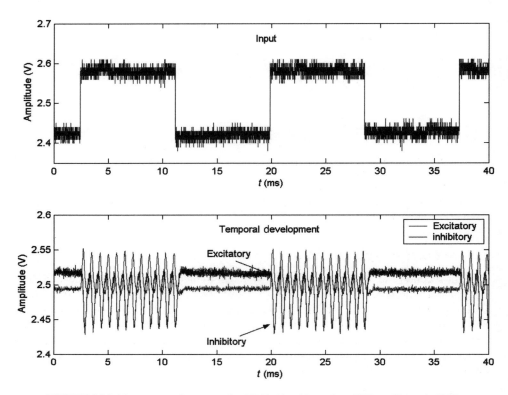

FIGURE 9.26. Measurement from an analog VLSI chip with a reduced KII set (Xu *et al.*, 2003).

9.6.5. COMPARISON OF DIGITAL SIMULATION AND HARDWARE DESIGN

To demonstrate the ability of digital simulation to facilitate analog VLSI design, we simulate in Neurosolutions® the blocks used in the hardware design. Figure 9.27 depicts a comparison between the behavior of the reduced KII sets for the impulse-invariant approximation and the analog circuit simulated using PSPICE. We see that the two systems give similar results on the qualitative dynamical behaviors. Figure 9.28 shows the transient waveforms of the two systems under external input control. $P(t)$ is set to 50 mV, while $K_{mg} = |K_{gm}| = 2.5$. Both systems have basically the same qualitative (transitions from the fixed point to oscillation controlled by the external stimulus) properties. Their quantitative properties (amplitudes of wave forms) have small differences because both simulation environments use approximations to the models and initial conditions may not be exactly the same. The results indicate that digital simulation could very well predict both the qualitative and the quantitative behaviors in the designed hardware. Moreover, digital simulation in NeuroSolutions® runs much faster than circuit simulation does. For the simulations shown in Figure 9.28, it took NeuroSolutions® about 0.2 s to achieve the result, whereas the circuit simulations had a run time of around 3 min.

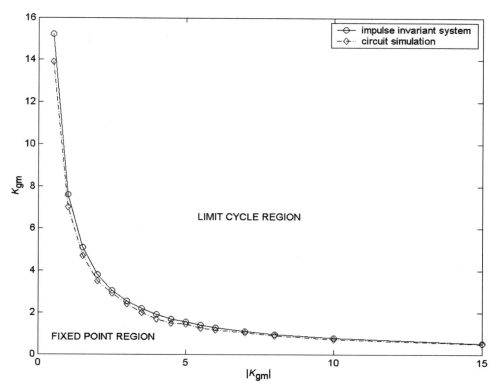

FIGURE 9.27. Boundaries between oscillatory and convergent dynamics for impulse response in the parameter space of the reduced KII set. Comparison between digital simulation and analog circuit simulation.

9.7. CONCLUSIONS

The major goal of this chapter is to introduce nonlinear dynamics of neural systems by the overviewing of the framework being developed to study, simulate, design, fabricate, and test biologically plausible information-processing paradigms. This chapter emphasizes showing the highly multidisciplinary nature of the enterprise, and the gains that can be achieved by synergistically employing knowledge from neurobiology, system neuroscience, dynamics, digital signal processing, artificial neural networks, and circuit and analog VLSI design.

There are two themes in our work that we believe are very important for successfully designing neuromorphic information processing systems. The first one is that the brain is the ultimate real-time machine exploiting the spatiotemporal dynamics of billions of synapses and neurons for information processing. As such we need to go beyond formal systems to preserve the plausibility of the biological computation. Dynamics enables us to simulate the capacity of biological brains to create information by virtue of positive Lyapunov exponents, which allows us to go beyond the algebraic framework of Von Neumann computation to

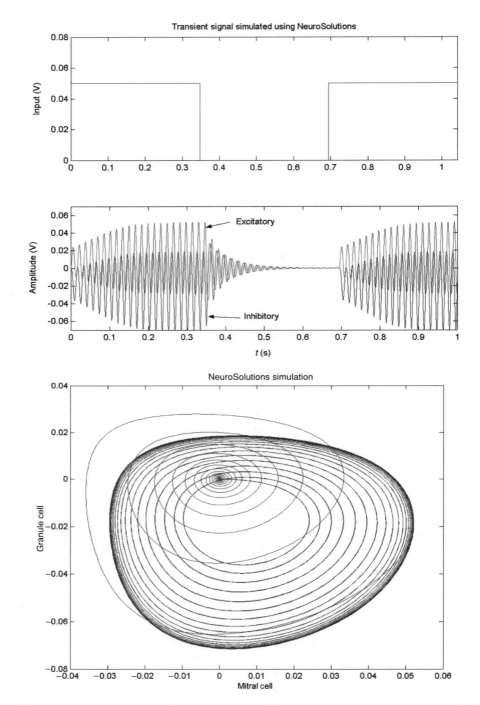

FIGURE 9.28. Transient signal simulations in Neurosolutions® and PSPICE. In both cases, $P(t)$ is set to 50 mV, while $K_{mg} = |K_{gm}| = 2.5$. (a) and (b) show the results obtained from digital simulation using Neurosolutions®; (c) and (d) show the circuit simulations.

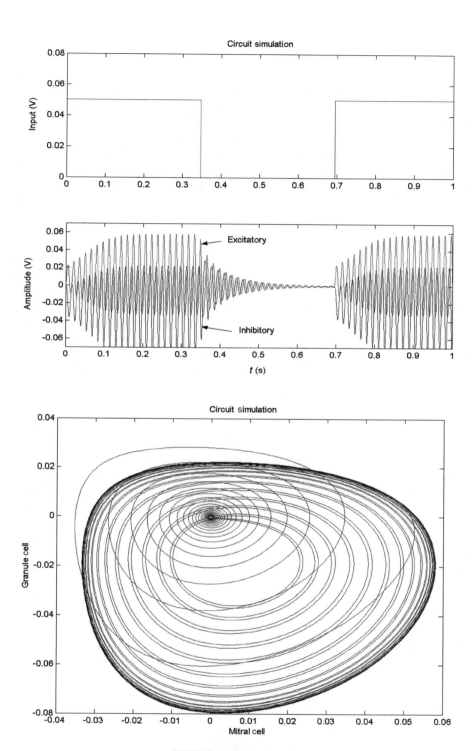

FIGURE 9.28. (*Continued*)

directly compute with functions. Moreover, dynamics provide, just like logic, independence with respect to the implementation medium. Once we capture the dynamical behavior of a cell assembly, we can implement the same behavior in the simulator (through digitization) as well as in the silicon.

The second aspect is to determine the most appropriate hierarchical level of analysis to study brain computation. The complexity and heterogeneity of the brain discourages a global approach and, at the other end of the granularity spectrum, the neuron seems to be too fine (and unreliable) a detail to explain computation. The dynamics of functional cell assemblies captured by the Katchalsky hierarchy was utilized here as the building block in our model. The modularity and reusability of this approach is impressive and simplifies all the aspects of this project, from the theoretical dynamical studies, the simulation environment, to the circuit design stage.

We are still very far from creating an olfactory cortex in silicon. However, we have advanced our understanding in several important aspects. First, we have now a first-principle understanding of the dynamics of the reduced KII set, which will hopefully lead to a full description of the bifurcation dynamics of the KII set, the building block of the olfactory bulb, and of all the other constituents of the olfactory system. Although the reduced KII set may be biologically unrealistic,[1] the mathematical simplicity achieved maybe a necessary step to understand for computational purposes the KII dynamics, which is crucial to unravel the nonconvergent dynamics of the KIII model. They have also been useful in helping us choose the operating points of the KII and preserve as much as possible the required dynamics even in the case of large parameter variations produced by chip fabrication.

Our efforts of creating a fast and efficient simulation environment to test both our theoretical dynamical predictions and the hardware are already paying off. The simulator will also be instrumental in studying information-processing characteristics of this dynamical computational model. This part of the project is the less developed so far, and the one where more rapid progress can be achieved in the short term, because the building blocks are all completed and tested. Still a remaining issue is the read-out of information from the spatiotemporal patterns of activity. The distributed nature of the information also poses a great challenge.

Finally, we have designed and fabricated all the building blocks necessary for building the KII set. The dynamical behavior measured from the chip is qualitatively the same as the simulation. In fabricated chips, we should expect more deviations and offsets in every system block due to the fabrication process, design mismatches, and testing environments, etc. Normally, a large portion of the offsets and noise could be modeled into circuit simulation under worst-case consideration by intentionally mismatching components, adding extra circuits, and simulating different working conditions. Also, various techniques are used to reduce systematic offsets so that actual chip measurement will follow the circuit simulation as accurately as possible. For every circuit simulation or observed offsets in the chip, we

---

[1] The rationale for forming a KII set from a KI_e and KI_i set (four K0 sets) instead of two K0 sets is to model an active neural population: at any moment as having some neurons that are transmitting while most are receiving. The partitioning remains constant over time, but each subset of neurons is constantly being renewed from the other subset. It is a necessary feature in making a lumped approximation at the mesoscopic level for the myriad neurons in the population at its microscopic level. This bipartite division also approximates the predominance of local connection density in the small-world network topology, and for simulating the $1/f^2$ PSD_T that characterizes the scale-free network (Wang and Chen, 2003) in extended KII layers.

could provide a corresponding mathematical model to be used in the digital simulation. In this way, we have the advantages of obtaining fast simulation results and verifying any theoretical conclusions more efficiently. This could greatly reduce the effort needed to perform trial-and-error procedures in the analog VLSI design and increase productivity.

The implementation of a KII network and the full KIII model in a chip are still not within our reach. The big difficulty is the size of the interconnection required for full connectivity (interconnection grows with the square of the processing elements). We have addressed this issue by creating a discrete state model for the dynamics, and a multiplexing scheme that reduces the growth of the interconnectivity (Freeman, 1988; Tavares, 2001; Xu *et al.*, 2003). But even this is too much for a network that can handle realistic pattern-recognition tasks. Therefore we are investigating spiking network architectures to implement the building blocks of the KII set.

ACKNOWLEDGMENT

This work was partially supported by grant ONR N00014-1-1-0405.

REFERENCES

Arnold, V., 1973, *Ordinary Differential Equations*, MIT Press, Cambridge.

Braun, M., 1993, *Differential Equations and Their Applications*, Springer-Verlag, New York.

Churchland, P., 1986, *Neurophilosophy: Towards a Unified Science of the Mind/Brain*, MIT Press, Cambridge.

de Vries, B., 1991, *Temporal Processing with Neural Networks—The Development of the Gamma Model*, Ph.D. Dissertation, University of Florida, Gainesville, Florida.

Freeman, W., 1975, *Mass Action in the Nervous System*, Academic Press, New York.

Freeman, W. J., 1988, Pattern learning and recognition device, United States Patent #4,748,674, May 31.

Freeman, W., Yao, Y., and Burke, B., 1988, Central pattern generating and recognizing on olfactory bulb: A correlation learning rule, *Neural Network* 1:277–288.

Furth, P. M., and Ommani, H. A., 1997, Low-voltage highly-linear transconductor design in subthreshold CMOS, in: *Proceedings of the 40th Midwest Symposium on Circuits and Systems*, Sacramento, CA, 1:156–159.

Grossberg, S., 1972, Pattern learning by functional-differential neural networks with arbitrary path weights, in: *Delay and Functional Differential Equations and Their Applications* (Schmitt, ed.), Academic Press, New York, pp. 121–160.

Haken, H., 1996, *Principles of Brain Function*, Springer-Verlag, New York.

Hopfield, J., 1982, Neural networks and physical systems with emergent collective computational abilities, *Proc. Natl. Acad. Sci.* 79:2554–2558.

Hopfield, J., 1984, Neurons with graded response have collective computational properties like those of two-state neurons, *Proc. Natl. Acad. Sci.* 81:3088–3092.

Hoppensteadt, F., 2000, *Analysis and Simulation of Chaotic Systems*, 2nd ed., Springer-Verlag, New York.

Jordan, D. W., and Smith, P., 1987. *Nonlinear Ordinary Differential Equations*, 2nd ed., Clarendon Press, Oxford.

Kaplan, D., and Glass, L., 1995, *Understanding Nonlinear Dynamics*, Springer-Verlag, New York.

Kelso, S., 1995, *Dynamic Patterns: The Self-Organization of Brain and Behavior*, MIT Press, Cambridge.

Kozma, R., and Freeman, W. J., 2001, Chaotic resonance-methods and applications for robust classification of noisy and variable patterns, *Int. J. Bifurc. Chaos* 11(6):1607–1629.

Kuznetsov, Y., Kuznetsov, L., and Marsden, J., 1998, *Elements of Applied Bifurcation Theory*, 2nd ed., Springer-Verlag, New York.

Lambert, J., 1991, *Numerical Methods for Ordinary Differential Systems: The Initial Value Problem*, Wiley, New York.

Llinas R., 2001, *I of the Vortex: From Neurons to Self*, MIT Press, Cambridge.

Mead, C., 1989, *Analog and Neural Systems*, Addison-Wesley, MA.

Nicolelis, M. (Ed.), 2001,*Advances in Neural Population Coding, Progress in Brain Research*, Vol. 130, Elsevier, Amsterdam.

Oja, E., 1982, A simplified neuron model as a principal component analyzer, *J. Math. Biol.* **15:**267–273.

Oppenheim, A. V., and Schafer, R. W., 1989, *Discrete Time Signal Processing*, Prentice Hall, Englewood Cliffs, NJ.

Principe, J., Tavares, V., Harris, J., and Freeman, W., 2001, Design and implementation of a biologically realistic olfactory cortex in analog VLSI, *Proceedings of the IEEE* **89**(7):1030–1051.

Principe, J., de Vries, B., and Oliveira, P., 1993, The gamma filter: A new class of adaptive IIR filters with restricted feedback, *IEEE Trans. Signal Process.* **41**(2):649–656.

Rao, R., Olshausen, B., and Lewicki, M. (eds.), 2002, *Probabilistic Models of the Brain*, MIT Press, Cambridge.

Tavares, V., 2001, *Design and Implementation of a Biologically Realistic Olfactory Cortex Model*, Ph.D. Dissertation, University of Florida, Gainesville, FL.

Wang, X., and Chen, G., 2003, Complex networks: Small-world, scale-free and beyond, *IEEE Circ. Syst. Mag.* **3(1):**6–20.

Xu, D., Deng, L., Harris, J., and Principe, J., 2003, Design of a reduced KII set and network in analog VLSI, *Proc. 2003 Int. Symp. Circ. Syst.* **5:**837–840.

Xu, D., and Principe, J., 2003, Dynamical analysis of a reduced KII set in the olfactory system, submitted to *IEEE Transactions on Neural Networks*.

10

CIRCUIT MODELS FOR NEURAL INFORMATION PROCESSING

Ting Ma, Ying-Ying Gu, and Yuan-Ting Zhang*

Joint Research Center for Biomedical Engineering, Department of Electronic Engineering,
The Chinese University of Hong Kong, Hong Kong, P. R. China

10.1. INTRODUCTION

Neural information processing is of great importance in understanding the principle of natural operation of nervous systems, in developing biologically inspired signal processing techniques, and in designing medical devices for the treatment of diseases. A basic neural information processing system is concerned primarily with signal generation at the biophysical level, information encoding at the communication system level, and integration for decision making and control at the central computation level. Complex spike sequences or action potential trains that reflect both the intrinsic dynamics of neurons and the temporal characteristics of the stimuli are the information-carrying signals in the nervous system. A large number of penetrating and useful models have been developed by many investigators to understand the complexities of the nervous system with particular relevance for information processing at various levels. The specific objectives of this chapter are to review concisely some of the important neural circuit models, to promote a balanced understanding of the interplay between the dynamics and temporal characteristics of action potential trains, and their effects on neural information processing.

10.2. CIRCUIT MODELS FOR SINGLE NEURONS

Circuit models can be used to describe the natural electrical properties of single neurons and thereby have the potential for addressing fundamental questions of neuronal coding and signal transmission within the nervous system. Awareness of the key notions of single neurons is necessary for constructing models. In this section, we will first brief on the

---

* Address for correspondence: Department of Electronic Engineering, Joint Research Center for Biomedical Engineering, The Chinese University of Hong Kong, Hong Kong, P. R. China; e-mail: ytzhang@ee.cuhk. edu.hk.

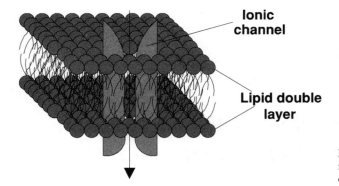

FIGURE 10.1. A patch of neuronal membrane with double-layer lipid and embedded proteins.

elementary background of single neurons and then introduce some important neuronal circuit models.

10.2.1. A SIMPLE CIRCUIT MODEL FOR PASSIVE NEURONAL MEMBRANES

Membrane potential of excitable neurons can be significantly altered and serves as the primary vehicle for information transmission in the nervous system. The structure of the neuronal membrane can be represented by a patch of double-layer lipid, within which there are embedded proteins (Kandel *et al.*, 1995), as shown in Figure 10.1. This relative insulating configuration with the selectivity and permeability of a membrane makes the neuron maintain some electrical properties, which make neural signaling in an electrical form possible.

The backbone of the membrane is made up of double-layer lipid with extremely low conductivity to charges. The double-layer lipid, which separates the internal and the external charges, therefore functions as a capacitance. The membrane capacitance, C_m, can be calculated by multiplying the capacitance of unit membrane area, c_m (F/cm^2), with the total membrane area.

The proteins spanning the lipid layer of the membrane act as the ionic channels through which ions can travel from one side of the membrane to the other under some gating mechanism (Kandel *et al.*, 1995). The gating of ionic channels can be described by the conductance to ionic current. The ionic channel is highly selective to ions, and the permeability varies with the state that the membrane maintains. The membrane conductance G_m is the lumped effect of all the ionic channels and is usually given by the multiplication of the membrane area with the unit conductance g_m. As we will see later, the permeability and gating mechanisms of ionic channels are crucial in electrical signal generation and transmission in the nervous system.

At steady state, the separation of charges on both sides of the membrane is maintained because ions cannot move freely across the membrane, and give rise to an electrical potential difference across the membrane, which is called the resting membrane potential, V_{rest}. Conventionally, the potential outside the neuron is defined as zero so that V_{rest} is negative. When V_{rest} is determined by two or more species of ions, the influence of each species is determined both by its concentration inside and outside the neuron, and the permeability of the membrane to that ion. The resting membrane potential can be quantified by the Goldman equation (Kandel *et al.*, 1995):

$$V_{rest} = \frac{RT}{F} \ln \frac{P_K[K^+]_{out} + P_{Na}[Na^+]_{out} + P_{Cl}[Cl^-]_{out}}{P_K[K^+]_{in} + P_{Na}[Na^+]_{in} + P_{Cl}[Cl^-]_{in}} \tag{10.1}$$

where R is the thermodynamic gas constant, T is the absolute temperature, F is the Faraday constant (96485.309 C/mol), $[i]_{out}$ is the extracellular concentration of ion i, $[i]_{in}$ is the intracellular concentration of ion i, and P_i is the membrane permeability for ion i.

The resting potential results from two characteristics of a neuron. One is the concentration gradients established by the sodium–potassium pump, a kind of ionic channel, keeping the sodium concentration low (about 10 times lower than outside) and the potassium concentration high (about 20 times higher than outside) within the neuron. The other is the resting membrane's high permeability to potassium and its relatively low permeability to sodium. At the resting state, these passive ionic currents flowing into and out of the cell are balanced, so that the membrane remains constant at its resting value, which typically is −70 mV. For most animal neurons, the only important ions are K^+, Na^+, and Cl^-. So, Eq. (10.1) lists these three, but there is no inherent limitation in the number of paired concentrations. Notice that the Goldman equation includes the permeability factor p of each ion and the temperature effect T; it would yield an accurate model of the transmembrane voltage at any particular ionic concentration and temperature. It can be observed from Eq. (10.1) that the Goldman equation does not provide any information about the changes in the membrane potential in response to a stimulus.

An equilibrium electrical circuit can describe the electrical properties of the passive membrane when we account for membrane capacitance, membrane conductance, and the resting potential, as shown in Figure 10.2. This circuit model provides an intuitive

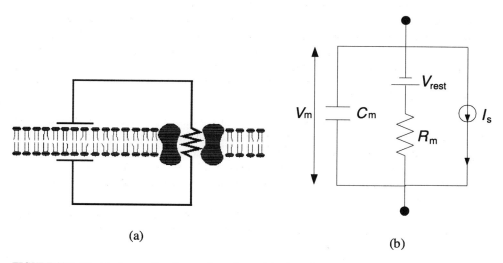

(a)

(b)

FIGURE 10.2. Electrical properties of a neural membrane: (a) schematic representation of a small patch of neuron membrane (Utah, 2002) and (b) an equilibrium electrical circuit with stimulus modeled by an ideal current source.

understanding of how current flow generates signals. By applying Ohm's law and Kirchhoff's law of circuit theory, the dynamics of the circuit can be described as

$$\tau_m \frac{dV_m(t)}{dt} = -V_m(t) + V_{rest} + \frac{I_s(t)}{G_m} \tag{10.2}$$

where $\tau_m = C_m/G_m$ is the time constant (C_m is the lump capacitor of the membrane, and G_m is the lumped conductance of the membrane), V_m represents the membrane potential, V_{rest} is the membrane resting potential at steady state, and $I_s(t)$ refers to the stimulus current.

Equation (10.2) provides the mathematical description of a neuronal model for the electrical properties of membrane potential, the equilibrium state of a neuron, etc. Obviously, the membrane potential is equal to the resting value at steady state when $I_s(t) = 0$ and V_m is constant. When a neuron is activated, the membrane potential is constantly driven away from its resting value by ionic currents flowing across the membrane. When the displaced potential from its resting value is sufficient up to a threshold, an action potential will be generated. However, what the gating schemes of ionic channels are, and how these schemes function in neuronal signaling, cannot be explained by this simple circuit. To answer these questions, we need more comprehensive models describing the relationship between membrane potential and ionic activities. In the next section, the Hodgkin–Huxley (HH) model will give an explicit description of this relation in terms of voltage-gated ionic channels.

10.2.2. EQUIVALENT CIRCUIT MODEL FOR ACTIVE NEURONS

The Hodgkin–Huxley model (Hodgkin and Huxley, 1952) is the most widely accepted formula for voltage-dependent ionic processes underlying the action potential. Hodgkin and Huxley established HH formula based on a large number of experiments on the giant axon of squid. The significance of their contribution is that, without having any knowledge of the underlying ionic channels, they described the membrane permeability changes using a quantitative model that is extremely influential in investigation and in the modeling of neuronal activities. They postulated the phenomenological model of the events underlying the generation of active potential in the squid giant axon under the following assumptions (Hodgkin and Huxley, 1952):

1. The action potential involves three ionic currents carried by sodium, potassium, and leakage ions. The sodium and potassium conductances are functions of time and membrane potential, whereas the leakage conductance can be regarded as constant.
2. The voltage-dependent ionic conductance can be represented by a maximum conductance multiplied by a coefficient indicating the fraction of the maximum ionic channel actually open.
3. Each species of ion has its own reversal potential owing to the balance concentration difference of the ions between inside and outside of the membrane. The reversal potentials can be taken as constants.

Figure 10.3 shows the circuit of a patch of neuronal membrane as equivalent circuits integrated with different functional ionic channels, where V_m is the membrane potential; C_m

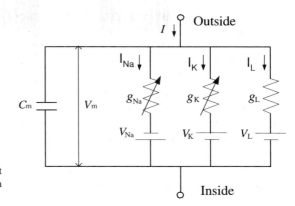

FIGURE 10.3. Electrical equivalent circuit of a patch of neuronal membrane of squid axon (Hodgkin and Huxley, 1952).

is the membrane unit capacitance and can be taken as a constant; G_{Na}, G_K, and G_L are the unit conductances for modeling the membrane permeability of Na^+, Ka^+, and leakage currents; I_{Na}, I_K, and I_L are the ionic current densities carried by Na^+, K^+, and leakage ions; and V_{Na}, V_K, and V_L are the reversal potentials of Na^+, K^+, and leakage ions. With the voltage-clamp technique, Hodgkin and Huxley (1952) provided the first complete description of the membrane current density in terms of voltage-dependent ionic mechanisms underlying an action potential.

$$I(t) = C_m \frac{dV(t)}{dt} + I_K(t) + I_{Na}(t) + I_L(t) = C_m \frac{dV(t)}{dt} + \bar{G}_K n^4(t)[V_m(t) - V_K]$$
$$+ \bar{G}_{Na} m^3(t) h(t)[V_m(t) - V_{Na}] + G_L[V_m(t) - V_L] \qquad (10.3)$$

where $\bar{G}_{Na}$ and $\bar{G}_K$ refer to the maximum unit conductances of the Na^+ and K^+ channels respectively, n is the gating factor for activating K^+ conductance (or channel), and m and h are gating factors for activating and inactivating Na^+ conductance (or channel) respectively. In Eq. (10.3), potentials are given in mV, current density in $\mu A/cm^2$, unit conductance in m·mho/cm^2, unit capacitance in $\mu F/cm^2$, and time in m·s. The gating factors n, m, and h are dimensionless coefficients and vary with time after a change of membrane potential. The detailed calculation of n, m, and h can be found in Hodgkin and Huxley (1952). Let x stand for m, n, or h. For a fixed voltage V_m, the differential equation for x dynamic against time can be written in the form (Gerstner and Kistler, 2002)

$$\frac{dx(t)}{dt} = -\frac{1}{\tau_x(V_m)}[x - x_\infty(V_m)] \qquad (10.4)$$

and the corresponding solution is

$$x(t) = x_\infty(V_m) - \{[x_\infty(V_m) - x_0 \exp[-(\alpha_x(V_m) - \beta_x(V_m))t]\} \qquad (10.5)$$

where x_0 stands for the boundary condition $x_0 = x(t = 0)$; $\alpha_x(V_m)$ and $\beta_x(V_m)$ are rate constants of x, which vary with voltage but not with time and have dimensions of [time]$^{-1}$.

It is obvious that variable x approaches the values $x_\infty(V_m)$ with a time constant $\tau_x(V_m)$. The asymptotic values $x_\infty(V_m)$ and the time constant $\tau_x(V_m)$ are given by the transformation $x_\infty(V_m) = \alpha_x(V_m)/[\alpha_x(V_m) + \beta_x(V_m)]$ and $\tau_x(V_m) = \alpha_x(V_m)/[\alpha_x(V_m) + \beta_x(V_m)]^{-1}$, respectively. On the basis of the results of the experiment on the squid axon, Hodgkin and Huxley (1952) established the relationships between the membrane voltage and the values of $\tau_x(V_m)$ and $x_\infty(V_m)$, respectively.

Physiologically, before stimulation, the resting voltage is about -70 mV, and the sodium and potassium channels are almost fully inactive with $G_K \gg G_{Na}$. Consider the effect of injection and of the inward current to the membrane (see Figure 10.3), that is, the membrane capacitance will be charged up, depolarizing the membrane in the process. As V_m approaches the threshold at about -50 mV, the Na^+ channels begin to open, and the resultant flow of Na^+ further depolarizes the membrane. Because of the depolarization of the membrane, the driving potential for K^+, $V - V_{rest}$, also increases. Consequently, conductance of the Na^+ channels further increases, causing the neuron to fire an action potential. According to the HH model, during the initial phase of the action potential, increased potassium activation n tends to bring the membrane potential toward V_K by increasing I_K, whereas increased sodium inactivation h decreases the amount of sodium conductance. Both processes drive the membrane potential back toward V_{rest}. Because the total I_{Na} quickly falls to zero but I_K persists longer at small amplitudes, the membrane potential is depressed below its resting level, that is, the hyperpolarization. At the low potential, potassium activation switches off, returning the membrane to its initial resting potential. Figure 10.4 shows an intracellular action potential recorded by Hodgkin and Huxley (1952), and the calculated action potential with the changing conductances by the HH formula. The experimental data of the parameters of the model can be found in Hodgkin and Huxley (1952).

As the first complete description of the excitability of a single cell, the HH model describes the current changes in terms of the opening or closing state of ionic channels, whose probability of being open is predicted by statistical approaches. It gives a universal mechanism of neuronal excitability that links the microscopic level of various ionic channels to the macroscopic level of currents and action potentials (Hausser et al., 2000). The great insights we can gain from the HH model include the following:

1. Separating the total conductance change into independent components. The action potential calculated from the HH conductance-based formula matches the electrical signal recorded in the axon remarkably well.
2. Describing the membrane permeability change with time and potential by gating factors. This description provides a basic approach to model voltage-dependent ionic channel conductance. For example, the Ca^{2+} current has been modeled following the HH formula to illustrate its contribution to the regulation of neuron firing patterns at synapses (Schutter and Smolen, 1998).
3. Providing a stereotype process of the generation and propagation of a neuronal action potential.

However, because the HH model was tested through experiments on the large axon of a squid at the given temperature of 6.3°C, to what extent this model is valid for different neurons or for the same axon under different conditions is still not clear. In different

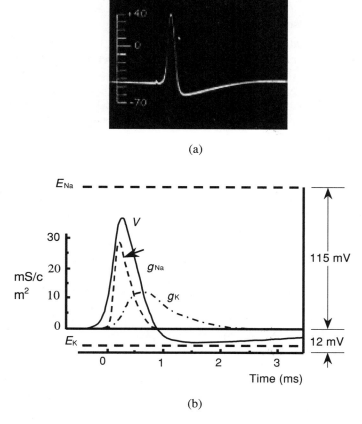

(a)

(b)

FIGURE 10.4. (a) Intracellular recording of an action potential from squid axon by Hodgkin and Huxley (Hausser, 2000), and (b) separated ionic conductances underlying the action potential in the HH model (Ram, 2000).

applications, the following factors should be taken into consideration: (1) neuronal configuration, (2) variety of voltage-gated ionic channels, and (3) influence of temperature. For example, a model for warm-blooded neurons is important in the study of the vertebrate and mammalian nervous systems. Hodgkin and Katz (1949) and Frankenhaeuser and Moor (1963) have reported the influence of temperature on the ionic channel activities and adjusted the HH model by a thermal factor k as follows:

$$\frac{d(x)}{dt} = [-(\alpha_x + \beta_x)x + \alpha_x]k \tag{10.6}$$

where x stands for the gating variables and k is a thermal coefficient. The acceleration factor k depends on the difference of the actual temperature T and the laboratory temperature T_0. A special constant Q_{10} is introduced to describe the acceleration in membrane behavior when the temperature is increased by $10°C$. Hodgkin and Huxley found that $Q_{10} = 3$ for squid membrane so that $k = Q_{10}^{(T-T_0)/10}$ (Hodgkin and Huxley, 1952), where T is the actual temperature.

It should be pointed out that the HH model in Eq. (10.3), describing the space clamp of a neuronal axon membrane with equal and uniform reaction, can represent well local bioelectrical behaviors without considering the detailed geometric structures of neurons. The action potential propagation along the axon can be described by cable theory (Rall and Agmon-Snir, 1998), combined with which a compartment model can describe the membrane potential distribution of a neuron with inhomogeneous structures. This will be discussed in the next subsection.

10.2.3. COMPARTMENT MODEL

Besides the individual ionic current and their interplays, it is proposed that the detailed geometric structures of neurons and the inhomogeneous ionic channel distributions on the neuron membrane will also affect the neuron firing and propagating properties. The compartment model considers a neuron with irregular-shaped dendrites and axons for the effects of the geometric structure. In a compartment model, the neuron is separated into regions or compartments, and the continuous membrane potential is modeled by a discrete set of values representing the potentials of different compartments. This scheme assumes that each compartment is small enough to neglect the variation of its property.

The development of the compartment models is based on the HH model and cable theory. Because cable theory has been very well explored (Rall and Agmon-Snir, 1998), we will brief on some of its salient points. In cable theory, the neural axons and the extended dendrite trees can be treated the same way as cable structures, where the cables can have different diameters and membrane properties. The thin tubes of dendrites and axons can be modeled as a cable with transmission capacitance and resistance, which refers to "core conductors" because both the intracellular cytoplasmic core and the extracellular fluid are ionic media that conduct electric current. For short lengths, the resistance to electrical current flow across the membrane is much greater than the resistance along the interior core, or along the exterior; the electric current inside the core conductor therefore tends to flow parallel to the cylinder axis for a considerable distance before a significant fraction leaks across the membrane. Formulating this concept mathematically leads to the cable equation (Segev *et al.*, 1998):

$$\lambda^2 \left(\partial^2 V/\partial x^2\right) - \tau \left(\partial V/\partial t\right) - V = 0 \qquad (10.7)$$

or, in terms of dimensionless variables, $X = x/\lambda$ and $W = t/\tau$, as

$$\partial^2 V/\partial X^2 - \partial V/\partial W - V = 0 \qquad (10.8)$$

where V represents the voltage difference across the membrane (interior minus exterior) as a deviation from its resting value (i.e., $V = V_{\text{interior}} - V_{\text{exterior}} - V_{\text{rest}}$), x is the distance along the axis of the membrane cylinder, λ is the length constant of the core conductor, t represents time, and τ is the membrane time constant (sometimes also expressed as τ_{m}) of the passive membrane. Although x and λ are expressed in centimeters or micrometers and t

and τ are expressed in seconds or milliseconds, the ratios $X = x/\lambda$ and $W = t/\tau$ represent dimensionless variables that are proportional to distance and time, respectively.

An action potential is generated at the hillock and propagated along the axon of a neuron. In developing compartment models, the models used for passive cables and for the HH model can be combined to simulate the shape and ionic distribution characteristics of a target neuron. A model of the central nervous system (CNS) neuron (Rattay, 1998) will be given to illustrate the process under the compartmental modeling scheme. The model of a neuron consists of the following subunits: dendritic tree, cell body (soma), initial segment (beginning of the axon), a myelinated axon with nodes and internodes, and an unmyelinated branching terminal part with a large number of synaptic endings. In the compartment model, the soma is assumed to have a spherical shape, and all the other elements are cylinders, as shown in Figure 10.5. The myelinated axon is divided into compartments of length Δx, with one compartment per node. The membrane currents of internodes are neglected, and the membrane is active only in the node with length L. The current flow for the nth compartment consists of the capacity current, the ionic current i_{ion}, and the current to the neighboring elements. The membrane potential within the nth compartment can be written in the form

$$C_{m,n}\frac{dV_n}{dt} = \frac{D_n\Delta x}{4\rho_n L}\left(\frac{V_{n-1} - 2V_n + V_{n+1}}{\Delta x^2} + \frac{V_{e,n-1} - 2V_{e,n} + V_{e,n+1}}{\Delta x^2}\right) - i_{ion,n}, \qquad (10.9)$$

where $C_{m,n}$ is the membrane capacitance; V_n is the reduced membrane potential, i.e., at rest $V_n = V_{i,n} - V_{e,n} - V_{rest}$; D_n is the diameter of the axon; ρ_n is the axoplasmatic resistivity; and $i_{ion,n}$ is the ionic current of the nth compartment, which can be defined by the classical HH formula (see Section 10.2.2). If we replace the $G_{m,n}$ in Figure 10.5(c) by a constant resistance, then the axon part will be a pure passive cable as introduced in Eq. (10.7). Because every compartment has its individual shape and geometric and electrical parameters, the values of ρ_n, D_n, $C_{m,n}$, and $i_{ion,n}$ can be different from compartment to compartment. The current flow to the center point of the nth compartment of the neuron can be calculated from the voltage between the points of the circuit in Figure 10.5(c). The terms in the following equation are derived from capacitance and ion currents across the membrane, and the inner axonal currents to the left and right neighbor elements (Rattay, 1998),

$$C_{m,n}\frac{d(V_{i,n} - V_{e,n})}{dt} + I_{ion,n} + \frac{V_{i,n} - V_{i,n-1}}{R_n/2 + R_{n-1}/2} + \frac{V_{i,n} - V_{i,n+1}}{R_n/2 + R_{n+1}/2} + \cdots = 0 \qquad (10.10)$$

where the suspension points are written for terms similar to the last one that has to be added in cases with more than two neighbor elements, e.g., at the soma or in other branching regions. Eq. (10.10) can also be rewritten as

$$\frac{dV_n}{dt} = \left[-I_{ion,n} + \frac{V_{n-1} - V_n}{R_{n-1}/2 + R_n/2} + \frac{V_{n+1} - V_n}{R_{n+1}/2 + R_n/2} + \cdots \right.$$
$$\left. + \frac{V_{e,n-1} - V_{e,n}}{R_{n-1}/2 + R_n/2} + \frac{V_{e,n+1} - V_{e,n}}{R_{n+1}/2 + R_n/2} + \cdots \right] \bigg/ C_{m,n} \qquad (10.11)$$

For the final compartment, the last term of Eq. (10.10) and the corresponding terms in Eq. (10.11) are canceled, as there is no current to the $(n + 1)$th compartment.

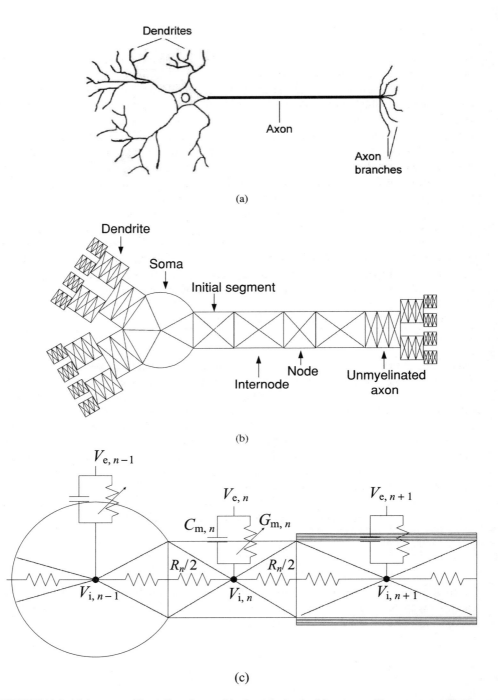

FIGURE 10.5. (a) A neuron with myelinated axon, (b) schematic sketch of the neuron with compartment (Rattay, 1998), and (c) a part of the compartment model of the neuron, including soma, a node, and an internode, with equivalent electrical circuits of each compartments. $V_{i,n}$ and $V_{e,n}$ are the inside and outside potential of the nth compartment. $C_{m,n}$ and $G_{m,n}$ are the membrane capacitance and conductance of the nth compartment ($n = 0, 1, 2, \ldots$). R is the core resistance of the axon, and can be generally taken as constant (Rattay, 1998).

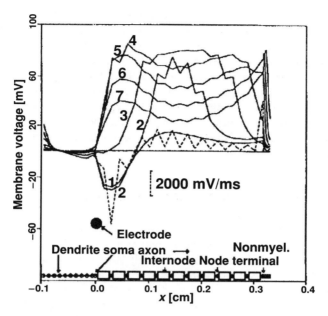

FIGURE 10.6. Voltage distribution of the modeled neuron evoked by a positive 100 μs/5 mA stimulation pulse (Rattay, 1998).

In Rattay's simulation (1998), the modeled neuron consists of 40 compartments (for $n = 40$), and is evoked by a positive 100-μs/5-mA stimulating pulse. The HH model is adopted for describing unmyelinated axon nodes and synapses. Values for geometric parameters of individual compartment and electrical parameters of membrane are adopted from empirical data. The calculation results of voltage distribution along the neuron are shown in Figure 10.6. The bottom plot indicates the positions of the center points of the 40 compartments where membrane voltages are calculated. The full lines show snapshots of the computed membrane voltage in the time interval of 50 μs, marked by numbers 1–7. In Figure 10.6, the dynamic of each compartment is represented by the curves from the time intervals of 1–7 at each corresponding point. The variation of the potential distribution along the neuron illustrates the propagation of an action potential. For the whole axon, the membrane potential can be represented three-dimensionally, for it is a function of both the time course and the distance of the compartment from the first node. This compartment model enriches our understanding of the main effects of electric and magnetic stimulation on CNS neurons. The activating function shows how structural and electrical factors can have a great impact on the overall excitatory response. For example, the zigzag shape of the curves in Figure 10.6 demonstrates the irregular influence of the substructures; therefore at least one compartment per internode should be included.

One significant advantage of the compartment model is that it places no limitation of the neuronal membrane geometry so that it faithfully embodies information both for the anatomical structure and the physiological properties of a neuron. Arbitrarily complex structures and some morphological inhomogeneties can be represented in the topology of the compartment connection. It also allows great flexibility in the level of solution for different compartments or for various types of neurons. Generally, compartment models can mainly serve three purposes (Segev *et al.*, 1998): (1) integrating the available morphological and physiological data in order to interpret experimental results (e.g., Rall, 1967); (2) suggesting different possibilities prior to experimental exploration; and (3) exploring the computational roles of the different compartments of a neuron in its various activities. It should be pointed out that the construction of a compartment model depends on the obtained experimental data that characterize the morphology and physiology of the target neurons. The interwoven improvements in experiment and modeling must make our understanding of the neuron more and more complete.

10.3. MODELS FOR NEURONAL RATE CODING

Detailed conductance-based neuronal models, such as the HH model, can reproduce local electrophysiological properties of single neurons to a high degree of accuracy. However, the neuronal code is closely linked to the seemingly stochastic or random characteristics of neuronal firing. Instead of completely regulated spaced action potentials, which are all-or-none events, firing rates carry the most important neural communication information. When the action potentials are taken to be identical and only their occurrence times are under study, simplified models are necessary for the analysis of the information carried by the train of neuronal firing events by converting the timing data into time series. Integrate-and-fire (IF) model and integral pulse frequency modulation (IPFM) model achieve this conversion by setting a fixed rule rather than by computing the membrane potential on the basis of explicit ionic processes. Both models simply stipulate that as long as the membrane potential V_m of a neuron remains below a threshold value, the dynamics are those of the passive membrane. Once the membrane potential reaches a threshold value, V_{th}, a stereotyped pulse, instead of a signal bearing the waveform of the action potential, will be fired. The membrane potential will be reset to the resting potential, V_{rest}, after occurrence of an impulse.

10.3.1. AN INTEGRATE-AND-FIRE CIRCUIT MODEL

An IF neuron can be modeled in an excitatory environment with stimulus from presynaptic neurons, as shown in Figure 10.7. For simplification, all active membrane conductances are ignored and the entire membrane conductance is modeled as a single passive leakage term, $i_m = G_m(V_m - V_{rest})$. Neuronal conductance is maintained approximately constant until the perturbation of membrane potential is beyond the threshold, by the time that an impulse, representing an event of action potential, will be generated. With these approximations, the modeled neuron acts as an electric circuit consisting of a capacitor C_m in parallel with a conductance G_m driven by stimulation current I_s. On the basis of the equilibrium circuit (see Figure 10.2 and Eq. (10.2)), when the stimulation current of the postsynaptic neuron is the integration of all synaptic driving currents from presynaptic neurons (see Figure 10.7), we have

$$\begin{cases} \tau_m \dfrac{dV_m}{dt} = -V_m(t) + V_{rest} + \dfrac{1}{G_m} \sum_k I_{syn}(t), & V_m < V_{th} \\ \text{impulse} & V_m \geq V_{th} \end{cases} \tag{10.12}$$

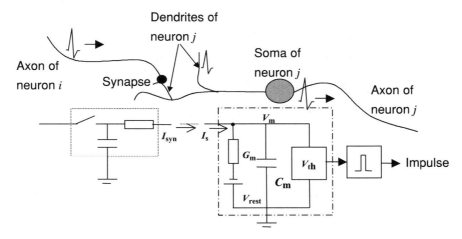

FIGURE 10.7. Schematic diagram of a leaky integrate-and-fire model: within the dot line is the circuit for the synapse and within the dot–dash line is the circuit for postsynaptic neural soma. (Adapted from Gerstner and Kistler, 2002)

where $\tau_m = C_m/G_m$ (the membrane time constant of neuron), V_m is the membrane action potential, and k is the number of presynaptic neurons which supply the stimulation currents to the postsynaptic neuron.

Actually, the neuron can be activated either by external current or by synaptic input from presynaptic neurons. Because the model does not involve the details of ionic currents shaping action potentials, the output spike is modeled as an identical impulse, which can be described by the delta function. The firing rate is one of the most important characteristics of the IF neuron output, carrying the neural coding information to corresponding stimulus. For a simple example, when the input of the models is constant current $I_s(t) = I_0$, we assume that a spike has occurred at t_1 and generated again at t_2. For simplicity, the reset potential is set such that $V_{rest} = 0$. The trajectory of the membrane potential can be obtained by integrating Eq. (10.12) with the initial condition $V_m(t_1) = V_{rest} = 0$ and the ending condition $V_m(t_2) = V_{th}$. The analytic solution of the membrane potential will be

$$V_m(t) = \frac{1}{G_m} I_0 \left[1 - \exp\left(-\frac{t_2 - t_1}{\tau_m} \right) \right] \tag{10.13}$$

Solving Eq. (10.13) for the time interval $T = t_2 - t_1$ yields

$$T = \tau_m \ln \frac{I_0}{I_0 - G_m V_{th}} \tag{10.14}$$

After the spike at t_2 the membrane potential is again reset to $V_{rest} = 0$ and the integration process starts again. On the basis of Eq. (10.14), the firing rate can be expressed by

$$r = \frac{1}{T} = \frac{1}{-\tau_m \ln\left(1 - \frac{G_m V_{th}}{I_0} \right)} \tag{10.15}$$

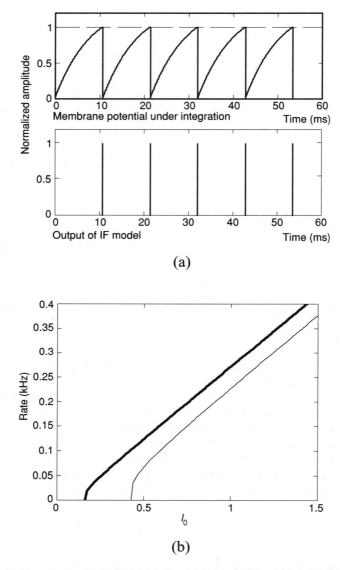

FIGURE 10.8. (a) Time course of action potential of an integrate-and-firing neuron driven by constant input current $I_0 = 0.58$ nA, $V_{th} = 16$ mV, $C_m = 0.2$ nF, $G_m = 26.3$ nS. The potential $V_m(t)$ is normalized by threshold V_{th}. (b) The firing rate of an IF neuron as a function of a constant driving current I_0. G_m is 10 nS for the thicker solid line and 26.3 nS for the thinner solid line. Other parameters are the same as in (a). (Adapted from Gerstner and Kistler, 2002)

Figure 10.8(a) shows the membrane potential integral process and the impulse train as an output of the IF model. The relationship between firing rate and the intensity of constant-stimulation current is given in Figure 10.8(b).

It can be observed from Figure 10.8(b) that the firing rate is not only affected by the input intensity, but also influenced by the membrane time constant of neuron, τ_m, which is determined by the membrane capacitor, C_m, and conductance, G_m. In other words, the membrane properties, such as the membrane permeability, could be important factors in determining neuronal firing, and the RC circuit in the IF model can reflect these influences on the firing rate well, as shown in Figure 10.8(b). By simplifying action potentials into identical impulses, the information of the input intensity and the firing rate is contained in the resulting sequence of spikes of IF model. Some mathematical approaches are already developed for analyzing the information lying in the output pulse train of such simplified models for spiking neuron. Detailed description and analysis of spiking neurons can be found in Koch (1999) and Gerstner and Kistler (2002).

10.3.2. INTEGRAL PULSE FREQUENCY MODULATION (IPFM) MODEL

10.3.2.1. Firing Rate Estimation Based on the IPFM Model

The rate–intensity function is the main focus of neural coding, where *intensity* refers to the stimulation amplitude and *rate* refers to the action potential repetition rate or firing rate, which serves as a mechanism to convey the neural information. As discussed in Section 10.3.1, the IF model, describing the input–output behavior of a single neuron, can transform the intensity information of a stimulus into a pulse train with the firing rate related to the input. Bearing the similar scheme of neuronal potential generation, the IPFM model achieves the "integrate" phase by an unspecified integration operator, instead of the RC circuit in the IF model.

Reconciling the potential generation pattern in the IF model, the IPFM model converts a nonnegative time-continuous input signal into a modulated information-carrying train of events as shown in Figure 10.9. In the IPFM model, any two consecutive event-occurrence times t_i and t_{i+1} satisfy the relationship (Bayly, 1968)

$$V_{th} = \int_{t_i}^{t_{i+1}} (m_0 + m_1(t))\, dt \qquad (10.16)$$

where $m_1(t)$ is the information or modulating input signal, m_0 is the constant input signal, t_i is the incidence of the ith impulse, and V_{th} is the integration threshold. The nonnegative input signal $m(t) = m_0 + m_1(t)$ is integrated, and whenever the integrated value exceeds a threshold V_{th}, an identical spike is generated and the integrator is reinitialized.

To simulate the action potential sequence generation process and various coding schemes of neurons, different neuronal models can be built by choosing a specific integrator with different parameters reflecting neuronal electrical behaviors to replace the unspecified integral operator as shown in Figure 10.9(a). The integration process is determined by membrane electrical properties, anatomical structure of neuron, etc. The membrane threshold V_{th} depends on neuronal geometry and the membrane voltage depolarization level at the site of origin required for action potential initiation. Depending on applications, the input signals to the IPFM model can be different in forms, representing the input variation, such as stretch, force, pressure, synaptic current, etc. In the neuron case, for example, the integrated

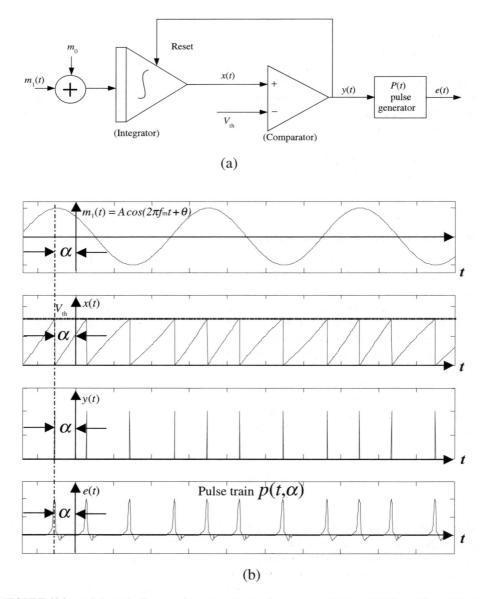

FIGURE 10.9. (a) Schematic diagram of the integral pulse frequency modulation (IPFM) model; and (b) the output of the IPFM model when the modulating input signal is given by $A \cos(2\pi f_m t + \theta)$ (Bayly, 1968).

signal is the neuronal membrane potential, which rises until the threshold is reached and consequently a pulse is generated. The resetting function corresponds to the reestablishment of resting membrane voltage over much of the neuron surface after the passage of a neuronal spike or an action potential (Bayly, 1968).

By converting the input modulating signal mimicking physical process (pressure, current, etc.) into a time series, the output of an IPFM model clearly reflects the important features, such as firing rate, of a neuron with close relevance to the physical input. Considering $m(t) = m_0$ with the input modulating signal $m_1(t) = 0$ as a simple example, the expression of the firing rate can be easily derived as $r_0 = m_0/V_{th}$ according to Eq. (10.16). It is clear that the firing rate r_0 is proportional to the dc input component m_0 and inversely proportional to the threshold V_{th}. This result is consistent with that obtained by the IF model, as illustrated in Figure 10.8(b). More generally with $m(t) = m_0 + A\cos(2\pi f_m t + \theta)$ ($A \leq m_0$ such that $m(t)$ is ensured nonnegative) as the input signal to the IPFM model, Eq. (10.16) becomes (Gu et al., 2003)

$$V_{th} = \int_{t_i}^{t_{i+1}} (m_0 + A\cos(2\pi f_m t + \theta))\,dt = m_0(t_{i+1} - t_i)$$

$$+ \frac{A}{\omega_m}[\sin(2\pi f_m t_{i+1} + \theta) - \sin(2\pi f_m t_i + \theta)] \tag{10.17}$$

Considering that

$$\sin a - \sin b = 2\sin\left(\frac{a-b}{2}\right)\cos\left(\frac{a+b}{2}\right)$$

we have

$$V_{th} = m_0(t_{i+1} - t_i) + \frac{A}{\pi f_m}\sin(\pi f_m(t_{i+1} - t_i))\cos(\pi f_m(t_{i+1} + t_i) + \theta) \tag{10.18}$$

This can be further simplified by the Taylor series of the sinusoidal function given by

$$\sin x = x - \left(\frac{x^3}{3!}\right) + \left(\frac{x^5}{5!}\right) - \left(\frac{x^7}{7!}\right) + \cdots \tag{10.19}$$

The difference between $\sin x$ and x is less than 4% when $x < 0.5$. For x with values less than 0.75, the deviation is not greater than 10% and for x with values closer to 1, the difference can become as high as 16%. In most physiological situations, the time interval between two consecutive firings is less than or close to 1. For example, in the nervous system the interpulse intervals are normally ranged from the order of 10 ms to the order of 100 ms, while in a neurocardiac case, the normal range of heart rate is 60 to 100 beats per minute (U.S. National Library of Medicine, 2003). If the frequency of the modulation signal can be suitably restricted such that $x = \pi f_m \Delta t_i$ is much less than 1, the first-order approximation of $\sin(\pi f_m \Delta t_i)$ by $\pi f_m \Delta t_i$ will be reasonable. In practice, physiological series of point events, e.g. neural spike trains or series of heartbeats, are often influenced by slow modulation, such as respiration, which has a frequency f_m around 0.3 Hz (Pallas-Areny et al., 1989). Hence, the resulting firing rate sequence is approximated by

$$\frac{1}{t_{i+1} - t_i} \approx \frac{m_0}{V_{th}} + \frac{A}{V_{th}} \times \cos(2\pi f_m \tilde{t}_i + \theta) \tag{10.20}$$

where $\tilde{t}_i = (t_i + t_{i+1})/2$.

In order to facilitate the discussions, the parameter $\tilde{t}_i$ is replaced by the continuous time variable t. Therefore the approximated firing rate can be expressed as

$$r = \frac{1}{T} = \frac{1}{t_{i+1} - t_i} \approx \frac{1}{V_{\text{th}}}[m_0 + A \times \cos(2\pi f_{\text{m}}t + \theta)] \tag{10.21}$$

which shows the relationship between the firing rate and the input intensity of the modulation signal.

As schematically depicted in Figure 10.10(a), the time interval between two consecutive firings will be lengthened as $m_1(t)$ decreases, and vice versa. Figure 10.10(b) gives the approximated firing rate as a function of time under different thresholds. It is obvious that small values of threshold result in high firing rates for the given intensity, and vice versa. This result supports the experimental observations on the relation between the rate and the intensity reported before (Winter *et al.*, 1990; Yates, 1990). When the intensity of the input stimulation (synaptic current, stretch, or pressure, etc.) increases, the time for the integrate-to-fire process will be shortened, which results in a corresponding increase in firing rate. Conversely, when the fixed threshold V_{th} representing the depolarization level of membrane potential required for potential initiation increases, the integration process will last for a longer time, resulting in a decrease in firing rate.

10.3.2.2. *Spectral Analysis of the IPFM Process*

Spectral analysis is a very important tool for studying the intrinsic properties of signals in the frequency domain. The spectral analysis of the IPFM process could provide insights into the mechanisms of encoding and decoding in nervous systems. For simplicity, we assume that the impulses have a unit surface area. Using Parseval relation and Bessel

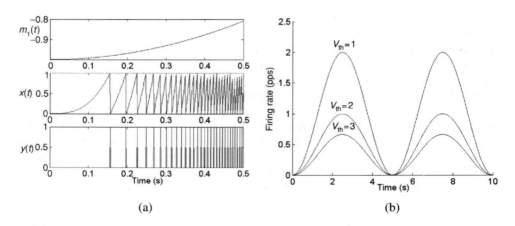

(a)   (b)

FIGURE 10.10. (a) The output of the IPFM model when the modulating input signal is given by $A\cos(2\pi f_{\text{m}}t + \theta)$, and the parameters are as follows: $A = 1$, $f_{\text{m}} = 0.2$ Hz, $\theta = \pi$, $m_0 = 1$, $V_{\text{th}} = 1$. (b) The approximated firing rate as a function of time for different threshold values, with the other conditions remaining unchanged.

function, Bayly (1968) derived the expression of the IPFM series in its time domain:

$$y(t) = \frac{m_0}{V_{th}} + \frac{A}{V_{th}} \cos(2\pi f_m t + \theta) + \frac{2m_0}{V_{th}} \sum_{k=1}^{\infty} \sum_{n=-\infty}^{\infty} b_{k,n} \cos[2\pi(kf_0 + nf_m)t + \phi_{k,n}]$$

(10.22)

In Eq. (10.22), $b_{k,n} = J_n\left(\frac{kAf_0}{f_m}\right)\left(1 + \frac{nf_m}{kf_0}\right)$ and $\phi_{k,n} = 2k\pi f_0\alpha + n\theta + \omega$ with $\omega = \frac{kAf_0}{f_m} \sin(2\pi f_m\alpha - \theta)$ where J_n is the Bessel function of the first kind of the order n, α denotes an arbitrary initial time instant as mentioned before, f_m is the frequency of the modulating signal, and f_0 is the pulse mean firing rate (or carry frequency) proportional to the dc signal component m_0. According to Eq. (10.22), the complex two-sided IPFM spectrum $Y(f)$ can be written as

$$Y(f) = \frac{m_0}{V_{th}}\delta(f) + \frac{A}{2V_{th}}\delta(f \pm f_m)\exp(\mp j\theta)$$

$$+ \frac{m_0}{V_{th}} \sum_{k=1}^{\infty} \sum_{n=-\infty}^{\infty} b_{k,n}\delta\{f \pm (kf_0 + nf_m)\}\exp(\mp j\phi_{k,n}) \qquad (10.23)$$

where $\delta(f)$ is the Dirac function, and the double symbols $\pm$ and $\mp$ are used to shorten the expression.

If the ratios A/V_{th} and f_m/f_0 are properly restricted, the spectrum of the IPFM signal will have the form shown in Figure 10.11 (Bayly, 1968). From Figure 10.11 and Eq. (10.23), three clusters of frequency components can be identified as follows:

(1) The dc component represents the average value of the pulse train. It is interesting to note that this dc component is, as a matter of fact, the mean firing rate given by m_0/V_{th} as shown in Eq. (10.21).
(2) The modulation or input information related signal component at the modulating frequency, f_m, next to the dc component, is given by $A/2V_{th}$, which is proportional to the ac component of the firing rate function defined by Eq. (10.21). Demodulation

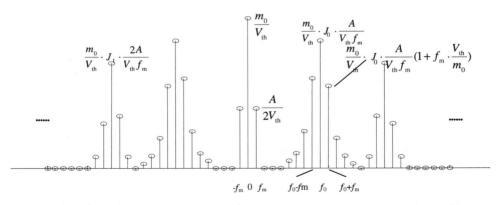

FIGURE 10.11. The spectrum of the IPFM signal. Component labels indicate the magnitudes of corresponding cosine components of $p(t, \alpha)$.

is, therefore, possible through the estimation of the firing rate by low-pass filtering (LPF) without any harmonic distortions if the threshold V_{th} is known.

(3) The components at harmonics of the mean sampling rate kf_0 are surrounded by a cluster of side components at sums and differences of mean firing rate and modulating frequency, $kf_0 \pm nf_m$. These harmonic components and sidebands are attributed to the nonlinearity of the IPFM model, specifically the threshold operation for the integrate-to-fire process. The amplitudes $b_{k,n}$ are a function of the firing rate A/V_{th} and the relative modulation frequency f_m/f_0, as well as the indexes k and n. The side frequencies are spread asymmetrically in amplitude around the harmonics kf_0, because of the second factor $b_{k,n}$. The spectrum contains components at all harmonics of the carrier frequency or the mean firing rate, each accompanied by an infinite set of adjacent components (or sidebands).

Theoretically, the main objective of the demodulation process is to recover the input modulating information, which is $m_1(t) = A \cos(2\pi f_m t + \theta)$ in the given example. A demodulation method using low-pass filtering with a proper cut-off frequency is illustrated in Figure 10.12, where the main requirement on the filter is that its cut-off frequency is high with respect to the frequency modulation and low with respect to the carrier frequency (Bayly, 1968). Figure 10.12(a) gives the schematic diagram of the demodulation of an IPFM pulse train by low-pass filtering with the actual output of $(A/V_{th})|G| \cos(2\pi f_m t + \theta + \phi) + \text{Noise}$, where $|G|$ is the gain of the low-pass filter. It should be noticed that in order to fully recover the input signal, the knowledge of V_{th} is necessary when the gain of the low-pass filter is given. However, as mentioned before, the

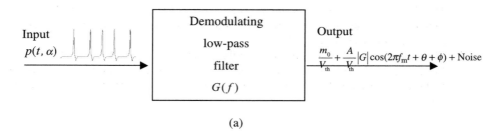

(a)

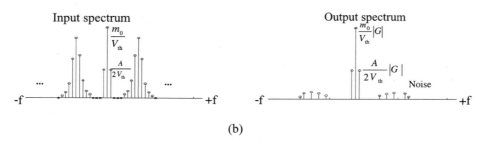

(b)

FIGURE 10.12. (a) Schematic diagram of the IPFM demodulation by low-pass filtering, and (b) the corresponding input and output spectra. Distortion at the output occurs when the first carrier component at f_0 and its lower sidebands are not completely removed by the LPF. $|G|$ is the gain of the filter.

threshold very much depends on the neuronal geometry and the membrane voltage depolarization level at the site of origin required for pulse initiation, and is therefore practically difficult to obtain without extensive exploration of the physiological mechanism of the pulse generation in the neurons. The qualitative aspects of the thresholds *in vivo* are at present not captured, although much has been speculated on this topic. Therefore, it is physiologically implausible to recover precisely the input modulation amplitude or intensity information by simple low-pass filtering because of the difficulty in estimating the exact threshold value V_{th} in the complicated living biological systems.

However, demodulation by low-pass filtering is of special importance for studying the firing rate characteristics of a neuron. In comparison with the approximated firing rate by Eq. (10.21), the output of the IPFM demodulator without noise is actually the instantaneous firing rate with both the dc component proportional to m_0 and the ac component to the modulating signal, as shown in Figure 10.12(b). Therefore, the demodulation by low-pass filtering is possible in reality for the estimation of the firing rate, which is related to the intensity of input signals.

As elaborated by Bayly (1968), because the firing characteristics of a neuron can be reconstructed by simple low-pass filtering, it is natural to expect that the nervous system would take advantages of this fact at its demodulation sites to decode and translate neural pulse train into the information of intensity-related firing rate, thus obtaining the description of stimulation. The response of a synapse in combination with neuronal membrane to a nerve impulse input usually exhibits a low-pass filtering characteristic. Indeed, in a nervous system a synapse can be modeled as an RC filter (Hodgkin and Rushton, 1946; Harmon, 1961; Lewis, 1963; Petz and Gerstein, 1963). Moreover, the low-pass properties of a neuromuscular junction and a muscle in demodulating efferent nerve messages are also well known (Fatt and Katz, 1951; Partridge, 1965). Therefore, it is apparent that the IPFM model is an appropriate tool not only for studying the biological impulse generation and modulation processes in neurons, but also for understanding the demodulation process in neuromuscular systems.

Also, there is evidence that the major features of the spectrum describing an IPFM process are similar to that observed in the nervous system. Bayly (1968) had investigated the spectrum of a relaxation oscillator having an approximate integration accomplished by an RC circuit and found it to be identical to those of IPFM, as shown in Figure 10.11. Hence, the spectral description of Eq. (10.23) indicates (1) change in the carrier rate or average pulse frequency is the most probable information-carrying parameter of nerve impulse trains, because it is represented by the spectral component at the modulation frequency and thus is theoretically recoverable by simple low-pass filtering; (2) modulation frequencies and modulation depths must be properly restricted for meaningful information transmission; and (3) multiple pathways or channels can give an improvement in the signal-to-distortion ratio over that possible on any one of the channels alone (Bayly, 1968; Zhang *et al.*, 1990).

10.3.2.3. *Effects of Neuronal Dynamics*

However, the specific characteristics of the pulse contain the important information of dynamic changes in physiological systems, including the variations of ionic currents in the underlying ionic channels and the changes in the membrane permeability, and are

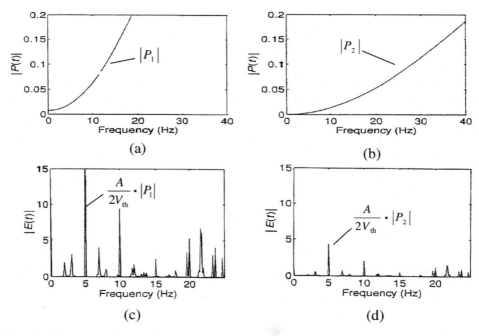

FIGURE 10.13. The spectrums of the frequency-modulated pulses multiplied by the spectrums of different action potentials.

therefore nonnegligible for a better understanding of the coding of neural information (Gu and Zhang, 2003). As a result, the ultimate output spectrum of the IPFM process integrated with neural dynamics is determined by the multiplication of the action potential spectrum with the spectrum given by Bayly (1968).

As illustrated in Figure 10.13, with different action potentials P_1 and P_2, the information signal ac component at $f_m = 5$ Hz may be distorted to different extents. From the point of view of the demodulation process, the interaction of the action potential with the IPFM process results in a scaling factor and it complicates the process of recovering any original information. Hence, demodulation by a single low-pass filtering can no longer be satisfactory to recover either the simplified firing rate function (Gu *et al.*, 2003), or the original input modulation signal. In order to solve this problem, prewhitening techniques can be considered to offset the influence of the "dynamic waveform" before low-pass filtering.

10.4. MODELING RATE–INTENSITY FUNCTION IN AN AUDITORY PERIPHERY

The previous sections dealt with circuit models for the generation of the action potentials in the living neuron responding to electrical stimuli and for the information coding at the single-neuron level. This section will focus on the studies of the rate behavior of an auditory system in response to a real physiological input such as an acoustic stimulus. A great deal of the established knowledge about sound perception has yielded a remarkable understanding of the biophysical mechanisms in the auditory periphery. The knowledge provides a basis for constructing models of acoustic information processing, which includes understanding human nature, speech encoding, and neuroprosthesis for the hearing impaired. Sound perception involves a number of mechanisms operating in cascade, dealing with sensory coding, electrical–mechanical signal transformation, and auditory nerve adaptation, etc. The code in an auditory system describes the manner of neural activities, which represent acoustic information. In the auditory nerve system, the sound waves, in a vibration pattern, are encoded in trains of action potentials. Therefore, the relationship between the temporal and spectral properties of acoustic signals and the discharge pattern of auditory nerve fibers is a key characteristic in the functional study of the auditory system. As the neural information is such sound pressure encoded in the firing rate, the input–output relationship of the auditory system can be described by the response rate versus stimulus frequency and intensity function. Therefore, the rate–intensity function at different frequencies becomes an important characteristic in the study of auditory information processing.

Circuit models of acoustic signal processing not only can precisely represent the activities of the auditory system, assisting people in understanding the mechanisms in sound perception, but also can be easily implemented by electronic circuits, which can be used for speech processing and neuroprosthesis of hearing damage. The models can also be evaluated by the known biophysical characteristics, such as rate–intensity functions obtained from experimental data. A system model will be introduced in this section, which emphasizes on functional description at the cellular level and circuit representation at the system level. Before the illustration of both models, some salient biophysical background of auditory periphery will be introduced in the next subsection.

10.4.1. BIOPHYSICS OF AUDITORY PERIPHERY

As shown in Figure 10.14 (Rice University School Science Project, 2003), the input of an auditory system is the sound wave, and the output is the temporal rate properties of auditory nerve responses. The signal goes through the auditory pathway as follows: Sound impinges on the outer ear and causes the eardrum to vibrate. These vibrations are transmitted via the middle ear to a fluid-filled tube called the cochlea in the inner ear, and ultimately cause the vibratory displacement of the basilar membrane (BM). The motion of the BM vibrates a series of tiny cilia of inner hair cells (IHCs), which stimulate action potentials of auditory nerve (AN) fibers by releasing neurotransmitters (Brugge, 1992). Another kind of

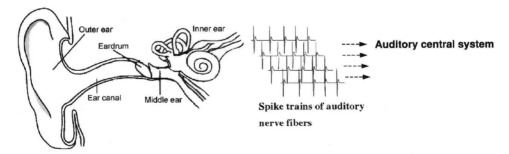

FIGURE 10.14. Information transmission pathway for acoustic signal.

hair cell, the outer hair cell (OHC), also senses the motion of BM, functioning as an active mechanism in acoustic information processing.

According to the information transmission characteristics, the auditory system is a multilayer nonlinear processor of acoustic information. It is believed that the mechanical–electrical signal transformation is achieved at the cochlea, where BM, IHC, and AN play a major role in the mammalian auditory system. The motion response of BM varies with the sound pressure level. The IHC, paired with the BM segment, translates this mechanical signal to chemical neurotransmitter release, which gives rise to AN firing. Meanwhile, OHC, amplifying the motion of BM to IHCs through their voltage-dependent length changes, would determine the dynamic range of IHCs.

Rate–intensity function is an important measure of the nonlinear transformation in the auditory system. The most basic properties of this nonlinear transformation originating in the IHC and AN complex are as follows: (1) signals are encoded in IHC receptor potentials in the analog form; (2) they are transformed in the auditory nerve into electrical pulses; and (3) these pulses are transmitted to auditory central nervous system. Biophysically, the firing rate of the auditory fibers is determined by the intracellular potential of IHC. However, each IHC can innervate 10–20 auditory nerve fibers with different spontaneous rates (SRs), leading to variation in rate thresholds: high-SR fibers are more sensitive to acoustic stimulation than low-SR fibers (Schoonhoven *et al.*, 1997). It is suggested that the dynamic range is an increasing function of the threshold at CF or a decreasing function of the discharge rate (Schalk and Sachs, 1980). Correspondingly, low-SR fibers have higher rate thresholds and a larger dynamic range, whereas for higher SR fibers the reverse is true, as shown in Figure 10.15.

In many auditory nerve fibers, especially those with a high spontaneous discharge rate, the firing rate saturates at some 40 dB or more above threshold. Hence, such fibers cannot encode sound over the entire dynamic range of hearing. For other fibers with low spontaneous rate, the discharge may not saturate, so that the dynamic range is extended (Brugge, 1992). The shape of a given fiber's CF rate–intensity (RI) function is strongly related to both its threshold at CF and its discharge rate. It has been found that there are three types of monotonic CF RI functions in the mammalian auditory nerve: saturating, sloping–saturating, and straight (Winter *et al.*, 1990). All these reports help researchers understand the nature of hearing and construct circuit models on the basis of these biophysical findings.

10.4.2. IHC MODEL AND RATE–INTENSITY FUNCTION

According to the cascade signal processing in auditory periphery, a system model was introduced recently by Sumner *et al.* (2002). The model as shown in Figure 10.16 is built on the basis of previous physiological findings and models with six modules: middle ear, basilar membrane, hair cell body, calcium kinetics, transmitter recycling, and auditory nerve. Module 1 is a second-order linear bandpass Butterworth filter (Nuttall and Dolan, 1996), modeling the response of middle ear. Module 2, a "dual-resonance nonlinear" (DRNL) filter architecture (Meddis *et al.*, 2001) representing the filtering of the BM. The input of module 1 is sound pressure (μPa) and the output of model 2 is BM motion velocity, which is the input of IHC and AN complex. Modules 3, 5, and 6 of the IHC and AN complex follow a generic hair cellular schematic circuit model and the Meddis model (Meddis, 1986,

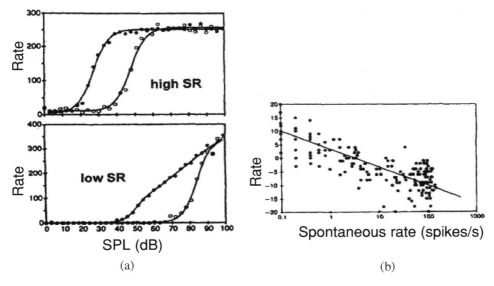

FIGURE 10.15. Experimental data (Schoonhoven *et al.*, 1997) for (a) low-threshold (upper), saturation rate–intensity function observed in guinea pig cochlear nerve fibers with an SR of typically 10 sp/s and above ("high"); and for high-threshold (lower), sloping–saturation rate–intensity function, with SR below 1 sp/s ("low"). The black dots represent measurements at the characteristic frequency (CF) of the fiber, and the circles represent the low-frequency tail of their frequency thresholds. (b) The relation between rate threshold and spontaneous discharge rate in the same set of data. The regression line has a slope of −0.29.

1988), respectively, for the rate response of auditory nerve fibers. The model combines subcomponents, which are faithful to physiology as is practicable and known (Sumner *et al.*, 2002). The membrane potential of the IHC cell body can be formulated by an electrical analog circuit (Figure 10.16, module 3; Shamma *et al.*, 1986), and can be mathematically described by

$$C_{\mathrm{m}}\frac{\mathrm{d}V(t)}{\mathrm{d}t} + G(u)(V(t) - E_t) + G_k(V(t) - E'_k) = 0 \qquad (10.24)$$

where $V(t)$ is the intracellular IHC potential; $C_{\mathrm{m}} = C_{\mathrm{A}} + C_{\mathrm{B}}$ is the cell membrane capacitance, $G(u)$ is the total apical conductance, G_k is the voltage-invariant basolateral membrane conductance; E_t is the endocochlear potential; and $E'_k = E_k + E_t R_p/(R_t + R_p)$ is the reversal potential of the basal current E_k (mostly potassium; Corey and Hudspeth, 1983) corrected for the resistance (R_t, R_p) of the supporting cell. The total apical conductance $G(u)$ is determined by the transduction conductance with all channels open, the passive conductance in the apical membrane, and the displacement of the IHC cilia, which is a function of BM motion velocity.

The IHC intracellular potential changes would give rise to a variation of Ca^{2+} concentration at the synapse, controlling the release of neurotransmitters into the synaptic cleft. Following HH formulism, the calcium ion current can be modeled as a function of receptor

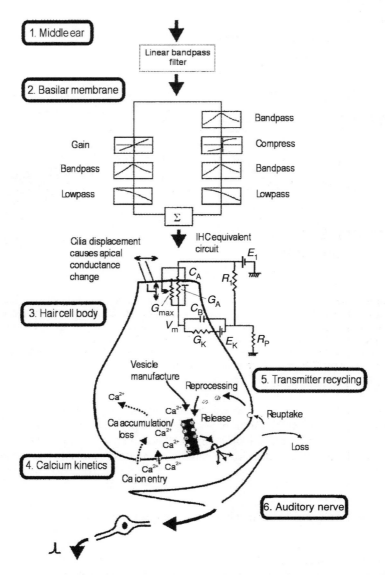

FIGURE 10.16. Schematic diagram of an inner-hair cell model embedded in a complete composite auditory periphery (Sumner *et al.*, 2002).

potential (Hudspeth and Lewis, 1988; Kidd and Weiss, 1990) as

$$I_{Ca}(t) = G_{Ca}^{max} m_{I_{Ca}}^3(t)[V(t) - E_{Ca}] \tag{10.25}$$

where E_{Ca} is the reversal potential for calcium, G_{Ca}^{max} is the calcium conductance in the vicinity of the synapse, with all the channels open, and $m_{I_{Ca}}(t)$ is the fraction of calcium channels that are open. The rate of generation of action potentials by the AN fiber is in accordance with the neurotransmitter release rate. The probability of the neurotransmitter release is modeled as a function of the Ca^{2+} concentration:

$$k(t) = \max \left\{ \left([Ca^{2+}]^3(t) - [Ca^{2+}]^3_{thr} \right) z, 0 \right\} \tag{10.26}$$

where $[Ca^{2+}]_{thr}$ is a threshold constant, z is a scalar for converting calcium concentration levels into release rate. The full description of the model is referred to Sumner's work (Sumner *et al.*, 2002). According to the Meddis model, the firing rate of the auditory nerve fiber is proportional to the amount of neurotransmitters in the synaptic cleft. Once the neurotransmitter release probability $k(t)$ is known, combined with other neurotransmitter processes, such as the reuptake process, the concentration of the neurotransmitter can be obtained by the differential equations of the Meddis model. The details accounting for these processes can be accessed in Meddis (1986, 1988) and Hewitt and Meddis (1991). Sumner's model describes the acoustic signal processing in the auditory pathway, especially the signal transformation at IHC from mechanical motion into chemical transmitter release rate, which determines the AN fiber firing rate. As mentioned before, the rate–intensity function is an important characteristic of acoustic signal processing in the auditory system, and is measurable *in vivo* and *in vitro* for animals. The distinguished advantage of Sumner's model lies in its good performance in reproducing the rate–intensity function of different fiber types. Figure 10.17 shows the comparison of the rate–intensity functions of three fiber types (saturation, sloping-saturation, and straight) between the model simulation results and the experimental data collected from guinea pigs by Winter *et al.* (1990).

It has been reported by Winter *et al.* (1990) that AN fibers with a high spontaneous rate (HSR) have low acoustic threshold and a steep rate–intensity function curve at CF that almost completely saturates within 20–30 dB of the threshold. Median spontaneous rate (MSR) fibers have a "sloping-saturation" rate–intensity function. Low spontaneous rate (LSR) fibers have almost no spontaneous activity with high threshold, and have a straight rate–intensity function without saturation. Yates (1990) concluded that these different shapes of rate–intensity functions at best frequency are due to the linear and compressive responses of BM activities. However, in the model simulation, only three transmitter release parameters are varied to produce the six model functions, which agree with the experimental data of guinea very well. These three parameters are the maximum calcium conductance G^{max}_{Ca}, $[Ca^{2+}]_{thr}$, and M, the maximum number of neurotransmitter that can be held in the free transmitter pool in Meddis model (Meddis, 1986). In addition, the rate–intensity functions are plotted at other frequencies. Figure 10.18 shows the RI function shape changes by adjusting aforementioned three synapse parameters.

It can be observed from Figure 10.18(a) that a BF stimulus gives rise to a typical HSR saturation function with a narrow dynamic range. However, when stimuli at other frequencies, the RI function for the same fiber can be sloping-saturation. As shown in Figure 10.18(c), MSR fiber also has similar variation in the RI response at different frequencies. Yates (1990) explained this partly due to the BM response, which is nonlinear at BF and linear away from BF. Sumner *et al.* (2002) investigated the influence of the neurotransmitter release function on rate–intensity function, as shown in Figure 10.19. It can be observed from Figure 10.19 that with the conductance increasing, the RI function changes from "straight"

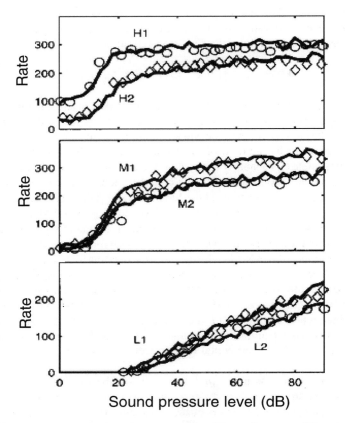

FIGURE 10.17. Auditory nerve rate–intensity responses for different fiber types. Unconnected symbols are animal data from Sumner *et al.* (2002). Solid lines are the responses of the model. All fits employ the same DRNL parameters.

to "sloping saturation," and finally to "saturation." At the same time, the threshold drops, the spontaneous rate increases, and the synapse saturates at low stimulus intensities. Both other two factors cannot govern RI function alone. Sumner *et al.* (2002) concluded that calcium conductance alone determines, at least qualitatively, the full range of RI function types, spontaneous rates, and thresholds, whereas $[Ca^{2+}]_{thr}$ affects the rate response mainly at low intensity and M is to scale the release rate linearly across the entire dynamic range.

   Although RI function variation cannot be fully explained in biophysical meaning, Sumner's model, with good performance in RI function reproduction, can assist researchers in identifying the possible key factors, thereby leading to a deeper understanding of the auditory perception. Further investigation can be conducted under the model guideline. A latest application of this model has been in the study of synapse adaptation of the IHC and AN complex (Sumner *et al.*, 2003). The basic adaptation characteristics in gerbils in Westerman's findings (Westerman and Smith, 1984, 1985, 1988) can be well reproduced by this model. One interesting result from this recent study is that the model parameters

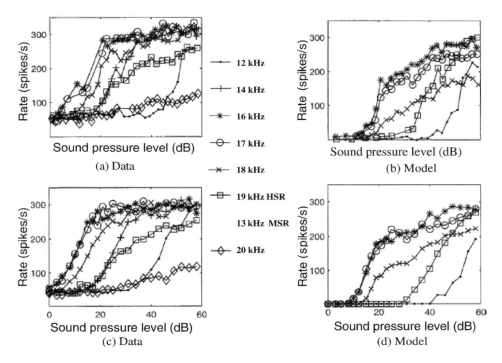

FIGURE 10.18. Rate–intensity functions at both CF and off-CF (Sumner *et al.*, 2002). (a) HSR fiber RI function from experiment; (b) response of model in accordance with (a); (c) MSR fiber RI function from experiment; and (d) response of model in accordance with (c).

determining rate characteristics also affect the adaptation characteristics, without change of the adaptation mechanism. What was obtained from the model may offer clues to understand the mechanism of the synapse adaptation at the IHC and AN synapses.

10.5. CONCLUSION

The voltage-dependent conductance-based circuit model developed by Hodgkin–Huxley in 1952 is still a milestone in neural electrophysiology. This HH model is still very important not only for its ability to describe quantitatively ionic channel activities underlying the neural dynamics at the single-neuron level, but also for its contribution to a better understanding of the neural information processing at the nervous system level. The HH conductance-based circuit model reproduces the electrophysiological data remarkably well and has become a stereotype for constructing more realistic models of the nervous system that can express the intrinsic neural dynamical behaviors. After a review of the HH model, this chapter focused on a compartment model, which combines the conductance-based circuit model with the cable equation for describing the membrane potential distribution and propagation along a complex neuron with inhomogeneous geometry.

However, these models requiring detailed knowledge of the kinetics of the individual ionic currents are difficult to construct and their dynamic behaviors are basically irrelevant

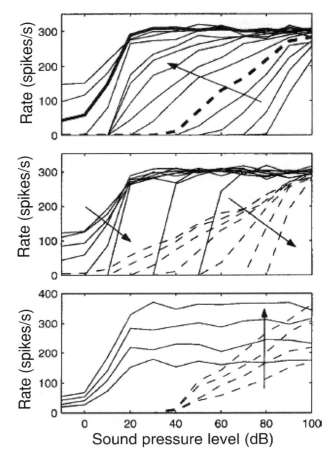

FIGURE 10.19. Effect of varying synapse parameters on rate–intensity functions (Sumner *et al.*, 2002). (a) The effect of increasing G_{Ca}^{max}, from 1.5 to 10 nS in the direction of the arrow. Thick solid line is for $G_{Ca}^{max} = 8$ nS and thick dashed line is for $G_{Ca}^{max} = 2.5$ nS. (b) The effect of increasing $[Ca^{2+}]_{thr}$ from 0 to 18×10^{-11} in the direction of the arrow, for two different values of G_{Ca}^{max} (2.5 nS for dotted lines and 8 nS for solid lines). (c) The effect of increasing M from 5 to 13 in the direction of the arrow, for the same two values of G_{Ca}^{max} as in (b).

with external stimuli. In other words, action potentials which are basically identical for a specified neuron are all-or-none events, the detailed waveform of which may carry no information. Therefore, rather than the detailed dynamics of the action potential itself, there must be other mechanisms responsible for encoding the information related to the stimuli, such as sounds, pressures, lights, currents, etc. Firing rate function of neurons has been found to be the major vehicle carrying neural information. The IF and IPFM models, capturing some of the neuronal behaviors, but at a much reduced complexity, are very powerful tools in studying neural coding aspects, including the rate and intensity relationships. These IF and IPFM models highlight two special features responsible for neural coding: a membrane threshold and an integration operator. Although IF and IPFM models do not incorporate the detailed neural dynamics, the effect of adaptation and membrane threshold can be included. Indeed, these simple models with a handful of parameters can be used to simulate a complex neural system for investigating the neural coding process (Kong, J. *et al.*, 1998, 1999). These IF and IPFM models represent a reasonable tradeoff between simplicity and faithfulness and the key attributes of neural coding. In this chapter, with a single sinusoidal signal as the input modulation signal to the IPFM model, the corresponding spectrum is studied from the demodulation point of view. It is found that the output of the IPFM demodulator without noise is actually the approximated instantaneous firing rate rather than the original input modulation signal as reported before. Therefore, this theoretical analysis provides an insight into the quantitative evaluation of the mean firing rate and the rate variability at the physiological modulating frequency. The results could be useful in exploring the underlying relevant coding mechanisms of a biological living system.

A circuit model for the rate–intensity function in auditory periphery was discussed at the end of this chapter to illustrate applications of modeling techniques to neural information processing at the system level. Rate–intensity function, characterizing the acoustic information transmission in the auditory system, can be remarkably reproduced by detailed models of mechanical to electrical signal transmission at the IHC–AN complex. The HH scheme was adopted as a framework for modeling the ionic channel activities while describing calcium activities at the IHC and AN nerve synaptic cleft to determine neurotransmitter release, which would determine the AN firing rate and thereby assist in understanding the neural coding by acoustic information in the auditory system. Furthermore, the knowledge of neural information processing mechanisms in the biological nervous system would also be beneficial for the development of bio-model-based signal processing techniques with various clinical applications.

REFERENCES

Bayly, E. J., 1968, Spectral analysis of pulse frequency modulation in the nervous systems, *IEEE Trans. Biomed. Eng.* **15**(4):257–265.

Brugge, J. F., 1992, An overview of central auditory processing, in: *The Mammalian Auditory Pathway: Neurophysiology* (N. Popper and R. F. Richard, eds.), Springer-Verlag, New York, pp. 1–33.

Corey, D. P., and Hudspeth, A. J., 1983, Kinetics of the receptor current in bullfrog saccular hair cells, *J. Neurosci.* **3**(5):962–976.

Fatt, P., and Katz, B., 1951, An analysis of the endplate potential recorded with an intracellular electrode, *J. Physiol.* **115**:320–370.

Frankenhaeuser, B., and Moor, L. E., 1963, The effect of temperature on the sodium and potassium permeability changes in myelinated nerve fibers of *Xenopus laevis*, *J. Physiol.* **169**:431–437.

Gerstner, W., and Kistler, W. M., 2002, *Spiking Neuron Models: Single Neurons, Populations, Plasticity*, Cambridge University Press, New York.

Gu, Y. Y., and Zhang, Y. T., 2003, Effects of neural dynamics on information decoding, *Proceedings of the Chinese Biomedical Engineering Conference*, Oct. 2003, pp. 62–63.

Gu, Y. Y., Zhang, Y. T., and Ma, T., 2003, A theoretical analysis of neural firing rate function based on an integral pulse frequency modulation model, *Proceedings of the IEEE EMBS Asia Pacific Conference on Biomedical Engineering*, Oct. 2003, pp.144–145.

Harmon, L. D., 1961, Studies with artificial neurons: Properties and functions of an artificial neuron, *Kybernetic* **1**:102–107.

Hausser, M., Spruston, N., and Stuart, G. J., 2000, Diversity and dynamics of dendritic signaling, *Science* **290**(5492):739–744.

Hausser, M., 2000, The Hodgkin–Huxley theory of the action potential, *Nat. Neurosci.* **3**(Suppl.):1165.

Hewitt, M. J., and Meddis, R., 1991, An evaluation of eight computational models of mammalian inner hair-cell function,*J. Acoust. Soc. Am.* **90**:904–917.

Hodgkin, A. L., and Huxley, A. F., 1952, A quantitative description of membrane current and its application to conduction and excitation in nerve, *J. Physiol.* **117**:500–544.

Hodgkin, A. L., and Katz, B., 1949, The effect of temperature on the electrical activity of the giant axon of the squid, *J. Physiol.* **109**:240–249.

Hodgkin, A. L., and Rushton, W. A. H., 1946, The electrical constants of a crustacean nerve fiber, *Proc. Royal Soc. Ser. B (London)* **144**:444–479.

Hudspeth, A. J., and Lewis, R. S., 1988, Kinetic analysis of voltage- and ion-dependent conductances in saccular hair cells of the bull-frog, *Rana catesbeiana, J. Physiol.* **400**:237–274.

Kandel, E. R., Schwartz, J. H., and Jessell, T. M., 1995, *Essentials of Neural Science and Behavior*, Appleton & Lange, Norwalk.

Kidd, R. C., and Weiss, T. F., 1990, Mechanisms that degrade timing information in the cochlea, *Hearing Res.* **49**(1–3):181–207.

Koch, C., 1999, *Biophysics of Computation: Information Processing in Single Neurons*, Oxford University Press, New York.

Kong, J., Zhang, Y. T., and Lu, W. X., 1998, Dynamical structures of integral pulse frequency modulation (IPFM) model with its applications to physiological systems, in: *20th Annual International Conference of the IEEE Engineering in Medicine and Biology Society*, No. 6 (1998), pp. 3028–3031.

Kong, J., Zhang, Y. T., and Lu, W. X., 1999, Dynamical behaviours of integral pulse frequency modulation (IPFM) model with a periodically-varied threshold, in: *Proceedings of the First Joint BMES/EMBS Conference*, Oct. 1999, pp. 1000.

Lewis, E. R., 1963, The locus concept and its application to neural analogs, *IEEE Trans. Bio-Med. Electron.* **10**:130–137.

Meddis, R., 1986, Simulation of mechanical to neural transduction in the auditory receptor, *J. Acoust. Soc. Am.* **79**(3):702–711.

Meddis, R., 1988, Simulation of auditory–neural transduction: Further studies, *J. Acoust. Soc. Am.* **83**(3):1056–1063.

Meddis, R., O'Mard, L. P., and Lopez-Poveda, E. A., 2001, A computational algorithm for computing nonlinear auditory frequency selectivity, *J. Acoust. Soc. Am.* **109**:2852–2861.

Nuttall, A. L., and Dolan, D. F., 1996, Steady-state sinusoidal velocity responses of the basilar membrane in guinea pig, *J. Acoust. Soc. Am.* **99**(3):1556–1565.

Pallas-Areny, R., Colominas-Balague, J., and Rosell, F. J., 1989, The effect of respiration-induced heart movements on the ECG, *IEEE Trans. Biomed. Eng.* **36**(6):585–590.

Partridge, L. D., 1965, Modifications of neural output signals by muscles: A frequency response study, *J. Appl. Physiol.* **20**(1):150–156.

Periodic Stimulus and the Single Cell, 2002, Utah University, Salt Lake City; http://www.math.utah.edu/~eric/research/talks/pacing/ODE1.html.

Petz, E. E., and Gerstein, G. L., 1963, An RC model for spontaneous activity of single neuron, *MIT Res. Lab. Electron.* **QPR-71**:249–257.

Ram, J. L., 2000, *The Axon Potential Simulator*, Department of Physiology, Wayne State University, http://sun.science.wayne.edu/~jram/axon_potential_simulator.htm.

Rall, W., 1967, Distinguishing theoretical synaptic potentials computed for different soma-dendritic distributions of synaptic input, *J. Neurophysiol.* **30**(5):1138–1168.

Rall, W., and Agmon-Snir, H., 1998, Cable theory for dendritic neurons, in: *Methods in Neuronal Modeling* (C. Koch and I. Segev, eds.), MIT Press, Cambridge, MA, pp. 27–92.

Rattay, F., 1998, Analysis of the electrical excitation of CNS neurons, *IEEE Trans. Biomed. Eng.* **45**(6):766–773.

Schalk, T. B., and Sachs, M. B., 1980, Nonlinearities in auditory-nerve fiber responses to bandlimited noise, *J. Acoust. Soc. Am.* **67**(3):907–913.

Schoonhoven, R., Prijs, V. F., and Frijns, J. H., 1997, Transmitter release in inner hair cell synapses: A model analysis of spontaneous and driven rate properties of cochlear nerve fibres, *Hearing Res.* **113**(1–2):247–260.

Schutter, E. D., and Smolen, P., 1998, Calcium dynamics in large neuronal models, in: *Methods in Neuronal Modeling: From Ions to Networks* (C. Koch and I. Segev, eds.), MIT press, Cambridge, MA, pp. 211–250.

Segev, I., Burke, R. E., and Hines, M., 1998, Compartmental Models of Complex Neurons, in: *Methods in Neuronal Modeling: From Ions to Networks* (C. Koch and I. Segev, eds.), MIT Press, Cambridge, MA, pp. 93–136.

Shamma, S. A., Chadwick, R. S., Wilbur, W. J., Morrish, K. A., and Rinzel, J., 1986, A biophysical model of cochlear processing: Intensity dependence of pure tone responses, *J. Acoust. Soc. Am.* **80**(1):133–145.

Sumner, C. J., Lopez-Poveda, E. A., O'Mard, L. P., and Meddis, R., 2002, A revised model of the inner-hair cell and auditory-nerve complex, *J. Acoust. Soc. Am.* **111**(5):2178–2188.

Sumner, C. J., Lopez-Poveda, E. A., O'Mard, L. P., and Meddis, R., 2003, Adaptation of revised inner-hair cell model, *J. Acoust. Soc. Am.* **113**(2):893–901.

Westerman, L. A., and Smith, R. L., 1984, Rapid and short-term adaptation in auditory nerve responses, *Hearing Res.* **15**(3):249–260.

Westerman, L. A., and Smith, R. L., 1985, Rapid adaptation depends on the characteristic frequency of auditory nerve fibers, *Hearing Res.* **17**(2):197–198.

Westerman, L. A., and Smith, R. L., 1988, A diffusion model of the transient response of the cochlear inner hair cell synapse, *Hearing Res.* **83**(6):2266–2276.

Winter, I. M., Robertson, D., and Yates, G. K., 1990, Diversity of characteristic frequency rate–intensity functions in guinea pig auditory nerve fibers, *Hearing Res.* **45**(3):191–202.

Yates, G. K., 1990, Basilar membrane nonlinearity and its influence on auditory nerve rate–intensity functions, *Hearing Res.* **50**(1–2):145–162.

Zhang, Y. T., Parker, P. A., and Scott, R. N., 1990, Study of the effects of motor unit recruitment and firing statistics on the signal to noise ratio of myoelectric control channel, *Med. Biol. Eng. Comput.* **28**(3):225–231.

11

NEURAL SYSTEM IDENTIFICATION

Garrett B. Stanley*

Division of Engineering and Applied Sciences, Harvard University,
Cambridge, Massachusetts

11.1. INTRODUCTION

One could argue that all scientific problems can be described in terms of two fundamental objectives: identification and control. Much of the current body of research in basic neuroscience revolves around the problem of identification, although not formally posed as such. The problem of identification is that of cause and effect. For example, in considering the relationship between two synaptically connected neurons, how does the presynaptic action potential cause the postsynaptic potential? At a more macroscopic level, how do the photons of light entering the eye cause the neuronal population activity in the visual pathway of the brain? Due to the overwhelming complexity of the nervous system, it is in fact difficult to think of threads of investigation that are not in some way reliant on identification or modeling at the systems level. The concept of system identification goes beyond simply reporting experimental observations. In many cases, the input can be controlled, and the goal is to identify a functional relationship between stimulus and response that will enable prediction of the response of the system to subsequent arbitrary inputs. Failure in prediction exposes previous misconceptions about the underlying dynamics, often leading to more intelligently designed experiments, and so on. Herein lies the true value of the identification process in this largely empirical field of science.

The field of system identification grew out of the statistics and engineering literature in the 1960s, motivated by the need to predict and control the behavior of complex systems (Box and Jenkins, 1976). Subsequently, there have been a number of general references that range from the applied (Jenkins and Watts, 1968; Ljung, 1987) to the theoretical (Brillinger, 1981; Söderström and Stoica, 1987) ends of the spectrum, as well as those that explicitly focus on the identification of physiological systems (Marmarelis and Marmarelis, 1978). Central to the framework of system identification is the idea that complex systems can be represented as a *black box*, as opposed to more simplistic physical systems whose dynamics suggest relationships based on first principles. As more is learned about a complex system,

---

*Address for correspondence: 321 Pierce Hall, 29 Oxford St., Harvard University, Cambridge, Massachusetts 02138; e-mail: gstanley@deas.harvard.edu.

the goal then becomes to reduce the relationship to smaller and smaller black boxes that represent the individual components of the system, until a sufficient level of detail has been achieved. Neuronal processing of information is complex at all levels, from the microscopic interaction between the pre- and postsynaptic cell, to the macroscopic interactions between the large populations of neurons involved in sensory processing and motor response. Since the 1960s, system identification techniques have been formally applied to the processing of neuronal information in a number of studies (Perkel *et al.*, 1967; DeBoer and Kuyper, 1968; Marmarelis and Naka, 1973a,b; Brillinger *et al.*, 1976), and has subsequently become a commonly utilized tool in systems neuroscience. The goals of this chapter are to introduce some important perspectives and techniques for system identification, and to present concrete examples of system identification strategies employed in sensory processing in the central nervous system and neural control in the peripheral nervous system.

The remainder of the chapter is presented in the following manner. Section 11.2 provides an introductory review of techniques in system identification. The basic characteristics of dynamical systems are discussed, as well as means for estimating the fundamental quantities relating input and output, in both time and frequency domains. The background on system identification, however, is provided for continuous signals. Neurons communicate information through discrete pulses, or action potentials. Section 11.3, therefore, provides the necessary details on the nature of measured neuronal signals and how they can be represented mathematically, as well as some basic correlation measures for this type of data. Section 11.4 presents examples of neuronal system identification in the context of the processing of visual information. Specific examples will be given for cases in which neuronal activity is well predicted by quasi-linear models estimated from stimulus–response data and cases in which the properties of the pathway that continuously change in response to a changing environment can be estimated using adaptive estimation strategies. Section 11.5 provides examples of the identification of neuronal processing mechanisms in the mammalian somatosensory (touch) pathway, specifically highlighting interactions that are not well explained through linear techniques. Finally, in contrast to the time-domain estimation presented in previous sections, Section 11.6 provides an example of the frequency domain identification of the dynamics related to the peripheral nervous control of cardiac function.

11.2. SYSTEM IDENTIFICATION

Before presenting issues related to the identification of dynamics related to neuronal processing, it is first necessary to discuss preliminaries regarding system identification in general. The following discussion will focus on discrete-time sampling of continuous-valued signals, but will, in subsequent sections, be extended to capture the discrete nature of neuronal spiking.

11.2.1. DYNAMICAL SYSTEMS

In general, problems of system identification revolve around a conceptual picture of an input–output dependence, as shown in Figure 11.1. The output or response $r(t)$ is in some way a function of the input or stimulus $s(t)$ (present and past), but is also influenced

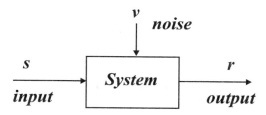

FIGURE 11.1. Input–output transformation.

by an uncontrolled, and typically unobserved, noise process $v(t)$ that describes all behavior of $r(t)$ not directly explained by $s(t)$. The following questions are then applicable. Is the relationship between s and r linear or nonlinear? Is the relationship between s and r time-invariant or time-varying? In other words, will an experiment conducted at time t yield the same results as an experiment conducted at $t + \Delta t$? For the rest of this section, we will assume that the properties are time-invariant, but will revisit this issue in the discussion of adaptation in the visual system in section 11.4.

When encountering a system in which little is known about the underlying input/output relationship, it is often prudent to begin with nonparametric models that minimally constrain the description of the dynamics. A general nonparametric nonlinear relationship between input, s, and output, r, can be expressed as a Volterra series (Volterra, 1959; Marmarelis and Marmarelis, 1978):

$$r(t) = g_0 + \sum_k g_1(k)s(t-k) + \sum_k \sum_j g_2(k, j)s(t-k)s(t-j) + \text{HOTS} + v(t)$$

$$(11.1)$$

where g_n is the "nth-order kernel," HOTS represents *higher order terms*, and $v(t)$ is the noise process (assumed additive). The term g_0 is representative of the component of the mean of $r(t)$ not related to the input, g_1 is representative of first-order dynamics, g_2 of second-order interactions, and so on. Note that this is analogous to a Taylor series expansion describing arbitrary static nonlinear relationships. The following discussion will focus on nonparametric estimation, where the input/output relationship takes on no more specificity than that of the expression in Eq. (11.1), but there are numerous references that can be consulted for parametric estimation for more specific functional forms (Cramér, 1946; Rao, 1973).

11.2.2. ESTIMATION

For the given model structure, the problem of system identification reduces to the minimization of a cost function that depends on observed data and the set of parameters θ:

$$\hat{\theta} = \arg \min_{\theta} J(\text{data}, \theta) \qquad (11.2)$$

where

$$\theta \triangleq [\, g_0 \quad g_1(0) \quad g_1(1) \quad \cdots \quad g_2(0, 0) \quad g_2(1, 0) \ldots \quad]^{\mathrm{T}}$$

where $J(\cdot)$ is a cost function, T represents matrix transpose, $\hat{\ }$ represents estimate, and θ is a vector containing elements of the model. The goal then is to find the set of parameters

θ that minimize the cost function for a given data set. The following measures are presented in the context of the class of minimization problems based on least-squares cost functions.

11.2.2.1. Time Domain Measures

There are a number of statistical measures that are important to define for the identification process, several of which are presented here.

Covariance

The auto-covariance function is a symmetric quantity that refers to how a signal, s, covaries with itself (i.e., relationship between signal at different time points):

$$C_{ss}(k) \triangleq E\{(s(t) - \mu_s)(s(t+k) - \mu_s)\}$$

where $E\{\cdot\}$ denotes statistical expectation, and μ_s denotes the mean of s. This function can be estimated in the following manner:

$$\hat{C}_{ss}(k) = \frac{1}{N - |k| - 1} \sum_t (s(t) - \hat{\mu}_s)(s(t+k) - \hat{\mu}_s) \qquad -\frac{N}{4} < k < \frac{N}{4}$$

where N is the length of the data set, and $\hat{\mu}_s$ is the mean estimated from data. The covariance estimates should not be computed for lags (shifts) greater than a quarter of the data length to avoid significant errors resulting from the limited number of data points involved in the estimate (Chatfield, 1989). The interpretation is that for a given value of k, the signal s is shifted by k points, and the point-by-point product of the shifted observations is formed, revealing periodicity or temporal structure in the signal. Suppose that $s(t)$ is an independent, identically distributed (IID) process (and therefore uncorrelated). The approximate 95% confidence intervals on an independent process are represented by the band $\pm 2\hat{C}_{ss}(k)/\sqrt{N}$. This band is often superimposed on the plot of the auto-covariance function to evaluate whether the observed correlation structure deviating from zero is statistically significant. A boot-strapping technique can alternately be employed, which involves repeatedly computing covariances from the randomized (temporally shuffled) data set to estimate the confidence bands on an uncorrelated process (Perkel *et al.*, 1967).

When analyzing the relationship between input $s(t)$ and output $r(t)$, it is often useful to consider the cross-covariance function:

$$C_{sr}(k) \triangleq E\{(s(t) - \mu_s)(r(t+k) - \mu_r)\}$$

which can be estimated in a manner similar to the auto-covariance estimate presented previously. For the case where $s(t)$ and $r(t)$ are independent, and individually uncorrelated IID processes, the band $\pm 2\sqrt{\hat{C}_{ss}(0)\hat{C}_{rr}(0)}/\sqrt{N}$ represents the approximate 95% confidence interval. We can thus assess to what extent the two processes are significantly correlated (Chatfield, 1989).

Kernel Estimation

In some very special cases, the input–output relationship is explained well by a linear system (i.e., the first-order kernel). In this case, the cross-covariance between the input and the output is related to the auto-covariance of the input through the first-order kernel (Papoulis, 1984):

$$C_{sr}(k) = \sum_{m=0}^{L} g_1(m) C_{ss}(k - m) \tag{11.3}$$

Using the fact that $C_{ss}(k) = C_{ss}(-k)$, this relationship can be written in a structured matrix form as

$$
\begin{bmatrix} C_{sr}(0) \\ C_{sr}(1) \\ \vdots \\ C_{sr}(L) \end{bmatrix}
=
\begin{bmatrix}
C_{ss}(0) & C_{ss}(1) & \cdots & C_{ss}(L) \\
C_{ss}(1) & C_{ss}(0) & \cdots & C_{ss}(L-1) \\
\vdots & \vdots & \ddots & \vdots \\
C_{ss}(L) & C_{ss}(L-1) & \cdots & C_{ss}(0)
\end{bmatrix}
\begin{bmatrix} g_1(0) \\ g_1(1) \\ \vdots \\ g_1(L) \end{bmatrix}
$$

or $C_{sr} = C_{ss} \cdot g_1$ in shorthand notation. We can then solve for g_1:

$$\hat{g}_1 = C_{ss}^{-1} C_{sr}$$

where the first-order kernel is simply the auto-covariance between input and output, normalized by the correlation structure of the input. Note that for a linear system, the first-order kernel is also known as the *impulse response* of the system, as it represents the response to an impulse input. The first-order kernel, when estimated in this manner, is precisely the solution to the least-squares problem. The cost function of Eq. (11.2) is the sum of the squared errors:

$$\hat{g}_1 = \arg \min_{g_1} \sum_{k} \left(r(k) - \sum_{m} g_1(m) s(k - m) \right)^2$$

If the input is uncorrelated (white), then C_{ss} will be a diagonal matrix, with σ_s^2 (the variance of s) along the diagonal. We can therefore write $\hat{g}_1(k) = C_{sr}(k)/\sigma_s^2$. In this case, the first-order kernel is simply the cross-covariance between input and output, normalized by the variance of the input. The application of this estimation technique will be discussed in the context of processing the visual pathway in section 11.4.

Higher-Order Interactions

It should be noted that even if the system is not strictly linear, and thus has nonzero kernels beyond the first order, the use of an orthogonal input, such as Gaussian white noise, makes possible the sequential estimation of kernels of different order (Lee and Schetzen, 1965; Marmarelis and Marmarelis, 1978). The higher-order kernels, in this case, can be estimated from higher-order correlations formed from averaging the products of the signals at varying delays. See Marmarelis and Marmarelis (1978) for a detailed discussion of this issue. This topic will again be revisited in the discussion of the nonlinear encoding of tactile information in section 11.5.

11.2.2.2. Frequency Domain Measures

It is often the case that the input–output relationships are better described (or estimated) in the frequency domain, as opposed to the time domain. Several of the related measures are presented here.

Spectra

The power spectrum of a signal, s, is defined as the Fourier transform of the auto-covariance:

$$S_{ss}(\omega) \triangleq \sum_{k=-\infty}^{\infty} C_{ss}(k)e^{-j\omega k} \in \mathbb{R}$$

which can be estimated from the Fourier transform of the estimated auto-covariance, where $\omega = 2\pi f$ is the frequency in radians per second. The interpretation of the power spectrum is that it is a frequency-dependent variance measure, describing how the variance is distributed across different frequency bands. The Fourier transform of the auto-covariance estimate is a rather raw estimate for the spectrum for finite observations. It is therefore prudent to modify this estimate of the spectrum by multiplying by a smoothing window w (Ljung, 1987):

$$\hat{S}_{ss}(\omega) = \sum_{k=-\infty}^{\infty} w(k)\hat{C}_{ss}(k)e^{-j\omega k}$$

Note that multiplication by a smoothing window in the time domain is equivalent to convolving with the smooth function in the frequency domain. See Ljung (1987) for a detailed discussion of smoothing windows and the corresponding effects on the spectral estimates. Confidence bands can be generated for a white process, and used as a test for whiteness of the observed process. An alternate use is to place bands around the actual estimate $\hat{S}_{ss}(\omega)$ so that the spectrum can be statistically compared to other spectra (Jenkins and Watts, 1968; Brillinger, 1981).

The cross-spectrum between two processes $s(t)$ and $r(t)$ is defined as the Fourier transform of the cross-covariance:

$$S_{sr}(\omega) \triangleq \sum_{k} C_{sr}(k)e^{-j\omega k} \in \mathbb{C}$$

which can be estimated from the Fourier transform of the estimated cross-covariance. Similar confidence bands exist as above for assessing correlations as a function of frequency (Brillinger, 1981).

Transfer Function

For linear systems, the transfer function between input and output is defined as the Fourier transform of the impulse response:

$$G(\omega) \triangleq \sum_{k} g_1(k)e^{-j\omega k} \in \mathbb{C}$$

which can be obtained in practice from the impulse response estimate, previously discussed. An alternate method of computing the transfer function estimate is derived from the relationship between the auto-covariance of the input and the cross-covariance between input and output. Because convolution in the time domain is equivalent to multiplication in the frequency domain, the relationship in Eq. (11.3) yields

$$S_{sr}(\omega) = G(\omega)S_{ss}(\omega)$$

Estimates of the quantities S_{ss} and S_{sr} then directly yield an estimate $\hat{G}$. The impulse response, g_1, can be obtained by taking the inverse Fourier transform of the resulting transfer function G. This has obvious advantages, because the time-domain estimation of g_1 involves matrix inversion, as previously described, and the frequency domain estimation simply involves the Fourier transform, which can be implemented efficiently using a variety of FFT algorithms (Press *et al.*, 1992). Estimation in the frequency domain will again be discussed in the context of neural control of cardiac function in section 11.6.

11.3. REPRESENTATIONS OF NEURONAL ACTIVITY

The neuron is the fundamental building block of our nervous system. Neurons encode information about the outside world through subthreshold membrane potentials and sequences of discrete electrical events. Various measurement strategies thus exist that produce a myriad of representations of the underlying neural activity, from the micro- to the macroscopic spatial scales (Eichenbaum, and Davis, 1998). The measures we will focus on here relate to the supra-threshold spiking activity of the neurons, with respect to both the precise timing and rate at which they occur in response to exogenous stimuli.

11.3.1. SPIKE TIMES

When observing the activity of a single neuron over time, the neuron undergoes a sequence of action potentials that are typically uniform in their magnitude and shape. By representing the times of these events as delta functions, or *spikes*, highly localized in time, we describe this as a *spike train*. For a more extensive discussion regarding this topic, see Dayan and Abbott (2001).

Suppose that the events occur at time t_i, with $i = 1, 2, \ldots, n$. The spike train can be represented as a continuous signal, with Dirac delta functions at the times of the events:

$$\rho(t) = \sum_{i=1}^{n} \delta(t - t_i)$$

The response, $\rho(t)$, is therefore a sequence of impulses at the event times. Such a train of events is shown in Figure 11.2a.

Intensity Measures

For discrete processes, such as the firing of a neuron, often termed *point processes*, it is natural to describe the process through intensity measures. We now turn our attention to

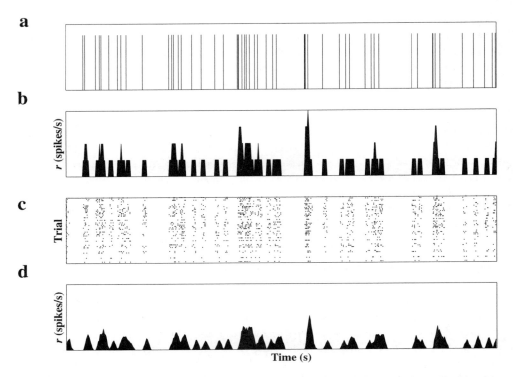

FIGURE 11.2. The spike train and various transformations. (a) A neuronal spike train, over 1 s. (b) The firing rate obtained from the single spike train in (a), by convolving with a rectangular window of width 10 ms. (c) Raster plot of repeated trials, showing the variability of the response. (d) Firing rate obtained from repeated trials, by summing across trials and convolving with a rectangular window of width 10 ms.

correlation measures for point processes (Brillinger *et al.*, 1976; Brillinger, 1992). Let the mean intensity of the process x be defined as

$$m_x \triangleq \lim_{h \to 0} \frac{\Pr\{x \text{ event in } [t, t+h)\}}{h}$$

where $\Pr\{\cdot\}$ denotes probability, and h is a binwidth, which goes to 0 in the limit. For an observation over $[0, T)$, we can estimate this as the number of events in $[0, T)$, divided by the total time interval T. The auto-intensity function, much like the auto-covariance function for continuous processes, is defined as

$$m_{xx}(u) \triangleq \lim_{h \to 0} \frac{\Pr\{x \text{ event in } [t, t+h) | x \text{ event at } t - u\}}{h} \tag{11.4}$$

Note that in practice, the binwidth h is finite, making the measures somewhat resolution-dependent. The heuristic description of the above measure is that given a spike at time $t - u$, it is the probability of observing a spike in a small window of width h at time t, normalized by the width of the window. Now, in addition to the process x, consider another point process y. The cross-intensity function, much like the cross-covariance function for continuous processes, is defined as:

$$m_{xy}(u) \triangleq \lim_{h \to 0} \frac{\Pr\{y \text{ event in } [t, t+h) | x \text{ event at } t - u\}}{h} \tag{11.5}$$

For all of the above estimators, there are corresponding error bounds that are presented in a number of references (Brillinger *et al.* 1976; Brillinger, 1992). These measures will be revisited in section 11.5 in the discussion of nonlinear encoding of tactile information.

11.3.2. FIRING RATE

Suppose that we are interested in the rate, r, at which spikes occur over some experimental trial interval T, which is equivalent to the mean intensity. This can be written as

$$r = \frac{1}{T} \int_0^T \rho(\tau) \, d\tau$$

For example, in Figure 11.2a, we observe 54 spikes in 1 s, resulting in a mean rate of 54 Hz. We can extend this to a more general expression that represents the firing rate in small fixed intervals of width Δt:

$$r(k \Delta t) = \frac{1}{\Delta t} \int_{k \Delta t}^{(k+1) \Delta t} \rho(\tau) \, d\tau \qquad k = 0, 1, 2, \ldots \tag{11.6}$$

This is essentially just binning and counting spikes within each bin. We will refer to this hereafter as simply $r(k)$, where the sampling interval is implicit. Note that this measure depends strongly on the size and placement of the bins. More generally we can express the rate as a convolution of the original spike train with a smoothing window $w(t)$. A rectangular function $w(t)$ of width Δt and height $1/\Delta t$ gives the number of events that occur in the interval $[t, t + \Delta t]$, normalized by the interval length Δt, which we will denote as the *firing rate*, shown in Figure 11.2b for $\Delta t = 10$ ms. Although this approach alleviates the problem of bin placement, it introduces correlation into the process $r(t)$, because adjacent bins overlap and the same spikes are counted in both.

11.3.3. NEURONAL VARIABILITY

The discussion thus far might lead one to believe that an individual neuron has very complex dynamics that simply need to be fully characterized. However, when presented with the same inputs over repeated trials, most neurons exhibit a significant degree of variability in their response. To illustrate this, Figure 11.2c shows a raster plot of the activity across different trials. For each trial, the spike times are plotted as a dot at the time of occurrence, whereas each row of the plot represents a different trial. This is an effective way of plotting neuronal data, to see repeatable structures in the activity and the variability across trials. The firing rate in this case can be determined by aligning the responses of each trial relative to the beginning of the individual trial, and averaging across the trials in a temporal bin. More precisely, let $t_{i,k}$ denote the ith spike in the jth trial, where $j = 1, 2, \ldots, N$ and $i = 1, 2, \ldots, n_j$. Further, let $\rho_j(t) = \sum_i \delta(t - t_{i,j})$ denote the spike train of the jth trial, where t is relative to the beginning of the jth trial. We then have

$$r(k\Delta t) = \frac{1}{\Delta t} \int_{k\Delta t}^{(k+1)\Delta t} \rho(\tau)d\tau \quad \text{where} \quad \rho(t) = \frac{1}{N} \sum_{j=1}^{N} \rho_j(t)$$

As before, the firing rate can also be obtained by convolving $\rho(t)$ with a smoothing window, the results of which are shown in Figure 11.2d, again for $\Delta t = 10$ ms.

We now turn to a series of examples that utilize the preceding perspectives and techniques.

11.4. NEURONAL ENCODING IN THE VISUAL PATHWAY

Humans are incredibly visual creatures. As a result, the visual pathway of the mammalian brain has been the focus of a significant amount of research in sensory coding. The visual pathway therefore serves as a good basis upon which to discuss the systems approaches presented thus far. Photons of light from the outside world enter the eye through the lens, and fall upon the back of the eye, or retina. Photoreceptors transduce the photons into electrical signals, which are propagated through the layers of the retina to the retinal ganglion cells, serving as the output layer of the retina. Action potentials that originate in the retina travel down axon bundles (the optic nerve), projecting to the visual region of the thalamus (the lateral geniculate nucleus, or LGN). The LGN then projects to the primary visual cortex (Hubel, 1995). In the early visual pathway, each neuron encodes information about a restricted region of visual space, which is generally referred to as the receptive field (RF) of the cell. More precisely, however, we will refer to the spatiotemporal receptive field (STRF) as the functional manner in which visual information is integrated over space and time to give rise to neuronal activity in the pathway, which is an explicit description of the input/output properties of the stages of processing in the pathway.

The fundamental task of identification problems in the visual pathway lies in the characterization of the relationship between a temporally continuous stimulus and the discrete process of the firing activity of the neuron at various locations in the pathway. One approach is to relate the continuous input (modulated light intensity) to the firing "rate" of the neuron, allowing the quantification of the relationship between the input and its modulatory effects on the rate of action potentials generated. Let the firing rate of the neuron be denoted as $r(k)$, which represents the number of events occurring in the interval $(k\Delta t, (k + 1)\Delta t]$ normalized by the interval width, Δt, as previously described in Eq. (11.6). Although the transformation to rate does simplify the relationship a great deal, the remaining relationship between the stimulus, which takes on values both positive and negative relative to the mean level, and the strictly nonnegative firing rate is nontrivial. Linear models are insufficient to capture such a transformation, and thus generally must be described through a higher-order kernel expansion, as in Eq. (11.1).

An alternate functional model to describe the relationship between the stimulus and the firing rate of a neuron is the linear–nonlinear (LN) cascade, which incorporates a linear

(L) system followed by a static nonlinearity (N) (Movshon *et al.*, 1978; Tolhurst and Dean, 1990; Ringach *et al.*, 1997; Meister and Berry, 1999; Chichilnisky, 2001; Stanley, 2002), as shown in Figure 11.3. The output of the linear element is expressed as a convolution of the temporal stimulus, s, with the first-order kernel g_1, and an integration over the visual space:

$$x(k) = \sum_i \sum_j \sum_m g_1(i, j, m)s(i, j, k - m)$$

where m ranges from 0 to L, and $L\Delta t$ is the filter length, which can be interpreted as the temporal integration window of the cell. The firing rate of the neuron is then a static nonlinear function of the intermediate signal x, so that $r(k) = f(x(k))$, representing rectifying and saturating properties of the encoding. The output of the nonlinearity is then driving an inhomogeneous Poisson process to produce the discrete neuronal events. In contrast to the complex nature of a higher-order kernel representation, the LN system provides a relatively simple means for describing the inherent nonlinearity in neural encoding (Hunter and Korenberg, 1986; Greblicki, 1992; Paulin, 1993; Korenberg and Hunter, 1999; Chichilnisky, 2001).

11.4.1. ESTIMATION OF THE STRF

Reverse-correlation techniques, which refer to the cross-correlation between stimulus and response for white noise stimuli, have been used extensively to characterize the dynamics of neurons in the retina, LGN, and primary visual cortex (Jones and Palmer, 1987; Reid *et al.*, 1991; Mao *et al.*, 1998), where there is an implicit assumption of the underlying LN cascade. The reverse-correlation technique closely mirrors the linear estimation techniques presented in section 11.2.2.1. Let the parameter vector be defined as $\theta \triangleq [g_1(0)\ g_1(1) \dots g_1(L-1)]^T$ to represent the first-order kernel. Here we represent a single visual pixel for simplicity, but the representations can easily be reformulated to represent a two-dimensional array, the elements of which can be restructured into the vector notation here. For this simple case, the response can then be written $r(k) = f(x(k))$, where $x(k) = \theta^T \varphi(k)$ and $\varphi(k) \triangleq [s(k)\ s(k-1) \dots s(k-L+1)]^T$ is the time history of the stimulus. As discussed previously, the parameter vector for a first-order kernel can be estimated from the cross-covariance between the stimulus and the output of the linear element, normalized by the auto-covariance structure of the stimulus, or $\hat{\theta} = C_{ss}^{-1}C_{sx}$, where C_{ss} is the structured auto-covariance matrix of the stimulus s, and C_{sx} is the cross-covariance vector between the stimulus and the output of the linear block, x. For Gaussian inputs to the static nonlinearity, the only effect of the nonlinearity is a scaling of the cross-covariance, and the estimate of the parameter vector can be expressed as a function of the stimulus, s, and response, r:

$$\hat{\theta} = C_{ss}^{-1}C_{sx} = A \cdot C_{ss}^{-1}C_{sr} \tag{11.7}$$

where A is a constant of proportionality. For half-wave rectification with Gaussian inputs, this scaling will be approximately 2 (Stanley, 2002). Figure 11.3 shows the estimate of a low-pass biphasic first-order kernel at a single pixel at the center of the RF from cat LGN X cell response to a spatiotemporal binary stimulus (m-sequence). See Stanley *et al.* (1999) for

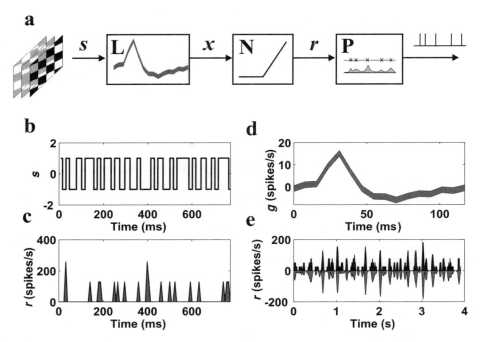

FIGURE 11.3. Linear–nonlinear–Poisson (LNP) model of visual encoding: (a) the linear–nonlinear–Poisson cascade. The first-order kernel was computed from the random binary sequence (b) and the corresponding firing rate (in 7.7-ms bins) of an X cell in the LGN (c). The temporal kernel at the center of the receptive field is shown in (d); the band represents ±2 SD around estimate. The actual (positive) and predicted (negative) firing rates (in 40-ms bins) from the full spatiotemporal receptive field model are shown in (e). (Adapted with permission from Stanley, 2002)

details of experimental methods. A segment of the m-sequence stimulus (at 128 Hz) at the center pixel is shown in Figure 11.3b, and the corresponding firing rate (in 7.7-ms bins) is shown in Figure 11.3c. The kernel estimate for the center pixel computed over the entire trial is shown in Figure 11.3d. The band represents two standard deviations around the estimator (Stanley, 2002). The uncertainty in the estimation (due to noise and unmodeled dynamics) is useful in comparing the encoding properties in different physiological states. Using the complete spatiotemporal kernel, the firing rate of the cell is predicted in Figure 11.3e; the actual firing rate is shown in the dark shaded (positive) region, whereas the predicted firing rate is shown in the gray shaded region, reflected about the horizontal axis for comparison. The cascade of the linear system with the static nonlinearity is a good predictor of the neuronal response for this relatively linear cell, under rather rigid conditions of stationary input statistics.

11.4.2. ADAPTIVE ESTIMATION

One of the major assumptions of the reverse-correlation technique is that the dynamics of the underlying functional mechanisms in the visual pathway are time-invariant, resulting in spatiotemporal receptive field properties that remain unchanged with time. As early as the retina, however, adaptation mechanisms act on a continuum of time scales to adjust encoding dynamics in response to changes in the visual environment (Enroth-Cugell and Shapley, 1973; Shapley and Victor, 1978). Adaptation mechanisms have also been identified in the lateral geniculate nucleus (LGN) (Ohzawa et al., 1985) and cortical area V1 (Albrecht et al., 1984). An adaptive approach for the estimation of encoding properties from *single* trials can be utilized to specifically address time-varying neural dynamics (Brown et al., 2001; Stanley, 2002; Lesica et al., 2003). A recursive estimate of the kernel at time t is posed as the least-squares estimate based on information up to time t (Ljung, 1987). In this context, the recursive STRF estimation can be reformulated as (Stanley, 2002)

$$\hat{\theta}_t = \arg \min_\theta \sum_{k=L}^{t} \lambda^{t-k}(r(k) - f(\theta^T \varphi(k)))^2$$

where $\lambda \in [0, 1]$ is a weighting parameter, and the subscript t denotes that the estimate is a function of time. The estimate of the kernel at time t can be written as the estimate at time $t - 1$ plus a correction based on new information arriving at time t:

$$\hat{\theta}_t = \hat{\theta}_{t-1} + \text{update} \tag{11.8}$$

where the update depends on the input–output covariance structure (Ljung and Söderström, 1983). For the explicit formulation for this example, see Stanley (2002). This estimate downweights past information in an exponential manner as controlled by the size of λ, which is often called the "forgetting" factor. As the encoding properties vary over time, the input/output covariance changes properties, which is accounted for by a single weighting parameter, telescoping backwards in time; the result is an exponential down-weighting of past information.

X cells in the cat LGN were stimulated from rest with the spatiotemporal m-sequence at 128 Hz and 100% contrast over several minutes, inducing an adaptation to the sudden increase in contrast (or variance). Figure 11.4 illustrates typical results obtained using the adaptive estimation approach. For this ON cell, the kernel at the center of the RF exhibits a clear decrease in response magnitude over the first 30 sec of the trial. The frequency response at the beginning and end of the trial is shown in Figure 11.4b. In addition to changes in magnitude, the kernel shows a sharpening in bandwidth over the same time course, where the bandwidth is the range of frequencies for which the transfer function magnitude was greater than 50% of the peak value. Figures 11.4c through f illustrate the spatial RF map for an OFF cell at the peak in temporal response at 8-s intervals over the stimulus trial. The RF map is normalized by the peak of the center pixel at each time slice. The extent of the spatial RF in Figure 11.4c–f appears to increase because the magnitude of the center pixel shows a greater decrease than that of the off-center pixels over the course the trial. The point to emphasize here is that these changes in encoding dynamics would not be discovered through traditional reverse-correlation techniques that assume a fixed relationship between stimulus and response over the trial.

Time-varying encoding properties are a ubiquitous characteristic of all sensory pathways. Adaptation has been studied for some time and is known to dramatically affect the

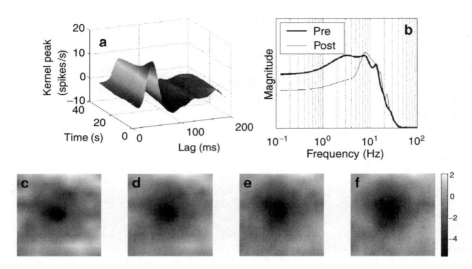

FIGURE 11.4. Evolution of adaptation in the visual pathway. Continuous tracking of adaptation in LGN X cells. (a) The temporal kernel between stimulus and firing rate for the center of the receptive field is plotted as a function of the time since stimulus onset (seconds) for an ON cell in the LGN. (b) The magnitude of the transfer function of the system at the beginning (thick) and end (thin) of the trial. Note that the magnitudes are normalized to emphasize the change in bandwidth. (c–f) Spatial RF of an OFF cell at peak in temporal kernel at 8, 16, 24, and 32 s after stimulus onset (region is approximately 1.8° of visual space). Dark represents minimum values, bright maximum values. The RF map is normalized by the peak of the center pixel at each time slice. (Adapted with permission from Lesica *et al.*, 2003)

temporal and spatial dynamics, whereas anecdotal evidence of modulatory effects from other brain regions on encoding properties has been reported for a number of stages in the visual pathway. However, it is not yet known what these phenomena imply for the coding strategies of the sensory pathway as a whole. The approach presented here is a first step toward capturing the evolution of spatiotemporal receptive field properties over a range of time scales, so that the pathway may be better understood in the context of the continually changing natural environment.

11.5. NONLINEAR ENCODING IN THE SOMATOSENSORY PATHWAY

The previous example focused on encoding mechanisms that are quasi-linear in their behavior. In many cases, however, neuronal encoding can be strongly nonlinear, and thus not well explained through linear methods. An example of such nonlinear encoding in the rat somatosensory pathway is presented in this section. Rats and other rodents have arrays of facial whiskers (vibrissae) that are vital for survival; neonates deprived of their vibrissa exhibit severely impaired behavioral development (Carvell and Simons, 1996). Rats have also been shown to be able, by actively palpating objects, to discriminate between very similarly textured surfaces based on vibrissa exploration alone (Guic-Robles *et al.*, 1989 Carvell and Simons, 1990), illustrating the exquisite sensitivity of this sensory modality.

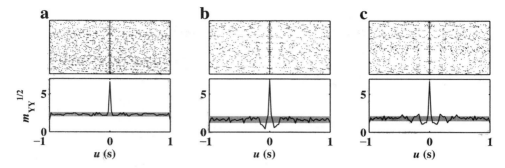

FIGURE 11.5. Correlation structure of spontaneous activity in the barrel cortex. (a–c) For typical cells from three classes of observed activity, raster of spikes surrounding each spike of the trial (top), and the square-root auto-intensity function (bottom) for the spontaneous activity, with the null band of an uncorrelated process shown with the gray band ($h = 25$ ms). The horizontal axis represents time relative to individual spikes. (Adapted with permission from Stanley and Webber, 2003)

In general, neurons in the vibrissa-related region of somatosensory cortex (tradition-ally referred to as the *barrel cortex*) do not have high spontaneous firing rates (Brumberg *et al.*, 1996). Nonetheless, the statistical properties of the spontaneous activity can reveal much about the underlying functional properties of the cell. Suppose that we denote the neuronal spike train as a discrete process y. The auto-intensity function $m_{yy}(u)$, as defined in Eq. (11.4), for the spontaneous activity of three characteristic cell types is shown in Figure 11.5. The top panels of Figures 11.5 a–c represent the raster of spikes surrounding each spike of the trial for each cell, and the bottom panels represent the corresponding square-root auto-intensity function (Stanley and Webber, 2003). It should be noted that these raster plots are different from traditional raster plots in that each row does not repre-sent a separate trial, but instead represents a collection of the spikes that occur relative to a different individual spike *within* the same trial. The gray bar represents a 95% confidence interval around the mean level on an uncorrelated process. The first cell, shown in Figure 11.5a, exhibits an intensity that is statistically different from the mean only at zero lag. This correlation structure is shared by the (memoryless) homogeneous Poisson process, which also exhibits an auto-intensity that is equal to the mean rate for nonzero lags and has a sharp positive peak at zero lag. The second cell, shown in Figure 11.5b, exhibits suppressive lobes for lags between 50 and 150 ms that extend out of the confidence band. The third cell, shown in Figure 11.5c, exhibits a suppressive lobe for lags between 50 and 150 ms, an excitatory lobe between 170 and 200 ms, and an additional suppressive lobe from 270 to 310 ms. The postexcitatory suppression revealed by the auto-intensity measure has significant bearing on how the cell would respond to exogenous stimulation at the periphery. In particular, punctate stimulation at the periphery induces postexcitatory suppression that influences the response to a subsequent stimulus in a strongly nonlinear manner.

11.5.1. THE IMPULSE RESPONSE AND NONLINEAR ENCODING

When a whisker is mechanically deflected with a step input to mimic contact with an object, the barrel neurons typically respond with a transient excitatory response after a short

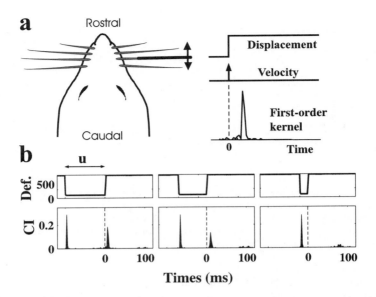

FIGURE 11.6. Higher-order dynamics in tactile encoding. (a) Whisker deflection paradigm. A step input in deflection is an impulse in velocity, producing a transient cortical response $r(t)$, which is equivalent to the cross-intensity (CI) function or first-order kernel. (b) Cross-intensity functions (spikes/ms) for caudal followed by rostral stimulation at latencies of 100, 70, and 20 ms.

latency of 5 to 15 ms, as shown in Figure 11.6a. It has been shown that the neurons in the barrel cortex are primarily sensitive to velocity, rather than displacement. The stimulus can therefore be thought of as an impulse in velocity, and the neural activity is then the impulse response:

$$r(t) = \int_\tau g_1(\tau)s(t - \tau)\,\mathrm{d}\tau = \int_\tau g_1(\tau)\delta(t - \tau)\,\mathrm{d}\tau = g_1(t)$$

In this case, the response is the first-order kernel. Note that the first-order kernel shown in Figure 11.6a was estimated from multiple repetitions of the velocity impulse, and averaging across the trials. This is equivalent to the cross-intensity measure m_{sy}, previously defined in Eq. (11.5).

What is not clearly exhibited is the known postexcitatory suppression that was observed in the auto-intensity measures in Figures 11.5b and c. Temporal interactions between paired tactile stimuli have been previously utilized to infer underlying levels of postexcitatory suppression (Simons, 1985; Kyriazi et al., 1994; Fanselow and Nicolelis, 1999; Stanley and Webber, 2003).The resulting behavior of a typical cortical neuron is shown in Figure 11.6b. If the time interval u between the first and second deflection is long (100 ms), there is still a robust response to the second deflection (left panels). However, when the interval is decreased, the response to the second stimulus is attenuated (middle panels), until eventually disappearing altogether for very short latencies (right panels). The response following the second stimulus is different from what would be predicted by the superposition of the responses to the two stimuli, and thus reflects second-order dynamics:

$$r(t) = g_1(t) + g_2(t, t - u) \qquad t > 0$$

where the first-order effect associated with the stimulus at $t - u$ is negligible because of the transient nature of the response. Alternately, the relationship can be expressed in terms of the first-order conditioned kernel (Klein, 1992):

$$r(t) = g_1(t, u) \tag{11.9}$$

which depends on both the time t and the time interval between the first and second stimulus u, where $g_1(t, u)$ becomes smaller as u is decreased. This is a means by which the second-order kernel can be embedded in the first-order representation.

The strongly nonlinear dynamics are important in behavioral contexts, where the animal is using its vibrissae to palpate object surfaces, inducing spatial and temporal patterns of tactile input. The approach presented here can be extended to more complex patterns or sequences that may be encountered in the animal's natural environment.

11.6. NEURAL CONTROL OF CARDIAC FUNCTION

The identification problems thus far have been discussed in terms of time domain estimation, and have arisen from phenomenology related to sensory pathways. An example is presented here in which frequency domain techniques are utilized to identify the dynamics of neural influence on the variability of cardiac rate. In healthy humans, the sino-atrial (SA) node acts as the pacemaker for the heart. Through an upward drift in electrical potential, these cells spontaneously depolarize to a threshold potential, at which point they rapidly depolarize, or "fire" as a group. This event is followed by a reset, which marks the start of a new cycle. Firing initiates the spread of electrical activity throughout the heart, followed by contraction of cardiac muscle. The R-wave of an electrocardiogram (ECG), which is readily localized in time, provides a convenient marker from which periods between SA node firings can be extracted. The spontaneous depolarization of SA nodal cells has an intrinsic rate that is regulated by direct input from the sympathetic and parasympathetic branches of the peripheral, or autonomic nervous system (ANS). Neural impulses arriving from the sympathetic branch tend to increase the mean HR, whereas impulses from the parasympathetic branch have the opposite effect. By this means, the ANS regulates HR. Generally speaking, heartbeats do not occur with exact regularity, but rather exhibit random variations around a mean rate. As a result, the beat-to-beat intervals measured from the ECG have a stochastic component, which may be termed "heart rate variability" (HRV) (Berger *et al.*, 1986). Interestingly, normal individuals show much greater HRV than those whose ANS function is attenuated due to aging, disease states, or pharmacologic blockade (Akselrod *et al.*, 1981, 1985, Appel *et al.*, 1989; Malliani *et al.*, 1991). It has been found that HRV, while random, exhibits a correlation structure in time, which can be associated with various periodicities of modulation of HR. For example, activity of the higher respiratory centers has been shown to modulate HR at the respiratory frequency via the parasympathetic branch of the ANS (Akselrod *et al.*, 1981; Liao *et al.*, 1995; Stanley *et al.*, 1996, 1997), which will be the focus of discussion here. This phenomenon is known as respiratory sinus arrhythmia (RSA). Continuous measures of lung volume serve as a noninvasive probe of central rhythm activity, and are well correlated with heart rate variability in the related frequency bands.

11.6.1. INPUT-DRIVEN THRESHOLD MODEL

In contrast to the nonparametric models previously described, we turn our attention to a mechanistic model that is a combination of parametric and nonparametric elements, as illustrated in Figure 11.7a. Note that the spectra are normalized to have a maximum of 1, and are therefore unitless, as are the transfer function magnitudes. In the model, the occurrence of R-waves is denoted by "spikes" at times $\tau_0, \tau_1, \tau_2, \ldots$, and the time interval between R-waves is $T_k = \tau_k - \tau_{k-1}$. These events arise from an integrate-and-fire mechanism that functionally represents the spatial and temporal summation of inputs at the SA node. When the output of the integrator reaches a threshold of 1, an event (heartbeat) occurs, and the integrator is reset to 0. The constant β represents the mean rate at which the SA node depolarizes. The additional input to the integrator consists of filtered versions of

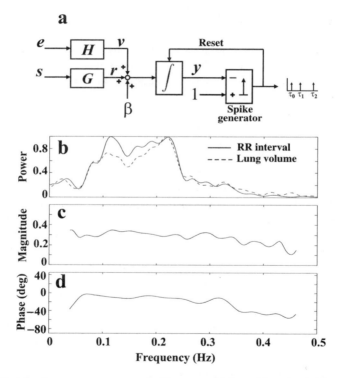

FIGURE 11.7. Peripheral nervous system control of heart rate. (a) Input driven threshold model. (b) Interval spectrum (solid) and lung volume spectrum (dotted), both normalized for comparison. (c, d) Estimated transfer function magnitude and phase, respectively. (Adapted with permission (© 2000 IEEE) from Stanley et al., 2000)

the respiratory signal s and a noise term e. For the time interval between two events, τ_{k-1} and τ_k, the integration reaches the unity threshold:

$$1 = \int_{\tau_{k-1}}^{\tau_k} [v(t) + r(t) + \beta]\, dt = \int_{\tau_{k-1}}^{\tau_k} [v(t) + r(t)]\, dt + \beta T_k = \Phi_k + \beta T_k \quad (11.10)$$

where r is a linearly filtered version of the zero-mean respiratory input s, and v is the output of a linear filter whose input is Gaussian white noise. Φ_k represents the sum of the noise activity and the respiratory influence, integrated over the interval from τ_{k-1} to τ_k. The input s is restricted to be zero mean, and therefore $E\{\Phi_k\}$ will also be 0, producing a mean R–R interval of $\mu_T \approx 1/\beta$. β in this case is representative of a combination of the intrinsic firing rate of the SA node and the mean levels of parasympathetic and sympathetic activity. It can be shown (Stanley et al., 2000) that the relationship between the spectrum of the respiratory input s and the sequence of R–R intervals can be approximated as

$$S_{TT}(\omega) \approx \mu_T^3 \{|H(\omega)|^2 + |G(\omega)|^2 S_{ss}(\omega)\} \quad (11.11)$$

Given the observed sequence of R–R intervals, the measured respiratory input s, and the previously identified noise model [see Stanley et al. (2000) for a discussion of noise model estimation], the respiratory-related transfer function magnitude, $|G(\omega)|$ can be estimated. The phase angle $\angle G(\omega)$ can also be reconstructed using additional arguments (Stanley et al., 2000).

Figure 11.7b shows the lung volume and R–R interval spectra for a typical data set from a previously published study (Stanley et al., 1996), in which the subject was breathing at randomly spaced intervals over a relatively broad frequency band. The corresponding transfer function magnitude and phase estimates over frequency bands of high coherence are shown in Figure 11.7c and d. The transfer function magnitude tends to be rather flat over frequency bands of sufficient excitation. The phase suggests that there are increased delays between central rhythm activity and the influence on heart rate variability at higher temporal frequencies.

This example provides a hybrid framework that is mechanistic, yet nonparametric, in contrast to the previous examples. The implementation in the frequency domain provides computational efficiency, but is also a relatively natural representation due to the frequency-band specific influence of the respiratory activity.

11.7. SUMMARY

In summary, an overview of the perspectives and techniques related to system identification have been presented and utilized within the context of several well-defined problems in the nervous system. Importantly, several of the problems presented here involved combinations of nonparametric and parametric representations that reflect an important point to emphasize. In the analysis of complex systems, where little is known about the underlying dynamics, nonparametric techniques provide a relatively unconstrained means by which to characterize and categorize the properties of the system. However, as more is learned about the underlying dynamics, particular nonlinear features of the relationships can be exploited to greatly simplify the characterizations, often into linear-like descriptions, for which many computational tools exist. Finally, the motivation for the application of system identification within the nervous system has been posed here as that of basic science, or exploration. However, as in engineering, the problem of system identification in neuroscience eventually will be oriented toward control applications, in the context of engineered prosthetics designed to enhance neural function impaired due to trauma or disease.

ACKNOWLEDGMENTS

The author would like to thank Nicholas A. Lesica and Roxanna M. Webber for helpful comments in the preparation of this manuscript.

REFERENCES

Akselrod, S., Gordon, D., Madwed, J., Snidman, N., Shannon, D., and Cohen, R., 1985, Hemodynamic regulation: Investigation by spectral analysis, *Am. J. Physiol.* **249**:H867–H875.

Akselrod, S., Gordon, D., Ubel, F., Shannon, D., Bargar, A., and Cohen, R., 1981, Power spectrum analysis of heart rate fluctuation: A quantitative probe of beat-to-beat cardiovascular control, *Science* **213**(10): 220–222.

Albrecht, D. G., Farrar, S. B., and Hamilton, D. B., 1984, Spatial contrast adaptation characteristics of neurones recorded in the cat's visual cortex, *J. Physiol.* **347**:713–739.

Appel, M., Berger, R., Saul, J., Smith, J., and Cohen, R., 1989, Beat to beat variability in cardiovascular variables: Noise or music? *J. Am. Coll. Cardiol.* **14**:1139–1148.

Berger, R., Akselrod, S., Gordon, D., and Cohen, R., 1986, An efficient algorithm for spectral analysis of heart rate variability, *IEEE Trans. Biomed. Eng.* **33**:900–904.

Box, G., and Jenkins, G., 1976, *Time Series Analysis: Forecasting and Control*, rev. ed., Holden-Day. San Francisco, CA.

Brillinger, D., 1981, *Time Series Analysis*, expanded ed., McGraw-Hill, New York.

Brillinger, D. R., 1992, Nerve cell spike train data analysis: A progression of technique, *J. Am. Stat. Soc.*, **87**(418):260–271.

Brillinger, D. R., Bryant, H. L., and Segundo, J. P., 1976, Identification of synaptic interactions, *Biol. Cybernetics.* **22**:213–228.

Brown, E. N., Nguyen, D. P., Frank, L. M., Wilson, M. A., and Solo, V., 2001, An analysis of neural receptive field plasticity by point process adaptive filtering, *Proc. Natl. Acad. Sci. USA* **98**:12261–12266.

Brumberg, J. C., Pinto, D. J., and Simons, D. J., 1996, Spatial gradients and inhibitory summation in the rat whisker barrel system, *J. Neurophysiol.* **76**:130–140.

Carvell, G. E., and Simons, D. J., 1990, Biometric analysis of vibrissal tactile discrimination in the rat, *J. Neurosci.* **10**:2638–2648.

Carvell, G. E., and Simons, D. J., 1996, Abnormal tactile experience early in life disrupts active touch, *J. Neurosci.* **15**:2750–2757.

Chatfield, C., 1989, *The Analysis of Time Series: An Introduction*, Chapman and Hall, London.

Chichilnisky, E. J., 2001, A simple white noise analysis of neuronal light responses, *Network* **12**:199–213.

Cramér, H., 1946, *Mathematical Methods of Statistics*, Princeton University Press, New Jersey.

Dayan, P., and Abbott, L. F., 2001, *Theoretical Neuroscience*, MIT Press, Cambridge.

DeBoer, E., and Kuyper, P., 1968, Triggered correlation, *IEEE Trans. Biomed. Eng.* **15**:169–179.

Eichenbaum, H., and Davis, J., 1998, *Neuronal Ensembles: Strategies for Recording and Decoding*, John Wiley & Sons, New York.

Enroth-Cugell, C., and Shapley, R., 1973, Adaptation and dynamics of cat retinal ganglion cells, *J. Physiol.* **233**:271–309.

Fanselow, E. E., and Nicolelis, M. A. L., 1999, Behavioral modulation of tactile response in the rat somatosensory system, *J. Neurosci.* **19**:7603–7616.

Greblicki, W., 1992, Nonparametric identification of wiener systems. *IEEE Trans. Inform. Theory* **38**:1487–1493.

Guic-Robles, E., Valdivieso, C., and Guajardo, G., 1989, Rats can learn a roughness discrimination using only their vibrissal system, *Behav. Brain Res.* **31**:285–289.

Hubel, D., 1995, *Eye, Brain, and Vision*, Scientific American Library, New York.

Hunter, I. W., and Korenberg, M. J., 1986, The identification of nonlinear biological systems: wiener and hammerstein cascade models, *Biol. Cybernet.* **55**:135–144.

Jenkins, G., and Watts, D., 1968, *Spectral Analysis and Its Applications*, Holden-Day, San Fransisco, CA.

Jones, J. P., and Palmer, L. A., 1987, The two-dimensional spatial structure of simple receptive fields in cat striate cortex, *J. Neurophysiol.* **58**:1187–1211.

Klein, S. A., 1992, Optimizing the estimation of nonlinear kernels, in: *Nonlinear Vision*, R. B. Pinter and B. Nabet, ed., CRC Press, Boca Raton, FL, pp. 109–170.

Korenberg, M. J., and Hunter, I. W., 1999, Two methods for identifying Wiener cascades having noninvertible static nonlinearities, *Ann. Biomed. Eng.* **27**:793–804.

Kyriazi, H. T., Carvell, G. E., and Simons, D. J., 1994, OFF response transformations in the whisker/barrel system, *J. Neurophysiol.* **172**(1):392–401.

Lee, Y. W., and Schetzen, M., 1965, Measurements of the Wiener kernels of nonlinear systems by cross-correlation, *Int. J. Control* **2**:237–254.

Lesica, N. A., Boloori, A. S., and Stanley, G. B., 2003, Adaptive encoding in the visual pathway, *Network Comput. Neural Syst.* **14**:119–135.

Liao, D., Barnes, R., Chambless, L., Simpson, R., Sorlie, P., and Heiss, G., 1995, Age, race, and sex differences in autonomic cardiac function measured by spectral analysis of heart rate variability—the ARIC study, *Am. J. of Cardiol.* **76**(12):906–912.

Ljung, L., 1987, *System Identification: Theory For the User*, Prentice-Hall, New Jersey.

Ljung, L., and Söderström, T., 1983, *Theory and Practice of Recursive Identification*, MIT Press, Cambridge, MA.

Malliani, A., Pagani, M., Lombardi, F., and Cerutti, S., 1991, Cardiovascular neural regulation explored in the frequency domain, *Circulation* **84**:482–492.

Mao, B. Q., MacLeish, P. R., and Victor, J. D., 1998, The intrinsic dynamics of retinal bipolar cells isolated from tiger salamander, *Visual Neurosci.* **15**:425–438.

Marmarelis, P. Z., and Marmarelis, V. Z., 1978, *Analysis of Physiological Systems: The White-Noise Approach*, Plenum Press, New York.

Marmarelis, P., and Naka, K. I., 1973a, Non-linear analysis and synthesis of receptive field responses in the catfish retina. I. Horizontal cell-ganglion chains, *J. Neurophysiol.* **36**:605–618.

Marmarelis, P., and Naka, K. I., 1973b, Non-linear analysis and synthesis of receptive feld responses in the catfish retina. II. One-input white-noise analysis, *J. Neurophysiol.* **36**:619–633.

Meister, M., and Berry, M., 1999, The neural code of the retina, *Neuron* **22**:435–450.

Movshon, J. A., Thompson, I. D., and Tolhurst, D. J., 1978, Spatial summation in the receptive fields of simple cells in the cat's striate cortex, *J. Physiol.* **283**:53–77.

Ohzawa, I., Sclar, G., and Freeman, R. D., 1985, Contrast gain control in the cat's visual system, *J. Neurophysiol.* **54**:651–667.

Papoulis, A., 1984, *Probability, Random Variables, and Stochastic Processes*, 2nd ed., McGraw-Hill, New York.

Paulin, M. G., 1993, A method for constructing data-based models of spiking neurons using a dynamic linear-static nonlinear cascade, *Biol. Cybernet.* **69**:67–76.

Perkel, D., Gerstein, G., and Moore, G., 1967, Neuronal spike trains and stochastic point processes. I. The single spike train, *Biophys. J.* **7**(4):391–418.

Press, W. H., Teukolsky, S. A., Vetterling, W. T., and Flannery, B. P., 1992, *Numerical Recipes in C: The Art of Scientific Computing*, Cambridge University Press, Cambridge.

Rao, C. R., 1973, *Linear Statistical Inference and Its Applications*, John Wiley & Sons, New York.

Reid, R. C., Soodak, R. E., and Shapley, R. M., 1991, Directional selectivity and spatiotemporal structure of receptive fields of simple cells in cat striate cortex, *J. Neurophysiol.* **66**:505–529.

Ringach, D., Sapiro, G., and Shapley, R., 1997, A subspace reverse-correlation technique for the study of visual neurons, *Vision Res.* **37**:2455–2464.

Shapley, R., and Victor, J. D., 1978, The effect of contrast on the transfer properties of cat retinal ganglion cells, *J. Physiol.* **285**:275–298.

Simons, D. J., 1985, Temporal and spatial integration in the rat SI vibrissa cortex, *J. Neurophysiol.* **54**:615–635.

Söderström, T., and Stoica, P., 1987, *System Identification*, Prentice-Hall, New Jersey.

Stanley, G. B., 2002, Adaptive spatiotemporal receptive field estimation in the visual pathway, *Neur. Comput.* **14**:2925–2946.

Stanley, G. B., and Webber, R. M., 2003, A point process analysis of sensory encoding, *J. Comput. Neurosci.*, in press.

Stanley, G., Li, F., and Dan, Y., 1999, Reconstruction of natural scenes from ensemble responses in the lateral geniculate nucleus, *J. Neurosci.* **19**(18):8036–8042.

Stanley, G. B., Poolla, K., and Siegel, R. A., 2000, Threshold modeling of autonomic control of heart rate variability, *IEEE Trans. Biomed. Eng.* **49**(9):1147–1153.

Stanley, G., Verotta, D., Craft, N., Siegel, R., and Schwartz, J., 1996, Age and autonomic effects on interrelationships between lung volume and heart rate, *Am. J. Physiol.* **270**:H1833–H1840.

Stanley, G., Verotta, D., Craft, N., Siegel, R., and Schwartz, J., 1997, Age effects on interrelationships between lung volume and heart rate during standing, *Am. J. Physiol.* **273**:H2128–H2134.

Tolhurst, D. J., and Dean, A. F., 1990, The effects of contrast on the linearity of the spatial summation of simple cells in the cat's striate cortex, *Exp. Brain Res.* **79**:582–588.

Volterra, V., 1959, *Theory of Functionals and of Integral and Integro-Differential Equations*, Dover Publications, New York.

12

SEIZURE PREDICTION IN EPILEPSY

Wim van Drongelen,* Hyong C. Lee, and Kurt E. Hecox

Department of Pediatrics, The University of Chicago, Chicago, Ilinois

12.1. INTRODUCTION

Epilepsy is the second most common serious neurological disease after stroke. This disease affects approximately 50 million people worldwide and 50–70 cases per 100,000 in the developed countries. In approximately 40% of patients with so-called partial seizures, current medications are unable to control their symptoms. One of the most devastating aspects of epilepsy is the anxiety and apprehension associated with the inability to predict when a seizure will occur. The inability to predict the time of seizure onset also implies the need for continuous medication therapy with the associated continuous side effects. For a number of years investigators and commercial-interest groups have sought methods for the early detection and anticipation of seizures so that "discontinuous" therapies could be introduced (e.g., Milton and Jung, 2003). At the heart of most predictive efforts is the description and analysis of the cerebral electrical activity reflected in the electroencephalogram (EEG). The brain electrical activity of a patient with epilepsy shows abnormal and often rhythmic discharges during the seizure. This activity pattern is called an electrographic seizure. Between such electrographic seizures, short discharges (spikes) are also frequently observed in the EEG of these patients. Identification of these activity patterns in clinical practice has typically been a subjective process. The introduction of computer-based instrumentation and analysis to the field of electroencephalography made evaluation of automated spike and seizure detection techniques possible (e.g., Gotman and Gloor, 1976; Gotman et al., 1979; Gotman, 1982). During the 1980s, the EEG during seizure activity was characterized using more complex measures such as those derived from chaos theory (e.g., Babloyantz and Destexhe, 1986; van Erp, 1988). There were a number of "early" reports of the successful application of frequency-domain template analyses and auto-regressive models to the problem of seizure prediction (e.g., Viglione and Walsh, 1975; Rogowski et al., 1981). Unfortunately, the performance of these methods was either difficult to evaluate, or the average anticipation time was only a few seconds. An interval of several seconds could fall within the uncertainty bounds of the clinical judgment against which the predictions were

* Address for correspondence: The University of Chicago, Department of Pediatrics, MC 3055, 5841 S Maryland, Chicago, Ilinois 60637; e-mail: wvandron@peds.bsd.uchicago.edu.

compared. Since the 1990s more successful attempts have been made to apply techniques from nonlinear dynamics to characterize, detect, and anticipate imminent seizure activity in electrophysiological recordings (e.g., Iasemidis *et al.*, 1990; Casdagli *et al.*, 1996; Elger and Lehnertz, 1998; Le Van Quyen *et al.*, 1998, 1999; Lehnertz and Elger, 1998, 1999; Hively *et al.*, 1999; Andrzejak *et al.*, 2001b; Jerger *et al.*, 2001; Savit *et al.*, 2001; van Drongelen *et al.*, 2003a). Although these studies suggest the feasibility of seizure detection and prediction, it is also clear that the applied methodology has limitations. One of these limitations is that we do not yet know the underlying processes against which prediction algorithms should be validated. In spite of a vast amount of electrophysiological studies in the field of epilepsy, the exact mechanisms responsible for initiating or stopping seizures are unknown, meaning there is no generally accepted "gold-standard" for the detection of the preseizure state. Therefore, prediction algorithms explore electrophysiological data sets that include seizures, and the seizure-prediction capability is assigned to a particular algorithm *a posteriori*, if it recognizes a change in the electrical activity prior to seizure onset.

What determines the predictability of processes? This is an important question in science, since the purpose of many experiments is to search for a cause–effect relationship: a relationship where the past and present determine the future state of an experimental system. Although a system such as a swinging pendulum behaves predictably, developments in the stock markets or the weather are not associated with high levels of predictability. One might conclude, therefore, that simple systems or systems with deterministic processes are predictable, whereas involvement of more complex processes puts that predictability at risk. *This conclusion, however, would be incorrect!* It can be shown that even simple and deterministic processes, for instance a time series where each point x_t depends on the previous obeying the logistic equation

$$x_t = a(1 - x_{t-1})x_{t-1} \tag{12.1}$$

can exhibit behavior that is stable, oscillatory, or very poorly predictable. Examples for different values of parameter a in Eq. (12.1) are shown in Figure 12.1. The number of final states are associated with the value of variable a, and vary between one (Figure 12.1A) and many (Figure 12.1C). If we study Eq. (12.1) for different values of a, we can produce a so-called final state diagram (Figure 12.2). This shows

   —that the behavior of the logistic equation converges to a single value for $a < 3$,
   —that there is a stable periodic behavior with two values for $3 < a < 3.4495$, and
   —that subsequently there are four, then eight values, etc.

A description of the final state, or bifurcation diagram, in Figure 12.2 shows a transition from stable to chaotic behavior for $a > 3.569$. This last transition, where discrete steps to higher frequencies evolve into an unpredictable regime is called the period-doubling route to *a chaotic state*. Here we have a simple deterministic system (meaning that there is a single future value x_{t+1} associated with the present state x_t) showing chaotic behavior. Furthermore, the logistic equation is not an exceptional case: many more examples can be found. The seminal example, a simplified and deterministic model of a weather system, showed similarly dramatic unpredictability (Lorenz, 1963). We can compare these unpredictable processes to tossing a coin, rolling a dice, or drawing a numbered lotto ball, in that they all show random behavior that can be characterized by measuring the probability of a certain

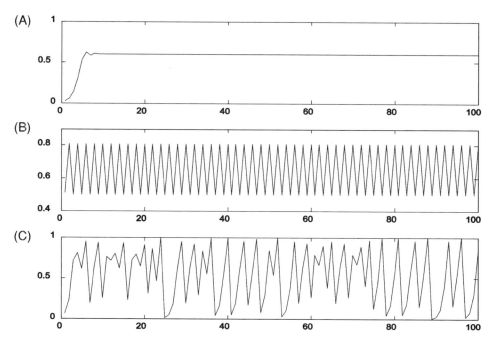

FIGURE 12.1. Three examples of a time series created with the logistic equation (Eq. (12.1)): (A) the time series converges to a single value for $a = 2.50$; (B) for $a = 3.24$ there is oscillatory behavior between two states; (C) chaos at $a = 4$.

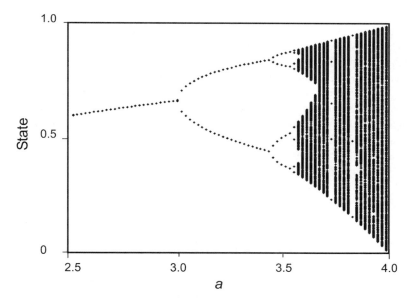

FIGURE 12.2. One of the icons of chaos: the final state diagram showing the period-doubling route to chaos. Final states plotted against the value of a in the logistic equation (Eq. (12.1)). The logistic equation (a quadratic iterator) transitions to oscillatory behavior at the bifurcation $a = 3$. For $a > 3.569\ldots$ the system transitions to chaotic behavior. Interestingly Feigenbaum (1983) discovered that the ratio of two successive ranges over which the period doubles, is a constant universally encountered in the period-doubling route to chaos (Feigenbaum's number: $4.6692\ldots$). A Matlab script to create the final state diagram can be found in Appendix 2.

outcome. In principle, if one precisely knew all the positions and mechanical parameters of the elements in a lotto drawing, one would be able to calculate the outcome. In spite of this "in-principle predictability," randomness seems inherently associated with these types of deterministic processes, and from these examples one can conclude that even a simple process can behave unpredictably. On the other hand, some complex phenomena such as tides that depend on many other processes (position of the moon, the wind, details in the coastline, etc.) can be a fairly predictable process again. From the examples above, we learn that the level of complexity in a time series does not necessarily correspond with the level of complexity of the underlying process.

Let us generalize and reconsider the components involved in prediction. At first sight, it seems that predicting the future of a process requires an algorithm or rule to generate a future state from knowledge of the present and the past (such as the logistic equation). However, there are problems both with knowing the present or past and with the computational aspect of predicting the future. There is uncertainty that prevents us from knowing all aspects of the (present) state of a dynamic system. In addition, any knowledge or computation is associated with a degree of precision, and precision limits the exact knowledge of the initial and subsequent states of an evolving process. This seems a trivial problem, but it is fairly serious, because it appears that in systems with *nonlinear dynamics*, minute perturbations (of the order of magnitude of a rounding error of a computer or even smaller) can be associated with a huge difference in the predicted outcomes. This difference can grow disproportionately toward the same order of magnitude as the predicted values: i.e. the evolution and outcome of certain types of processes may depend critically on initial conditions. This dependence is sometimes referred to as the "butterfly effect": as was pointed out by Lorenz, a perturbation as small as the flap of the wings of a butterfly could make a difference in the development of a tornado. Of course, sensitivity to perturbations also exists in *linear systems*. However, the error in a linearly evolving process grows proportionally with the predicted values.

Chaos theory in mathematics deals with systems like those described in the examples above. *Aperiodic behavior*, *limited predictability*, and *sensitivity to initial conditions* in the dynamics of deterministic systems are all hallmarks of chaos. One of the prerequisites for systems to be able to demonstrate chaotic behavior is *nonlinearity*. As demonstrated with the example in Figures 12.1 and 12.2, nonlinear dynamics, even in very simple processes, can be responsible for a transformation from an equilibrium or orderly oscillatory behavior to chaos. Peitgen *et al.* (1992), Elbert *et al.* (1994), Kaplan and Glass (1995), and Kantz and Schreiber (1997) provide excellent introductions to nonlinear dynamics and chaos theory with numerous practical examples.

12.2. PROCESSES UNDERLYING THE ELECTROENCEPHALOGRAM

In most current studies, the basis of detection, anticipation, and prediction of seizures in epilepsy is the electroencephalogram (EEG) or the electrocorticogram (ECoG). Both of these signals reflect the electrical activity of the brain. The EEG is a signal that can be measured from electrodes placed at standard locations on the surface of the scalp (American Electroencephalographic Society, 1994). The ECoG is recorded with surgically implanted cortical electrodes; their positions vary with each patient because the brain areas with suspected pathology determine the locations. Both EEG and ECoG represent summed electrical activity of the underlying networks of nerve cells. The compound neural activity recorded by a surface electrode is assumed to be a *linear* summation of all the activity sources in the brain. Under the simplifying assumption that the brain behaves as a volume conductor with homogeneous conductivity, one can calculate the field potential from all the neurons as the sum of all currents (Nunez, 1981), each weighed by the inverse of the distance between the current source and the position at which the potential is measured. In other models, the EEG activity is modeled as a dipole representing electric activity of a group of cortical neurons. In case of a dipole, the decline of the potential is proportional to the inverse squared distance between dipole and measurement position.

Nonlinear processes are manifest both in intrinsic neuronal properties and in the coupling between nerve cells. One simplified model of the neuron consists of a network representing the cell's membrane-bound components: membrane resistance, membrane capacitance, and several potential sources due to different ion concentrations inside and outside the neuron. Hodgkin and Huxley (1952) were the first to describe this model based on their measurements of the squid giant axon, where they showed that the membrane resistance is modified by the membrane potential in a nonlinear fashion. Their model has been shown to apply to a wide variety of nerve cells and has been applied to study oscillatory processes in neuronal models (e.g., Traub and Llinas, 1979; Lytton and Thomas, 1999; van Drongelen *et al.*, 2002). Although this kind of detailed knowledge of neuronal function is available at the cellular level, the relationship between clinical recordings and single unit cellular physiology is far from understood. Because of the size of the clinical recording electrodes (on the order of mm^2 or cm^2), the EEG signal from a single electrode represents the average of more than 1,000,000 neurons. This indicates that clinical data is "blind" to details of processes at a small scale, though these small-scale processes are likely to play a role in the onset of seizure activity. On the other hand, single-cell or small-network data obtained from microelectrode studies is not easily related to more global multiunit or population activity. Despite the enormous challenge involved in relating activity "derived across levels," there may be good reason to persist in addressing the challenge. First, more refined measurements in the epileptic focus may elucidate some of the early seizure onset processes that so far have not been observed in the large-scale clinical electrophysiology. Second, data sets from large-scale recordings may contain hidden, useful information reflecting small-scale processes, and appropriate signal-processing tools may make these small-scale events visible.

12.3. ELECTROGRAPHIC SEIZURE ACTIVITY

Defining the onset of a seizure assumes a clear definition of what constitutes a seizure. Surprisingly, subtle interpretation issues still arise from this definitional perspective. The most widely accepted key elements of a clinical seizure are a change in observable behavior associated with a diminished adaptive response to the environmental input, and electrical abnormalities in the cortex. It is the combination of these elements and not the isolated elements that allow separation of seizures from movement disorders, fainting, migraine, sleep, and other nonepileptic paroxysmal events.

Another important term is the electrographic seizure, a much more restricted concept. In contrast to clinically defined seizures, electrographic seizures do *not* have to involve changes in observable behavior nor loss of adaptive skills, but are defined instead by paroxysmal abnormal cortical electrical activity. Clinical and electrographic elements do not generally appear at the same time. Most often, there is an electrographic onset followed by the appearance of behavioral changes associated with diminished responsiveness to external input. The time-shift between elements is usually greater than several seconds, but less than 1 min. Typically, we relate anticipation and prediction relative to the electrographic seizure onset. It should also be noted that the scalp EEG recordings do not always change during a seizure. This may occur, for example, in partial seizures when the area of the activated region is too small, when the location is distant (e.g., orbital–frontal), or when the electrophysiological changes are atypical.

Normal EEG is described in terms of the rhythms that occur "spontaneously" as well as events that interrupt this background. The most common EEG rhythms are the delta (δ: 0–4 Hz), theta (θ: 4–8 Hz), alpha (α: 8–12 Hz), and beta (β: 12–30 Hz) bands. Recently, there has been an increased interest in higher frequency components (γ, ω, ρ, σ, with rhythms up to about 1 kHz). EEG patterns are also variable across subjects, and show age and state dependence. In general, they may appear very complex, but during an epileptic seizure the pattern is often a rhythmic bursting activity (For an overview of electrographic seizure patterns, see Commission on Classification and Terminology of the International League Against Epilepsy, 1981; Engel, 2001). During a typical seizure a variety of electrical patterns may emerge with time. Patterns of low-voltage desynchronization, rhythmic high-voltage signals of varying frequency, generalized voltage attenuation, or high amplitude bursts can be observed. Four examples of different types of recordings during a seizure onset from three different patients are shown in Figure 12.3. An idealized diagram of one type of EEG activity around a seizure, commonly indicated as the electrographic seizure, is shown in Figure 12.4. The *ictal* period is the epoch between seizure onset and offset (in this simplified example we assume that the electrographic seizure activity and the clinical seizure coincide). A recent summary of terminology in epilepsy can be found in the commission report from the International League Against Epilepsy (2001). A set of associated terms based on the term *ictus* is currently in use to describe the different states in patients with epilepsy. The epoch in between seizures is defined as *interictal*, and during this period two hypothetical states occur: the *preictal* (prior to seizure onset) and *postictal* (after seizure offset). It is hypothesized that underlying processes during the preictal interval are essential for causing the seizure to start. This process leading to the seizure onset is also referred to as *ictogenesis*. Detecting ictogenesis can lead to anticipation of an imminent seizure; quantification of the ictogenesis may lead to a prediction of the seizure onset time. In most current studies there is detection of a preictal process that satisfies the anticipation criterion: i.e., an imminent seizure can be anticipated but the onset time is not predicted (Ebersole, 2001). When using "anticipation" in the following text, we imply both anticipation and prediction, unless stated differently. In Figure 12.4 we make a distinction between the alarm (2, Figure 12.4) and the "formal" onset of the preictal period (1, Figure 12.4): the former is determined by a variable exceeding a prespecified threshold, and the latter indicates the "true" onset of the physiologic process leading to the seizure.

The anatomic location of seizure onset is critical because the normal electrophysiological processes, the cellular relationships, the frequency of cell types, and the behavior

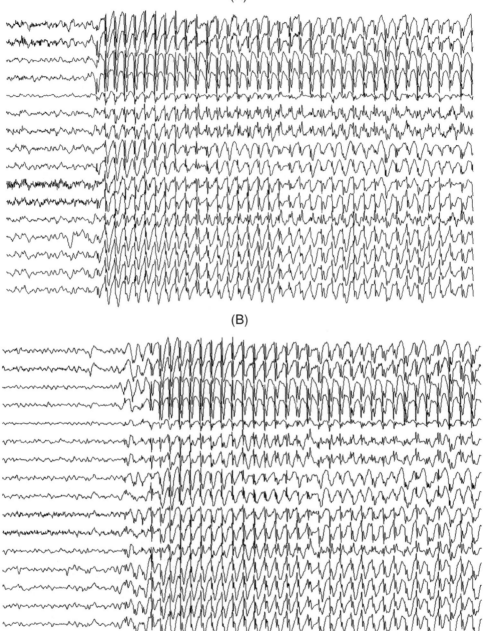

FIGURE 12.3. Examples of 15-s EEG/ECoG epochs around seizure onset. (A), (B) Two generalized seizures recorded from the scalp. Both (A) and (B) are from the same patient to show the stereotypical aspect of the sudden seizure onset. (C) Seizure onset from a patient with a mixed seizure disorder. Data was recorded from the surface of the cortex. Only few channels show involvement in seizure activity. (D) An example of a complex partial seizure recorded from the cortical surface. Initially the seizure activity can be observed in few electrodes; subsequently it propagates to a wider area.

(C)

(D)

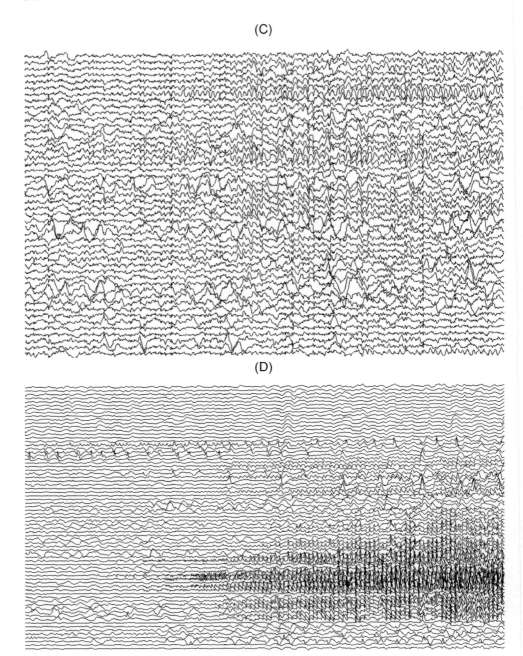

FIGURE 12.3. (*Continued*)

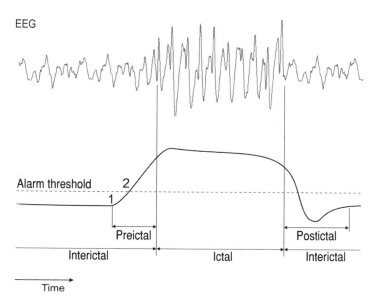

FIGURE 12.4. Schematic representation of the different stages involved in seizure anticipation and detection. The upper trace is the EEG and the lower trace depicts an idealized time series of an extracted metric. The interval between seizures is interictal. The preictal state is the hypothetical state in which processes leading to the seizure onset start (1). The anticipation epoch starts where the metric exceeds the alarm threshold (2). The epochs during the seizure and the recovery after the event are indicated by the ictal and post-ictal epochs respectively.

emerging as a consequence of the seizure all depend on their location in the central nervous system. The brain is usually subdivided into the brain stem, the cerebellum, and cerebrum. A large part of the cerebrum is the cortex, which is divided into left and right hemispheres, each consisting of four lobes (the frontal, temporal, parietal, and occipital lobes). An alternate subdivision of the cortex is into paleo-, archi-, and neocortex, thought to have developed during different stages in vertebrate evolution. In pediatric epilepsy, abnormal discharges can be measured in almost all areas of the brain but are often located in the neocortex. Seizure frequency in children can be high, i.e. several seizures per day or per hour. In most adults with epilepsy, the hippocampus (a component of the archicortex, located deep in the temporal lobe) is known to play a critical role.

12.4. TIME SERIES ANALYSIS AND APPLICATION IN EEG

In this section we present some of the methods that have been applied to anticipate seizures. In general, attempts have been made to extract metrics from the EEG that show behavior as depicted in the lower trace of Figure 12.4: that is, trends that occur prior to the onset of the electrographic event, signaling that a seizure is imminent. Work in this area has come from several groups worldwide exploring a variety of metrics. References describing a representative example from each of the groups can be found in Table 12.1.

TABLE 12.1. Examples of Seizure Anticipation Methods and the Associated Performance

Method	Reference	Anticipation Interval (min)
Linear methods		
Linear decomposition	Jerger *et al.*, 2001	1–5
Power	Litt *et al.*, 2001	>60
Variance	McSharry *et al.*, 2003	~5
Nonlinear methods		
Attractor dissimilarity	Hively *et al.*, 1999	30–60
Lyapunov exponent	Iasemidis and Sackellares, 1997	5–30
Correlation dimension based	Lehnertz and Elger, 1998	5–30
Complexity loss (*L*)		
Correlation integral similarity	Le Van Quyen *et al.*, 1999	5–30
Order-2 Kolmogorov entropy (*KE*)	van Drongelen *et al.*, 2003a	5–30
Multichannel		
Cross-correlation	Lange *et al.*, 1983	5–30
Phase synchrony	Mormann *et al.*, 2000	5–30
Adaptive seizure prediction	Iasemidis *et al.*, 2003	>60

Note: The methods are grouped in linear, nonlinear, and multichannel analysis sets. This summary is not an overview of the seizure anticipation field, for which we refer to a 2001 issue of *J. Clin. Neurophysiol.* (Vol. 18) and a 2003 issue of *IEEE Trans. Biomed. Eng.* (Vol. 50).

12.4.1. LINEAR METHODS

The most common step in most types of time series analysis is to explore linear methods. Techniques using the fast Fourier transform (FFT), linear filters, and linear decomposition have been incorporated in studies of seizure anticipation (e.g., Jerger *et al.*, 2001; Litt *et al.*, 2001; van Drongelen *et al.*, 2003a). One of the simplest methods applied in seizure anticipation is the calculation of the total power (Po) from the demeaned EEG time series. This metric is defined as the sum of squares of the sampled points (x_i) in a window divided by the number of sample points (N) in that window:

$$P\text{o} = \frac{1}{N} \sum_{i=1}^{N} x_i^2 \tag{12.2}$$

The window can be shifted over the recorded time series to obtain a power index over time. Litt *et al.* (2001) and van Drongelen *et al.* (2003a) applied this power measure to long-term recordings obtained from adult patients and pediatric epilepsy patients, respectively. The total power index detected an increase in energy in the epileptic focus caused by subclinical seizures and bursts. For the adult population, increased episodes of power were found several hours prior to a seizure onset (Litt *et al.*, 2001) (Table 12.1). In the children with epilepsy, the measure was successful in anticipating seizure in two out of five cases, with maximum anticipation times of up to 45 min (van Drongelen *et al.*, 2003a). McSharry *et al.* (2003) applied signal variance to predict seizure onset (Table 12.1).

12.4.2. NONLINEAR METHODS

This section will focus on the application of nonlinear dynamics to recorded time series. An important development in the analysis of dynamical systems is the so-called embedding procedure. Embedding of a time series $x_t(x_1, x_2, x_3, \ldots, x_N)$ is done by creating a set of vectors X_i such that

$$X_i = [x_i, x_{i+\Delta}, x_{i+2\Delta}, \ldots, x_{i+(m-1)\Delta}] \qquad (12.3)$$

where Δ is the delay in number of samples and m is the dimension of the vector. When embedding a time series one must make a decision about the dimension m of X_i and the delay Δ, such that each vector represents values that show the topological relationship between subsequent points in the time series. The value of m should not be too large so that the first and last values in the epoch are practically uncorrelated. On the other hand, the number of samples in the embedded vector should be large enough to cover the dominant frequency in the time series. The evolution of the system can now be depicted as the projection of the vectors X_i in multidimensional space, often referred to as *phase space* or *state space*. If the multidimensional evolution converges to a subspace in the phase space, this subspace is the attractor of the system. For a correct representation of the attractor, the embedded dimension must be larger than the dimension of the attractor. The construction and characterization of the attractor plays a major role in the analysis of time series. As was proven mathematically, the attractor of a single variable (e.g., the EEG or ECoG) can characterize the system that generated the time series (Takens, 1981). In the analysis of dynamics, measures that describe the attractor are used as an index for the system's state. Measures that are commonly used to describe the attractor in phase space are *dimension, entropy*, and *Lyapunov exponents*. For the dimension and entropy measures, several "flavors" exist and a multitude of algorithms have been developed over the past decades. Measures can be subdivided into a group that quantifies the attractor's spatial characteristics and a group that quantifies dynamics of trajectories in phase space.

Examples of time series and a two-dimensional embedding are shown in Figure 12.5. The upper time series (Figure 12.5A) is an example of the excursion of a pendulum and the associated embedding, indicating a strict relationship between past and future points. The middle example (Figure 12.5B) shows a random time series, where the embedded vector shows no specific relation between successive points. The lower example (Figure 12.5C) is the logistic equation (Eq. (12.1)). Interestingly the time series generated by the random process and the logistic iterator do not seem that different. However, by plotting x_t versus x_{t-1}, one can see that one time series shows a random relationship and the next has a fairly simple attractor characterized by a quadratic relationship from Eq. (12.1). The time series embedding in Figure 12.5D is characterized by more complex relationships of a type often referred to as a strange attractor. This strange attractor represents a more complex geometry than the curved line in the quadratic relationship, but is more confined than the random process. Both time series in Figure 12.5D and E are examples of a Henon map, a classic chaotic iterator that defines the coevolution of two variables x_t and y_t. Both plots in Figure 12.5D and E show x_t, but with only slightly different initial conditions: (0,0) in Figure 12.5D and $(10^{-5}, 0)$ in Figure 12.5E. The difference between the two closely related time series in Figure 12.5D and E is shown in Figure 12.5F, clarifying the sensitivity to a small perturbation (in this example, 10^{-5}) that was mentioned earlier. Initially the difference between the two time series is small, but after 25 points the difference grows disproportional and the error is of the same order of magnitude as the amplitude of each of the time series in Figure 12.5D and E. This phenomenon illustrates the point that even knowing the initial

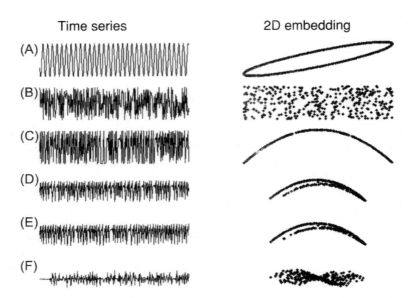

FIGURE 12.5. Examples of time series (left column) and embedding in two dimensions (right column). (A) sinusoidal signal; (B) random signal; (C) time series determined by the logistic equation ($x_t = 4x_{t-1}[1 - x_{t-1}]$; $x_0 = 0.397$); (D, E) two examples of a Henon map ($x_t = y_{t-1} + 1 - ax_{t-1}^2$; $y_t = bx_{t-1}$, $a = 1.4$, $b = 0.3$). The initial conditions differ between (D) and (E): $x_0 = 0$, $y_0 = 0$ and $x_0 = 10^{-5}$, $y_0 = 0$, respectively; (F) the difference between (D) and (E) shows that initially both time series develop in a similar fashion (difference $\to$ 0). However, after 25 iterations the difference in initial condition causes a different evolution in each time series. A Matlab script to create this figure can be found in Appendix 2.

condition at a precision of 10^{-5} results in *poor predictability of a chaotic process*: i.e., the values may deviate considerably after only a few time steps.

Several studies have explored the use of embeddings to characterize EEG time series in epilepsy (Hively *et al.*, 1999; Protopopescu *et al.*, 2001; Hively and Protopopescu, 2003). The distance (L), as well as the χ^2 statistic were computed by using a reference set (Q) and a test set (R). In these studies, a reference set was selected from a "normal" interictal epoch of EEG, and compared with the time series under investigation. The L and χ^2 metrics reflect the dissimilarity between the reference and test attractors, and Hively *et al.* (1999) report anticipation intervals up to 38 min (Table 12.1).

12.4.2.1. Lyapunov Exponent

To begin with a trivial statement: an attractor would not be an attractor if there weren't attraction of trajectories into its space. On the other hand an attractor wouldn't represent a chaotic process if neighboring trajectories didn't diverge exponentially. The Lyapunov exponent describes attraction (convergence) or divergence of trajectories in each dimension of the attractor. We indicate the exponent in the ith dimension as λ_i, describing the rate at which the distance between two initially close trajectories changes over time as an exponent: e^{λ_i}. A value of $\lambda_i > 0$ indicates there is divergence and $\lambda_i < 0$ indicates convergence in the ith dimension. In two dimensions, the sum of two exponents determines how a surface in the ith and $(i + 1)$th dimension evolves: $e^{\lambda_i} e^{\lambda_{i+1}} = e^{\lambda_i + \lambda_{i+1}}$. In three dimensions, three

Lyapunov exponents describe the evolution of a cube, and the sum of all Lyapunov exponents indicate how a so-called hypercube evolves in a multidimensional attractor. In order to show sensitivity to the initial condition, the largest Lyapunov exponent determined in an attractor of a chaotic process must be > 0. Therefore the characterization of EEG signals by the Lyapunov exponent is usually focused on the largest exponent. The largest exponent describes the expansion along the principal axis (p_i) of the hypercube over a given time interval t. Formally, the exponent (λ_i) is calculated as

$$\lambda_i = \lim_{t \to \infty} \frac{1}{t} \log_2 \left[\frac{p_i(t)}{p_i(0)} \right] \tag{12.4}$$

Wolf et al. (1985) developed an algorithm to estimate the largest Lyapunov exponent in a measured time series. The point nearest to the starting point of the embedded time series is found, and trajectories from this and the starting point are followed during a fixed interval. The initial distance d_0 and the distance after the time interval d_1 are measured. If the distance d_1 is smaller than a preset value, the procedure is repeated. Figure 12.7 shows part of the EEG attractor in Figure 12.6, with an example of two initially close trajectories and their start and end positions. If the distance between the end positions (d_1) grows larger than the preset value, an attempt is made to rescale the distance by searching for a new point closer to the reference trajectory. The rescaling algorithm described by Wolf et al. (1985) was revised for application to the EEG and ECoG time series by Iasemidis et al. (1990). This procedure is repeated k times to cover the measured attractor from t_0 to t_k, and the largest Lyapunov exponent $(\lambda_{\max})$ is calculated as

$$\lambda_{\max} = \frac{1}{t_k - t_0} \sum_{i=1}^{k} \log_2 \left[\frac{d_i}{d_{i-1}} \right] \tag{12.5}$$

In the early 1990s, Iasemidis, Sackellares, and coworkers applied nonlinear analysis to both the scalp EEG and ECoG before and after seizure onset. This group focused mainly

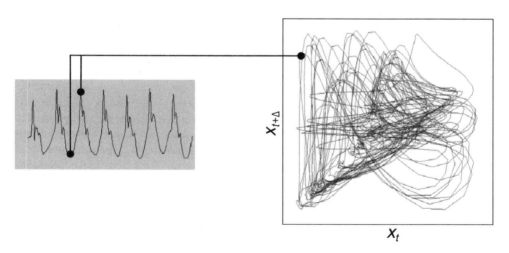

FIGURE 12.6. An example of 2-D embedding of an EEG signal. Two points of the time series are plotted as one single point in a two-dimensional state space diagram. By embedding all subsequent pairs in the same manner, a 2-D section of the attractor is obtained.

on the largest Lyapunov exponent or the comparison of the exponent with other metrics (e.g., Iasemidis *et al.*, 1990, 1993, 1997, 2003; Iasemidis and Sackellares, 1991). Anticipation times of up to about 10 min were reported by Iasemidis and Sackellares (1991) (Table 12.1).

12.4.2.2. *Kolmogorov Entropy*

Another variable that can characterize the dynamics of an attractor is order-2 Kolmogorov entropy (Grassberger and Procaccia, 1983b; Schouten *et al.*, 1994a). Order-2 Kolmogorov entropy is a measure of the rate at which information about the state of a system is lost, and it can be estimated by the examination of two initially close orbits in an attractor. The time interval (t) required for the orbits to diverge beyond a set distance satisfy a distribution:

$$C(t) \propto e^{-KEt} \tag{12.6}$$

where KE is the Kolmogorov entropy. Schouten *et al.* (1994a) found an efficient maximum-likelihood method of estimating KE. Their method assumes a time series of N points that is uniformly sampled at intervals of t_s; under these assumptions, Eq. (12.6) becomes

$$C(b) = e^{-KEt_s b} \tag{12.7}$$

where b represents the number of time steps for pair separation. They then show that the maximum likelihood estimate of the Kolmogorov entropy KE_{ml}, in bits per second, can be written as

$$KE_{ml} = -\frac{1}{t_s}\left[\log_2\left(1 - \frac{1}{b_{avg}}\right)\right] \tag{12.8}$$

where b_{avg} is the average number of steps required for close pairs to diverge. Not only does this method converge relatively quickly, but Schouten *et al.* (1994a) also provide a way of estimating the standard deviation for KE_{ml} (again in bits per second) by using a set of M escape steps $\{b_1, \ldots, b_M\}$, giving $\sigma = 1/(\ln(2)t_s\sqrt{b_{avg}(b_{avg} - 1)M}) \propto 1/\sqrt{M}$ for a given data set. A listing of a routine to estimate KE_{ml} according to the procedure described by Schouten *et al.* (1994a) can be found in Appendix 1.

To collect the necessary bs, methods of choosing nearby independent points as well as determining the divergence threshold are needed. Schouten *et al.* (1994a,b) suggest estimating these from the data in the following way. First, the data is demeaned and normalized to the average absolute deviation of the demeaned data $x_{abs} = (1/N)\sum_{i=1}^{N}|x_i|$, where N is the number of sample points; x_{abs} is then used as an estimate of the divergence threshold. Second, the number of cycles in the time series is estimated as 1/2 of the number of zero crossings; this is used to calculate the number of samples/cycle m, which is used as the independence criterion.

The algorithm proceeds by selecting a pair of samples in the data at randomly chosen time steps i and j; if they are separated by at least m time steps ($|i - j| \geq m$), then they are considered independent. The largest of m absolute differences between pairs of values starting at i and j constitutes the maximum norm: $d = \max(|x_{i+k} - x_{j+k}|)$ for $0 \leq k \leq m - 1$; if $d \leq x_{abs}$, the samples are considered nearby. Finally, having found a pair of randomly chosen, nearby, independent data points, the number of steps b needed for them to diverge (such that $|x_{i+m-1+b} - x_{j+m-1+b}| > x_{abs}$) can be added to the set used to calculate b_{avg}.

The above thresholds for determining independence and divergence work reasonably for many data sets, but Schouten *et al.* (1994b) stress that x_{abs} and m are heuristics that provide reasonable guidelines; they may yield better results if modified by a factor of order unity. We have applied their default estimates to 30-s segments of EEG sampled at 400 Hz and have found them to work well. Roughly 1–10% of randomly selected pairs in our data contribute to b_{avg} (i.e., are nearby and independent), so $M = 10,000$ (corresponding to ~1% spread in the KE estimate) requires testing 10^5–10^6 random points.

Both the value of the largest Lyapunov exponent and the KE relate directly to predictability, and the procedures for looking into the evolution of trajectories in these two metrics are fairly similar. For the KE estimation the interpoint distance is set and the time of evolution is measured, whereas for estimation of the largest Lyapunov exponent it is the other way round. For the KE estimation, close trajectories are selected randomly, for the Lyapunov exponent the procedure (described by Wolf *et al.*, 1985) covers the attractor from start to end. Large values of both measures indicate an important divergence in trajectories that are initially close. As in the example in Figure 12.5D, E, and F, small perturbations or inaccuracies in the initial state or in the calculation of subsequent values in a time series will create large differences after only a few iterations, thus limiting the potential for accurate prediction over a longer interval.

Application of the Kolmogorov entropy to EEG and ECoG was explored by Hively *et al.* (1999), Protopopescu *et al.* (2001), and van Drongelen *et al.* (2003a). These studies report seizure anticipation intervals of up to ~30 min for both adult and pediatric patients (Table 1).

12.4.2.3. Attractor Dimension

Measures of dimensionality are used to characterize the geometry of an attractor in space. Several flavors of the dimension metric are currently in use. An overview of the relationships between the different dimension measures (the so-called Renyi dimensions) would be beyond the scope of this chapter and can be found in Peitgen *et al.* (1992). Theoretically, central among these measures is the capacity dimension D_Cap of an attractor, which can be estimated with a box-counting algorithm. This procedure determines the space that is occupied by the attractor in terms of the number of boxes $N(s)$ with size s in which points of the attractor are located (Figure 12.7). For different sizes of s, the value of $N(s)$ scales according to a power law:

$$D\_Cap = \lim_{s \to 0} \frac{\log N(s)}{\log(1/s)} \qquad (12.9)$$

For instance, a cube of size 1 m $\times$ 1 m $\times$ 1 m can be subdivided into 1000 small cubes of 0.1 m $\times$ 0.1 m $\times$ 0.1 m, and 10,00,000 small cubes of 0.01 m $\times$ 0.01 m $\times$ 0.01 m, etc. In this example the number of small cubes versus the inverse of the size (s) scales as $(1/s)^3$,

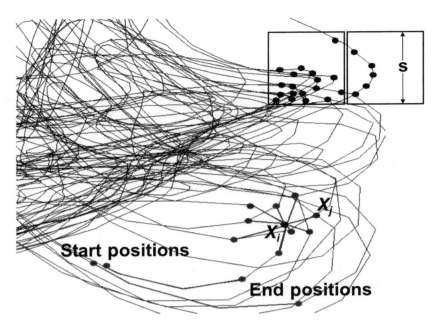

FIGURE 12.7. Simplified representation of metrics that characterize properties of an attractor. These trajectories are a detail of Figure 12.6. Counting the occupation of the attractor by using boxes of size s, is the basis for the estimation of the capacity dimension (Eq. (12.9)). Estimation of the correlation dimension is based on counting pairs (within a set small distance) X_i, X_j on the attractor (Eq. (12.10)). Both the largest Lyapunov exponent and the order-2 Kolmogorov entropy describe how initially close trajectories diverge. The largest Lyapunov exponent measures the ratio between the distances over a given time interval (Eqs. 12.4 and 12.5). The order-2 Kolmogorov entropy measures the time over which trajectories diverge beyond a set distance (Eqs. 12.6 and 12.7).

the power being the capacity dimension of the cube. Similar procedures can be applied for a line or a square, generating powers of 1 or 2, respectively. Applying the same box counting and scaling procedure for more irregular structures, such as an attractor embedded in a cube, which doesn't fill up the whole space entirely, can generate a noninteger value in between 2 and 3 for the dimension. The smaller the size of the box in the counting procedure, the more precisely can the area, volume, etc., covered by the attractor be described. Unfortunately, a requirement for a reliable small-box count is an attractor that is known in great detail, i.e. many points that are available to characterize the attractor's space. For measured time series, such large data sets are often not available. The use of larger boxes is easier to accomplish but reflects the attractor's space less precisely. Therefore, the capacity dimension is not attractive for application to measured time series, in spite of the efficient algorithms that are available (e.g., Ghorui *et al.*, 2000). Another measure that is related to D_Cap is the information dimension. This measure relates to the entropy, the distribution, and the local density of the attractor's points in space. In box-counting terms, one counts the number of boxes occupied in space and weighs the box by the number of points it includes. Like capacity dimension, the computational burden of estimating information dimension prevents it from being frequently used in experimental work. The most popular measure is the so-called correlation dimension (Grassberger and Procaccia, 1983a, Schouten *et al.*, 1994b).

A metric derived from the so-called correlation integral is as follows:

$$C(s) = \left[\frac{1}{N(N-1)} \right] \sum_{i \neq j} \Theta \left(s - |X_i - X_j| \right)$$

(12.10)

With Θ = Heaviside step function, and N = the number of points. The term $|X_i - X_j|$ denotes the distance between the points in state space. The summation (Σ) and the Heaviside function count the vector pairs (X_i, X_j) with an interpoint distance smaller than s, because $\Theta(\ldots)$ is 1 if this distance is smaller than s, and 0 in all other cases. The value of $C(s)$ is a measure of the number of pairs of points (X_i, X_j) on the reconstructed attractor whose distance is smaller than a set distance (Figure 12.7). For a large number of points (N) and small distances (s), $C(s)$ scales according to a power law $C(s) \propto s^{D\_Cor}$, where D_Cor is the correlation dimension of the attractor.

Most procedures for calculating nonlinear metrics from data rely on building up the value from a randomly sampled subset of the data; they therefore become more accurate as the size of the sampled subset increases. Like uncertainties in a Monte Carlo integral, there is statistical uncertainty in these approaches that depends only on the number of sampled points and not on any error in the data. The relevant convergence properties can usually be estimated using the same assumptions used to derive the method itself, so their effect on accuracy is predictable. A second source of uncertainty is the classic measurement noise; this is more troublesome because measurement error is usually not well characterized. One attempt to assess the impact of measurement noise in an estimate of the correlation dimension is given by Schouten et al. (1994b). They model additive noise that is strictly bounded by a value s_n and modify the above scaling law. When the maximum norm is used to measure distance, this yields:

$$C(s) = \left[\frac{s - s_n}{s_0 - s_n} \right]^{D\_Cor}$$

(12.11)

For $s_n < s < s_0$, where s_0 represents the distance above which the power law scaling for correlation dimension breaks down; Schouten et al. (1994b) recommend that this threshold be set to the average absolute deviation of the data. The effect is to reject all distances smaller than s_n (since, within the noise, these points are coincident) and correct the rest of the distances by s_n. Fitting data to Eq. (12.11) will generally yield a larger D_Cor than the uncorrected scaling law, implying that estimates of D_Cor that do not take noise into account will be biased low.

Neuronal complexity loss (L^*), a metric derived from the correlation dimension, was introduced by Lehnertz, Elger, and coworkers. The neuronal complexity loss was obtained from the temporal changes of the estimated correlation dimension. The value of L^* equals the surface between the correlation dimension plot and an arbitrarily determined upper limit D_u (e.g., $D_u = 10$) over a fixed time interval (e.g., 25 min). One of the early descriptions of neuronal complexity loss can be found in Lehnertz and Elger (1995), and a statistical evaluation of seizure anticipation using this metric was first performed by Lehnertz and Elger (1998). The latter study indicates that a drop in correlation dimension may occur up to 25 min prior to seizure onset (Table 12.1). An overview of this group's work can be found in Lehnertz et al. (1999, 2001). More recently, this group has become interested in measures of nonlinearity and nonstationarity (Andrzejak *et al.*, 2001a,b; Rieke *et al.*, 2003).

In an initial study on interdependencies in the EEG, Le Van Quyen *et al.* (1998) described the performance of linear and nonlinear correlations. Subsequently, a similarity index was introduced and evaluated by this group (Le Van Quyen *et al.*, 1999, 2000, 2001a,b). A unique aspect of this group's approach is the pretreatment of the EEG signal. Most groups use the raw sampled time series as the input to the algorithms, whereas Le Van Quyen and coworkers transform the raw data prior to embedding. First the times T_n of threshold crossings (e.g., zero crossings) are determined; subsequently the associated intervals $I_n = T_{n+1} - T_n$ are used as the basis for embedding. The rationale for this preprocessing is to make their measures less dependent on large-amplitude signals, such as during interictal spikes. The similarity index used by this group is calculated as the correlation between a reference set (based on an epoch of interictal EEG) and the time series under investigation. Correlation is determined with the correlation integral for two signals, similar to the integral for a single set in Eq. (12.10). For a given distance (s in Eq. (12.10)), the integral counts the common points between the reference set S_{ref} and the test set S_t. A cross-correlation value $C(S_{ref}, S_t)$ is found and normalized as $C(S_{ref}, S_t)/[C(S_{ref}, S_{ref})C(S_t, S_t)]$. Seizure anticipation intervals of up to several minutes were found by this group (Table 12.1).

12.4.3. MULTICHANNEL-BASED METHODS

Metrics described in the previous sections can be applied to both single and multichannel time series. Because clinical recordings typically contain multiple channels, specific algorithms to analyze interchannel relationships have been applied to the EEG. Iasemidis *et al.* (2003) created a multichannel version for the Lyapunov exponent, and Jerger *et al.* (2001) combine the extracted metrics from multiple channels for an optimal anticipatory effect. Because seizure activity is typically associated with high levels of synchrony between the EEG data at different locations, some of these techniques focus on measures of similarity between channels. Coherence and cross-correlation are examples of techniques that have been used to find linear interchannel relationships (e.g., Lange *et al.*, 1983; Towle *et al.*, 1999). Nonlinear equivalents to detect channel interaction are frequently based on the mutual information concept (e.g., Mormann *et al.*, 2000; Le Van Quyen *et al.*, 2001a; Chavez *et al.*, 2003). Typical anticipation intervals are found in Table 1. Interestingly, not all recent work shows an increased synchrony close to and at seizure onset. This seems to be supported by the observation of low levels of synchrony between neurons during experimental seizures in *in vitro* preparations (Netoff and Schiff, 2002; van Drongelen *et al.*, 2003b).

12.4.4. SURROGATE TIME SERIES

An important question when applying nonlinear time series analysis to recorded data is the nature of the underlying process. For example, the application of sophisticated nonlinear dynamic systems tools to a time series is not appropriate if there is no underlying nonlinear process. To determine whether a data set contains nonlinearities, several methods were developed in which surrogate data sets were generated and compared against a measured data (e.g., Pijn *et al.*, 1991; Theiler *et al.*, 1992; Kaplan and Glas, 1995; Casdagli *et al.*, 1996; Stam *et al.*, 1998; Andrzejak *et al.*, 2001a). The idea is to compute one of the nonlinear measures (e.g., the ones described in the previous paragraphs) from both the measured time series and the surrogate time series generated by some linear model. The linear model generates these surrogate time series on the basis of the measured data. Subsequently, the values of the nonlinear measure obtained from the real data and a set of surrogate time series are compared. The null hypothesis is that the value of the computed nonlinear measure can be explained from the linear model, and if the null hypothesis is rejected, a nonlinear process may have generated the original data. The procedure to obtain surrogate data depends on the null hypothesis at hand. If the null hypothesis is that the data originates from a random process, a random shuffle of the data is sufficient to generate a surrogate time series. Another commonly applied null hypothesis is to assume that the underlying process is stationary, linear, and stochastic. A commonly applied technique to obtain surrogate time series satisfying this hypothesis is to compute the fast Fourier transform (FFT) followed by a randomization of the phase. The inverse FFT generates a surrogate time series with the same power spectrum. Methods of surrogate time series comparison provide a relatively robust technique for the critical task of demonstrating underlying nonlinearity, a prerequisite for existence of chaos. Unfortunately, in a similar manner, objective tests to demonstrate an underlying chaotic process in the measurement do not exist.

12.5. EVALUATION AND FUTURE DIRECTIONS

Prediction of events is difficult, both in science and in daily life. Unfortunately, prediction difficulties also include activity patterns generated in the brain from patients with epilepsy, the topic of this chapter. Considering the body of evidence we have summarized here, it is clear that at least some types of seizure seem to allow prediction or anticipation. However, a critical attitude toward this evidence is appropriate. First, one might question the applicability of most of these signal-processing techniques to nonstationary EEG signals (Manuca and Savit, 1996; Manuca *et al.*, 1998; Rieke *et al.*, 2003). Proof of the existence of deterministic chaos in EEG may be impossible, and this has led to an informal interpretation of the nonlinear measures (Lehnertz and Elger, 1998), which may distance the use of these techniques from their theoretical foundations. Consequently, after the initial enthusiasm about the feasibility of detecting the preictal state, the questions have let most researchers to take a more careful approach.

First, the *general* feasibility of the approach itself has been questioned. On theoretical and experimental grounds, Lopes da Silva *et al.* (2003) subdivide seizures into different types: those that *can* and those that *cannot* be predicted. In the latter group of seizures there is a random transition between the "normal" and ictal attractors. In this category, both attractors are so close that a small perturbation may cause the brain to jump between interictal and ictal states. An illustration of how this process might work is shown in Figure 12.8. This example is based on the EEGs in Figure 12.3A and B, showing the onsets of two similar seizures from the same patient. In Figure 12.8, these EEGs are embedded in 2-D plots as $x(t + 18)$ versus $x(t)$; in this example the 18 data points represent a delay of 45 ms. The 2-D section of the "normal" EEG attractor forms an area (red) that is located in the center of a hole in the seizure attractor (blue). The transition to the seizure state (black trajectory in

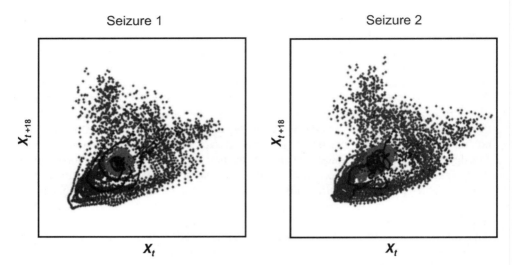

FIGURE 12.8. An example to demonstrate the different routes to a generalized epileptic seizure. The left and right plot show 2-D embedded data sets around both seizure onsets shown in Figure 12.3(A) and (B). The red dots (the doughnut hole) represent the 2-D section of the inter-ictal attractor, the blue ones (the warped doughnut) for the seizure state. It can be seen that both sections of the attractors are very close, and that the routes to seizure onset (black dots) are different in both cases.

Figure 12.8) is clearly different in both cases. The finding that routes other than the normal signal to the seizure onset exist indicates that (in this case) a random perturbation may have caused the transition between the two attractors. As a result, we were unable in this case to predict these seizures with a set of the linear and nonlinear methods: power, variance, linear decomposition, correlation dimension, and Kolmogorov entropy. The best one can do to quantify this type of seizure is to describe the domain of the ictal and interictal states in state space, and to collect statistics of state transitions. Milton *et al.* (1987) found that some seizures in adults, including the generalized seizure type, could be statistically described using a Poisson process.

Second, the effort in seizure anticipation is shifting from feasibility studies to a statistical evaluation of anticipation algorithms. The statistical validation in most published studies is based on small and/or discontinuous data sets. A large-scale study where anticipation algorithms are exercised for several days or weeks on a large patient group is required to better establish a sensitivity–specificity function for the prediction and anticipation techniques known to date. Results from initial statistical studies are favorable, in the sense that a number of seizure types may be anticipated with the associated reasonably low false detection rates. Lehnertz *et al.* (2001), for example, reported statistics of dimension drops prior to seizure onset for a group of 59 patients. Although this statistical comparison between interictal and preictal behavior of the dimension was based on discontinuous data sets, statistically significant drops were found in 67% of the seizures in patients with mesial temporal lobe epilepsy, and in 29% of the seizures in patients with neocortical lesional epilepsy. Iasemidis *et al.* (2003) report results of an adaptive seizure prediction algorithm on continuous data sets up to several days. Their method predicts 82% of seizures and has a false detection rate of 0.16 seizures per hour. Although such large-scale studies on continuous recordings are required, the endeavor is far from trivial since the estimation of most metrics is computationally intensive, and there are typically 21–256 candidate channels to explore in every EEG or ECoG. However, such a lengthy evaluation is inevitable, because knowledge of true anticipation alarms against false positive and false negative alarm rates is critical for practical application where a therapeutic intervention is based on seizure anticipation. To address anticipation performance in a pediatric population, our group recently implemented a software package for the analysis of long-term data sets. Results obtained with monitoring the Kolmogorov entropy are shown in Figures 12.9 and 12.10. Each figure shows the time course of the Kolmogorov entropy for an epoch of about 24 h and a detail of one of the events. In this case, the generalized seizures in Figure 12.9 (indicated with S) are typically associated with a drop in entropy, but as can be seen in the detail (Figure 12.9B), no prediction of the onset is possible. The data in Figure 12.10, however, show a different pattern. A relatively noisy evolution of the entropy signal is disrupted by episodes where the entropy rises, followed by a decrease. At the end of the entropy decrease, an event occurs. This event can either be a clinical seizure (e*, Figure 12.10) or an epoch of bursting (e, Figure 12.10). A detail of the first seizure (Figure 12.10B) shows that the trend of entropy increase starts about 45 min prior to seizure onset, and the trend of the entropy decrease around 15 min before the event starts: a consistent pattern that allows anticipation of the upcoming event. Both records shown in Figures 12.9 and 12.10 also show evidence of false detection. In the example in Figure 12.9, the four negative peaks not associated with seizure are associated with chewing or movement artifacts in the signal. In the example in Figure 12.10, we have three correct positive detections (e*) and at least four false positive ones (e). It is important to note that as long as the physiology of the preictal state remains unknown, it is not easy to define false positive detections in a principled way (Lopes da Silva *et al.*, 2003). One might assume that certain preictal processes do not lead to a seizure in all cases. Under that assumption, detection of a preictal state that is not followed by a seizure may be a true and correct detection. From a practical standpoint one might argue that any detection of a preictal state not followed by a seizure is a false detection. On the other hand, no real seizures were missed by the trend in Figure 12.10. A low false negative detection rate is critical in a situation where one moves from continuous to discontinuous therapy. In this case and many other real-life situations, one might be willing to tolerate a reasonable level of false detection if it is associated with a high sensitivity of the algorithm. One approach to further increase sensitivity and selectivity is the combination and individualization of algorithms (e.g., D'Allessandro *et al.*, 2003).

Detection of preictal states can be used as a guide for further study of the mechanisms of epileptogenesis, as well as for therapeutic intervention. The required anticipation interval for seizures relates to the type of therapy at hand. Treatment on the basis of anticonvulsants requires an anticipation interval that is sufficient for the delivery of medication at the site of seizure onset (probably 1 to several minutes). However, for effective electrostimulation to prevent seizure onset or propagation, a shorter interval or even early onset detection may be sufficient. A description of one of the earliest attempts of the application of electrostimulation in a clinical setting is described in Chkhenkeli (2003). In cases where prediction is impossible, detailed knowledge of "normal" and ictal attractors may allow control of the brain activity without the necessity of prediction. In the example shown in Figure 12.8, the goal of an effective control would be to keep the system as much as possible in the center

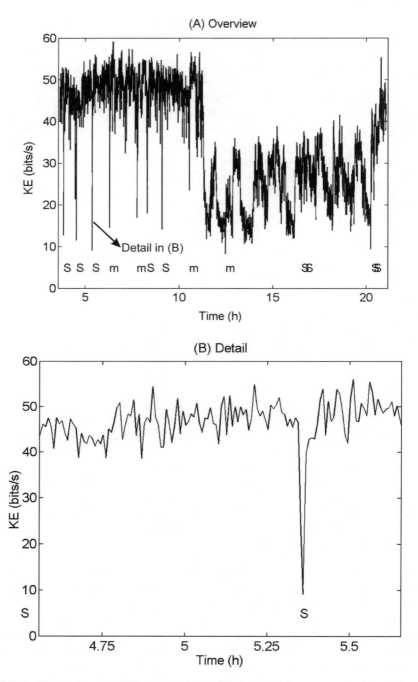

FIGURE 12.9. A long-term trend of Kolmogorov entropy in a patient with generalized seizures (A) and a detail of the third seizure (B). S indicates seizures, m indicates movement artifacts. Seizures are associated with decreased entropy. However, anticipation of seizure onset isn't feasible with the trend shown.

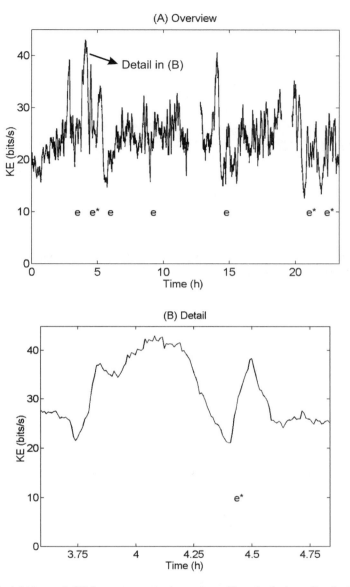

FIGURE 12.10. A 24-hr trend of Kolmogorov entropy in a patient with a mixed seizure disorder (A) and a detail of the first clinical seizure (B). Bursting events are indicated by e, clinical seizures by e*. Event onset is preceded by a rise and decrease in entropy.

of the normal attractor. This might prevent a small perturbation from causing a transition to the neighboring ictal attractor.

Future seizure prediction will likely continue to focus on aspects of practical application, including the issues discussed above. Evaluation, fine-tuning, plus combination of existing and new algorithms will generate increased reliability of prediction/anticipation procedures for different types of seizure and may make its application to therapeutic intervention feasible. From the point of view of basic science, we expect that attempts to translate the predictive techniques to the level of small networks and single neurons may lead to a better understanding of ictogenesis. On the therapeutic side, increased knowledge of the effects of electrostimulation paradigms and intracerebral delivery of medication are required to develop a robust and clinically viable automated seizure-prevention device.

ACKNOWLEDGEMENT

This work was supported by the Falk Medical Trust Foundation. We thank Drs. F. Elsen, V. L. Towle, J. D. Hunter, and M. Carroll for the useful discussions and comments, and S. Nayak and K. van Drongelen for their help in preparing the manuscript.

12.6. APPENDIX 1: C FUNCTION TO CALCULATE MAXIMUM LIKELIHOOD KOLMOGOROV ENTROPY

```
#include <stdio.h>
#include <stdlib.h>
#include <math.h>

#define KOLMOG_TEST_LIMIT   (1000)

/*
 *KEML:
 *
 * Returns max likelihood estimate of Kolmogorov entropy as
   described in
 * Schouten, Takens, and van den Bleek 1994 Phys. Rev. E v49, #1
 * (from now on, STB94a), or -1 on error. The estimated standard
   * deviation is stored in the target of kolmog_err_ptr.
 *
   * Input data are assumed to be demeaned.
 * Note: should seed the C random number generator before calling
 *
 * n_samples      number of data pts
 * samples             pointer to data
 * dt                  time between sample pts (sec)
 * correl_steps  estimated min steps between indep pts (m in STB94a)
 * kolmog_pairs        number of pairs to be used for KE estimate
 * l0                  divergence threshold of selected pairs; STB94a
   *                    recommend using avg absolute deviation of data.
   *                    Typically, data are normalized to this, so
                        l0 = 1.0
 * kolmog_err_ptr      output: estimated std dev of KE (bps) is placed
                        here
 */
```

```
double KEML(int n_samples, double *samples, double dt,
      int correl_steps, int kolmog_pairs, double 10,
      double *kolmog_err_ptr)
{

double avg_b;        // avg # escape steps
double kolmog_bps;       // KE (bits/sec)
double rej_b=0;          // sum of all rejected escape steps
int    total_b=0;        // sum of all escape steps
int    tested_pairs=0;   // # random pairs tested
int    pairs_left;       // # pairs needed to finish
int    n_rejected=0;     // # pairs w/b<correl_steps
                                    //||max dist always<=10

int    test_limit;       // test & exit from pathological runs

/* If we process test_limit pairs, flag an error and return */
test_limit=KOLMOG_TEST_LIMIT*kolmog_pairs;

/* Set number of pairs needed for estimate */
pairs_left=kolmog_pairs;

/* Accumulate ML estimate of KE w/requested # of pairs */
while (pairs_left && tested_pairs++<test_limit) {

      int    i,j;    // Starting indices of new sample pair
      int    s;      // Index at which dist > 10 for 1st time
      int    b;      // # of steps needed to diverge

      /* Randomly choose indices of new pair */
      i=rand()%(n_samples-correl_steps);
      j=rand()%(n_samples-correl_steps);

      /* Ensuring that i>=j makes life simpler */
      if (i<j) {
       int    temp=i;

      i=j;
      j=temp;
      {

      /* Initialize divergence index to largest index */
      s=i;

    /*
     * Only indices separated by at least correl_steps elems
     * represent valid independent vectors
     */
if (i-j>=correl_steps) {
```

```
      /* Get # of steps before max norm dist > 10 */
      while (s<=n_samples && fabs(samples[s]-samples[j])<=10) {
      s++;
      j++;
      }
      /*
      * Comparing to the first index in eqn 12.10 of STB94a,
      * s=i+m-1+b and correl_steps=m, so set b=s-(i+m-1)
      */
      b=s-(i+correl_steps-1);

      /*
      * Accept if max norm is <= 10 for correl_steps (b>0)
      * && evolves to > 10 before the end of data (s<=n_samples)
      */
      if (b>0 && s<=n_samples) {
      pairs_left--;
      total_b+=b;
      } else {
      n_rejected++;
      rej_b+=(b+correl_steps);
      }

}

}     // end of while loop

/*
* Calculate max likelihood KE from eq. 20 in STB94a; note
* comments on normalizing entropy estimate for the sampling
* rate in the last paragraph of Section V. If no valid
* pairs were found, flag an error by setting KE=-1
*/
if (total_b>0) {
        int     pairs_used=kolmog_pairs-pairs_left;

avg_b=total_b;
     avg_b=avg_b/pairs_used;
kolmog_bps=-(log(1.0-1.0/avg_b))/(dt*log(2.0));
     *kolmog_err_ptr=1.0/
                 ( dt
                     *log(2.0)
                     *sqrt(pairs_used
                             *avg_b
                             *(avg_b-1.0)
                       )
                 );
} else {
        avg_b=-1;
        kolmog_bps=-1;
```

```
}

/* Print out some diagnostic statistics */
fprintf(stderr,
        "# KEML: %d pairs tested, %d rejected, avg rej_b=%f\n",
        tested_pairs, n_rejected, rej_b\n_rejected);
fprintf(stderr, "# KEML: avg_b=%f\n", avg_b);
fprintf(stderr, "# KE (bps) = %f\n", kolmog_bps);
fprintf(stderr, "# KE error (bps) = %f\ n", *kolmog_err_ptr);

return(kolmog_bps);

}
```

12.7. APPENDIX 2: MATLAB SCRIPTS TO CREATE FIGURES 12.2 AND 12.5

```
%  Figure 12.2
clear
xn=0.01;
figure;
hold;
for a=2.5:.02:4;              % range for coefficient a
for k=1:.1:200               % iterate for 200 steps
xn=a*xn*(1-xn);              % logistic equation
if (k>100)                   % Do not show initial val-
ues <100
plot(a,xn,'k.');             % Plot the data points
end;
end;
end;
xlabel('a')                  % Provide labels and title
ylabel('state')
title('Feigenbaum Diagram from Logistic Equation')

%  Figure 12.5
clear;
le=400;
figure;
for k=1:le;x(k)=sin(k/(le/le-3));end;        %  Sine map
subplot(6,2,1); plot(x, 'k');axis('off');   % time series
subplot(6,2,2); hold;for n=2:length(x); plot(x(n-1),
x(n),'k.');end; axis('off');
                                             % phase space
for k=1:le;x(k)=rand(1); end;                %  Random map
```

```
subplot(6,2,3); plot(x,'k');axis('off');        % time series
subplot(6,2,4); hold;for n=2:length(x); plot(x(n-1),x(n),
'k.');end; axis('off');
                                        % phase space
x(1)=.397; for k=2:1e;x(k)=4*x(k-1)*(1-x(k-1));end;
                                        %   Logistic map
subplot(6,2,5); plot(x, 'k');axis('off');       % time series
subplot(6,2,6); hold;for n=2:length(x);
plot(x(n-1),x(n),'k.');end; axis('off');
                                        % phase space
x(1)=0; y(1)=0;for k=2:1e;x(k)=y(k-1)+1-1.4*x
(k-1)^ 2;y(k)=0.3*x(k-1);end;
                                        %   Henon map 1
subplot(6,2,7); plot(x, 'k');axis('off');       % time series
subplot(6,2,8); hold;for n=2:length(x);
plot(x(n-1),x(n), 'k.');end; axis('off');
                                        % phase space
xx(1)=1e-5; y(1)=0;for k=2:1e;xx(k)=y(k-1)+1-1.4*xx
(k-1)^ 2;y(k)=0.3*xx(k-1);end;
                                        %   Henon map 2
subplot(6,2,9); plot(xx,'k');axis('off');       % time series
subplot(6,2,10); hold;for n=2:length(xx);
plot(xx(n-1),xx(n), 'k.');end; axis('off');
                                        % phase space
xxx=xx-x;                               % Effect initial condition
subplot(6,2,11); plot(xxx, 'k');axis('off'); % time series
subplot(6,2,12); hold;for n=2:length(xxx);
plot(xxx(n-1),xxx(n),'k.');end; axis('off');
                                        % phase space
```

REFERENCES

American Electroencephalographic Society, 1994, American Electroencephalographic Society. Guideline thirteen: Guidelines for standard electrode position nomenclature, *J. Clin. Neurophysiol.* **11**:111–113.

Andrzejak, R. G., Lehnertz, K., Mormann, F., Rieke, C., David, P., and Elger, C. E., 2001a, Indications of nonlinear deterministic an infinite-dimensional structures in time series of brain electrical activity: Dependence on recording region and brain state, *Phys. Rev. E* **64**:061907-1–061907-7.

Andrzejak, R. G., Widman, G., Lehnertz, K., Rieke, C., David, P., and Elger, C. E., 2001b, The epileptic process as nonlinear deterministic dynamics in a stochastic environment: An evaluation on mesial temporal lobe epilepsy, *Epilepsy Res.* **44**:129–140.

Babloyantz, A., and Destexhe, A., 1986, Low-dimensional chaos in an instance of epilepsy, *Proc. Natl. Acad. Sci. USA* **83**:3513–3517.

Casdagli, M. C., Iasemidis, L. D., Sackellares, J. C., Roper, S. N., Gilmore, R. L., and Savit, R. S., 1996, Characterizing nonlinearity in invasive EEG recordings from temporal lobe epilepsy, *Physica D* **99**:381–389.

Chavez, M., Le Van Quyen, M., Navarro, V., Baulac, M., and Martinerie, J., 2003, Spatio-temporal dynamics prior to neocortical seizures: Amplitude versus phase coupling, *IEEE Trans. Biomed. Eng.* **50**:571–583.

Chkhenkeli, S. A., 2003, Direct deep-brain stimulation: First steps towards the feedback control of seizures, in: *Epilepsy as a Dynamic Disease* (J. Milton and P. Jung, eds.), Springer-Verlag, Berlin, pp. 249–261.

Commission on Classification and Terminology of the International League Against Epilepsy, 1981, Proposal for revised clinical and electroencephalographic classification of epileptic seizures, Epilepsia **22**: 489–501.

Commission Report International League Against Epilepsy, 2001, Glossary of descriptive terminology for ictal semiology: Report of the ILAE task force on classification and terminology, *Epilepsia* **42**: 1212–1218.

D'Allessandro, M., Esteller, R., Vachtsevanos, G., Hinson, A., Echauz, J., and Litt, B., 2003, Epilepsy seizure prediction using hybrid feature selection over multiple intracranial EEG electrode contacts: A report of four patients, *IEEE Trans. Biomed. Eng.* **50**:603–614.

Ebersole, J. S., 2001, Editorial: The last word, *J. Clin. Neurophysiol.* **18**:299–300.

Elbert, T., Ray, W. J., Kowalik, Z. J., Skinner, J. E., Graf, K. E., and Birbaumer, N., 1994, Chaos and physiology: Deterministic chaos in excitable cell assemblies, *Physiol. Rev.* **74**:1–47.

Elger, C. E., and Lehnertz, K., 1998, Seizure prediction by non-linear time series analysis of brain electrical activity, *Eur. J. Neurosci.* **10**:786–789.

Engel, J., 2001, A proposed diagnostic scheme for people with epileptic seizures and with epilepsy: Report of the ILAE task force on classification and terminology, *Epilepsia* **42**:796–803.

Feigenbaum, M. J., 1983, Universal behavior in nonlinear systems, *Physica D* **7**:16–39.

Ghorui, S., Das, A. K., and Venkatramani, N., 2000, A simpler and elegant algorithm for computing fractal dimension in higher dimensional state space, *Pramana* **54**:L331–L336.

Gotman, J., 1982, Automatic recognition of epileptic seizures in the EEG, *Electroencephalogr. Clin. Neurophysiol.* **54**:530–540.

Gotman, J., and Gloor, P., 1976, Automatic recognition and quantification of interictal epileptic activity in the human scalp EEG, *Electroencephalogr. Clin. Neurophysiol.* **41**:513–529.

Gotman, J., Ives, J. R., and Gloor, P., 1979, Automatic recognition of inter-ictal epileptic activity in prolonged EEG recordings, *Electroencephalogr. Clin. Neurophysiol.* **46**:510–520.

Grassberger, P., and Procaccia, I., 1983a, Characterization of strange attractors, *Phys. Rev. Lett.* **50**:346–349.

Grassberger, P., and Procaccia, I., 1983b, Estimation of the Kolmogorov entropy from a chaotic signal, *Phys. Rev. A* **28**:2591–2593.

Hively, L. M., Gailey, P. C., and Protopopescu, V. A., 1999, Detecting dynamical change in nonlinear time series, *Phys. Lett. A* **258**:103–114.

Hively, L. M., and Protopopescu, V. A., 2003, Channel-consistent forewarning of epileptic events from scalp EEG, *IEEE Trans. Biomed. Eng.* **50**:584–593.

Hodgkin, A. L., and Huxley, A. F., 1952, A quantitative description of membrane current and its application to conduction and excitation in nerve, *J. Physiol. (London)* **117**:500–544.

Iasemidis, L. D., 2003, Epileptic seizure prediction and control, *IEEE Trans. Biomed. Eng.* **50**:549–558.

Iasemidis, L. D., Sackellares, J. C., Zaveri, H. P., and Williams, W. J., 1990, Phase space topography and the Lyapunov exponent of electrocorticograms in partial seizures, *Brain Topogr.* **2**:187–201.

Iasemidis, L. D., and Sackellares, J. C., 1991, The evolution with time of the spatial distribution of the largest Lyapunov exponent on the human epileptic cortex, in: *Measuring Chaos in the Human Brain* (D. Duke and W. Pritchard, eds.), World Scientific, Singapore, pp. 49–82.

Iasemidis, L. D., Sackellares, J. C., and Savit, R. S., 1993, Quantification of hidden time dependencies in the EEG within the framework of nonlinear dynamics, in: *Nonlinear Dynamical Analysis of the EEG* (B. H. Jansen and M. E. Brand, eds.), World Scientific, Singapore, pp. 30–47.

Iasemidis, L. D., Principe, J. C., Czaplewski, J. M., Gilmore, R. L., Roper, S. N., and Sackellares, J. C., 1997, Spatiotemporal transition to epileptic seizures: A nonlinear dynamical analysis of scalp and intracranial EEG recordings, in: *Spatiotemporal Models in Biological and Artificial Systems* (F. H. Lopes da Silva, J. C. Principe, and L. B. Almeida, eds.), IOS Press, Amsterdam, pp. 81–88.

Iasemidis, L. D., Shiau, D.-S., Chaovalitwongse, W., Sackellares, J. C., Pardalos, P. M., Principe, J. C., Carney, P. R., Prasad, A., Veeramani, P., and Tsakalis, K., 2003, Adaptive epileptic seizure prediction system, *IEEE Trans. Biomed. Eng.* **50**:616–627.

Jerger, K. K., Netoff, T. I., Francis, J. T., Sauer, T., Pecora, L., Weinstein, S. L., and Schiff, S. J., 2001, Early seizure detection, *J. Clin. Neurophysiol.* **18**:259–268.

Kantz, H., and Schreiber, T., 1997, *Nonlinear Time Series Analysis*, Cambridge University Press, Cambridge, UK.

Kaplan, D., and Glass, L., 1995, *Understanding Nonlinear Dynamics*, Springer-Verlag, New York.

Lange, H. H., Lien, J. P., Engel, J., and Crandall, P. H., 1983, Temporo-spatial patterns of pre-ictal spike activity in human temporal lobe epilepsy, *Electroencephalogr. Clin. Neurophysiol.* **56**:543–555.

Le Van Quyen, M., Adam, C., Baulac, M., Martinerie, J., and Varela, F. J., 1998, Nonlinear interdependencies of EEG signals in human intracranially recorded temporal lobe seizures, *Brain Res.* **792**:24–40.

Le Van Quyen, M., Martinerie, J., Baulac, M., and Varela, F., 1999, Anticipating epileptic seizures in real time by a non-linear analysis of similarity between EEG recordings, *NeuroReport* **10**:2149–2155.

Le Van Quyen, M., Adam, C., Martinerie, J., Baulac, M., Clemenceau, S., and Varela, F., 2000, Spatio-temporal characterizations of non-linear changes in intracranial activities prior to human temporal lobe seizures, *Eur. J. Neurosci.* **12**:2124–2134.

Le Van Quyen, M., Martinerie, J., Navarro, V., Baulac, M., and Varela, F. J., 2001a, Characterizing neurodynamic changes before seizures, *J. Clin. Neurophysiol.* **18**:191–208.

Le Van Quyen, M., Martinerie, J., Navarro, V., Boon, P., D'Have, M., Adam, C., Renault, B., Varela, F. J., and Baulac, M., 2001b, Anticipation of epileptic seizures from standard EEG recordings, *Lancet* **357**:183–188.

Lehnertz, K., and Elger, C. E., 1995, Spatio-temporal dynamics of the primary epileptogenic area in temporal lobe epilepsy characterized by neuronal complexity loss, *Electroencephalogr. Clin. Neurophysiol.* **95**:108–117.

Lehnertz, K., and Elger, C. E., 1998, Can epileptic seizures be predicted? Evidence from nonlinear time series analyses of brain electrical activity, *Phys. Rev. Lett.* **80**:5019–5023.

Lehnertz, K., Widman, G., Andrzejak, R., Arnhold, J., and Elger, C. E., 1999, Is it possible to anticipate seizure onset by non-linear analysis of intracerebral EEG in human partial epilepsies? *Rev. Neurol.* **155**:454–456.

Lehnertz, K., Andrzejak, R. G., Arnhold, J., Kreuz, T., Mormann, F., Rieke, C., Widman, G., and Elger, C. E., 2001, Nonlinear EEG analysis in epilepsy: Its possible use for interictal focus localization, seizure anticipation, and prevention, *J. Clin. Neurophysiol.* **18**:209–222.

Litt, B., Esteller, R., Echauz, J., D'Alessandro, M., Shor, R., Henry, T., Pennell, P., Epstein, C., Bakay, R., Dichter, M., and Vachtsevanos, G., 2001, Epileptic seizures may begin hours in advance of clinical onset: A report of five patients, *Neuron* **30**:51–64.

Lopes da Silva, F. H., Blanes, W., Kalitzin, S. N., Parra, J., Suffczynski, P., and Velis, D. N., 2003, Dynamical diseases of brain systems: Different routes to epileptic seizures, *IEEE Trans. Biomed. Eng.* **50**:540–548.

Lorenz, E. N., 1963, Deterministic non-periodic flow, *J. Atmos. Sci.* **20**:130–141.

Lytton, W. W., and Thomas, E., 1993, Modeling of thalamocortical oscillations, in: *Cerebral Cortex*, Vol. 13 (P. Ulinski *et al.* eds.), Plenum Publishers, New York, pp. 479–509.

Manuca, R., and Savit, R., 1996, Stationarity and nonstationarity in time series analysis, *Physica. D* **99**:134–161.

Manuca, R., Casdagli, M. C., and Savit, R. S., 1998, Nonstationarity in epileptic EEG and implications for neural dynamics, *Math. Biosci.* **147**:1–22.

McSharry, P. E., Smith, L. A., and Tarassenko, L., 2003, Comparison of predictability of epileptic seizures by a linear and a nonlinear method, *IEEE Trans. Biomed. Eng.* **50**:628–633.

Milton, J. G., Remillard, G. M., and Anfermann, F., 1987, Timing of seizure recurrence in adult epileptic patients: A statistical analysis, *Epilepsia* **28**:471–478.

Milton, J., and Jung, P. (eds.), 2003, *Epilepsy as a Dynamic Disease*, Springer-Verlag, Berlin.

Mormann, F., Lehnertz, K., David, P., and Elger, C. E., 2000, Mean phase coherence as a measure for phase synchronization and its application to EEG of epilepsy patients, *Physica D* **144**:358–369.

Netoff, T. I., and Schiff, S. J., 2002, Decreased neural synchronization during experimental seizures, *J. Neurosci.* **22**:7297–7307.

Nunez, P. L., 1981, *Electric Fields of the Brain: The Neurophysics of EEG*, Oxford University Press, Oxford.

Peitgen, H. O., Jurgens, H., and Saupe, D., 1992, *Chaos and Fractals: New Frontiers in Science*, Springer-Verlag, New York.

Pijn, J. P., Neervan, J. V., Noest, A., and Lopes da Silva, F. H., 1991, Chaos or noise in EEG signals: Dependence on state and brain site, *Electroencephalogr. Clin. Neurophysiol.* **79**:371–381.

Protopopescu, V. A., Hively, L. M., and Gailey, P. C., 2001, Epileptic event forewarning from scalp EEG, *J. Clin. Neurophysiol.* **18**:223–245.

Rieke, C., Mormann, F., Andrzejak, R. G., Kreuz, T., David, P., Elger, C. E., and Lehnertz, K., 2003, Discerning nonstationarity from nonlinearity in seizure-free and preseizure EEG recordings from epilepsy patients, *IEEE Trans. Biomed. Eng.* **50**:634–639.

Rogowski, Z., Gath, I., and Bental, E., 1981, On the prediction of epileptic seizures, *Biol. Cybern.* **42**:9–15.

Savit, R., Li, D., Zhou, W., and Drury, I., 2001, Understanding dynamic state changes in temporal lobe epilepsy, *J. Clin. Neurophysiol.* **18**:246–258.

Schouten, J. C., Takens, F., and van den Bleek, C. M., 1994a, Maximum-likelihood estimation of the entropy of an attractor, *Phys. Rev. E* **49**:126–129.

Schouten, J. C., Takens, F., and van den Bleek, C. M., 1994b, Estimation of the dimension of a noisy attractor, *Phys. Rev. E* **50**:1851–1861.

Stam, C. J., Pijn, J. P., and Pritchard, W. S., 1998, Reliable detection of nonlinearity in experimental time series with strong periodic components, *Physica D* **112**:361–380.

Takens, F., 1981, Detecting strange attractors in turbulence, in: *Lecture Notes in Mathematics, Dynamical Systems and Turbulence*, Vol. 898 (D. A. Rand and L. S. Young, eds.), Springer-Verlag, Berlin, pp. 366–381.

Theiler, J., Eubank, S., Longtin, A., Galdrikian, B., and Farmer, J. D., 1992, Testing for nonlinearity in time series: The method of surrogate data, *Physica D* **58**:77–94.

Towle, V. L., Carder, R. K., Khorasani, L., and Lindberg, D., 1999, Electrographic coherence patterns, *J. Clin. Neurophysiol.* **16**:528–547.

Traub, R. D., and Llinas, R., 1979, Hippocampal pyramidal cells: Significance of dendritic ionic conductances for neuronal function and epileptogenesis, *J. Neurophysiol.* **42**:476–496.

van Drongelen, W., Hereld, M., Lee, H. C., Papka, M. E., and Stevens, R. L., 2002, Simulation of neocortical activity, *Epilepsia* **43**(Suppl. 7):149.

van Drongelen, W., Nayak, S., Frim, D. M., Kohrman, M. H., Towle, V. L., Lee, H. C., McGee, A. B., Chico, M. S., and Hecox, K. E., 2003a, Seizure anticipation in pediatric epilepsy: Use of Kolmogorov entropy, *Pediatr. Neurol.* **29**:207–213.

van Drongelen, W., Koch, H., Marcuccilli, C., Pena, F., and Ramirez, J. M., 2003b, Synchrony levels during evoked seizure-like bursts in mouse neocortical slices, *J. Neurophysiol.* **90**:1571–1580.

van Erp, M. G., 1988, *On Epilepsy: Investigations on the Level of the Nerve Membrane and of the Brain*, Thesis, Leiden, The Netherlands.

Viglione, S., and Walsh, G., 1975, Epileptic seizure prediction, *Electroencephalogr. Clin. Neurophysiol.* **39**:435–436.

Wolf, A., Swift, J. B., Swinney, H. L., and Vastano, J. A., 1985, Determining Lyapunov exponents from a time series, *Physica D* **16**:285–317.

13

RETINAL BIOENGINEERING

Robert A. Linsenmeier*

Departments of Biomedical Engineering, and Neurobiology and Physiology,
Northwestern University, Evanston, Illinois

13.1. INTRODUCTION

The retina is a tiny piece of neural tissue, with each retina being only about 250 μm thick at the thickest point and weighing less than 100 mg in humans. The retina's importance is out of proportion to its size for two reasons. First, the retina has long served as a model for understanding complex parts of the nervous system, and it has attracted a great deal of attention from neuroscientists from all fields, including bioengineers. Many of its properties hold up well *in vitro*, and it is accessible to microelectrodes both *in vivo* and *in vitro*. It has a modest number of principal cell types, and the total number of output neurons (ganglion cells) in each eye is about 1 million in humans, and much less in other mammals, numbers that are almost manageable by comparison with the outputs of other parts of the central nervous system. The retina can be studied while it responds to its natural input, light, which can be controlled easily. For deeper neural structures, one often has to make the choice between studying responses to electrical stimulation, which is unnatural, or responses to natural inputs from other locations in the nervous system that may be difficult to control or completely characterize. The retina is also simpler than many areas of the brain because there is no significant feedback from the brain to the retina. In short, no other region of comparable complexity provides the advantages for study that the retina does, and this has allowed bioengineers to make considerable progress in understanding the retina in quantitative ways.

The second reason for the importance of the retina is its role in human lifestyles and performance, coupled with its sensitivity to disease. A large part of the brain is devoted to visual processing, and all of this relies on the transduction and initial visual processing steps that occur in the retina. Both our ability to receive information about the world and our mobility within it are ordinarily strongly dependent on vision. Unfortunately, the retina is a rather fragile tissue, and a number of genetic, vascular, and metabolic diseases interfere with its function. Just as engineers can contribute to understanding normal retinal function,

*Address for correspondence: Department of Biomedical Engineering, Northwestern University, 2145 Sheridan Rd, Evanston, Illinois 60208-3107; e-mail: r-linsenmeier@northwestern.edu.

they can help unravel the etiology of disease and assist in providing solutions to some of the many blinding diseases.

Diagnosing the problems of the diseased retina and repairing or providing substitutes for its functions are obviously within the purview of design-oriented neural engineers. However, there is also a large body of work by retinal bioengineers who have been engaged with other types of neuroscientists in measuring and modeling normal retinal function. Understanding retinal neural mechanisms will ideally provide some information for the design of artificial retinas.

This chapter focuses on aspects of retinal bioengineering related to mathematical modeling of neural responses and the retinal microenvironment, and aspects related to replacement of retinal function. These topics are at the intersection of bioengineering and neuroscience. Because of space limitations and the focus of this book, this review omits topics at the intersection of bioengineering and optics, which comprise another type of retinal bioengineering. Optical imaging of blood flow and the retinal structure are important in the identification of pathology and delivery of treatments. Laser-based techniques are used for both diagnostic applications, as in the case of the scanning laser ophthalmoscope, and therapeutic ones, as in panretinal photocoagulation.

13.2. THE NEURAL STRUCTURE AND FUNCTION OF THE RETINA

This review of retinal structure and function is necessarily brief, and more comprehensive views of the retina can be found in many books, including those by Rodieck (1973, 1998), Dowling (1987), McIlwain (1996), Oyster (1999), and a website by Kolb *et al.* (2002). The retina is the innermost of three layers comprising the posterior part of the eye (Figure 13.1). It lies inside the choroid, which has a vascular role, and the sclera, the fibrous coat that provides most of the structural rigidity. However, the eye maintains its shape only because secretion of fluids from the ciliary body (just behind the lens) keeps the intraocular pressure at about 15 mm Hg above atmospheric. The retinas of all vertebrates share basic structural and physiological similarities, but we will concentrate on the retina of mammals, especially cats and primates, which serve as the most directly relevant models for understanding the human retina (Figure 13.2).

13.2.1. PHOTORECEPTORS

At the back of the retina are the photoreceptor cells, which contain many stacked disks in their outer segments. The disk membranes contain the visual pigment, which absorbs light and begins the process of transducing light into electrical signals. In vertebrates, light leads to a hyperpolarization of the photoreceptors, as described more fully below. Photoreceptors in humans fall into two classes, rods and cones. Rods mediate vision (scotopic conditions) over about 6 log units of illumination, from the threshold of less than 0.001 quanta per second per rod up to about the illumination at dawn and twilight (Rodieck, 1998). The amplitude of rod responses then saturates, and cones gradually take over and are responsible for vision under the rest of the approximately 10 log units of illumination over which we have vision (Shapley and Enroth-Cugell, 1984; Rodieck, 1998) (photopic conditions). Still, in order

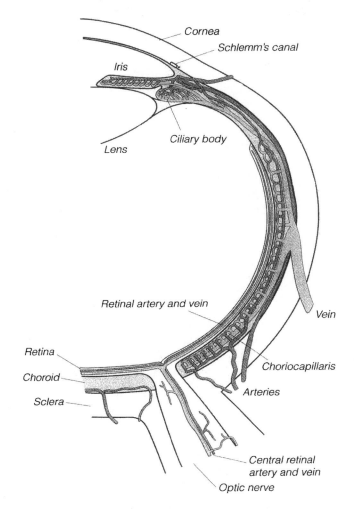

FIGURE 13.1. Structure of the three coats of the vertebrate eye, the sclera on the outside, the choroid in the middle, and the retina adjacent to the vitreous humor. The major arteries and veins are also shown. The central retinal artery enters through the optic nerve and feeds the capillaries of the retinal circulation within the retina. The central retinal vein drains this circulation, leaving the eye through the optic nerve. The completely separate choroidal circulation is fed by short posterior ciliary arteries that penetrate the sclera near the optic nerve, distribute in a capillary bed called the choriocapillaris, and is then drained by the vortex veins. (Reprinted from Rodieck, 1998, with permission)

to cover this entire range adequately, both rods and cones (and subsequent neurons) must adapt, or adjust their sensitivity as mean illumination changes, because the dynamic range at any given time for a rod or a cone is only about 2 log units (e.g., Rodieck, 1998). The transition region where rods and cones may both be involved is called the mesopic range.

There are four standard visual pigments in humans, one in rods and the other three in cones. All have the same light-absorbing component, the chromophore retinal, which

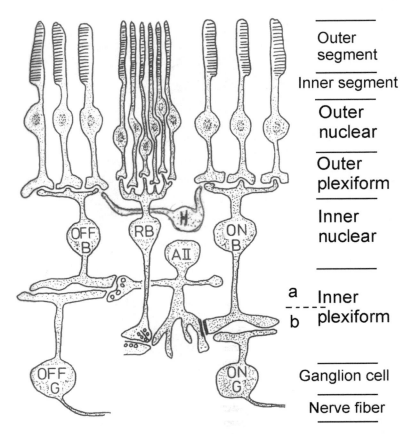

Outer
segment

Inner segment

Outer
nuclear

Outer
plexiform

Inner
nuclear

a   Inner
---- plexiform
b

Ganglion cell

Nerve fiber

FIGURE 13.2. Cell types and lamination of the mammalian retina. At the top are the photoreceptors, which comprise about half of the retinal thickness. The thinner photoreceptors are rods and the thicker ones are cones. They are intermixed in most of the retina. At the outer plexiform layer (OPL), rods contact rod bipolars (RB) and cones contact two main types of cone bipolars (OFF B and ON B) as described in the text. Horizontal cells (H) also make synapses in the OPL, receiving input from cones and providing output to other cones. Horizontal cell processes are also found in the rod–RB synaptic complex. Cell bodies of bipolar cells, horizontal cells and amacrine cells (A) are found in the inner nuclear layer. Connections of bipolars and amacrines to ganglion cells are found in the inner plexiform layer in separate sublaminae for the ON and OFF systems. Ganglion cell bodies are found in the ganglion cell layer, and their axons run in the nerve fiber layer, which is adjacent to the vitreous humor. There are a few subtypes of cone bipolars, and 10 to 20 subtypes each of amacrine and ganglion cells. The connections that are shown are the principal ones needed to explain the circuitry of Figure 13.3. Not shown are interplexiform cells, whose cell bodies are in the inner nuclear layer, and project from the inner plexiform to the outer plexiform, and the Müller cells, the principal glial cell of the retina, which spans all the layers except the outer and inner segments. (Modified from Wässle and Boycott, 1991, with permission of the American Physiological Society)

is derived from vitamin A, but they vary slightly in the protein, called opsin, to which the chromophore is attached. All the pigments respond to light over a wavelength range of more than half the complete visual spectrum (400 to 750 nm) but the slight differences in the proteins modify the spectral sensitivity of the pigment, so that rods and the three types of cones, now generally referred to as short-, middle- and long-wavelength (or S, M, and L), each have maximum absorbances at different wavelengths. Comparison between the outputs of different cones by second-order neurons is required to extract a wavelength signal and discriminate color. Although full color vision requires all three cone types, many humans, especially males, function reasonably well with only one or two cone types. Thus, it is vision at high illuminations, rather than color vision, that is the critical function of cones in humans.

13.2.2. *RETINAL CIRCUITS*

Photoreceptors make their synapses in the outer plexiform layer, the first of the two synaptic layers in the retina (Figure 13.2). At this location rods and cones project to separate subtypes of bipolar cells, which carry the signals forward, and to horizontal cells, which make lateral inhibitory connections back to other photoreceptors. The cell bodies of bipolar cells, horizontal cells, and amacrine cells (along with Müller cells, the principal glial cell of the retina) form the inner nuclear layer, and their outputs connect to several classes of ganglion cells at the inner plexiform layer. The ganglion cell bodies along with some "displaced" amacrine cells are located in the ganglion cell layer (GCL). Because of the different requirements for visual information going to different locations in the brain, ganglion cells are of several different physiological types, which are correlated with different anatomical types (e.g., Troy and Shou, 2002). It is believed that each of the major types of ganglion cells tiles the retina, providing several overlapping representations of the visual world (Wassle and Boycott, 1991). Because of the need to transmit signals over long distances, ganglion cells and some amacrine cells (e.g., Bloomfield, 1996) fire action potentials. Other retinal neurons ordinarily do not support action potentials, but instead control their transmitter release by graded potential changes.

Each ganglion cell sends an unmyelinated axon toward the optic disc (also called the optic nervehead) in the nerve fiber layer. The axons then pass through a modified part of the sclera called the lamina cribrosa at the optic disc. Past the lamina cribrosa, the axons become myelinated, and go on to higher structures (e.g., McIlwain, 1996; Troy and Shou, 2002). The most important of these are (1) the lateral geniculate nucleus of the thalamus, which is the major relay station for signals that travel to visual cortex to mediate visual perception, (2) the superior colliculus in the midbrain, which uses visual input to guide eye movements, and (3) the suprachiasmatic nucleus, which is the circadian pacemaker and uses visual input to synchronize the clock to the light–dark cycle.

There are several pathways by which photoreceptor signals reach the ganglion cells. Mammals have two main classes of bipolar cells, ON and OFF, subserving the cones, and another type, all ON bipolars, subserving the rods. As with other retinal neurons, ON and OFF refer to the type of stimulus that depolarizes the cell. An increase in illumination leads to a depolarization of ON bipolars and a decrease in illumination depolarizes OFF bipolars. The ON cone bipolar cells are sometimes called depolarizing cone bipolars, and OFF bipolars are sometimes called hyperpolarizing cone bipolars. All photoreceptors hyperpolarize with illumination, so they can be regarded as "off cells," although they are never actually called "off cells." Connections from cones to OFF bipolars therefore preserve the sign of the responses, and are fundamentally excitatory connections, whereas connections from cones

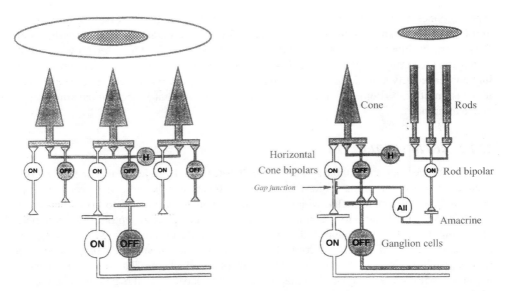

FIGURE 13.3. Cellular connections underlying the center and surround pathways of cone-driven ganglion cells (left) and rod-driven ganglion cells (right) in the mammalian retina. The ovals at the top represent a spot centered on the receptive field of cones or rods, and also on the relevant bipolars or ganglion cells. Shaded cells hyperpolarize in response to light, whereas clear cells depolarize. Thus, if a shaded cell connects to a clear cell, there is a sign-reversing or inhibitory synapse. In the cone pathways, the ON–OFF dichotomy arises at the OPL and is preserved in the IPL. In the rod system the situation is more complex because all rod bipolars are depolarizing (ON). The AII amacrine cells project to off-center ganglion cells through a sign-reversing synapse, and to on-center ganglion cells through a gap junction with ON bipolars. Note that on the right, the stimuli are too weak to stimulate the cones themselves. (Reprinted from Schiller, 1992, with permission from Elsevier)

to ON bipolars require a sign reversal, which implies an inhibitory connection (Figure 13.3). Photoreceptor to horizontal cell connections are excitatory, so horizontal cells are OFF cells. At the inner plexiform layer, where bipolars connect to ganglion cells, the ON/OFF separation of response types is preserved by segregated excitatory connections of ON cone bipolar cells to ON-center ganglion cells, and OFF cone bipolars to OFF-center ganglion cells (Schiller, 1992), although there is some evidence for more complexity (e.g., Belgum et al., 1982; Chen and Linsenmeier, 1989). These excitatory connections occur in separate sublamina of the inner plexiform layer, so both the "axons" of bipolar cells and dendrites of ganglion cells have to find the correct sublamina (Figures 13.2 and 13.3) (Nelson et al., 1978; Wassle and Boycott, 1991).

Rod signals converge onto the same ganglion cells as do cone signals, but, surprisingly, rod bipolars do not actually contact ganglion cells. Interposed in the pathway from rod bipolars to ON-center ganglion cells is an AII amacrine cell connecting to an ON cone bipolar via gap junctions. Interposed in the pathway from rod bipolars to OFF-center ganglion cells is the same AII amacrine cell connecting via chemical synapses to the OFF-center ganglion cell and OFF cone bipolar (Figures 13.2 and 13.3) (Kolb and Famiglietti, 1974; Wassle and Boycott, 1991). The connections shown in Figure 13.3 appear to be the most important ones under very strong and very dim illumination, but at intermediate levels of illumination, other signal pathways exist (Nelson, 1977; Smith *et al.*, 1986; Bloomfield and Dacheux, 2001).

As noted above, the rod "system" (i.e., the pathway from rods to rod-driven ganglion cells) is more sensitive to dim light than is the cone system. One reason is that the gain of the biochemical cascade inside the rod outer segment is greater, so one photoisomerization leads to a response of about 0.7 pA in primate rods and only about 0.033 pA in primate cones (Baylor *et al.*, 1984; Schnapf *et al.*, 1990). Thus, rods themselves are about 20 times more sensitive than cones. Because the increase of cone signals with illumination is shallower than the increase in rod signals, the half saturation value for primate cones is about 100 times greater than the half saturating intensity of rods (Baylor, 1987). An additional reason for the higher sensitivity of the rod system is that there is a greater convergence of rod signals at both the outer and inner plexiform layers. Near the center of the area centralis in the cat retina, for instance, about 30 rods make synaptic contact with a single rod bipolar cell, whereas only 4 cones contact each cone bipolar. Then, at the inner plexiform layer about 100 rod bipolars converge through interneurons onto a beta-type (or "X type," see below) ganglion cell that receives input from only 4 cone bipolars (Sterling *et al.*, 1986). Essentially, weak signals in the rod system have a greater chance to sum to noticeable signals in ganglion cells.

13.2.3. RECEPTIVE FIELDS

One of the important concepts for understanding the processing that occurs in the retina is the idea of a receptive field. The receptive field of a neuron in the visual system is defined to be that portion of visual space within which light will influence the neuron's behavior. This is directly related to a particular region of the retina, so the receptive field can also be discussed in terms of an area or distance on the retinal surface. The region of visual space and its projection onto the retina can both be specified in terms of the visual angle, as indicated in Figure 13.4. One degree of visual angle is about 0.294 mm on the retinal surface in humans. For reference, a U.S. quarter held at arm's length roughly subtends 2.4° of visual angle. The concept of eccentricity is also important. If one looks straight at an object, it is said to be at a visual eccentricity of 0°. If one moves the quarter horizontally by five quarter diameters but still gazes straight ahead, the quarter is now at an eccentricity of about 12° off the visual axis of the eye.

To a first approximation, retinal receptive fields are circular, but their form and size change as one moves from photoreceptors to ganglion cells. Photoreceptors have simple small receptive fields, since it is mainly the light falling on that photoreceptor that influences its membrane potential. Bipolar and ganglion cell receptive fields are somewhat larger, because of the convergence of signals from the cells preceding them. As noted above, bipolars and ganglion cells are named for the influence of increased light falling in the middle of the receptive field, as shown in Figure 13.5A by firing patterns and a histogram of firing frequency for an ON-center ganglion cell. As also shown in Figure 13.5, a reduction in illumination in the middle of the receptive field of an ON-center ganglion cell leads to the opposite effect as an increase in illumination: a suppression of the response or hyperpolarization. An additional feature is the "surround" region of the receptive field, or

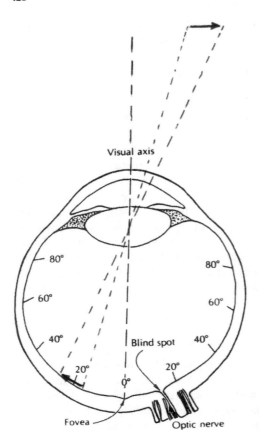

FIGURE 13.4. Illustration of the concepts of retinal eccentricity and visual angle. The fovea is defined to be at an eccentricity of 0°. The eye is viewing an object that subtends about 10° of visual angle, located at an eccentricity of about 20°. This is a top view of a left eye, because the blind spot (optic disk) is nasal to the fovea. (Modified from Cornsweet, 1970. Reprinted with permission from Elsevier)

the "surround mechanism" as first proposed by Kuffler (1953). Light falling outside of the middle of the receptive field region in a larger, concentric region has the opposite effect as light falling in the middle, antagonizing the effect of light on the center (Figure 13.5B). ON-center ganglion cells have OFF-surrounds and OFF-center cells have ON-surrounds. Surrounds are sometimes said to be "inhibitory," but this is not correct in the sense of synaptic inhibition, and it is preferable to refer to them as being "antagonistic." Increased activity in the surround pathway of an OFF-center ganglion cell depolarizes the ganglion cell, which is not an inhibitory action. Also, the surround pathway of an ON-center ganglion cell exerts a net inhibitory effect on firing when illumination is increased, but a net excitatory effect when illumination is decreased. The center and surround strengths are relatively well-balanced, as described below, so diffuse flashes, which stimulate both the center and surround, cause only a small change in the firing of ganglion cells, as shown in Figure 13.5B.

The first quantitative description of the receptive field of ganglion cells suggested that the influence of light was not uniform across the center (or surround), but in each case had a Gaussian weighting (Figure 13.6), so that a stimulus in the middle of the receptive field center would have a larger effect than one near the edge of the center (Rodieck, 1965; Rodieck and Stone, 1965). This idea is still accepted as valid for most of the ganglion cells

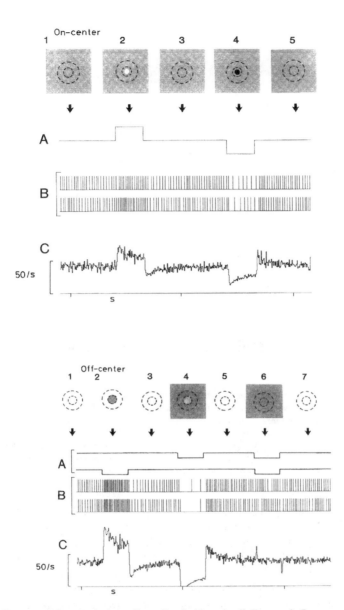

FIGURE 13.5. Responses of a cat retinal ganglion cell to flashing stimuli. *Top panels*: Responses of an on-center ganglion cell. The pictures show the spatial configuration of the stimulus with respect to the center (inner circle) and surround (outer circle) of the receptive field, trace A shows the time course of the stimulus, B shows the spike pattern in two repetitions of the stimulus, and C shows peristimulus time histograms (PSTHs) of the firing rate averaged over several presentations. The odd-numbered panels show times when the cell is subjected to a uniform gray background. In period 2, the firing rate is increased by presentation of a centered bright spot of light. In period 4 the centered stimulus is made dimmer than the background, causing the firing rate to decrease. The surround was not activated in this set of stimuli, but if an annulus of light brighter than the background had been presented to activate the surround, a response similar to that in panel 4 would have been observed. *Lower panels*: Responses of an off-center ganglion cell. The plan of the figure is the same, but now the dimming of a spot activates the center (panel 2), whereas the dimming of the surround suppresses firing (panel 4). Stimulation of both center and surround evoke only a small transient response from the cell. (Reprinted from Enroth-Cugell and Robson, 1984, with permission from the Association for Research in Vision and Ophthalmology)

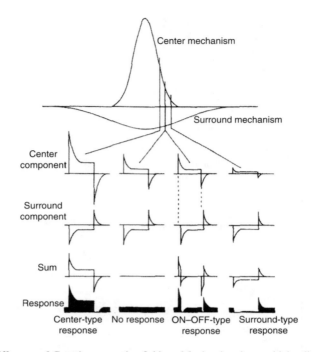

FIGURE 13.6. Difference of Gaussians receptive field model, showing the sensitivity distribution across the receptive field center, and the sensitivity distribution across the receptive field surround, which is shown below the horizontal line because it produces antagonistic responses. The rows of responses show hypothetical responses at three locations in the receptive field for an on-center cell for the receptive field center and surround, and representations of the ganglion cell membrane potential (sum) and firing rate (response), which would be a truncated version of the membrane potential because firing rate cannot be negative. (Reprinted from Rodieck and Stone, 1965, with permission from the American Physiological Society)

whose axons project to the lateral geniculate nucleus (Troy and Shou, 2002), and for a small fraction of the rest of the ganglion cells as well. Other ganglion cells (many W cells, see below) have receptive fields that cannot be described easily by the center-surround model.

It appears that horizontal cell feedback to photoreceptors in the outer plexiform layer is responsible for at least part of the surround signal, which is already present in bipolar cells at least in cold-blooded vertebrates (e.g., Kaneko, 1970; Hare and Owen, 1990). Less direct evidence suggests that the same pathway is important for photopic responses in mammals (Mangel, 1991). Evidence from rabbit AII amacrine cells indicates that their surround, and probably therefore the surround of ganglion cells in the scotopic range, is generated at the inner plexiform layer (Bloomfield and Xin, 2000).

13.2.4. ECCENTRICITY AND ACUITY

Retinal structure and function vary considerably with retinal eccentricity. On the optic axis of primates and humans is the fovea, a region about 5° in diameter (e.g., Oyster, 1999), in which the retina is thinner. Here the second- and third-order neurons are pushed aside, presumably for optical clarity, and the photoreceptors extend long lateral processes out to their bipolar cells. An area that is about 600 μm in diameter in the middle of the fovea contains only cones. This area has the best acuity and is ordinarily used for tasks like reading.

The concept of acuity is important in subsequent sections. The most obvious way to specify acuity is to consider the minimum spacing that is required between two points or lines in order that they can be seen as distinct objects rather than as a single object. This is set by the point spread function of the optics (see Oyster, 1999), but the eye is constructed so that the acuity that would be predicted by the photoreceptor spacing, without considering optical blur, is almost the same. That is, retinal anatomy is well matched to the best that the optics can do. Because these two ways of looking at acuity give essentially the same answer, we will discuss only the more intuitive concepts based on detector spacing.

If one had a pattern of dark and light lines of high contrast, i.e., a grating pattern, the minimum detectable spacing between light lines would be the spacing where two light lines were detected by two different photoreceptors, with another photoreceptor receiving less light between them. The minimal detectable line spacing is then twice the spacing between centers of the detector elements, which is about 2.5 μm in the fovea. (Photoreceptors are tightly packed, so this is also the diameter of one photoreceptor.) The maximum resolution in cycles of the grating that could be resolved per degree is the inverse of this minimal spacing in degrees. Therefore, if there are 300 μm across the retinal surface per degree of visual angle, the resolution limit or acuity should be about

$$R = (300 \, \mu m / deg) \times [1 \, cycle/(2 \times element \, spacing)] \tag{13.1}$$

The best acuity would then theoretically be about 60 cycles per degree, and this is not far from the actual acuity of a person with good vision (e.g., Cornsweet, 1970). A combination of factors gives the fovea the best acuity. First, the photoreceptors are smaller there, so a large number can be packed in. Second, foveal photoreceptors project through almost 1:1 connections (i.e., almost no convergence) through bipolar cells to ganglion cells. Third, the representation of this region in the visual cortex is large, so the detailed retinal information from this region is not lost. More peripherally, cone density falls and rod density rises, and the two are intermixed in most of the retina, so acuity decreases outside the fovea. The peripheral retina is, however, important in motion detection.

Acuity is often specified in terms of the visual ability of a person relative to the ability of a "normal" observer. The familiar "20/20" (or in metric units 6/6) vision means that an individual (numerator) can see at 20 feet what a "normal" person (denominator) can see at 20 feet. The features of the test stimulus in a standard eye chart that are just barely distinguishable (e.g., the gap that makes a "C" different from an "O") subtend 1 minute of arc at this distance. A person with 20/100 vision needs to be at 20 feet from the object to resolve what a "normal" person can see at 100 feet. In other words, the person with 20/100 vision would have to be five times closer to the object to achieve the same resolution as the normal person. If the poor acuity is due to optical imperfections in the eye, such as myopia (nearsightedness) or astigmatism, it can usually be corrected to 20/20 vision (or better 20/15 is not uncommon) with lenses. If poor acuity is due to disease of the retina or brain, or opacity of the lens, the same system of acuity designations is used, but the vision cannot be corrected optically. A person is legally blind if vision in the best eye, when best corrected, is no better than 20/200 or if the visual field is less than 20° in diameter. This is still useful vision, however, but is frequently called "low vision" rather than blindness. In clinical ophthalmology, there are several degrees of visual function worse than 20/200, the standard levels being "counting fingers," "hand motion," "light perception," and "no light perception," which describe exactly what one would expect from these designations.

13.3. VASCULATURE OF THE RETINA

Many retinal diseases are fundamentally vascular or have a vascular component, so it is important to consider the dual circulation of the retina. Wise *et al.* (1971), Bill (1984), and Alm (1992) provide reviews of this material. It is often useful to think of the retina as two domains. The outer retina, consisting primarily of photoreceptors, is supplied mainly by the choroidal circulation. The inner retina, consisting primarily of the second- and third-order neurons and glia, is supplied by the retinal circulation. Diseases that affect the vasculature typically cause blindness by affecting either the inner or outer retina, and then there may be secondary effects on the other region.

The choroidal circulation is behind the retina, separated from it by the retinal pigment epithelium (RPE) (Figures 13.1 and 13.7). The arterial inflow is via several short posterior ciliary arteries that enter the choroid near the optic nerve, and two long posterior ciliary arteries that travel within the sclera and enter the eye at about the equator (about half way between the edge of the cornea and the posterior pole) to supply the anterior choroid. The outflow from the choroid is via four vortex veins that exit the eye posterior to the equator (Wise *et al.*, 1971). The choroidal circulation has a very high flow rate, on the order of 1000 ml/100 g·min (e.g., Alm and Bill, 1972), but in cat and primate this is normally sufficient to supply only the photoreceptors, not other retinal neurons (e.g., Linsenmeier,

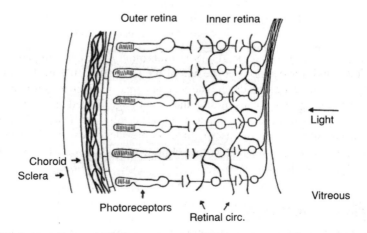

FIGURE 13.7. Schematic showing the relation of the choriocapillaris and the retinal capillaries to the retinal neurons. The choriocapillaris is separated from the photoreceptors by the retinal pigment epithelial cells. The retinal circulation only occupies the inner half of the retina, leaving the photoreceptors in an avascular region that is more than 100 μm thick. See also Figure 13.1.

1986; Linsenmeier and Padnick-Silver, 2000). The flow rate is high because the vessels are large and the resistance is low. The capillaries are 10 μm or more in diameter (Wise *et al.*, 1971). Choroidal capillaries are fenestrated, and the blood–retinal barrier, analogous to the blood–brain barrier, is provided by the tight junctions between RPE cells rather than by the choroidal endothelium. The RPE also pumps fluid from the retina into the choroid, which is important for keeping the retina attached to the back of the eye (Marmor, 1998), because there are no junctional complexes of any kind between the photoreceptors and RPE cells. Choroidal blood flow is controlled by the autonomic nervous system, and there is almost no matching of its flow rate to the metabolic needs of the outer retina, even though this circulation is of critical importance to the photoreceptors. The choroidal circulation exhibits some autoregulation in response to changes in arterial pressure, but less in response to changes in intraocular pressure (Kiel and Shepherd, 1992).

The retinal circulation supplies the inner half of the retina. In humans the central retinal artery enters the eye at the optic nervehead and then branches to form four retinal arteries that travel superficially in the retina. These in turn branch into a relatively typical arteriolar and capillary network that forms two layers in most of the inner retina, with the innermost layer in the ganglion cell layer, and the deeper layer extending as far as the outer plexiform layer. The venous drainage is via a central retinal vein that exits the eye next to the central retinal artery. The retinal circulation is similar to brain circulation, with a flow rate of 20 to 40 ml/100 g·min (e.g., Ahmed *et al.*, 2001). The retinal circulation has tight junctions between capillary endothelial cells that form the blood–retinal barrier. In contrast to the choroidal circulation, it has no autonomic control, but good autoregulation with respect to arterial pressure (Alm, 1992). It also responds to changing metabolic conditions in the inner retina. Blood flow increases during hypoxemia (e.g., Eperon *et al.*, 1975) decreases during hyperoxia (Riva *et al.*, 1983), and also increases in response to greater retinal neural activity (Bill and Sperber, 1990; VoVan Toi and Riva, 1994; Kiryu *et al.*, 1995).

13.4. MAJOR RETINAL DISEASES

It is estimated that there are nearly 15 million blind or visually impaired people in the United States and, for people over the age of 65, more than 10% of the population is legally blind. Age-related macular degeneration, glaucoma, and diabetic retinopathy are the most prevalent retinal diseases causing blindness, with age-related cataract as the fourth largest cause (Braille Institute, 2002). Glaucoma is especially significant because it affects many working-age adults. A few diseases with lower prevalence also need to be considered because of the potential for bioengineering solutions. In this discussion we move from genetic disorders to vascular disorders.

13.4.1. RETINITIS PIGMENTOSA

The most common hereditary cause of blindness is called retinitis pigmentosa (RP). It affects about 1 in 4000 (Heckenlively *et al.*, 1988) to 1 in 3000 (Saleem and Walter, 2002). There are more than 50 genetic defects in photoreceptor or RPE proteins that lead to loss of photoreceptors. These may be autosomal or sex-linked and can be dominant or recessive.

Many of these diseases cause loss of rods first, with cone degeneration following later (rod–cone degeneration), and a few act in the reverse manner (cone–rod degeneration). The inner retina seems largely normal during the time when photoreceptors are degenerating, but later there is often a loss of inner retinal neurons as well (e.g., Humayun *et al.*, 1999b). Interestingly, there is also a loss of retinal vasculature over time (Grunwald *et al.*, 1996; Penn *et al.*, 2000). Oxygen from the choroid is generally used by the photoreceptors, but when photoreceptors are sick or absent, they use less oxygen, allowing oxygen from the choroid to diffuse into the inner retina (Linsenmeier *et al.*, 2000), where the vasculature constricts and then becomes permanently damaged (Penn *et al.*, 2000).

The time course of RP is variable, with some types leading to blindness in adolescence, while others progress more slowly. There are several cases in which the genetic defect is known exactly, and some in which an animal model appears to be a good model for the human disease (e.g., LaVail, 1981). In some cases it is clear why the genetic defect kills photoreceptors. For instance, in the rd mouse (Farber and Lolley, 1973) and Irish Setter (Aguirre *et al.*, 1978) there is a reduced activity of PDE, which presumably increases the concentration of cGMP, keeping the plasma-membrane cation channels open and flooding the photoreceptor with more Na^+ and Ca^+ than can be pumped out effectively and causing apoptosis (Fox *et al.*, 1999). In most types of RP, however, the signaling pathway that connects the gene defect to cell death is not clear. A number of treatments are being investigated, of which the most successful appears to be gene transfer. In the Briard dog, as in humans with a type of RP called Leber's Congenital Amaurosis, the defect is a null mutation causing the absence of the protein RP65, an enzyme usually expressed in the RPE that is responsible for creation of all-trans retinal. Without retinal, rhodopsin is not functional, lipoid deposits build up in the RPE, and animals are blind. Gene transfer to the subretinal space using an adeno-associated virus has been able to restore photoreceptor structure, retinal electrical responses, and vision in dogs (Acland *et al.*, 2002; Narfstrom *et al.*, 2002).

13.4.2. MACULAR DEGENERATION

A more prevalent type of photoreceptor degeneration is age-related macular degeneration (AMD or ARMD). This occurs in at least 1 in 100 adults over the age of about 40, and its incidence is considerably higher in adults over the age of 65 (Klein, 1999). A hallmark of macular degeneration is the presence of extracellular deposits called drusen between the RPE and choroid. Ordinarily the outer segments of photoreceptors undergo continuous renewal, with new disks being synthesized at the base of the outer segment, and old disks being shed at the tip and phagocytosed by the RPE (Young, 1976). Drusen contain both lipids and proteins, and may be the waste product of incomplete digestion of the photoreceptor outer segments by the RPE. There is debate over whether drusen are causally related to photoreceptor loss, since they can be present with no visual symptoms in "dry" AMD. Dry AMD sometimes proceeds to the more severe "wet" or exudative form, and it may be that the larger the drusen are, the more they impede transport of both large and small molecules between the choroid and the retina. In the exudative form, which is responsible for 75% of cases with severe visual loss, there is choroidal neovascularization (CNV), in which choroidal vessels proliferate, break through the RPE, and enter the retina and vitreous (Abdelsalam *et al.*, 1999). Like all neovascularizations of the retina, these vessels are abnormal, and may cause traction on the retina, resulting in retinal detachment or other damage. There may be a genetic component of AMD, but the major risk factors are smoking and hypertension, and females are at higher risk than males (Klein, 1999). Carotenoids in the diet, especially lutein and xeaxanthin, may be partially protective (Cho *et al.*, 1999), suggesting that oxidative damage is a component of the disease. Laser treatment is effective against CNV. A number of the same strategies being investigated for RP, such as RPE and retinal cell transplantation, are being investigated for use in AMD.

13.4.3. GLAUCOMA

Glaucoma is a slow, neurodegenerative disease of a different type, because the affected cells are not photoreceptors but ganglion cells. The principal risk factor for glaucoma is elevated intraocular pressure (IOP), from its normal value of about 15 mm to 2 SD higher than the mean, about 22 mm Hg (Stamper and Sanghvi, 1996). Glaucoma is quite prevalent. It is estimated to occur in about 0.8 in 100 to about 3 in 100 Caucasians, and the incidence is higher in African Americans (Sassani, 1996). The elevation of IOP is generally caused by a decrease in the conductance, C (called the "outflow facility"), to flow of the aqueous humor out of the eye. About 75% of this flow is pressure dependent and occurs across a complex structure in the "angle" where the cornea meets the iris, and the rest is a pressure-independent uveoscleral outflow, U (Stamper and Sanghvi, 1996). The total flow out of the eye, F, must balance the active secretion of aqueous humor across the ciliary epithelium, which does not depend on IOP. The static relation between pressure and outflow of aqueous humor is then given by

$$F = (\text{IOP} - P_e)C + U \tag{13.2}$$

or

$$\text{IOP} = (F - U)/C + P_e \tag{13.3}$$

where P_e is the pressure in the veins outside the eye, the episcleral veins. Because F must match the inflow, any decrease in C is accompanied by an increase in IOP. Increased inflow of aqueous humor by active secretion at the ciliary body could also theoretically force an increase in F and contribute to elevating pressure as well, but this is rarely if ever the cause of elevated IOP. Treatments for glaucoma, however, typically do attempt to reduce secretion of aqueous humor by pharmacological means.

High IOP probably causes damage to the retina by compressing optic nerve fibers as they pass through the lamina cribrosa. This blocks axonal transport and causes retrograde degeneration of the ganglion cells (e.g., Oyster, 1999; Quigley *et al.*, 2000). As the nerve fibers are lost, a characteristic depression, or cupping, of the optic disk develops that is visible ophthalmoscopically. As the disease progresses, there is a loss of visual function. This is detectable with standard perimetry, in which the patient is asked to adjust the intensity of spots of light presented at different points in the visual field so that they are at a threshold intensity (i.e., just visible) (e.g., Johnson, 1996; Prince and Solomon, 1996). Glaucoma patients usually exhibit a loss of sensitivity (elevation of threshold) first in the mid-periphery of the nasal visual field (temporal retina), and the loss gradually progresses closer to the central visual field (e.g., Oyster, 1999). Unfortunately, standard automated perimetry does not detect early changes. It has been estimated that one can lose up to 50% of ganglion cells before the loss is detectable with standard visual testing (Quigley *et al.*, 1989; Harwerth *et al.*, 1999).

Although high IOP is the single most important risk factor, some individuals with elevated pressure (ocular hypertension) do not exhibit the retinal symptoms of glaucoma. On the other hand, some individuals with normal IOP, and others whose elevated IOP is lowered by drugs, still develop disc cupping and visual loss (Sassani, 1996). These individuals are classified as having normal tension glaucoma. There have been persistent theories that it is not just the mechanical effects of high intraocular pressure that matter, but vascular effects of the perfusion pressure in the optic nervehead circulation (e.g., Hayreh, 1978). Low arterial pressure or poor autoregulation of the circulation could have a similar effect to increased IOP and explain the cases of normal tension glaucoma, as well as some effects of high tension glaucoma. Unfortunately, experimental investigation of the circulation in this critical region is very difficult.

13.4.4. DIABETIC RETINOPATHY

In both insulin-dependent (type 1) and non–insulin-dependent (type II) diabetes, elevated blood glucose over many years can lead to microvascular complications in several organs, including the eye, the kidney, and peripheral nerves. Almost all diabetics with a disease duration of more than 20 years show some signs of retinopathy (Engerman *et al.*, 1982). The earliest clinical signs of retinopathy are microaneurysms and capillary leakage that are especially apparent in fluorescein angiograms (Engerman *et al.*, 1982; Kincaid, 1996). At the microvascular level, there is a loss of pericytes, a cell type that along with endothelial cells makes up the capillary wall, as well as a thickening of the basement membrane of retinal capillaries, and plugging of capillaries with leukocytes and platelets (e.g., Schroder *et al.*, 1991; Hatchell and Sinclair, 1995). Clinically the next stage is further fluid leakage including hemorrhage, and capillary nonperfusion in patches across the retina (Engerman *et al.*, 1982). With the loss of retinal capillaries comes loss of visual function. As the capillary dropout progresses, there is a growth of new, abnormal tufts of blood vessels. These can grow out into the vitreous, collapse, and shrink, and cause traction on the retina. Because the retinal attachment to the back of the eye is tenuous, this traction can detach the retina from the RPE, leading to blindness.

There is evidence that diabetics who take measures to tightly control their blood glucose have less retinopathy (Chase *et al.*, 1989) and that institution of good control can retard the progression of retinopathy (DCCT Research Group, 1993). In some patients, however, sudden tight control after years of poor control may worsen retinopathy (Grunwald *et al.*, 1994). Currently the treatment for diabetic retinopathy is panretinal photocoagulation, in which hundreds of small laser burns are made across the retina (e.g., Diabetic Retinopathy Research Group, 1976). The theory is that this allows oxygen from the choroid to diffuse into the inner retina, ending retinal hypoxia (Wolbarsht and Landers, 1980) and removing the stimulus for upregulation of vascular endothelial growth factor (VEGF) that is important in both increased vascular permeability and angiogenesis in diabetes (Adamis *et al.*, 1994; Kunz Mathews *et al.*, 1997). Another serious consequence of diabetes that may be related to the vascular effects is macular edema, for which laser photocoagulation and intraocular steroids are the main treatments.

13.4.5. VASCULAR OCCLUSIVE DISEASE

Like the brain, the retina is susceptible to vascular occlusive events that occur from thrombi or atherosclerosis in either the arteries or veins (e.g., Brown, 1999). These produce the retinal equivalent of strokes. The most serious type of occlusion is one that affects the central retinal artery, because this prevents circulation to the entire inner retina. Occlusion of a branch artery produces a scotoma (blind spot) in the region supplied by that vessel because there is no redundancy in the retinal circulation. Experimentally produced occlusions lead to irreversible damage to the primate retina if they last more than about 2 h (Hayreh and Weingeist, 1980), which is a much longer window than one has for the brain, possibly because the vitreous provides a small reservoir of oxygen and glucose. Many treatments have been attempted, but there are no accepted treatments for arterial vascular occlusion. Venous occlusions often produce multiple hemorrhages in the retina, but may resolve without permanent visual loss (Clarkson, 1994). Venous occlusions may, however, lead to neovascularization in either the retina or the iris (Hayreh et al., 1983; Pournaras, 1995), probably because VEGF is produced in the retina and diffuses to the anterior part of the eye (Adamis et al., 1996).

13.4.6. RETINAL DETACHMENT

A frequent result of proliferative diabetic retinopathy is the detachment of the retina. There are other causes for detachment as well, including trauma, severe myopia, detachments of the vitreous from the retina, and retinal holes of idiopathic origin (Michels et al., 1990). In all cases, fluid gains access to the subretinal space between the retina and the RPE, lifting the retina off and sometimes detaching large areas. The photoreceptors, being separated from the choroid and deprived of their main source of nutrition, undergo apoptotic cell death unless the retina is reattached by one of several surgical procedures that bring the retina and eye wall closer together (Michels et al., 1990). The quality of vision following these procedures depends on the time between detachment and reattachment and on whether the detachment had reached the macula.

13.5. ENGINEERING CONTRIBUTIONS TO UNDERSTANDING RETINAL PHYSIOLOGY AND PATHOPHYSIOLOGY

Engineering approaches to understanding retinal function date to the 1960s. The work has been done by a combination of physiologists, psychophysicists, and biomedical engineers who have constructed mathematical models of the retina with several goals in mind. These include providing a compact representation of a great deal of data, extracting parameters characterizing retinal function and then investigating how those parameters vary with independent variables such as adaptation level and retinal eccentricity, and devising systems models whose transfer functions are similar to those of the retina. There are models of many aspects of retinal function, and even more of visual function. It will not be possible to review all of this work, but the intention is to review some of the major analytical threads that constitute retinal bioengineering. First, models have been constructed of how light is transduced by the photoreceptors into an electrical signal. Second, some models take as their basis the transformations that occur at different stages of retinal processing and are grounded in the electrophysiological recordings from interneurons and the anatomical relations among them as seen in synaptic ultrastructure and dendritic and axonal arborizations. Third, there are a number of models describing the receptive field and response properties of retinal ganglion cells in cats and monkeys, based on extracellular recordings. These typically treat the entire retina as a black box, with patterns of light as the input and ganglion cell firing behavior as the output. Fourth, there are a number of models of the retinal microenvironment focusing on nutrient and ionic balance. In these four areas, the models are all based on data, rather than being totally theoretical constructs. Of course this does not mean that they are necessarily the best models or unique models, but they are constrained and at least have descriptive validity. These research areas are also ones in which sustained effort and refinement of models has taken place over many years. The models have led to conclusions pertinent to the human retina, although the data were often derived from other animals. Only the photoreceptor models are based directly on human data.

Several major threads are omitted by this choice of topics. Some of these are (1) models of information transfer through the tiger salamander retina (e.g., Werblin, 1991; Roska *et al.*, 2000), (2) cybernetic models that attempt to explain general properties of retinal responses, but not to account for data in a detailed way or to derive parameter values (e.g., Moreno-Diaz and Rubio, 1980; Oguztoreli, 1980), (3) models of spatiotemporal transfer properties of horizontal cells (e.g., Foerster *et al.*, 1977; Tranchina *et al.*, 1983), and (4) models of light and dark adaptation (e.g., Shapley and Enroth-Cugell, 1984).

13.5.1. PHOTORECEPTOR MODELS

Models of photoreceptor function have sought to quantify the relation between incident light and photoreceptor hyperpolarization. The biochemistry, physiology, and biophysics of photoreceptors have been reviewed in Baylor (1987), Yau (1994), and Pugh and Lamb (2000), among others. In darkness, photoreceptors have resting potentials that are depolarized relative to those of many other neurons (ca. -30 mV), because their outer segments have a cation-selective channel with a high Na^+ and Ca^{2+} conductance. More of these channels are open in the dark, keeping the cell depolarized. The number of channels that are open, and therefore the current entering the cell, is determined by the level of cGMP, with more cGMP leading to more open channels. Absorption of a photon causes activation of rhodopsin, which then activates another protein bound to the photoreceptor discs, called transducin. Transducin is a G-protein, which requires GTP binding for activity, and whose activity is terminated by hydrolysis of GTP. Transducin in turn activates a phosphodiesterase, which breaks down cGMP. Decreased cGMP closes the channels, decreasing the inflow of Na^+ and Ca^{2+} and causing hyperpolarization. (The synthesis of cGMP from GTP is controlled by light only indirectly, when levels of Ca^{2+} in the cytoplasm decrease and the activity of guanylate cyclase increases). The light-evoked hyperpolarization increases in amplitude with light, up to a saturating value. The dependence of response amplitude on illumination has often been characterized by Eq. (13.4), which is sometimes called the Naka–Rushton equation (Naka and Rushton, 1966) when it is used in vision:

$$R = R_{\max} \left(\frac{I}{I + \sigma} \right) \tag{13.4}$$

where R is the response amplitude at intensity I, R_{max} is the maximum amplitude, and σ is the illumination at half-saturation. Sometimes this does not rise steeply enough to fit the data, so a modified form is used in which n is greater than 1.0.

$$R = R_{max} \left(\frac{I^n}{I^n + \sigma^n} \right) \tag{13.5}$$

Amplitude–intensity data from many stages subsequent to the photoreceptors can also be well described by one of these equations. In addition to the change in response amplitude with illumination, the time course of the photoreceptor hyperpolarization also speeds up with increasing illumination, and the leading edge continues to become steeper even after amplitude saturation occurs (Figure 13.8). The current that flows into the outer segments is completed by current (mostly K^+) flowing out of the inner segments (Hagins *et al.*, 1970). The first recordings from individual primate photoreceptors were made by sucking the outer segments of isolated photoreceptors into a pipette and forcing the receptor current to flow through the electrode (Baylor, 1987).

In the intact retina, some of the receptor current flows out of the retina and across the wall of the eye. The voltage drop associated with this current produces a negative-going signal as large as several hundred microvolts that can be recorded between the vitreous

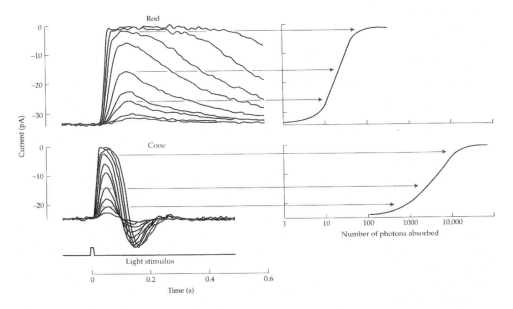

FIGURE 13.8. Responses of membrane current of a primate rod (top) and cone (bottom) outer segment, each in response to several brief stimuli of different intensities. Responses were recorded by sucking the outer segment of isolated photoreceptors into a pipette and recording the current. Inward current is reduced by light. The rod reaches saturation, with all channels closed, in the top two traces. These are essentially the impulse responses of the photoreceptors, and are the inverse of the voltage changes that would be observed with an intracellular electrode if the outer segment could be isolated. The cone responses are characteristically faster. The half-saturating intensity for the cones was 100 times that required for the rods. [Reprinted from Oyster, 1999 (after Baylor, 1987) with permission from Sinauer Associates]

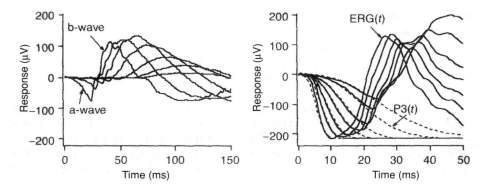

FIGURE 13.9. *Left*: Electroretinograms (ERGs) in response to flashes of several intensities in the dark-adapted human retina, showing the a-wave, originating from the photoreceptors and the b-wave, originating from the bipolars. *Right*: Fits of the photoreceptor model (P3(t)) described in the text ($n = 4$; $t_p = 189$ ms) to the early part of the ERGs. The intensities used for the right half of the figure were 2 to 4 log scotopic td-sec, which were higher than those on the left. (Reprinted from Hood and Birch, 1995, with permission from IEEE)

humor or cornea and a reference electrode. This makes it possible to record photoreceptor activity from the surface of the human eye as part of the electroretinogram (ERG). The ERG manifestation of the photoreceptor signal is often called P3, or PIII, because it was the third component of the ERG to disappear following treatment with ethyl alcohol or anoxia (Granit, 1933). If the stimulus is very bright, the initial part of P3 is observed in almost pure form as the "a-wave" of the ERG (Figure 13.9). The entire photoreceptor current is not usually observed in ERG recordings, because voltage-dependent conductances in the inner segment come into play, which make the photoreceptor voltage response different from the current response (Fain *et al.*, 1978; Schneeweis and Schnapf, 1995). In addition, activity from other types of retinal cells begins before the photoreceptors are done responding. The next major component is the b-wave, which is caused by the activity of bipolar and Muller cells (e.g., Robson and Frishman, 1995). However, for up to about 15 ms after a brief flash, the a-wave is thought to be a good reflection of the light-dependent current in the outer segment. Even though the ERG is a complex set of potential changes, its clear advantage is that it can be used to study retinal electrophysiology in the intact human eye, and potentially to investigate how disease processes affect different types of retinal neurons.

It should not be surprising that models of photoreceptor activity apply both to the signals from individual rods and to the leading edge of the a-wave of the ERG, as long as one uses diffuse light, which stimulates many photoreceptors. There have been two different approaches to modeling of the onset of photoreceptor activity that eventually converged to the same mathematical form. One model fitted families of a-wave responses to different brief flash intensities by an input–output analysis having a few characteristic parameters (Baylor *et al.*, 1984; Hood and Birch, 1990, 1995). The other attempted to characterize each of the known steps in transduction by an equation, and then to couple these individual equations into an overall model (Lamb and Pugh, 1992; Pugh and Lamb, 2000).

13.5.1.1. Input–Output Analysis of Rod Responses

The input–output analysis that describes the a-wave data (Hood and Birch, 1990) consists of an n-stage low pass filter for $r(t)$, the impulse-response function of the photoreceptors, in which the time-to-peak of the response is t_p.

$$r(t) = \left\{ \left(\frac{t}{t_p} \right) \exp\left[1 - \left(\frac{t}{t_p} \right) \right] \right\}^{(n-1)} \tag{13.6}$$

Here each response is normalized to a peak response of 1.0, so it does not depend on illumination. This is shown in Figure 13.9 for $n = 4$ and $t_p = 189$ ms, which provided good fits to data. The a-wave, called P3(i, t) in this analysis, depends on $r(t)$ and on the intensity of a brief flash of energy, i (in scotopic troland-seconds, a measure of light incident at the cornea). This involves a second stage, which is a saturating exponential nonlinearity:

$$P3(i, t) = \left[1 - \exp\left(\frac{-\ln 2ir(t)}{\sigma} \right) \right] R_{mP3} \tag{13.7}$$

Equation (13.7) introduces two new variables. R_{mP3} is the maximum amplitude of P3 in response to bright flashes. It is assumed to be just the sum of maximum responses of individual photoreceptors. The other new parameter is the semisaturation constant, σ, which reflects the sensitivity of the photoreceptor to light. It is the value of $ir(t)$ at which P3(i, t) $= R_{mP3}/2$. These two equations were used to fit families of a-waves, setting t_p and n fixed and extracting R_{mP3} and σ. The chosen value of n was 4, implying a four-stage filter. The parameter t_p is not directly observable in a-wave recordings, because other waves intrude before t reaches t_p, so t_p was set for human a-wave recordings to be the value observed in primate rods, 189 ms. Fits of this model for the human retina are shown in Figure 13.9.

13.5.1.2. Biochemically Based Analysis of Rod Responses

Lamb and Pugh (1992) derived an alternate model that was based on the biochemical steps in transduction. In this very detailed model, also presented in a simplified form by Breton et al. (1994), the dynamics of five major processes were considered: (1) activation of rhodopsin by light, (2) activation of transducin by rhodopsin, (3) activation of PDE by activated transducin, (4) hydrolysis of cGMP by activated PDE, and (5) channel closure caused by the fall in cGMP. Other models had taken similar approaches (Cobbs and Pugh, 1987; Forti et al., 1989), but this one started at the most molecular level. It was the first to explicitly consider that rhodopsin diffuses in the disc membrane to cause activation of many transducin molecules, which converts the step activation of rhodopsin by a flash into a ramp increase in transducin activity. Formally, it also allowed for the longitudinal diffusion of cGMP in the cytoplasm, although only isotropic conditions (i.e., illumination of the whole outer segment) were considered. It turned out that the overall gain of transduction was an important parameter that came from the model. This parameter has been called A in subsequent work, and was the product of the gains of steps 2 through 5 above. A was related to the characteristic time constant of transduction, τ_ϕ, by $A = \tau_\phi^{-2}$. In terms of timing, processes 2 and 4 above were found to contribute substantially to the time course of the flash responses, and the others were very fast by comparison. A small delay called t_{eff} was also needed to

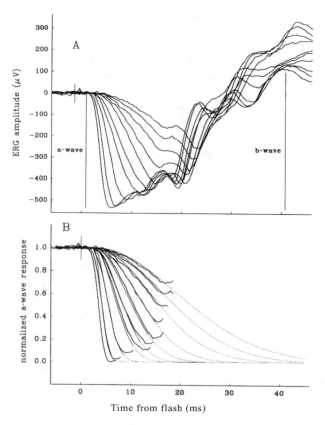

FIGURE 13.10. *Top*: The first 40 ms of human ERGs in response to brief flashes ranging from 402 to 128,000 photoisomerizations per rod. Each is the average of four stimulus presentations, except at the highest intensity, which is an average of two. *Bottom*: Fits (dotted line) of the model described in the text to the early parts of these responses. The same parameters, $A = 8.7$ s^{-2} and $t_{\text{eff}} = 2.7$ ms, were used to fit all responses. (Reprinted from Breton *et al.*, 1994, with permission from the Association for Research in Vision and Ophthalmology)

account for the onset of a noticeable change in PDE activity. The overall response was then

$$R(\Phi, t) = \left[1 - \exp\left(-\frac{1}{2}\Phi A[t - t_{\text{eff}}]^2\right)\right] R_{\text{m}} \qquad (13.8)$$

Here Φ is the intensity in isomerizations of rhodopsin, rather than scotopic troland-seconds. These two intensity units are related by a constant that depends on the optics and light-capturing efficiency of rods, and this is different for different animals. Although the Lamb–Pugh and Hood–Birch formulations look different, Hood and Birch (1995) showed that they had very similar forms if $t < t_{\text{p}}$, which are the only times at which either model can be applied. The Lamb and Pugh model was originally applied to salamander rod responses, but it does a good job of fitting a-waves in human ERGs as well (Breton *et al.*, 1994) (Fig. 13.10). For human rods, the amplification constant is about 100 times higher, which means that the responses develop about 100 times faster. The fits to the ERG a-wave are even better if one rectifies certain simplifying assumptions that were made originally (Cideciyan and Jacobson, 1996). These are (1) taking into account the photoreceptor membrane time constant, which was ignored originally because the responses modeled were current responses, (2) allowing the isomerizations to take place over a short interval rather than all at $t = 0$, and (3) recognizing that for high-intensity flashes, the response time course will be on the time scale of t_{eff}.

13.5.1.3. Paired Flash Analysis

One clear limitation of these analyses is that they were only intended to address photoreceptor behavior at the onset of light, and not the recovery as rhodopsin, and the subsequent steps are inactivated. There are several mechanisms to turn photoreceptors off, so only a partial model of the entire response has emerged (e.g., Pugh and Lamb, 2000). Fitting a complete model would require knowing the complete time course of the photoreceptor current, but this cannot generally be recorded in humans because responses from other retinal neurons contribute to the ERG. When one uses a test flash to evoke a sustained response from photoreceptors, one may get a complex ERG. However, if the test flash is followed at different intervals by a brief, bright "probe" flash designed to drive the photoreceptor current all the way to saturation (the "paired-flash technique"), one can determine how far the photoreceptors were from saturation before the probe flash, and this allows reconstruction of the complete photoreceptor response to the test flash (Figure 13.11) (Birch *et al.*, 1995; Pepperberg *et al.*, 1996, 1997). An extension of this work led to the development of a descriptive equation that characterized the complete time course of rod responses in mice, but this did not link biochemical steps to their electrical consequences (Hetling and Pepperberg, 1999). This model is similar in form to Eq. (13.7) above, but includes the dynamics of what is supposed to be the underlying single-photon response.

13.5.1.4. Diagnostic Value of a-Wave

The ERG has always had some diagnostic value, because it is the only objective measure of retinal neural function available for use in humans. Until the models allowed a deeper understanding of the waveforms, however, most of the diagnoses were simply based on the presence or absence of components in the responses to flashes of light. These may or may not have been optimal for revealing particular disease processes. With more quantitative models of the ERG, more detailed conclusions have become possible. To give just one example, it appears that in a particular type of retinitis pigmentosa, caused by a pro-23-his mutation in rhodopsin, the decrease in amplitude of the a-wave cannot be explained completely by loss of photoreceptors or disks, but must involve a decrease in the gain of transduction (Birch *et al.*, 1995). These patients also have a delayed recovery of rod responses.

13.5.2. POSTRECEPTOR ERG ANALYSES

13.5.2.1. b-Wave Analyses

If the photoreceptor models accurately described the time course of P3 for times out to the peak of the response, then the model fits could be subtracted from the entire

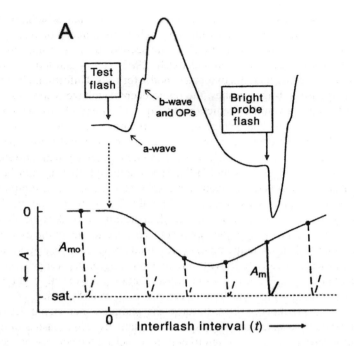

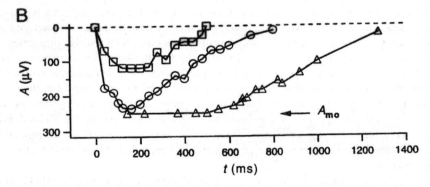

FIGURE 13.11. The paired flash technique for revealing the entire time course of photoreceptor responses. (A) Illustration of the method with hypothetical data: The top trace shows an ERG in response to the "test flash" of moderate intensity followed in approximately 200 ms by a "probe flash" designed to saturate the response of the rods. The presence of the b-wave makes it impossible to determine the time course or amplitude of the photoreceptor's response to the test flash alone. The lower graph shows amplitudes of the probe flashes given at different times before and after a test flash, which occurs at $t = 0$. The ERG in response to one probe flash, labeled A_m, is shown as a solid curve, and those at other test-prove intervals are dashed. A_{m0} is the probe flash amplitude when given alone. The curve connecting the data points is the reconstructed response of the rod to the test flash. (B) Reconstructions of rod responses using the method in part A: The test flashes were 11 scotopic td-s (squares), 44 scotopic td-s (circles) and 320 scotopic td-s (triangles). Probe flashes in all cases were 1.2×10^4 scotopic td-s. (Reprinted from Pepperberg et al., 2000, with permission of Elsevier)

ERG to reveal the time course of the remaining ERG components, which are dominated by the b-wave (also called P2) at high intensity. That "photoreceptor-free" part of the ERG could then be used to derive a model of the second level of retinal processing as well. This analysis was made by Hood and Birch (1992, 1995), and led to a three-stage model for the b-wave, rather than the two-stage model for rods. Unfortunately, they did not realize that in response to bright flashes, the photoreceptor voltage responses seen in the mammalian ERG depart from the predictions of Eqs. (13.7) and (13.8), which are based on the photoreceptor current. The voltage shows a rebound or recovery that does not seem to be present in the photoreceptor current (e.g., Kang Derwent and Linsenmeier, 2001). This recovery probably occurs because voltage-dependent inner segment currents become activated (Fain *et al.*, 1978; Barnes and Hille, 1989), although that has not been proven for mammals (Kang Derwent and Linsenmeier, 2001). This means that a subtraction of the photoreceptor activity on the basis of Eqs. (13.7) and (13.8) subtracts too much from the overall ERG, even if one works with times shorter than the time-to-peak of the photoreceptor impulse response. Consequently, another approach is needed to model the dynamics of the retina's second stage. Also, while the models presented above give very good fits, and are effectively being used clinically to provide an understanding of disease processes (Breton *et al.*, 1994; Birch *et al.*, 1995), this does not mean that they are exactly right. One might expect the ERG to have some inner retinal activity even at early times, i.e., before the b-wave (Robson and Frishman, 1996), because the photoreceptor signal activates other neurons. The a-wave is observed only in response to relatively bright flashes of light, when the photoreceptor response is fast enough to dominate, before the activity of second-order neurons comes into play, and when it is large enough to be detectable in the ERG. For moderate illuminations the earliest detectable activity is the positive-going b-wave, but for very weak illumination, when the b-wave is slow or absent, the earliest activity again becomes a negative response, the scotopic threshold response (STR), which originates in the inner retina (e.g., Steinberg *et al.*, 1991). These changes with illumination can be understood on the basis of convergence of signals in the retina. For the weakest flashes, responses from photoreceptors are so small that they are in the noise in ERG recordings, but they sum to give barely detectable responses from second-order neurons, and the signals then sum again at the inner plexiform layer to give measurable STRs. For this reason, the STRs dominate for very weak flashes. As one increases illumination, the responses of first the bipolars and then the photoreceptors become large enough and fast enough that they are detectable with less summation.

To begin to quantitatively address responses of second-order cells in the ERG, Robson and Frishman (1995, 1996) blocked activity of the cat inner retina, that is, post–bipolar cells, with intravitreal applications of the glutamate antagonist N-methyl-DL-aspartic acid (NMDLA). NMDLA has been shown to block activity at the inner plexiform layer, but not at glutamatergic synapses at the outer plexiform layer, so it should simplify the ERG. Then, if one blocks the b-wave with a glutamatergic blocker of the outer plexiform layer, one can isolate the b-wave by subtraction. In cats, the only agent required to block the rod bipolars is APB. When this was done, Robson and Frishman found that a good fit to the rising side of the b-wave could be obtained with a six-stage process, of which three are carried forward from the main activation steps in the photoreceptor, and the others are associated with the response of the bipolars. ON-bipolars respond to glutamate with a G-protein-mediated cascade that was expected to introduce three more stages. The rising side of the b-wave could be fitted, therefore, by

$$R_b(t) = kI (t - t_d)^5 \tag{13.9}$$

where $R_b(t)$ is the b-wave as a function of time, k is a constant, I is illumination, and t_d is a brief delay, less than 5 ms. This only fits the early part of the response, and there is still considerable work that could be done to model the entire b-wave and other ERG responses.

13.5.2.2. Multifocal ERG

The ERG provides an objective electrophysiological test of retinal function, but one of its disadvantages as a diagnostic tool has been the inability to determine what region of the retina generates the signal. The ERG typically represents summed activity across the retina, so if the temporal half of the retina were severely damaged, the ERG a- and b-wave amplitudes might be approximately halved relative to normal, but one could not from the ERG infer that the damage was in the temporal retina. For a major functional deficit ophthalmoscopic inspection might provide enough additional information to identify the site of the problem, but for more subtle changes this is not the case. One might expect that local stimulation of different parts of the retina with a spot of light would be able to elicit a corneal ERG from just the part of the retina stimulated. However, in practice, when a bright flash is presented on a dark background, the scatter of light in the eye and the sensitivity of the retina to small amounts of light have made it impossible to isolate corneal responses from different regions. Studies were carried out in the 1950s that showed that a perfectly normal ERG could be generated by flashing a light at the optic disk, which of course has no photoreceptors, and these studies emphasized the contribution of scattered light (Asher, 1951; Boynton and Riggs, 1951). One solution has been the multifocal ERG (e.g., Sutter and Tran, 1992; Palmowski *et al.*, 1997). In this technique, one presents a grid of approximately 100 flashing elements to a region 20 to 50° in diameter (Figure 13.12). The elements are hexagons of varying size, so the grid looks like a distorted honeycomb. Each element turns on and off with a pseudorandom sequence, uncorrelated with the behavior of any other element, and, since there is a reasonably high mean level of illumination (typically in the photopic range) in all elements over time, light scattered from bright to dark regions has less influence on the cells in the dark region than it would if the dark regions were dark-adapted. This means that signals are generated reasonably specifically in the part of the retina corresponding to each element. The signals resulting from this type of stimulation are invisible until a cross-correlation is done between the voltage and the pattern of stimulation for each element, and then one pulls out the component of the signal that is correlated with the activity of each element. One can look at different orders of these multifocal responses, but the first-order ones look very much like miniature ERGs. There is no new mechanistic model provided by this technique, but it is an application of engineering methods to derive more information.

13.5.3. GANGLION CELL MODELS

At the other end of the retina from photoreceptors are retinal ganglion cells. Models of the receptive field and response properties of ganglion cells generally do not connect with

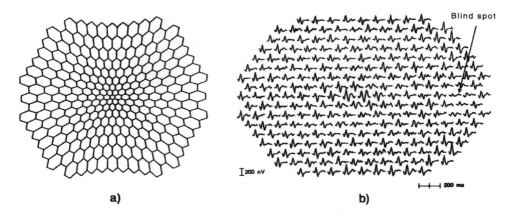

a)   **b)**

FIGURE 13.12. *Left*: Stimulus pattern used to elicit the multifocal ERG. This pattern has 241 elements that could be turned on and off independently, and subtended the central 23° of the visual field. *Right*: ERG responses at each location obtained by cross-correlation of the voltage with the stimulus pattern. Note that the largest signals, which come from the fovea, are about three orders of magnitude smaller than those in Figures 13.9 and 13.10. (Reprinted from Sutter and Tran, 1992, with permission of Elsevier)

photoreceptor or ERG models, or attempt to achieve a description of how the neural circuitry of the retina works. Instead they treat the retina as a black box receiving light inputs and generating neural outputs, similar to the approach of Hood and Birch to photoreceptors that has been discussed already, although the ganglion cell work started much earlier. The reason for this is twofold. First, these models have been designed to characterize the retinal output and understand the several channels of information that the retina provides to the rest of the brain, which has not required modeling the photoreceptors or interneurons. Second, it has been feasible to record from mammalian ganglion cells for more than 50 years, so a large database of ganglion cell behavior began to accumulate before other types of data on the retina. A great deal of this work has been on the cat retina, with more recent contributions on the monkey retina. The present analysis focuses on models. Shapley and Perry (1986), Kaplan (1991), and Troy and Shou (2002) have written more comprehensive reviews of the ganglion cell literature.

13.5.3.1. Systems Analysis

Systems analysis techniques began to be applied to the retina by Enroth-Cugell and Robson (1966). This engineering approach dominated much of the work on retinal physiology, not just the thinking of engineers. Enroth-Cugell and Robson (1966) used "grating" patterns whose contrast was sinusoidally modulated in one dimension in space, and temporally varied either sinusoidally or as a square wave in time. These patterns are characterized by a luminance profile L:

$$L(x, t) = L_{\mathrm{mean}} + L_1[\sin(2\pi kx + \phi)]M(t) \tag{13.10}$$

where L_1 is the sine wave amplitude, x is distance in degrees, k is spatial frequency, usually

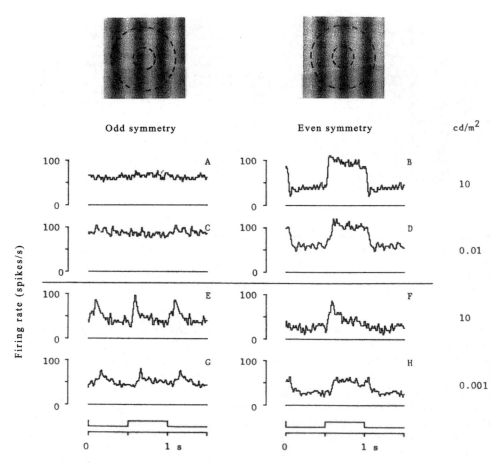

FIGURE 13.13. Difference in spatial summation between X and Y type cat retinal ganglion cells. At the top are sinusoidal grating patterns positioned in odd symmetry (spatial phase of 0° and 180°) and even symmetry (spatial phase of 90° and 270°) on the receptive field. The grating contrast reversed with the timing shown at the bottom of the figure. The top four histograms are from an X cell at two background levels separated by three log units, illustrating that summation is linear at both backgrounds, because there is no response to the grating in odd symmetry. The bottom four histograms are from a Y cell at two backgrounds separated by four log units, showing that the Y cell generates frequency-doubled responses at both backgrounds when the grating is in odd symmetry. The contrasts were A: 0.2, B: 0.2, C: 0.7; D: 0.3; E: 0.07; F: 0.03; G: 0.4; H: 0.2. The spatial frequency was one chosen to be above the peak of the contrast sensitivity curve for the fundamental. (Modified from Linsenmeier and Jakiela, 1979, with permission from Springer-Verlag GmbH & Co. KG)

expressed in cycles per degree of visual angle, ϕ is the phase of the grating with respect to the receptive field, and M is the sinusoidal or square-wave temporal reversal. Grating patterns are shown in Figure 13.13. For the "drifting" gratings that are also used for this type of work, the temporal modulation is produced by having the spatial phase continuously change at a frequency of f Hz:

$$L(x, t) = L_{\mathrm{mean}} + L_1 \sin(2\pi k x - f t) \qquad (13.11)$$

Contrast refers to the amplitude of the sine wave divided by the mean illumination, the Rayleigh contrast (Shapley and Enroth-Cugell, 1984):

$$C = L_1/L_{mean} = (L_{max} - L_{min})/(L_{max} + L_{min}) = (L_{max} - L_{min})/2L_{mean} \qquad (13.12)$$

where L is mean, maximum, or minimum luminance of the pattern. Sinusoidal patterns have become standard for this field because arbitrary patterns can be represented by the Fourier sum of such patterns. To the extent that the retina operates linearly, retinal responses to arbitrary stimuli can be predicted by knowing the spatial and temporal tuning curves of ganglion cells. Further, unlike flashing spots, gratings are effective stimuli for probing all levels of the visual system, including psychophysical analyses of human performance (e.g., Shapley and Lennie, 1985). An alternative to the use of sinusoidal gratings that does not assume linearity is the use of pseudorandom (e.g., Shapley and Victor, 1978) or white-noise stimuli (e.g., Citron et al., 1988), which can reveal nonlinear behavior, and we will consider below the insights that these have provided.

13.5.3.2. X and Y Cells

Enroth-Cugell and Robson (1966; reviewed in 1984) discovered that two prominent classes of ganglion cells in the cat retina, which are believed to make up most of the ganglion cells projecting ultimately to the visual cortex, could be discriminated by whether the light distribution in the receptive field was reported on linearly or nonlinearly by the ganglion cells. For the X cells, the linearity was quite remarkable. It was possible to position a high spatial frequency grating on the receptive field so that contrast reversal of the light and dark bars led to no response from the cell, even though the photoreceptors and bipolar cells must all have been producing responses (Figure 13.13). Shifting the phase of the grating with respect to the receptive field away from this "null position" yielded large responses from the cell at the fundamental frequency of contrast reversal. For Y cells, there was no null position; all positions of the grating evoked responses from the cell indicating that summation of light was nonlinear (Figure 13.13). As also shown in Figure 13.13, the X–Y distinction proved to be a fundamental property of the cells, independent of adaptation level (Linsenmeier and Jakiela, 1979). X and Y cells differ not only in spatial summation, but in receptive field size (e.g., Cleland and Levick, 1974; Linsenmeier et al., 1982) and soma and dendritic field size (e.g., Boycott and Wassle, 1974; Peichl and Wassle, 1979), with Y cell receptive field diameter being about three times as large as X cells at any eccentricity. The conduction velocity of Y cell axons is also faster because they have larger axons (Cleland and Levick, 1974; Stone and Fukuda, 1974).

Hochstein and Shapley (1976a,b) further analyzed Y cells, and showed that both X and Y cells had a linear response at the fundamental frequency of temporal modulation whose amplitude depended on the phase of the grating with respect to the receptive field (Figure 13.14). In addition, Y cells had an additional nonlinear response that could be characterized as a pure second-harmonic response that was independent of phase (Hochstein and Shapley, 1976a) and was most pronounced at high spatial frequencies. For Y cells the second harmonic was at least twice as large as the fundamental at some spatial frequency, and for X cells the second harmonic was always less than the fundamental, providing

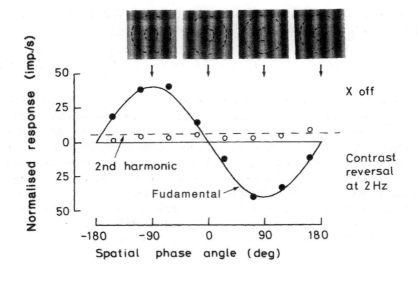

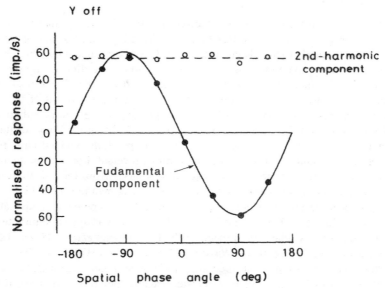

FIGURE 13.14. Spatial phase dependence of the linear (fundamental) and nonlinear (second harmonic) responses of an X cell (top) and a Y cell (bottom) to contrast-reversing gratings, as shown at the top. X cells have negligible second-harmonic responses, whereas Y cells have nonlinear responses that are present at all contrasts and that exceed those of the linear receptive field mechanisms at high spatial frequencies. (Reprinted from Enroth-Cugell and Robson, 1984, from the Association for Research in Vision and Ophthalmology)

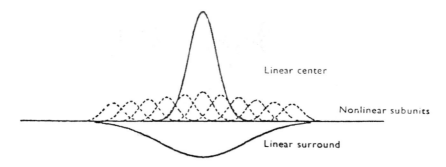

FIGURE 13.15. Modification of the Difference of Gaussians receptive field model to account for the nonlinear responses of Y cells. The data are consistent with the existence of a number of subunits, each smaller than the center, within the receptive field. Each subunit generates either a half-wave or full-wave rectified response that appears as a frequency doubling in response to stationary gratings, and may appear as an elevation of the mean rate of firing in response to a drifting grating. (Reprinted from Hochstein and Shapley, 1976b, with permission of the Physiological Society)

quantitative support for a true dichotomy between these cell types (Hochstein and Shapley, 1976a), rather than a range of properties. This work led to an important modification of the center-surround model of ganglion cells to include small, nonlinear subunits (Figure 13.15; Hochstein and Shapley, 1976b) that may arise from the behavior of amacrine cells (e.g., Frishman and Linsenmeier, 1982).

13.5.3.3. Difference of Gaussians Model of the Receptive Field

Enroth-Cugell and Robson (1966) also quantitatively described the spatial transfer functions of cat X cells, i.e., their contrast sensitivities as a function of spatial frequency, and Hochstein and Shapley (1976b) and Linsenmeier *et al.* (1982) did the same for the linear part of cat Y cell behavior. Contrast sensitivity is the reciprocal of the contrast needed to evoke a small fixed response from the cell at the fundamental frequency of contrast reversal or grating movement. This measure was adopted rather than response amplitude for two reasons. First, Enroth-Cugell and Robson were interested in linear behavior, so they wished to remain in the linear part of the response vs. contrast relationship. The responses they recorded, of 10 to 15 impulses per second in amplitude (Enroth-Cugell *et al.*, 1983), allowed them to insure that the response amplitudes were not saturated. Second, they wanted to be able to relate their findings to measures of human visual performance, which were beginning to use systems analysis techniques at about the same time. It is feasible to determine the minimum contrast at which a person sees a grating (i.e., the contrast sensitivity), but not the sizes of the neural responses in the human retina or brain. The results of measuring contrast sensitivity as a function of spatial frequency (Enroth-Cugell and Robson, 1966) were interpreted as the spatial frequency domain representation of the spatial "Difference of Gaussians" model of Rodieck (1965). The point weighting function, expressed in radial coordinates, assumes a linear addition of center (c) and surround (s), and is

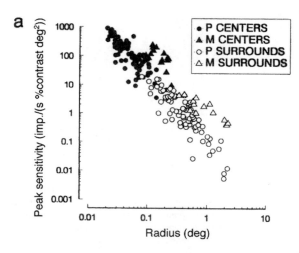

FIGURE 13.16. *Part A*: The characteristic parameters, radius and peak sensitivity for the center mechanism in the Difference of Gaussians model in spatial coordinates. Similar parameters define the surround. *Part B*: Symbols show the response of an on-center X cell to gratings of different spatial frequencies at a temporal drift rate of 2 Hz. Fits of the Difference of Gaussians model to these data yielded the solid curve, which was composed of the spatial frequency tuning curves for the center (C) and surround (S). The C curve and the solid curve are the same at high spatial frequencies, because high spatial frequencies are invisible to the surround. The receptive field profile in A was generated from the parameters obtained for this cell. (Reprinted from Linsenmeier *et al.*, 1982, with permission of Elsevier)

given by

$$W(r) = W_c(r) - W_s(r) = K_c \exp[-(r/r_c)^2] - K_s \exp[-(r/r_s)^2] \tag{13.13}$$

The corresponding spatial frequency representation is

$$S(v) = S_c(v) - S_s(v) = K_c \pi r_c^2 \exp[-(\pi r_c v)^2] - K_s \pi r_s^2 \exp[-(\pi r_s v)^2] \tag{13.14}$$

where W is the sensitivity as a function of radial position, and S is the contrast sensitivity (the reciprocal of the contrast required for a particular small response amplitude) at spatial frequency v. The Ks and rs are the maximum sensitivities and characteristic radii (at K/e) of the center and surround, as shown in Figure 13.16. This model fits the spatially linear parts of the responses of both X and Y cells (Enroth-Cugell and Robson, 1966; Hochstein and Shapley, 1976a; Linsenmeier *et al.*, 1982).

Monkey ganglion cells projecting to the LGN are generally designated M and P rather than X and Y. M cells project to the magnocellular (lower) layers of the lateral geniculate, and P cells project to the parvocellular (upper) layers. P cell receptive fields are smaller than those of M cells. All P cells have linear spatial summation, but M cells may have spatially linear or nonlinear behavior (DeMonasterio, 1978a; Kaplan, 1991; Croner and Kaplan, 1995). Thus, primate cells identified as X would include not only P cells, but some M cells as well. In addition, most primate ganglion cells have color opponency, meaning that both the

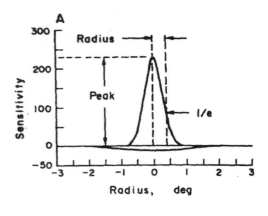

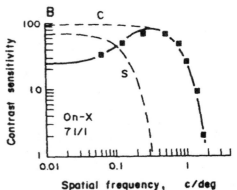

FIGURE 13.17. Relation between the peak sensitivity and size of the center and surround for primate M and P ganglion cells. The slope of the line for centers and surrounds of P and M cells is about −2 on this log−log plot, indicating that peak sensitivity is inversely proportional to the area of the center or surround. (Reprinted from Croner and Kaplan, 1995, with permission of Elsevier).

center and surround signals are spectrally tuned, apparently because they have inputs from only a subset of cone types (DeMonasterio and Gouras, 1975; DeMonasterio, 1978a). The Difference of Gaussians model developed for cat retinal ganglion cells also works well for concentrically organized primate ganglion cells (e.g., Croner and Kaplan, 1995), although there is evidence from experiments with chromatic stimuli that the surround mechanism of at least some P cells is absent in the middle of the receptive field (DeMonasterio, 1978b), which is not predicted by the model. The Difference of Gaussions model also works for cat and primate LGN cells (So and Shapley, 1979; Kaplan and Shapley, 1982). For both cat X and Y and primate M and P cells, this model is valuable because it allows an analysis of how the different receptive field parameters depend on eccentricity, and how they depend on each other. For instance, it is interesting that the larger the receptive field center, the lower the peak sensitivity (Figure 13.17) (Linsenmeier *et al.* 1982; Croner and Kaplan, 1995), and this tradeoff works in such a way that the integral under the center turns out to be almost independent of center radius. Also, despite adjustments in all the individual parameters characterizing the receptive field, the integrated strength of the surround relative to the center tends to be fairly tightly constrained (average of 0.73 in cat; Linsenmeier *et al.*, 1982) and 0.55 in monkey (Croner and Kaplan, 1995)).

13.5.3.4. Gaussian Center–Surround Models

The Difference of Gaussians model works when the center and surround are $180°$ out of phase, but this is true for only some temporal frequencies. In order to deal with the limitations of the original Difference of Gaussians Model, Derrington and Lennie (1982), Enroth-Cugell et al. (1983), Dawis et al. (1984), and Frishman et al. (1987) used models that can be called "Gaussian Center–Surround Models," which allowed the temporal phases of both center and surround to vary with temporal frequency. These have five to eight parameters, rather than the four of the Difference of Gaussians model. The response in the Gaussian Center–Surround Model of Frishman et al. (1987) had six parameters, allowing center and surround responsivity to vary with temporal frequency, ω,

$$R(v, \omega) = R_c(v, \omega) + R_s(v, \omega) \tag{13.16}$$

R is the responsivity of the cell or of the center or surround, a new term that means amplitude divided by contrast. It is used only when the response is small enough that it is in the linear part of the response vs. contrast function, and is functionally equivalent to sensitivity. R can be expressed in terms of magnitude and phase of the center and surround components:

$$|R(v, \omega)| \exp[i P(v, \omega)] = |R_c(0, \omega)| \exp\{i P_c(\omega) - [\pi v r_c(\omega)]^2\}$$
$$+ |R_s(0, \omega)| \exp\{i P_s(\omega) - [\pi v r_s(\omega)]^2\} \tag{13.17}$$

Here the quantities in the absolute value symbols represent the strengths of the center and surround. It turned out that not only the temporal phase, but also the center and surround radii, and the center and surround strength had to be allowed to vary with the temporal frequency (Dawis et al., 1984; Frishman et al., 1987). When this was combined with the fact that center and surround strength vary with spatial frequency, the overall behavior of ganglion cells depended strongly on temporal frequency. This can be seen in both temporal frequency tuning curves at selected spatial frequencies (Figure 13.18) and spatial frequency tuning curves at selected temporal frequencies, which were fitted by Eq. (13.17) (Figure 13.19).

13.5.3.5. More Complex Ganglion Cell Models

Unfortunately, although models can be fitted to individual spatial and temporal frequency tuning curves to investigate the parameter space, this does not mean that there is a comprehensive systems model that can predict spatiotemporal behavior completely, even for X cells. Chen and Freeman (1989) published a model in which each stage of processing was represented by cable equations, and either a feedforward or feedback loop was used to represent the interaction of center and surround. Although this model did fit the data reasonably well, it did not take advantage of the existing Gaussian models. Extensions of the Gaussian analyses have been made to investigate ganglion cell properties at different adaptation levels (Derrington and Lennie, 1982; Chan et al., 1992; Troy et al., 1993, 1999), but again there is not a comprehensive model. A great deal of additional work to understand the process of adaptation has been done on ganglion cells and other retinal neurons without the use of grating stimuli (e.g., Enroth-Cugell and Shapley, 1973; Shapley and Enroth-Cugell, 1984).

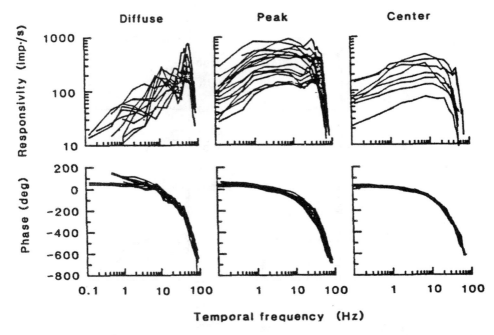

FIGURE 13.18. Dependence of the temporal tuning curve of cat X cells on the spatial properties of the stimulus. On the left are the amplitude and phase of the responses for 17 ON-center X cells when the stimulus was a diffuse field (i.e., zero spatial frequency) that stimulated both center and surround. In the center and right panels are similar temporal tuning curves for spatial frequencies at the peak of the spatial tuning curve, which may involve some surround, and at a spatial frequency above the peak, where the response is solely due to the center. Responsivity is response divided by contrast. Solid lines are fits to the Gaussian Center–Surround model described in the text. (Reprinted from Frishman *et al.*, 1987, with permission from the Rockefeller Institute of Medical Research)

In the work discussed so far, the stimuli were modulated at one temporal frequency at a time. A more general approach is to use white noise or a sum of discrete temporal frequencies as stimuli. It is then possible to use first-order responses (i.e., the response components at the input temporal frequencies) as an alternative way of investigating linear behavior. By measuring second- and higher-order components present in the responses, one could also investigate nonlinear behavior. Victor and Shapley (Victor *et al.*, 1977; Shapley and Victor, 1978; Victor and Shapley, 1979) took this approach and used a sum of six or eight sinusoids that were nearly incommensurate in temporal frequency (i.e., no individual test frequency was an integer multiple of another, and could not be created by a sum or difference of two others). This series of studies cannot be reviewed completely here, but it supported most of the fundamental conclusions about X and Y cells outlined above. One striking new result of their work, however, was the finding of a "contrast gain control" as shown in Figure 13.20 (Shapley and Victor, 1978). On the right are responses of a Y cell to individual sinusoidal stimuli at different contrasts, showing the intuitive result that the shape of the temporal tuning curve is independent of contrast. However, on the left a sum of sinusoids was used, and in this case the responses at low temporal frequency increase little with contrast, and it is only those at higher temporal frequencies that grow with contrast. Thus, the presence of

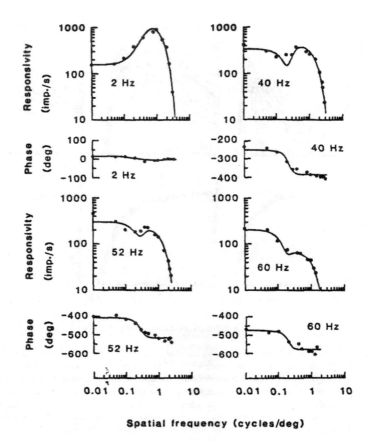

FIGURE 13.19. Dependence of the spatial tuning curve of cat X cells on temporal frequency for four temporal frequencies (2, 40, 52 and 60 Hz). (Reprinted from Frishman *et al.*, 1987, with permission from the Rockefeller Institute of Medical Research)

stimulus components at high temporal frequencies made the cell almost "ignore" increases in contrast at low temporal frequencies. This behavior was observed more strongly in Y cells than in X cells, but occurred in both. As shown in the lower part of the figure, the temporal phase of the response components also shifted with contrast.

13.5.3.6. W Cells

The X and Y cells comprise 40 to 60% of the ganglion cells in cats (Troy and Shou, 2002). Other ganglion cells have axons that all conduct more slowly than X cells, but they form a heterogenous group in terms of other properties. Most of these do not project to the visual cortex, but appear to subserve roles other than perception. These were called W cells by Stone and Fukuda (1974). One of the approximately seven types of W cells is the highly linear "Q cell" (Enroth-Cugell *et al.*, 1983; Troy *et al.*, 1995), also called sluggish-sustained (Cleland and Levick, 1974) or tonic W cells (Stone and Fukuda, 1974; Rowe and Cox, 1993). Their spatial summation is similar to that of X cells but they have

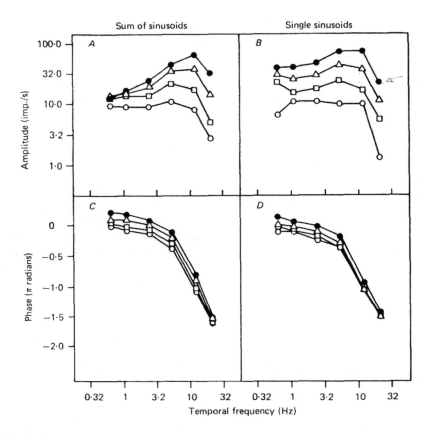

FIGURE 13.20. One manifestation of the contrast gain control. On the left are responses elicited from a Y cell when six sinusoidal stimuli were presented simultaneously at different temporal frequencies (shown on the abscissa). The points show the amplitude and phase of the fundamental at each temporal frequency. The stimulus was a stationary bar, half a degree in diameter, positioned to produce a maximal fundamental response. Each curve represents a different contrast (0.0125, 0.025, 0.05 and 0.10 per sinusoid from bottom to top). On the right are responses of the same cell when the same stimulus was presented to the cell, with each sinusoidal component presented separately. (Reprinted from Shapley and Victor, 1978, with permission of The Physiological Society)

receptive field centers similar in size to Y cells, and have lower peak sensitivity. All other W cells appear to have nonlinear spatial summation (e.g., Troy *et al.*, 1989; Rowe and Cox, 1993). Phasic W cells (sluggish transient cells of Cleland and Levick, 1974) have spatial summation similar to Y cells, but poor sensitivity to gratings, and most can be characterized by a Difference of Gaussians model. Directionally selective cells and ON–OFF ganglion cells have receptive fields that are not well described by a Difference of Gaussians (Rowe and Cox, 1993).

13.5.4. MODELS OF THE RETINAL MICROENVIRONMENT

A completely different set of engineering approaches has been used to study the retinal microenvironment and retinal metabolism. The microenvironment refers to the composition of the extracellular space surrounding the neurons, in terms of ion distributions, nutrient and waste product concentrations, and extracellular volume. These properties can be studied with intraretinal microelectrodes sensitive to ions (e.g., K^+, Ca^+, and H^+) and gases (O_2 and NO in particular). Diffusion models can then be fitted to the data to understand both the fluxes of these substances through the retina and cellular metabolism. This work is important because alterations in the microenvironment, caused either by vascular dysfunction or cellular dysfunction, are often the aspect of disease that leads to retinal cell death. In addition, these measurements can often give a different kind of insight into retinal cell physiology. However, unlike the modeling discussed earlier, where electrophysiological data provided almost all of the information on which the models were constructed, the microelectrode techniques are not the only way to study the microenvironment. A full understanding, which we will not attempt here, requires the use of many complementary techniques, including recordings of retinal activity, biochemical measurements of various metabolites, optical measurements of intracellular ion concentrations, histological measurements of deoxyglucose uptake and cytochrome oxidase, and measurements of blood flow. The microenvironment also includes molecules used to signal between cells, such as neurotransmitters and paracrine substances like melatonin, but in general there are no techniques available to measure these with spatial and temporal precision.

13.5.4.1. Oxygen

One of the important constituents of the microenvironment is oxygen. Research on retinal oxygenation has been reviewed by Hogeboom van Buggenum et al. (1996), Yu and Cringle (2001), and Wangsa-Wirawan and Linsenmeier (2003). Normally, the metabolism of the retina is limited by the presence of oxygen, which cannot be stored in tissue. Hypoxia, the lack of oxygen, clearly plays a role in diabetic retinopathy, retinopathy of prematurity, and retinal vascular disease, and may be involved in any situation where blood flow is compromised. Oxygen partial pressure, P_{O_2}, can be measured with oxygen-sensitive polarographic electrodes, which chemically reduce oxygen and yield a current proportional to the concentration of oxygen at the tip of the electrode. These electrodes have a spatial resolution approaching 1 μm and have response times of milliseconds. It is possible to map out the P_{O_2} as a function of position across the retina, and this has been done in several species. The animals fall into two categories: those with both a choroidal circulation and a circulation in the inner half of the retina, such as human, monkey, cat, and rat, and those without the retinal circulation, including rabbit and guinea pig. The gradient of oxygen across the retina of a cat under dark-adapted conditions is shown in Figure 13.21.

Oxygen moves only by simple diffusion, and it diffuses equally well through membranes as through intracellular and extracellular space, so the tissue can be modeled as homogeneous. In the most general terms, oxygen diffusion is described by:

$$Dk\nabla^2 P + Q = k\,\partial P/\partial t \qquad (13.18)$$

where D is the diffusion coefficient of oxygen (cm^2/s), k is oxygen solubility (ml O_2/ml retina/mm Hg), P is partial pressure (mm Hg or torr), Q is utilization of oxygen (ml O_2/100 g/min) and ∇P is the second spatial derivative of P (mm Hg/cm^2). This only applies in a region that can be assumed to have a homogeneous value of Q, so the challenge is to

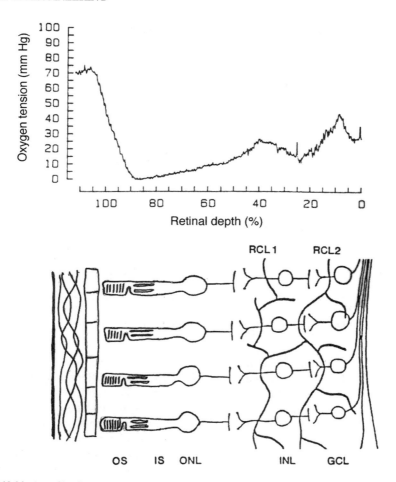

FIGURE 13.21. A profile of oxygen tension across the cat retina during dark adaptation. The recording was made with an oxygen microelectrode that was first advanced through the retina in steps to the choriocapillaris, and then was withdrawn continuously at 2 μm/s to the vitreous. Evidence of retinal capillaries is visible as peaks in the inner half of the retina. The correspondence to retinal layers is shown in the diagram at the bottom.

define a region where this can be applied, and specify appropriate boundary conditions. Most analyses performed to date have attempted to extract a value for Q at steady state, so the right side of the equation is set to zero.

Equation (13.18) can be applied to the outer half of the retina because this part of the retina can be taken to be an avascular slab of tissue, where oxygen supply is all from the boundaries, which are at the choriocapillaris and half way through the retina, where the retinal circulation begins. The curvature of the retina is negligible with respect to the thickness. In this slab, oxygen is assumed to diffuse only in one dimension, along the photoreceptors, because any lateral gradients are expected to be very small. Using these geometrical simplifications, one can fit models with different numbers of layers to oxygen

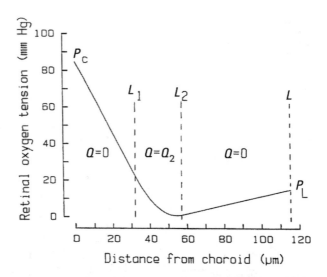

FIGURE 13.22. The structure of the oxygen model used to describe the profile of oxygen in the outer half of the retina. Only the middle layer, corresponding to the photoreceptor inner segments, has a nonzero oxygen consumption. The parameters that are adjustable during fitting are P_C, P_L, L_1, L_2, and Q_2.

profiles (P_{O_2} as a function of distance from the choroid) in order to determine how many layers are needed to fit the data and whether Q is different in different layers. The earliest model simulated the outer retina as one layer (Dollery *et al.*, 1969) before any intraretinal P_{O_2} recordings were available. Dollery *et al.* reached the somewhat surprising conclusion that although there was adequate oxygen at the boundaries of the outer retina, the P_{O_2} was likely to be almost zero somewhere in the tissue. The first intraretinal measurements were made in cats by Alder *et al.* (1983), and revealed steep gradients of oxygen in the outer retina during light adaptation, but did not find very low P_{O_2}s. Measurements in the dark-adapted cat retina, however, a condition in which metabolism was known to be higher in some animals (Sickel, 1972; Zuckerman and Weiter, 1980) supported the idea that part of the outer retina had a very low P_{O_2} (Linsenmeier, 1986; Linsenmeier and Braun, 1992). Subsequent work (Haugh *et al.*, 1990) led to a model for oxygen diffusion in the outer retina (Figure 13.22) that had three layers rather than the one used by Dollery *et al.* (1969). The solution to Eq. (13.18) under these conditions is

$$
\begin{aligned}
P_1(x) &= a_1 x + b_1 & 0 \leq x \leq L_1 \\
P_2(x) &= (Q_2/2Dk)x^2 + a_2 x + b_2 & L_1 \leq x \leq L_2 \\
P_3(x) &= a_3 x + b_3 & L_2 \leq x \leq L
\end{aligned}
\tag{13.19}
$$

where the constants a_i and b_i for each of the three layers are determined from the boundary conditions (Haugh *et al.*, 1990). The boundary conditions include specified P_{O_2} values at the choroid and at the outer–inner retinal border, about half way through the retina, as well as matching of P_{O_2}s and O_2 fluxes at L_1 and L_2, the borders between layers.

The fits of this model to data yielded values for the P_{O_2}s at the choroid and inner retinal boundary (P_C and P_L), the locations of the boundaries, L_1 and L_2, and for Q_2/Dk

in the middle layer, the only layer in which consumption was found to be necessary. The outermost layer corresponded to the outer segments, the middle layer corresponded to inner segments, and the innermost layer corresponded to the layers of cell bodies in the outer nuclear layer. As is well known, mitochondria are present only in the inner segments, so this model agreed with the anatomy. The metabolism of the consuming region in dark adaptation was very high, on the order of 20 ml O_2/100 g·min, which is four to five times the oxygen consumption of brain tissue. This high consumption, in combination with the distance of the inner segments from the choroid, is responsible for the very low P_{O_2}s observed in the inner segment layer. The low P_{O_2} suggests that photoreceptors would be at risk if arterial P_{O_2} is reduced (Linsenmeier and Braun, 1992), or if choroidal blood flow is reduced by elevated intraocular pressure (Yancey and Linsenmeier, 1989), or if the retina is detached (Linsenmeier and Padnick-Silver, 2000).

In general, the same equations cannot be applied to the inner retina, because there are vessels embedded in the tissue, reflected in the peaks in the inner retina in Figure 13.21 that make it impossible to reduce the geometry to a one-dimensional problem. The three-dimensional vascular geometry is difficult to measure, and there are no three-dimensional data to use in fits to a three-dimensional model. Cringle et al. (2002) have attempted to circumvent this problem and analyze the metabolism of the inner retina of rats by using an eight-layer model (five for the inner retina) and not attempting to derive parameters from those layers containing retinal capillaries. The other strategy is to block the circulation of the inner retina so that all of the oxygen is derived from the choroid, and provide enough oxygen in the choroid to supply the entire retina. In this case, another layer representing the inner retina can be added to the three-layer model described above (Braun et al., 1995). A model of the inner retina can also be used when the inner retina is avascular, as in guinea pigs (Cringle et al., 1996). The inner retina in these animals receives very little oxygen and has low oxygen consumption.

13.5.4.2. Ion Distribution

The tip of a microelectrode can be filled with a resin that is selectively permeable to a particular ion, allowing the recording of the Nernst potential for that ion across the resin. When placed in the retinal extracellular fluid (ECF), measurements of ion concentrations can be made with 1 μm resolution. A great deal of quantitative information leading to understanding of retinal neural activity has come from studies of the distributions in the retina of K^+ (e.g., Karwoski and Proenza, 1977; Oakley, 1977; Steinberg et al., 1980; Shimazaki and Oakley, 1984; Frishman et al., 1992), Ca^{2+} (e.g., Gold and Korenbrot, 1980; Dmitriev et al., 1990; Gallemore et al., 1994), and H^+ (e.g., Oakley and Wen, 1989; Yamamoto et al., 1992; Dmitriev and Mangel, 2000, 2001).

Unfortunately, the number of studies that have coupled ion measurements to quantitative diffusion models has been small. Modeling the transport of ions through the retina is complicated. The tissue cannot be treated as homogeneous, because ions diffuse only through the ECF, and require facilitated or active transport across membranes. In order to describe extracellular transport, the concepts of tortuosity of the extracellular space, λ, and fraction of the total volume that is extracellular, α, have to be introduced, so the general equation developed for ion diffusion in the brain by Nicholson and coworkers (e.g., Nicholson and Phillips, 1981; Nicholson and Rice, 1991) includes corrections for these factors:

$$(D/\lambda^2)\nabla^2 C + Q/\alpha = \partial C/\partial t \qquad (13.20)$$

Values for α are on the order of 0.1, and values for λ are on the order of 1.5. Once these modifications have been made to the diffusion equation, one can attempt to define production rates and fluxes in the extracellular space. For the ions, which are not actually produced or consumed, "production" is actually the extrusion of the ion from cells and its appearance extracellularly and "consumption" is the uptake of the ion by cells.

H^+ Distribution and Production

Using Eq. (13.20) as a basis, an analysis of H^+ diffusion and production has recently been done for the cat retina (Padnick-Silver and Linsenmeier, 2002), but only in the steady state. Like the oxygen model described above, the pH model was one-dimensional and required the same three layers to fit the data. An H^+ profile across the retina, obtained with an ion-selective H^+ electrode, and the corresponding fitted model are shown in Figure 13.23. The curvature of the profile is opposite to that of the oxygen profile, because H^+ is produced, whereas oxygen is consumed. Two layers, the inner segments and the outer nuclear layer, were found to produce H^+, leading to a minimal pH of about 7.2. H^+ production is believed to reflect the high rate of utilization of glycolytically derived ATP in the retina, which is needed even under aerobic conditions because the oxidative metabolism is limited by oxygen availability. H^+ production decreased during illumination and increased during hypoxemia (Padnick-Silver, 2000b; Padnick-Silver and Linsenmeier, 2002). An interesting finding that resulted from this modeling, and was not apparent from the data alone, was that the H^+ production rates were far below those that were expected on the basis of lactate production

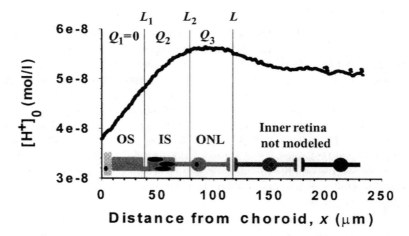

FIGURE 13.23. Gradients and modeling of hydrogen ion in the cat retina. The profile was recorded with an ion-selective H^+ microelectrode. For the model fitted to this data, the outer half of the retina was composed of three layers, of which layers 2 and 3 produced H^+ and the outer segments (layer 1) did not. Values of H^+ production derived from this model are believed to be underestimates of actual H^+ production. (L. Padnick-Silver, unpublished)

in the outer retina (Wang *et al.*, 1997), even though there should be a 1:1 stoichiometry between lactate and H^+. This meant that some H^+ is probably cleared or buffered rapidly, and the values of production derived from the model underestimate total H^+ production. Further, the buffering of H^+ may be of great importance in preventing the retina from having a pH in the range of 4 to 5. The importance of buffering is supported by experiments with carbonic anhydrase blockers (Wangsa-Wirawan *et al.*, 2001), which greatly acidify the retina, and other lines of work (Oakley and Wen, 1989). Whether pH plays a significant role in any retinal diseases is not known, but these results highlight the possibility that a failure of buffering mechanisms could lead to severe acidosis and contribute to cellular dysfunction.

Retinal Extracellular Volume

Another line of investigation has explored whether extracellular volume changes under any physiological or pathological conditions, because in brain slices, extracellular volume (α) was found to decrease during hypoxia (Rice and Nicholson, 1991). This would affect the diffusion of all molecules in the extracellular space. The technique of measuring changes in α in the brain (e.g., Nicholson and Phillips, 1981) involved using a micropipette to introduce an impermeant cation, such as tetramethylammonium (TMA^+), into the ECF. Its concentration was then followed over time with "K^+" microelectrodes, which, in the presence of TMA^+, become TMA^+ electrodes, because they are about 200 times more sensitive to TMA^+ than to K^+ (Li *et al.*, 1994a). Because TMA^+ is not produced or consumed and does not enter cells, its concentration changes are caused by ECF volume changes and by diffusion of TMA^+ away from the injection pipette. If one holds constant the amount of TMA^+ injected, then differences in the concentration vs. time curves before and after a manipulation, such as hypoxia, reveal differences in volume under the two conditions.

A modification of this approach was taken in the isolated frog (Huang and Karwoski, 1992) and chick retina (Govardovskii *et al.*, 1994; Li *et al.*, 1994a) and the intact cat retina (Li *et al.*, 1994b; Cao *et al.*, 1996). In this work a uniform initial concentration of TMA^+ could be achieved by adding it to the bathing solution of the isolated retina, or injecting enough in the cat vitreous to achieve an equilibrium ECF concentration of about 5 mM. During illumination, [TMA^+] was found to change in a way that was consistent with an increase in ECF in the subretinal space, but not the rest of the retina (Huang and Karwoski, 1992; Li *et al.*, 1994a,b). Pharmacological experiments suggested that this hydration of the subretinal space was probably initiated by the light-induced decrease in [K^+] in the subretinal space (e.g., Oakley, 1977). The decrease in [K^+] reduces the activity of a Na/K/Cl transporter at the apical membrane of the RPE. That transporter is a major driving force for water transport out of the retina, so water transport decreases and the space hydrates. In order to quantify the volume change, models were developed and fitted to the data. The model for the subretinal space of chick retina (Govardovskii *et al.*, 1994) was

$$\frac{\partial C(x,t)}{\partial t} = D_s \frac{\partial^2 C(x,t)}{\partial x^2} - \frac{C(x,t)}{\alpha} \frac{\partial \alpha}{\partial t} \tag{13.21}$$

where D_s is an apparent diffusion coefficient for the subretinal space that includes the tortuosity. The first term on the right represents the diffusion of TMA^+ into a region of lower concentration, and the second term reflects the change in volume with time. Figure 13.24

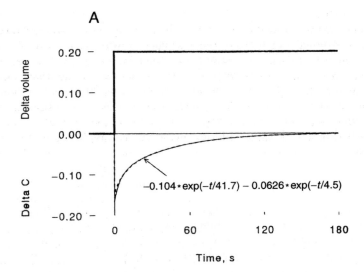

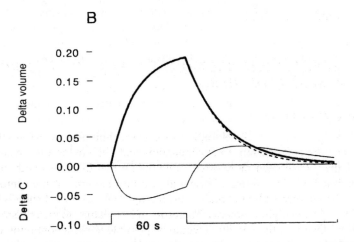

FIGURE 13.24. Model of the extracellular volume change of the subretinal space in the isolated chick retina–RPE–choroid preparation during illumination. Tetramethylammonium ion (TMA^+) was added to the bathing solution to produce a concentration of TMA^+ that was initially uniform across the retina. (A) The delta volume trace represents a step increase in subretinal extracellular volume, and the lower trace shows how TMA^+ concentration would be expected to change in response to this. The concentration decreases because TMA^+ is diluted, and then recovers because TMA^+ diffuses into the outer retina from the inner retina. (B) The diffusion response in A was deconvolved from a curve simulating an actual concentration change during illumination (delta C), yielding a derived volume change. Although this is a simulation, the Delta C curve closely matches actual concentration changes with light. (Reprinted from Govardovskii *et al.*, 1994, with permission from the Association for Research in Vision and Ophthalmology)

shows two situations. In A, a step change of volume (delta volume) was used as the input to the model and the resulting concentration change (delta concentration) showed a steep decrease followed by a recovery. The recovery was due to diffusion of TMA^+ from the inner retina into this increased volume. The point of the model was to extract the unknown $d\alpha/dt$ from the actual change in concentration, and a deconvolution procedure was necessary for this. An example is shown in Figure 13.24B. A curve of dC/dt similar to those actually observed is shown as delta C. When the calculated step response in A was deconvolved from this, the resulting delta volume was computed as the solid line. Here a 7% concentration change, which was the magnitude of the change observed, implied a 20% increase in α. In cats, a similar model suggested that illumination could increase α by 60% on average (Li *et al.*, 1994b). These are very large changes, and would require shrinkage of the RPE cells or photoreceptors. A limitation of the model is that it assumed that the light-evoked volume change was sustained during illumination, and that the transient nature of the TMA concentration change was due solely to diffusion of TMA into the subretinal space. If the model had allowed for recovery of the *volume* during sustained illumination, the derived volume changes would have been smaller. The estimated changes should therefore probably be regarded as upper limits. The failure to account for possible transience in the volume change could also explain the apparent inconsistency that the derived α in cat retina did not recover after the end of sustained illumination. Similar procedures showed that the subretinal space in cats shrank during hypoxemia by as much as a factor of 4 (Cao *et al.*, 1996), again probably an upper limit.

13.6. ENGINEERING CONTRIBUTIONS TO TREATMENT OF RETINAL DISEASES—VISUAL PROSTHESES

13.6.1. VISUAL PROSTHESES

The engineering models and electrophysiological techniques discussed above make one set of engineering contributions to understanding and treating disease. In addition, for about 60 years there have been more direct attempts to treat blindness with engineering solutions. The function of the retina is essentially to turn patterns of light into electrical signals that can be interpreted by the brain. Turning light into electrical signals is also the function of television and digital cameras. Therefore, it has seemed to many visionaries that it would be only a matter of time before it would be possible to use a nonbiological "visual prosthesis" to replace the function of either the photoreceptors or the entire retina. Electrical signals would be generated that could be fed into some stage of intact neural processing. The prosthesis could connect to functional bipolar cells or ganglion cells in the retina, or to the visual cortex. Intervention at the level of the lateral geniculate nucleus would also be theoretically possible, but this is a much less accessible neural structure.

13.6.1.1. Cortical Visual Prostheses

Cortical visual prostheses potentially have at least two advantages. First, they completely bypass the retina, which means that these devices do not necessarily have to be especially compact and, second, they may be able to restore sight in diseases like glaucoma and diabetic retinopathy, where the inner retina is severely damaged. The main efforts in this area following the early work of Brindley (e.g., Brindley and Rushton, 1974) have been those of Dobelle (2000) and Normann *et al.* (1999), reviewed in Margalit *et al.* (2002) and elsewhere in this book. Some of the challenges to be met in a cortical prosthesis are (1) providing an electrical signal that the brain can interpret, when the coding of natural signals in the cortex is not fully understood; (2) providing enough stimulating electrodes to achieve adequate spatial resolution; (3) overcoming the potential adaptation of the brain to repeated stimulation; and (4) long-term biocompatibility, both on the materials side, so that the electrodes remain functional, and on the biological side, to minimize infection and the gliosis that would insulate the electrodes from the neurons they are supposed to stimulate.

13.6.1.2. Retinal Visual Prostheses

Retinal visual prostheses are interchangeably known as "artificial retinas," and we focus on these here. Like cortical prostheses, they have advantages and disadvantages. Retinal prostheses could connect to bipolar cells or ganglion cells, where the response properties are relatively well understood. Thanks to 50 years of work on retinal ganglion cells in animals, the neural code in the retina is better understood than it is further along in the hierarchy of visual processing, and it is probably simpler. Retinal prostheses could also potentially allow the best use of remaining visual pathways. Assuming that the technical barriers can be overcome, the chief drawback that will remain is that they require the presence of at least some intact ganglion cells. This means that they are suitable only for photoreceptor diseases. The earliest target population will be people with retinitis pigmentosa (RP) who have reached the stage of complete blindness. These devices will not be useful in diabetic retinopathy, glaucoma, or vascular occlusive diseases, which, unfortunately, account for a larger fraction of blindness than RP does. Individuals with AMD may also be a candidate population, but they often retain some central vision (20/400) for a long time, and it is not clear at present that an implanted device would improve their vision (Rizzo *et al.*, 2001). In addition, AMD patients with choroidal neovascularization may have inner retinal damage, which would rule them out as candidates for prostheses.

Artificial retinas pose substantial challenges. First we will discuss some of the design requirements that have to be considered, and the quality of vision that can be expected. Then, after a description of each type of prosthesis, we will consider their biocompatibility (damage to the retina and the device, effects on retinal metabolism, heat transfer, and longevity and reliability) and their electrical coupling to neurons.

13.6.2. DESIGN GOALS

The goal in making an artificial retina is not to restore normal vision, but to restore enough vision for mobility, and then work toward better vision after that. The normal retina in a person with 20/20 vision has a resolution limit of about 2.5 μm and a corresponding acuity of about 60 cycles per degree for foveal viewing, as discussed above. However, even very nearsighted people can navigate the world without their glasses, and can read if the letters are large enough. A resolution of 3 cycles per degree would be about 20/400 vision, and this would require the spacing between the elements to be about 40 μm. A spacing of 60 μm would give an acuity limit of 2.5 cycles per degree according to Eq. (13.1). This is approximately the resolution that is being sought with the retinal prostheses. Incidentally, 2–3 cycles per degree is also the approximate functional resolution that cats have (Pasternak and Merigan, 1981) at high illumination, although the most central X-type ganglion cells seem to have a resolution closer to 10 cycles per degree (Cleland *et al.*, 1979).

One can also ask how much of the visual field needs to be restored. The normal visual field from the far temporal periphery to the far nasal periphery is about 150° of visual angle. This will be unachievable with any of the techniques currently under consideration, but a sort of tunnel vision subtending about 20° of visual angle would be a reasonable design goal. The largest devices are about 3 mm in diameter, and therefore cover about 10° of the retinal surface. If one holds a standard cardboard toilet paper tube (1.5″ in diameter; 4.5″ long) up to one's eye, the field of view is 19° in diameter. (Of course the other eye should be closed, because no one for a long time will receive binocular retinal prostheses.) If one is also very nearsighted, it becomes possible to appreciate the best that could be done with the retinal prostheses under development. Even though this is much reduced from normal vision, it is still better than total blindness, and mobility is possible if one moves the head to compensate for the loss of the extended visual field. Of course this demonstration uses intact visual pathways, so it assumes that the signaling from the prosthesis to the neurons is optimal. As discussed below, this is not currently the case.

There are other considerations besides the spatial ones. From what we know about ganglion cell behavior, it seems useful to filter out signals from large objects with low spatial frequency information, which means that the spatial response of a retinal prosthesis should ideally have a bandpass rather than a low-pass spatial frequency characteristic. Adequate temporal frequency response should be relatively easy to achieve, because humans have their best sensitivity to frequencies of 20 Hz or less (Cornsweet, 1970).

Assuming that retinotopic stimulation can be achieved, so that a pattern of retinal locations could be stimulated in a controlled way, another line of investigation becomes relevant. These are psychophysical studies that convert visual scenes into pixilated images, and assess the number of array elements that must be seen by a subject with normal visual pathways to allow mobility and reading of large print. These studies were performed mainly for the design of cortical prostheses, but are useful for retinal implant design as well. The result is that an array of 25×25 elements subtending an angle of 1.7° (a spacing of about 17 µm across about 425 µm of retinal surface) would be sufficient (Cha *et al.*, 1992a–c), and some functions could be attained with fewer elements (reviewed by Margalit *et al.*, 2002).

A gain control would also be extremely useful in a visual prosthesis. Our neurons encode a range of no more than about 2 log units—from 1 action potential per second to about 100 per second—but they manage to handle more than 12 log units of ambient illumination in the world, from starlight to bright sunlight. In the natural visual system this could be done by a compression that encoded the very large input range into the small neuronal output range, but this would make it difficult to detect small differences in illumination from a mean level, which is the situation we are in normally. The mean illumination is usually provided by a self-luminous source of a particular illumination (sun, moon, light bulb), and different illuminations reaching our eye depend on relatively small differences in the reflectance of objects that are illuminated by this source. Instead of using a compressive function, early elements in the visual system adapt to light. This means that their processing adjusts to the ambient level of illumination, so that the small response range shifts depending on the mean

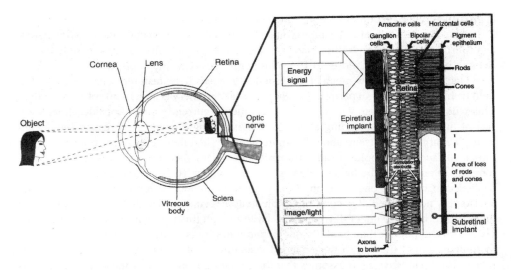

FIGURE 13.25. Schematic diagram showing the concept of epiretinal and subretinal prostheses. Epiretinal prostheses are electrode arrays that receive stimuli from an intraocular signal generator and are designed to activate retinal ganglion cells. Subretinal prostheses are positioned under the retina, receive light, and convert it to a current that stimulates bipolar cells. (Reprinted from Zrenner, 2002, with permission from copyright holder, A. Stett)

illumination. Information about the mean itself is less valuable, and is essentially thrown out. Digital cameras are quite good at this sort of sensitivity adjustment, and this function could be built into a prosthesis. Cameras are not as good as the eye at low illuminations, but prostheses that work over the range from normal room illumination to daylight would probably be sufficient.

13.6.3. SUBRETINAL AND EPIRETINAL PROSTHESES

Two fundamentally different types of retinal prostheses are under development, reviewed recently by Margalit *et al.* (2002) and Zrenner (2002) (Figure 13.25). The first type slides under the retina, into the subretinal space, so it is called a subretinal implant. These devices would be small (maximum of a few millimeters in diameter) and might be as unobtrusive as an intraocular lens. They would require no external power because there is no circuitry involved; the light itself causes the generation of a current designed to stimulate the neurons. The other type of retinal prosthesis would have a camera and a signal-processing apparatus outside the eye, and then transmit signals to an intraocular stimulator. The stimulator would control an array of electrodes that are placed on the retina in order to stimulate ganglion cells or optic nerve fibers. Because they would be placed on the surface of the retina, they are called epiretinal prostheses.

13.6.3.1. Subretinal Prostheses

The subretinal prostheses are silicon-based disks that are 50 to 100 μm thick and up to 3 mm in diameter. The artificial retina from Chow's group (Peyman *et al.*, 1998; Chow

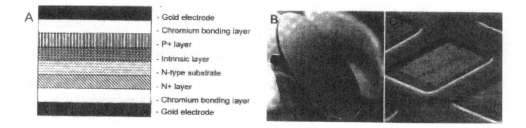

A
- Gold electrode
- Chromium bonding layer
- P+ layer
- Intrinsic layer
- N-type substrate
- N+ layer
- Chromium bonding layer
- Gold electrode

B

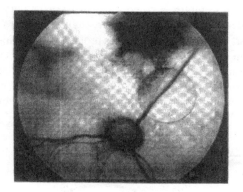

FIGURE 13.26. *Top left*: Lamination of a subretinal prosthesis. *Top middle*: Photograph of a 2-mm-diameter subretinal prosthesis. *Top right*: Detail of one element of the subretinal prosthesis. *Bottom*: Fundus photograph of the cat retina showing a subretinal prosthesis in place after 1 month and 3 months. Note that the retinal circulation appears intact over the prosthesis. (Reprinted from Chow *et al.*, 2001, with permission from IEEE)

et al., 2001) and the company Optobionics, contains microphotodiodes, essentially the same as solar cells, with a center-to-center spacing that is currently about 30 μm. This gives a density of about 1100 elements per mm², or 7600 electrodes in a 3-mm-diameter implant (Figure 13.26). They are made with standard photomasking and etching techniques from a sheet of multilayered material that can be assembled to make either PiN (positive–intrinsic–negative) or NiP junctions. In this device, called the artificial silicon retina, or ASR, the light-sensitive elements are 20 ×20 μm and are separated by 10 μm. Transparent gold or titanium nitride (TiN) electrodes are individually deposited on top of each photosensitive element. The reference electrode is a continuous sheet at the back of the device. When incident light stimulates an element of the array, it generates a current at the electrode that is supposed to stimulate adjacent bipolar or ganglion cells. In a PiN device, the front electrode generates a positive signal and in an NiP device the stimulating current is inverted. The electrodes in the ASR are the same size as the photodiodes. The other subretinal implant (Zrenner *et al.*, 1997, 1999) is called the microphotodiode array (MPDA). It is very similar, but TiN electrodes of 8 ×8 μm are applied to 20 ×20 μm diodes, so that the current density is higher, and positive and negative electrodes are built alternatingly into the device with the hope of stimulating neurons with both on and off polarities (Zrenner *et al.*, 1997). For testing stimulation parameters, arrays with fewer and larger electrodes have been used (Schwan, 2001). None of these devices are sufficiently sensitive to be driven solely by visible light, but they are also sensitive to near-infrared. Their spectral sensitivity ranges from 500 to 1100 nm. There is still a question of whether they will have high enough current densities to stimulate the remaining retinal neurons, as discussed below.

The outstanding advantages of these devices are their simplicity, ease of attachment to the eye, high density of electrodes, and the lack of an external camera and image-processing hardware and software.

13.6.3.2. Epiretinal Prostheses

Epiretinal prostheses would employ a camera or field sensor outside the eye, and send signals to a receiver placed intraocularly, in the vitreous or in an intraocular lens (Figure 13.27). The intraocular part of the device would then provide stimulating current through a cable to an electrode array in contact with the retinal surface. Although epiretinal devices are more complex than subretinal devices, they have at least two potential advantages. First, the extraocular signal-processing device could use the same gain control methods to adapt to mean illumination that are used in cameras at present, whereas subretinal implants are

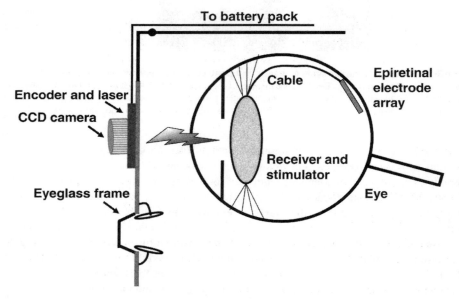

FIGURE 13.27. Components of an epiretinal system. A video camera, possibly in a pair of eyeglasses, is connected to an encoder. The encoder contains signal-processing hardware and software that encode the light into electrical signals for transmission into the eye. The lightning bolt represents laser energy that transmits these signals, as well as power for the intraocular portion of the prosthesis, into the eye. The receiver in the eye is located in a modified intraocular lens that replaces the natural lens. After performing further processing, the stimulator would send signals to the electrode array attached to the retina. The electrode array shown is the approximate size of ones currently being tested. An alternate mode of transferring signal and power into the eye would be with radiofrequency telemetry rather than a laser. (After Rizzo and Wyatt, 1999; Singer, 2002)

likely to work only within a limited range of high illuminations. Second, reprogramming and shaping of the signals sent to the intraocular portion of the device can be done following implantation, which would not be possible with the subretinal devices. In addition, parameters necessary for neuronal stimulation can be worked out in animal or pilot clinical experiments with individual electrodes that are similar to the ones that will actually be incorporated into the devices. One disadvantage of an epiretinal device is that it has to be attached to the retina with adhesives or tacks—more like rivets—that go through the retina and sclera. Other challenges include miniaturization, electronics packaging, and power supply.

There are three principal groups working on epiretinal devices. One effort (Humayun *et al.*, 1999a,b; Margalit *et al.*, 2002) includes at least two universities, four national labs, and a company called Second Sight (Singer, 2002). The extraocular part of the artificial retina consists of a camera and video-processing board, a telemetry encoder chip, a radio frequency amplifier, and a primary coil. Power to drive the stimulator, as well as signals, have to be sent to the eye. The intraocular unit consists of a secondary coil, a rectifier and regulator, a processor to decode the signals, a stimulator, and the electrode array (Margalit *et al.*, 2002). A 4×5 mm prosthesis with 16 electrodes was implanted in a patient in 2002, but the goal is to reach 1000 electrodes (Singer, 2002). A fundamentally similar system is being developed at Harvard and MIT (Wyatt and Rizzo, 1996; Rizzo and Wyatt, 1997; Grumet *et al.*, 2000). Wyatt and Rizzo (1999) propose, however, that optical signaling via a laser could carry both power and signal information to the intraocular device. The third device is being designed by the EPI-RET research group in Germany (Eckmiller, 1997; Walter *et al.*, 1999; Walter and Heimann, 2000.) The front end of this device is called a "retina encoder" and consists of up to 1000 elements with tunable filters mimicking the Difference of Gaussian receptive fields of ganglion cells discussed earlier (Eckmiller, 1997). A novel feature is that the properties of the filters could be "trained" by the individual user to achieve the best vision. As in the other epiretinal devices, the encoder would be outside the eye and would send signals to a "retinal stimulator." The electrode array of the stimulator would have platinum or TiN electrodes, 50 to 100 μm in diameter, mounted in a polyimide or silicon strip (Walter *et al.*, 1999; Walter and Heimann, 2000).

13.6.4. BIOCOMPATIBILITY

13.6.4.1. Health of Retina

The subretinal implant placed in normal rabbits led to loss of photoreceptors and a reduction of the density of neurons in the inner retina over the implant (Peyman *et al.*, 1998). This was to be expected, because there is no retinal circulation in rabbit, and the implant blocks diffusion of nutrients from the choroid. The damage was local, and the retina appeared normal away from implant. In cat, which has a retinal circulation, there were no significant differences in cell counts in the inner nuclear or ganglion cell layers (Chow *et al.*, 2001). The ERG in response to light in the implanted eye was slightly smaller than in the fellow eye, but the reduction was consistent with the area of photoreceptors destroyed by the implant, not with more extensive damage (Chow *et al.*, 2001). Similar results have been obtained in the Yucatan micropig after 14 months of implantation (Schwahn *et al.*, 2001) and in the RCS rat after 4 months (Zrenner *et al.*, 1999). There was a mild glial reaction

(Zrenner et al., 1999; Schwahn et al., 2001), but no glial proliferation. Some of the devices used in these studies were inactive, or produced potentials themselves but did not appear to stimulate the retina. There is a question of whether they will continue to be benign when the current output is large enough to clearly stimulate retinal cells. The best evidence for long-term (2.5-year) safety in humans is from implantation in 10 patients with RP (Chow et al., 2002, 2003).

Inactive epiretinal implants also appear to leave the rabbit retina in good condition, on the basis of light microscopy, and ERG and visual-evoked potential recordings, despite the introduction of titanium tacks to attach the arrays to the retina (Walter et al., 1999). These arrays were 5×3 or 10×2 flexible silicon or polyimide substrates with platinum electrodes. Similar results were obtained in dogs with inactive 5×5 arrays that were in place for up to a year (Majji et al., 1999).

The epiretinal implant will require a powered stimulator that will generate heat. It has been estimated that a 5-mW supply may generate enough power for 100 stimulating electrodes. This is likely to be acceptable, especially if it is located in the vitreous rather than at the retinal surface. At the retina, a 50-mW source over 1.4 mm^2 could be tolerated for only a second, whereas a 500-mW source in the vitreous caused no problems for over 2 h (Margalit et al., 2002).

13.6.4.2. Health of Implant

In cats, the amplitude of the response of a subretinal implant increased for 1–2 months after implantation, and began to fall at about 4 months after implantation (Chow et al., 2001). Loss of gold electrode probably accounted for this. This was related to the current that was produced, because it occurred in active but not inactive implants. Active epiretinal electrodes have not been left in place long enough to gauge implant damage.

13.6.5. COUPLING OF PROSTHESES TO NEURONS

To stimulate retinal neurons with either of the proposed devices, they must generate enough current to stimulate neurons, but not so much as to damage the electrodes or the retina. The many aspects of this problem have not been adequately addressed for retinal prostheses; however, a discussion of the basic principles is provided in Margalit et al. (2002), and work on this important topic is ongoing. Reports have appeared that both an epiretinal prosthesis (Singer, 2002) and a subretinal prosthesis (Chow et al., 2003) can restore vision in humans, which is an exciting development. This suggests that there is some coupling from the devices to the retina, but, on the basis of the data from animal studies reviewed below, this coupling is not yet optimal.

One issue for both types of prostheses is that the current required depends on the proximity of the current source to the neurons, since the current will take the path of least resistance, which would primarily be through the extracellular fluid if there is fluid between the implant and the neurons. For this reason, Zrenner's group wishes to have the subretinal implant in contact with neurons, and they have proposed coating the implant with molecules that promote adhesion. Coating implants with poly-L-lysine, poly-D-lysine, or laminin was effective in promoting adherence of isolated retinal cells for several weeks, but a comparison of coated and uncoated implants in vivo has not been made (Zrenner et al., 1997). Zrenner *et al.* (1997) also suggest that signals have to generate at least a 10-mV depolarization in the neurons, but there is no obvious basis for this.

13.6.5.1. Subretinal Stimulation

One of the problems with the subretinal electrodes is that the current pathways from the positive to the negative electrode have not been defined. If the current generated by each microphotodiode is transmitted to its attached electrode, and then flows around the edge of the chip to reach the large electrode at the back of the chip, then the current loops are not well localized. One then has to rely on the current density being high enough near the active element to stimulate some neurons, but low enough that it does not stimulate adjacent ones.

A rational approach to determine the electrical requirements of retinal prostheses would be to investigate the relation between the stimulating current and neuronal responses under conditions where stimulation parameters can be varied, and recordings from individual neurons can be made. This has been done by using isolated retinas from newly hatched chickens with damaged outer segments (Zrenner *et al.*, 1997) or RCS rats (Zrenner *et al.*, 1999). The retina was laid, ganglion cell side down, on a multielectrode array. The MPDA was then laid on top of the retina, forming a sandwich. In response to very intense (30 to 70 klux) whole-field stimuli, transient ganglion cell activity was recorded. A burst of spikes lasting about 50 ms occurred near the beginning of a 500-ms period of stimulation in chicks, but in RCS rats the stimulation caused a transient depression of spontaneous firing.

Direct electrical stimulation, rather than stimulation by microphotodiodes, has also been done. Zrenner *et al.* (1999) found that 10 μA of current could evoke a change in firing rate of ganglion cells. Schwahn *et al.* (2001) implanted an electrode array consisting of eight 100×100 μm electrodes (i.e., much larger than the microphotodiode electrodes) in the subretinal space of minipigs and rabbits, and recorded electrically evoked responses (EERs) from the visual cortex. This is not entirely comparable to the implantation of a subretinal prosthesis, because the outer nuclear layer and possibly other parts of the photoreceptors were still intact in these acute experiments, whereas they deteriorate relatively rapidly when an implant is chronically implanted. In two of the five pigs, stimulus pulses of 400 μs and 3 V (a charge transfer of 50 nC per electrode) evoked EERs having an amplitude of about 10 μV, comparable to the size of visual-evoked responses (VERs). However, stimuli greater than 2 V caused damage to the retina. The threshold stimulus was generally lower in rabbits, with 0.6 V (7 nC per electrode) producing an EER in 4 of 10 animals. Why only 40% of the animals generated responses is not clear, but the authors suggested that in some cases the fluid prevented good contact of the electrode with the neurons.

Although there is a lack of detailed analyses of current flows and electrical requirements, *in vivo* tests of the function of subretinal implants have been performed with electroretinographic techniques. They show that signals of 20–30 μV can be recorded in the cornea in rabbits, rats, and cats in response to intense IR illumination (Zrenner *et al.*, 1997; Peyman *et al.*, 1998; Chow *et al.*, 2001). These responses reach a peak and then decrease before illumination is terminated. Significantly, in tests done by Peyman *et al.* (1998) on rabbits, different waveforms were obtained before and after the rabbit was killed, suggesting that the ERG responses included both a direct response from the implant and a component of retinal activity evoked by the implant. It is not known which retinal cells might have generated the responses. In other studies, responses to infrared light were mainly dominated by the implant signal, and if a retinal response was generated by the implant, it was small (Zrenner *et al.*, 1997; Chow *et al.*, 2001). ERG techniques will not necessarily reveal retinal activity even if it exists, because the implant occupies only a small region in the eye, and because ganglion cell activity is not present under standard ERG stimulating and recording conditions. At one time, Chow's group believed that cortical responses had been obtained, but this appeared to be mainly due to a slight sensitivity of the cat retina to IR illumination (Peachey *et al.*, 2000).

RP patients in the phase I clinical trial of a subretinal implant have reported regaining some vision (Chow *et al.*, 2002, 2003). Some of these patients had remnant vision before the surgery however, and no rigorous tests of visual function have been reported yet that would definitively show that it is the implant that was responsible for the improved vision. One possibility is that the surgery itself, rather than the implant, was responsible for the improvement. Faktorevitch *et al.* (1990) found that trauma to the retina increased the production of growth factors and slowed the loss of photoreceptors in an animal model of RP, and Chow *et al.* (2003) acknowledge the potential of a neurotrophic effect. On the whole, evidence that the current generated by subretinal implants provides visual function is quite limited.

13.6.5.2. Epiretinal Stimulation

Considerable information about stimulation parameters has been obtained by the groups developing epiretinal prostheses. Currents of more than 100 µA are required to generate an EER from the cortex of rabbits in response to stimulating electrodes on the retinal surface (Rizzo and Wyatt, 1997), but ganglion cells laid on electrodes can be activated by currents below 2 µA (Grumet *et al.*, 2000). Using the types of electrodes proposed for the EPI-RET prosthesis, Walter and Heimann (2000) were able to record graded EERs in acute rabbit experiments in response to 10 to 150 µA/phase (stimuli were either monopolar or bipolar) and charge transfer as low as 0.1 to 0.3 nC/phase. The EERs were larger in response to bursts of 5 to 20 pulses at 1 kHz than with single pulses.

Epiretinal electrical stimulation has been shown to evoke visual percepts in humans. Rizzo and Wyatt found that volunteers were able to distinguish two spots of light when electrodes were separated by 2° of visual angle (Greenberg, 2000). Humayan *et al.* (1999) used up to 25 electrodes in a two-dimensional array in the eye of RP and AMD patients and were able to evoke a sensation of multiple spots of light. Importantly, the patients were able to identify simple shapes and letters when the electrodes were activated in order to produce patterns, and a stable, nonflickering perception was created with stimulation frequencies of 40 to 50 Hz. Limited data were presented, but this was the first attempt to address whether it will be possible to achieve the perception of more than just brief phosphenes.

One of the questions concerning epiretinal stimulation is whether the ganglion cell bodies can be stimulated, which might allow an array of elements to yield a retinotopic percept, or whether the unmyelinated ganglion cell axons closer to the stimulating electrodes would be stimulated, possibly making it difficult to produce an organized percept. Although there is no complete resolution to this yet, a computational model of ganglion cells suggests that the threshold for stimulating cell bodies compared with that for stimulating axons is lower enough for preferential cell body stimulation to be feasible (Greenberg *et al.*, 1999).

13.6.6. VASCULAR ISSUES

One of the important challenges of using a subretinal device is that when implantation of a prosthesis would be considered for a patient with RP, there is severe attenuation of the retinal circulation as discussed above. This circulation is presumably lost, because in the absence of photoreceptors the choroid can adequately fulfill the nutritional needs of the retina. However, if one then inserts a silicon-based layer under the retina, the choroid will be too far away to supply the inner retina effectively, and diffusion of nutrients from the choroid will be impeded, if not completely blocked, by the chip itself. It is conceivable that the retinal circulation could redevelop, but on the basis of the examples we have to date (diabetic and other retinopathies), we can conclude that angiogenesis in the adult retina always produces abnormal vessels that cause more problems than they solve. Furthermore, it is unlikely that vascular development could occur in a time frame that would prevent the inner retinal cells from dying. Subretinal electrodes containing spaces have been proposed. The circulation in and the nutrition of the retina would not be expected to be a problem for an epiretinal prosthesis, unless electrical stimulation increases the activity and metabolism of the inner retinal neurons so that the choroid cannot supply adequate nutrients.

13.7. OPPORTUNITIES

Although a great deal of progress has been made in retinal bioengineering, many avenues of development remain open. First, the modeling of retinal neural function in a way that ties it to the underlying physiological and anatomical substrates is not complete, and further work could have many benefits. Second, retinal electrophysiological testing and visual function testing have not reached optimal development for early detection of the major retinal diseases. Advances will rely on combining an understanding of retinal pathophysiology with improved instrumentation. Third, an understanding of the retinal microenvironment and its response to disease can potentially lead to improved treatments. Fourth, the prostheses under development still face major issues in achieving vision close to natural vision and demonstrating long-term efficacy. Finally, even if successful, the retinal prostheses under development will be useful for improving vision in only a few of the major blinding diseases, leaving many challenges for the future.

ACKNOWLEDGMENTS

I thank Dr. John Troy for useful discussions. The work in my laboratory was supported largely by NIH R01 EY05034.

REFERENCES

Abdelsalam, A., Del Priore, L., and Zarbin M. A., 1999, Drusen in age-related macular degeneration: Pathogenesis, natural course, and laser photocoagulation–induced regression, *Surv. Ophthalmol.* **44**:1–29.

Acland, G. M., Aguirre, G. D., Ray, J., Zhang, Q., Aleman, T. S., Cideciyan, A. V., Pearce-Kelling, S. E., Anand, V., Zeng, Y., Maguire, A. M., Jacobson, S. G., Hauswirth, W. W., and Bennett, J., 2002, Gene therapy restores vision in a canine model of childhood blindness, *Nat. Genet.* **28**:92–95.

Adamis, A. P., Miller, J. W., Bernal, M.-T., D'Amico, D. J., Folkman, J., Yeo, T. K., and Yeo, K. T., 1994, Increased vascular endothelial growth factor levels in the vitreous of eyes with proliferative diabetic retinopathy, *Am. J. Ophthalmol.* **118**:445–450.

Adamis, A. P., Shima, D. T., Tolentino, M. J., Gragoudas, E. S., Ferrara, N., Folkman, J., D'Amore, P. A., and Miller, J. W., 1996, Inhibition of vascular endothelial growth factor prevents retinal ischemia-associated iris neovascularization in a nonhuman primate, *Arch. Ophthalmol.* **114**:66–71.

Aguirre, G., Farber, D., Lolley, R., Fletcher, R. T., and Chader, G. J., 1978, Rod–cone dysplasia in Irish Setters: A defect in cyclic GMP metabolism in visual cells, *Science* **201**:1133–1134.

Ahmed, J., Pulfer, M. K., and Linsenmeier, R. A., 2001, Measurement of blood flow through the retinal circulation of the cat during normoxia and hypoxemia using fluorescent microspheres, *Microvasc. Res.* **62**:143–153.

Alder, V. A., Cringle, S. J., and Constable, I. J., 1983, The retinal oxygen profile in cats, *Invest. Ophthalmol. Visual Sci.* **24**:30–36.

Alm, A., 1992, Ocular circulation, in: *Adler's Physiology of the Eye: Clinical Application*, 9th ed. (W. M. Hart Jr., ed.), Mosby Year Book, St. Louis, pp. 198–325.

Alm, A., and Bill, A., 1972, The oxygen supply to the retina. II. Effects of high intraocular pressure and of increased arterial carbon dioxide tension on uveal and retinal blood flow in cats: A study with radioactively labeled microspheres, including flow determination in brain and some other tissues, *Acta Physiol. Scand.* **84**:306–319.

Asher, H., 1951, The electroretinogram of the blind spot, *J. Physiol.* **112**:40P.

Barnes, S., and Hille, B., 1989, Ionic channels of the inner segment of tiger salamander cone photoreceptors, *J. Gen. Physiol.* **94**:719–743.

Baylor, D. A., 1987, Photoreceptor signals and vision, *Invest. Ophthalmol. Visual Sci.* **28**:34–49.

Baylor, D. A., Nunn, B. J., and Schnapf, J. L., 1984, The photocurrent, noise, and spectral sensitivity of rods of the monkey *Macaca fascicularis*, *J. Physiol.* **357**:575–607.

Belgum, J. H., Dvorak, D. R., and McReynolds, J. S., 1982, Sustained synaptic input to ganglion cells of mudpuppy retina, *J. Physiol.* **326**:91–108.

Bill, A., 1984, Circulation in the eye, in: *Handbook of Physiology. The Cardiovascular System IV* (E. M. Renkin and C. C. Michel, eds.), American Physiological Society, Bethesda, MD, pp. 1001–1034.

Bill, A., and Sperber, G. O., 1990, Control of retinal and choroidal blood flow, *Eye* **4**:319–325.

Birch, D. G., Hood, D. C., Nusinowitz, S., and Pepperberg, D. R., 1995, Abnormal activation and inactivation mechanism of rod transduction in patients with autosomal dominant retinitis pigmentosa and the pro-23-his mutation, *Invest. Ophthalmol. Visual Sci.* **36**:1603–1614.

Bloomfield, S. A., 1996, Effect of spike blockade on the receptive-field size of amacrine and ganglion cells in the rabbit retina, *J. Neurophysiol.* **75**:1878–1893.

Bloomfield, S. A., and Dacheux, R. F., 2001, Rod vision: Pathways and processing in the mammalian retina, *Prog. Retin. Eye Res.* **20**(3):351–384.

Bloomfield, S. A., and Xin, D. Y., 2000, Surround inhibition of mammalian AII amacrine cells is generated in the proximal retina, *J. Physiol.* **523**:771–783.

Boycott, B. B., and Wassle, H., 1974, The morphological types of ganglion cells of the domestic cat's retina, *J. Physiol.* **240**:397–419.

Boynton, R. M., and Riggs, L. A., 1951, The effect of stimulus area and intensity upon on the human retinal response, *J. Exp. Psych.* **42**:217–226.

Braille Institute/Braille Press, 2000, Los Angeles (July 31, 2002); http://www.brailleinstitute.org.

Braun, R. D., Linsenmeier, R. A., and Goldstick, T. K., 1995, Oxygen consumption in the inner and outer retina of the cat, *Invest. Ophthalmol. Visual Sci.* **36**:542–554.

Breton, M. E., Schueller, A. W., Lamb, T. D., and Pugh, E. N., Jr., 1994, Analysis of ERG a-wave amplification and kinetics in terms of the G-protein cascade of phototransduction, *Invest. Ophthalmol. Visual Sci.* **35**:295–309.

Brindley, G., and Rushton, D., 1974, Implanted stimulators of the visual cortex as visual prosthetic devices, *Trans. Am. Acad. Ophthalmol. Otolaryngol.* **78**:741–745.

Brown, G. C., 1999, Arterial occlusive disease, in: *Vitreoretinal Disease: The Essentials* (C. D. Regillo, G. C. Brown, and H. W. Flynn, eds.), Thieme, New York, pp. 97–115.

Cao, W., Govardovskii, V., Li, J.-D., and Steinberg, R. H., 1996, Systemic hypoxia dehydrates the space surrounding photoreceptors in the cat retina, *Invest. Ophthalmol. Visual Sci.* **37**:586–596.

Cha, K., Horch, K. W., and Normann, R. A., 1992a, Simulation of a phosphene-based visual field: Visual acuity in a pixelized vision system, *Ann. Biomed. Eng.* **20**:439–449.

Cha, K., Horch, K. W., and Normann, R. A., 1992b, Mobility performance with a pixelized vision system, *Vision Res.* **32**:1367–1372.

Cha, K., Horch, K. W., Normann, R. A., and Boman, D. K., 1992c, Reading speed with a pixelized vision system, *J. Opt. Soc. Am.* **9**:673–677.

Chan, L. H., Freeman, A. W., and Cleland, B. G., 1992, The rod–cone shift and its effect on ganglion cells in the cat's retina, *Vision Res.* **32**:2209–2219.

Chase, H. P., Jackson, W. E., Hoops, S. L., Cockerham, R. S., Archer, G., O'Brien, D., 1989, Glucose control and the retinal and retinal complications of insulin-dependent diabetes, *JAMA* **261**:1155–1160.

Chen, E. P., and Freeman, A. W., 1989, A model for spatiotemporal frequency response in the X cell pathway of the cat's retina, *Visual Res.* **29**:271–291.

Chen, E. P.-C., and Linsenmeier, R. A., 1989, Centre components of cone-driven retinal ganglion cells: differential sensitivity to 2-amino-4-phosphonobutyric acid, *J. Physiol.* **419**:77–93.

Cho, E., Hung, S., and Seddon, J. H., 1999, Nutrition, in: *Age-Related Macular Degeneration* (J. W. Berger, S. L. Fine, and M. G. Maguire, eds.), Mosby, St. Louis, pp. 57–67.

Chow, A. Y., Pardue, M. T., Chow, V. Y., Peyman, G. A., Liang, C., Perlman, J. I., and Peachey, N. S., 2001, Implantation of silicon chip microphotodiode arrays into the cat subretinal space, *IEEE Trans. Neural Syst. Rehabil. Eng.* **9**:86–95.

Chow, A. Y., Peyman, G. A., Pollack, J. S., and Packo, K. H., 2002, Safety, feasibility, and efficacy of subretinal artificial silicon retina prosthesis for the treatment of patients with retinitis pigmentosa, Association for Research in Vision and Ophthalmology Abstracts, no. 2849. www.arvo.org.

Chow, A. Y., Packo, K. H., Pollack, J. S., and Schuchard, R. A., 2003, Subretinal artificial silicon retina microchip implantation in retinitis pigmentosa patients: Long term follow-up. Association for Research in Vision and Ophthalmology Abstracts, no. 4205. www.arvo.org.

Cideciyan, A. V., and Jacobson, S. G., 1996, An alternative phototransduction model for human rod and cone ERG a-waves: Normal parameters and variation with age, *Vision Res.* **16**:2609–2621.

Citron, M. C., Emerson, R. C., and Levick, W. R., 1988, Nonlinear measurement and classification of receptive fields in cat retinal ganglion cells, *Ann. Biomed. Eng.* **16**:65–77.

Clarkson, J. G., 1994, Central retinal vein occlusion, in: *Retina*, 2nd ed., Vol. 2 (S. J. Ryan, ed.), Mosby, St. Louis, pp. 1379–1385.

Cleland, B. G., Harding, T. H., and Tulunay-Keesey, U., 1979, Visual resolution and receptive field size: Examination of two kinds of cat retinal ganglion cell, *Science* **205**:1015–1017.

Cleland, B. G., and Levick, W. R., 1974, Brish and sluggish concentrically organized ganglion cells in the cat's retina, *J. Physiol.* **240**:421–456.

Cobbs, W. H., and Pugh, E. N., Jr., 1987, Kinetics and components of the flash photocurrent of isolated retinal rods of the larval tiger salamander, *Ambystoma tigrinum, J. Physiol.* **394**:529–572.

Cornsweet, T. N., 1970, *Visual Perception*, Academic Press, New York, pp. 387–392.

Cringle, S. J., Yu, D.-Y., Alder, V., Su, E.-N., and Yu, P. K., 1996, Oxygen consumption in the avascular guinea pig retina, *Am. J. Physiol.* **271**:H1162–H1165.

Cringle, S. J., Yu, D.-Y., Yu, P. K., and Su, E.-N., 2002, Intraretinal oxygen consumption in the rat *in vivo, Invest. Ophthalmol. Visual Sci.* **43**:1922–1927.

Croner, L. J., and Kaplan, E., 1995, Receptive fields of P and M ganglion cells across the primate retina, *Vision Res.* **35**:7–24.

Dawis, S., Shapley, R., Kaplan, E., and Tranchina, D., 1984, The receptive field organization of X-cells in the cat: Spatiotemporal coupling and asymmetry, *Vision Res.* **24**:549–564.

DCCT (Diabetes Control and Complications Trial) Research Group, 1993, The effect of intensive treatment of diabetes on the development and progression of long-term complications in insulin-dependent diabetes mellitus, *N. Eng. J. Med.* **329**:977–986.

DeMonasterio, F. M., 1978a, Properties of concentrically organized X and Y ganglion cells of macaque retina, *J. Neurophysiol.* **41**:1394–1417.

DeMonasterio, F. M., 1978b, Center and surround mechanisms of opponent-color X and Y ganglion cells of retina of macaques, *J. Neurophysiol.* **41**:1418–1434.

DeMonasterio, F. M., and Gouras, P., 1975, Functional properties of ganglion cells of the rhesus monkey retina, *J. Physiol.* **251**:167–195.

Derrington, A. M., and Lennie, P., 1982, The influence of temporal frequency and adaptation level on receptive field organization of retinal ganglion cells in cat, *J. Physiol.* **333**:343–366.

Diabetic Retinopathy Research Group, 1976, Preliminary report on the effects of photocoagulation therapy, *Am. J. Ophthalmol.* **81**:383–396.

Dmitriev, A. V., Govardovskii, V. I., Schwahn, H. N., and Steinberg, R. H., 1990, Light-induced changes of extracellular ions and volume in the isolated chick retina-pigment epithelium preparation, *Visual Neurosci.* **16**:1157–1167.

Dmitriev, A. V., and Mangel, S. C., 2000, A circadian clock regulated the pH of the fish retina, *J. Physiol.* **522**:77–82.

Dmitriev, A. V., and Mangel, S. C., 2001, Circadian clock regulation of pH in the rabbit retina, *J. Neurosci.* **21**:2897–2902.

Dobelle, W. H., 2000, Artificial vision for the blind by connecting a television camera to the visual cortex, *ASAIO J.* **46**:3–9.

Dollery, C. T., Bullpit, C. J., and Kohner, E. M., 1969, Oxygen supply to the retina from the retinal and choroidal circulations at normal and increased arterial oxygen tensions, *Invest. Ophthalmol.* **8**:588–594.

Dowling, J., 1987, *The Retina: An Approachable Part of the Brain*, Belknap Press, Cambridge, MA.

Eckmiller, R., 1997, Learning retina implants with epiretinal contacts, *Opthalmic Res.* **29**:281–289.

Engerman, R., Finkelstein, D., Aguirre, G., Diddie, K. R., Fox, R. R., Frank, R. N., and Varma, S. D., 1982, Ocular complications, *Diabetes* **31**(Suppl. 1):82–88.

Enroth-Cugell, C., and Robson, J. G., 1966, The contrast sensitivity of retinal ganglion cells of the cat, *J. Physiol.* **187**:517–552.

Enroth-Cugell, C., and Robson, J. G., 1984, Functional characteristics and diversity of cat retinal ganglion cells, *Invest. Ophthalmol. Visual Sci.* **25**:250–267.

Enroth-Cugell, C., Robson, J. G., Schweitzer-Tong, D. E., and Watson, A. B., 1983, Spatio-temporal interactions in cat retinal ganglion cells showing linear spatial summation, *J. Physiol.* **341**:279–307.

Enroth-Cugell, C., and Shapley, R. M., 1973, Adaptation and dynamics of cat retinal ganglion cells, *J. Physiol.* **233**:271–309.

Eperon, G., Johnson, M., and David, N. J., 1975, The effect of arterial PO2 on relative retinal blood flow in monkeys, *Invest. Ophthalmol.* **14**:342–352.

Fain, G. L., Quandt, F. N., Bastian, B. L., and Gerschenfeld, H. M., 1978, Contribution of cesium sensitive conductance increase to the rod photoresponse, *Nature* **272**:467–469.

Faktorevitch, E. G., Steinberg, R. H., Yasumura, D., Matthes, M. T., and LaVail, M. M., 1990, Photoreceptor degeneration in inherited retinal dystrophy delayed by basic fibroblast growth factor, *Nature* **347**:83–86.

Farber, D. B., and Lolley, R. N., 1973, Cyclic guanosine monophosphate: Elevation in degenerating photoreceptor cells of the C3H mouse retina, *Science* **186**:449–450.

Foerster, M. H., van de Grind, W. A., and Grusser, O.-J., 1977, Frequency transfer properties of three distinct types of cat horizontal cells, *Exp. Brain Res.* **29**:347–366.

Forti, S., Menini, A., Rispoli, G., and Torre, V., 1989, Kinetics of phototransduction in retinal rods of the newt *Triturus cristatus*, *J. Physiol.* **419**:265–295.

Fox, D. A., Poblenz, A. T., and He, L., 1999, Calcium overload triggers rod photoreceptor apoptotic cell death in chemical-induced and inherited retinal degenerations, *Ann. N. Y. Acad. Sci.* **893**:282–285.

Frishman, L. J., Freeman, A. W., Troy, J. B., Schweitzer-Tong, D. E., and Enroth-Cugell, C., 1987, Spatiotemporal frequency responses of cat retinal ganglion cells, *J. Gen. Physiol.* **89**:599–628.

Frishman, L. J., and Linsenmeier, R. A., 1982, Effects of picrotoxin and strychnine on non-linear responses of Y-type cat retinal ganglion cells, *J. Physiol.* **324**:347–363.

Frishman, L. J., Yamamoto, F., Borgula, J., and Steinberg, R. H., 1992, Light-evoked changes in $[K^+]_o$ in proximal portion of light-adapted cat retina, *J. Neurophysiol.* **67**:1201–1212.

Gallemore, R. P., Li, J.-D., Govardovskii, V. I., and Steinberg, R. H., 1994, Calcium gradients and light-evoked calcium changes outside rods in the intact cat retina, *Visual Neurosci.* **11**:753–761.

Gold, G. H., and Korenbrot, J. I., 1980, Light-induced calcium release by intact retinal rods, *PNAS* **77**:5557–5561.

Govardovskii, V. I., Li, J.-D., Dmitriev, A. V., and Steinberg, R. H., 1994, Mathematical model of TMA+ diffusion and prediction of light-dependent subretinal hydration in chick retina, *Invest. Ophthalmol. Visual Sci.* **35**:2712–2724.

Granit, R., 1933, The components of the retinal action potential in mammals and their relation to the discharge in the optic nerve, *J. Physiol.* **77**:207–240.

Greenberg, R. J., 2000, Visual prostheses: A review, *Neuromodulation* **3**:161–165.

Greenberg, R. J., Velte, T. J., Humayun, M. S., Scarlatis, G., and de Juan, E., 1999, A computational model of electrical stimulation of the retinal ganglion cell, *IEEE Trans. Biomed. Eng.* **46**:505–514.

Grumet, A. E., Wyatt, J. L., and Rizzo, J. F., 2000, Multi-electrode stimulation and recording in the isolated retina, *J. Neurosci. Methods* **101**:31–42.

Grunwald, J. E., Brucker, A. J., Braunstein, S. N., Schwartz, S. S., Baker, L., Petrig, B. L., and Riva, C. E., 1994, Strict metabolic control and retinal blood flow in diabetes mellitus, *Br. J. Ophthalmol.* **78**:598–604.

Grunwald, J. E., Maguire, A. M., and Dupont, J., 1996, Retinal hemodynamics in retinitis pigmentosa, *Am. J. Ophthalmol.* **12**:502–508.

Hagins, W. A., Penn, R. D., and Yoshikami, 1970, Dark current and photocurrent in retinal rods, *Biophys. J.* **10**:380–412.

Harwerth, R. S., Carter-Dawson, L., Shen, F., Smith, E. L., III, and Crawford, M. L. J., 1999, Ganglion cell losses underlying visual field defects from experimental glaucoma, *Invest. Ophthalmol. Visual Sci.* **40**:2242–2250.

Hatchell, D. L., and Sinclair, S. H., 1995, Role of leukocytes in diabetic retinopathy, in: *Physiology and Pathophysiology of Leukocyte Adhesion* (D. N. Granger and G. W. Schmid-Schoenbein, eds.), Oxford University Press, New York, pp. 458–466.

Haugh, L. M., Linsenmeier, R. A., and Goldstick, T. K., 1990, Mathematical models of the spatial distribution of retinal oxygen tension and consumption, including changes upon illumination, *Ann. Biomed. Eng.* **18**:19–36.

Hayreh, S. S., 1978, Pathogenesis of optic nerve damage and visual field defects, in: *Glaucoma, Conceptions of a Disease* (K. Heilman and K. T. Richardson, eds.), Saunders, Philadelphia, pp. 104–137.

Hayreh, S. S., Rojas, P., Podhajsky, P., Montague, C. R. A., and Woolson, R. F., 1983, Ocular neovascularization with retinal vascular occlusion, III. Incidence of ocular neovascularization with retinal vein occlusion, *Ophthalmology* **90**:488–506.

Hayreh, S. S., and Weingeist, T. A., 1980, Experimental occlusion of the central artery of the retina: IV. Retinal tolerance time to acute ischaemia, *Br. J. Ophthalmol.* **64**:818–825.

Heckenlively, J. R., Bouchman, J., and Friedman, L., 1988, Diagnosis and classification of retinitis pigmentosa, in: *Retinitis Pigmentosa* (J. R. Heckenlively, ed.), JB Lippincott., Philadelphia.

Hetling, J. R., and Pepperberg, D. R., 1999, Sensitivity and kinetics of mouse rod flash responses determined *in vivo* from paired-flash electroretinograms, *J. Physiol.* **516**(2):593–609.

Hochstein, S., and Shapley, R. M., 1976a, Quantitative analysis of retinal ganglion cell classifications, *J. Physiol.* **262**:237–264.

Hochstein, S., and Shapley, R. M., 1976b, Linear and nonlinear spatial subunits in Y cat retinal ganglion cells, *J. Physiol.* **262**:265–284.

Hogeboom van Buggenum, I. M., Van der Heijde, G. L., Tangelder, G. J., and Reichert-Thoen, J. W. M., 1996, Ocular oxygen measurement, *Br. J. Ophthalmol.* **80**:567–575.

Hood, D. C., and Birch, D. G., 1990, A quantitative measure of the electrical activity of human rod photoreceptors using electroretinography, *Visual Neurosci.* **5**:379–387.

Hood, D. C., and Birch, D. G., 1992, A computational model of the amplitude and implicit time of the b-wave of the human ERG, *Visual Neurosci.* **8**:107–126.

Hood, D. C., and Birch, D. G., Feb. 1995, Computational models of rod-driven retinal activity, *IEEE Eng. Med. Biol. Mag.*, pp. 59–66.

Huang, B., and Karwoski, C. J., 1992, Light-evoked expansion of subrietinal space volume in the retina of the frog, *J. Neurosci.* **12**:4243–4252.

Humayun, M. S., DeJuan, E., Jr., Weiland, J. D., Dagnelie, G., Katona, S., Greenberg, R., and Suzuki, S., 1999a, Pattern electrical stimulation of the human retina, *Vision Res.* **39**:2569–2576.

Humayun, M., Prince, M., DeJuan E., Jr., Barron, Y., Moskowitz, M., Klock, I. B., and Milam, A. H., 1999b, Morphometric analysis of the extramacular retina from postmortem eyes with retinitis pigmentosa, *Invest. Ophthalmol. Visual Sci.* **40**:143–148.

Johnson, C. A., 1996, Evaluation of visual function, in: *Duane's Foundations of Clinical Ophthalmology*, Vol. 2 (W. Tasman and E. A. Jaeger, eds.), Lippincott-Raven, Philadelphia, Chapt. 17, pp. 1–20.

Kang Derwent, J., and Linsenmeier, R. A., 2001, Intraretinal analysis of the a-wave of the electroretinogram in the dark-adapted intact cat retina, *Visual Neurosci.* **18**:353–363.

Kaplan, E., 1991, The receptive-field structure of retinal ganglion cells in cat and monkey: in *Vision and Visual Dysfunction, Vol. IV. The Neural Basis of Visual Function* (G. Leventhal, ed.), CRC Press, Boca Raton, FL, pp. 10–40.

Kaplan, E., and Shapley, R. M., 1982, X and Y cells in the lateral geniculate nucleus of macaque monkeys, *J. Physiol.* **330**:125–143.

Karwoski, C. J., and Proenza, C. J., 1977, Relationship between Muller cell responses, a local transretinal potential, and potassium flux, *J. Neurophysiol.* **40**:244–259.

Kiel, J., and Shepherd, A. P., 1992, Autoregulation of choroidal blood flow in the rabbit, *Invest. Ophthalmol. Visual Sci.* **33**:2399–2410.

Kincaid, M. C., 1996, Pathology of diabetes mellitus, in: *Duane's Foundations of Clinical Ophthalmology*, Vol. 2 (W. Tasman and E. A. Jaeger, eds.), Chapt. 18, Lippincott-Raven, Philadelphia, pp. 1–14.

Kiryu, J., Asrani, S., Shahidi, M., Mori, M., and Zeimer, R., 1995, Local response of the primate retinal microcirculation to increased metabolic demand induced by flicker, *Invest. Ophthalmol. Visual Sci.* **36**:1240–1246.

Klein, R., 1999, Epidemiology, in: *Age-Related Macular Degeneration* (J. W. Berger, S. L. Fine, and M. G. Maguire, eds.), Mosby, St. Louis, pp. 31–55.

Kolb, H., and Famiglietti, E. V., 1974, Rod and cone pathways in the inner plexiform layer of cat retina, *Science* **186**:47–49.

Kolb, H., Fernandez, E., and Nelson, R., 2002, Webvision: The organization of the retina and visual system. http://webvision.med.utah.edu.

Kuffler, S. W., 1953, Discharge patterns and functional organization of mammalian retina, *J. Neurophysiol.* **16**:37–68.

Kunz Mathews, M., Merges, C., McLeod, D. S., and Lutty, G. A., 1997, Vascular endothelial growth factor (VEGF) and vascular permeability changes in human diabetic retinopathy, *Invest. Ophthalmol. Visual Sci.* **38**:2729–2741.

Lamb, T. D., and Pugh, E. N., Jr., 1992, A quantitative account of the activation steps involved in phototransduction in amphibian photoreceptors, *J. Physiol.* **449**:719–758.

LaVail, M. M., 1981, Analysis of neurological mutants with inherited retinal degeneration, *Invest. Opthalmol. Visual Sci.* **21**:638–657.

Li, J.-D., Gallemore, R. P., Dmitriev, A., and Steinberg, R. H., 1994a, Light-dependent hydration of the space surrounding photoreceptors in chick retina, *Invest. Ophthalmol. Visual Sci.* **35**:2700–2711.

Li, J.-D., Govardovskii, V. I., and Steinberg, R. H., 1994b, Light-dependent hydration of the space surrounding photoreceptors in the cat retina, *Visual Neurosci.* **11**:743–752.

Linsenmeier, R. A., 1986, Effects of light and darkness on oxygen distribution and consumption in the intact cat retina, *J. Gen. Physiol.* **88**:521–542.

Linsenmeier, R. A., and Braun, R. D., 1992, Oxygen distribution and consumption in the cat retina during normoxia and hypoxemia, *J. Gen. Physiol.* **99**:177–197.

Linsenmeier, R. A., Frishman, L. J., Jakiela, H. J., and Enroth-Cugell, C., 1982, Receptive field properties of X and Y cells in the cat retina derived from contrast sensitivity measurements, *Vision Res.* **22**:1173–1183.

Linsenmeier, R. A., and Jakiela, H. G., 1979, Non-linear spatial summation in cat retinal ganglion cells at different background levels, *Exp. Brain Res.* **36**:301–309.

Linsenmeier, R. A., and Padnick-Silver, L., 2000, Metabolic dependence of photoreceptors on the choroid in the normal and detached retina, *Invest. Ophthalmol. Visual Sci.* **41**:3117–3123.

Linsenmeier, R. A., Padnick-Silver, L., Kang Derwent, J., Ramirez, U., and Narfstrom, K., 2000, Changes in photoreceptor oxidative metabolism in Abyssinian cats with a hereditary rod/cone degeneration, *Invest. Ophthalmol. Visual Sci.* **41**(4):S887 [ARVO Abstract].

Majji, A. B., Humayun, M. S., Weiland, J. D., Suzuki, S., D'Anna, S. A., and de Juan, E., 1999, Long-term histological and electrophysiological results of an inactive epiretinal electrode array implantation in dogs, *Invest. Ophthalmol. Visual Sci.* **40**:2073–2081.

Mangel, S. C., 1991, Analysis of the horizontal cell contribution to the receptive field surround of ganglion cells in the rabbit retina, *J. Physiol.* **442**:211–234.

Margalit, E., Maia, M., Weiland, J., Greenberg, R. J., Fujii, G. Y., Torres, G., Piyathaisere, D. V., O'Hearn, T. M., Liu, W., Lazzi, G., Dagnelie, G., Scribner, D. A., de Juan, E., Jr., and Humayun, M. S., 2002, Retinal prosthesis for the blind, *Survey Ophthalmol.* **47**:335–356.

Marmor, M. F., 1998, Mechanisms of retinal adhesiveness, in: *The Retinal Pigment Epithelium* (M. F. Marmor and T. J. Wolfensberger, eds.), Oxford University Press, New York, pp. 392–405.

McIlwain, J. T., 1996, *An Introduction to the Biology of Vision*, Cambridge University Press, Cambridge, UK.

Michels, R. G., Wilkinson, C. P., and Rice, T. A., 1990, *Retinal Detachment*, Mosby, St. Louis.

Moreno-Diaz, R., and Rubio, E., 1980, A model for non-linear processing in the cat's retina, *Biol. Cybernet.* **37**:25–31.

Naka, K.-I., and Rushton, W. A. H., 1966, S-cone potentials from luminosity units in the retina of fish (*Cyprinidae*), *J. Physiol.* **185**:587–599.

Narfstrom, K., Bragadottir, R., Redmond, T. M., Katz, M. L., Lei, B., Lai, C. M., and Rakoczy, E. P., 2002, Gene therapy in 6 dogs with RPE65 null mutation improves visual function: A short term study using clinical observations, electrophysiology and morphology, Association for Research in Vision and Ophthalmology, abstract number 4601. www.arvo.org.

Nelson, R., 1977, Cat cones have rod input: A comparison of response properties of cones and horizontal cell bodies in the retina of the cat, *J. Comp. Neurol.* **172**:109–136.

Nelson, R., Famiglietti, E. V., Jr., and Kolb, H., 1978, Intracellular staining reveals different levels of stratification for ON- and OFF-center ganglion cells in the cat retina, *J. Neurophysiol.* **41**:472–483.

Nicholson, C., and Phillips, J. M., 1981, Ion diffusion modified by tortuosity and volume fraction in the extracellular microenvironment of the rat cerebellum, *J. Physiol.* **321**:225–257.

Nicholson, C., and Rice, M. E., 1991, Diffusion of ions and transmitters in the brain cell microenvironment, in: *Volume Transmission in the Brain: Novel Mechanisms for Neural Transmission* (K. Fuxe and L. F. Agnati, eds.), Raven Press, New York, pp. 279–294.

Normann, R. A., Maynard, E. M., Rousche, P. J., and Warren, D. J., 1999, A neural interface for a cortical vision prosthesis, *Vision Res.* **39**:2577–2587.

Oakley, B., 1977, Potassium and the photoreceptor-dependent pigment epithelial hyperpolarization, *J. Gen. Physiol.* **70**:405–425.

Oakley, B., and Wen, R., 1989, Extracellular pH in the isolated retina of the toad in darkness and during illumination, *J. Physiol.* **419**:353–378.

Oguztoreli, M. N., 1980, Modelling and simulation of vertebrate retina, *Biol. Cybernet.* **37**:53–61.

Oyster, C. W., 1999, *The Human Eye: Structure and Function*, Sinauer Associates, Sunderland, MA.

Padnick-Silver, L., and Linsenmeier, R. A., 2002, Quantification of *in vivo* anaerobic metabolism in the normal cat retina through pH measurements, *Visual Neurosci.* **19**:793–806.

Padnick-Silver, L., 2000, *Characterization of Anaerobic Metabolism and the Effect of Acute Hyperglycemia in the Cat Retina through In Vivo pH and Oxygen Measurements*, PhD Thesis, Northwestern University.

Palmowski, A. M., Sutter, E. E., Bearse, M. A., Jr., and Fung, W., 1997, Mapping of retinal function in diabetic retinopathy using the multifocal electroretinogram, *Invest. Ophthalmol. Visual Sci.* **38**:2586–2596.

Pasternak, T., and Merigan, W., 1981, The luminance dependence of spatial vision in the cat, *Vision Res.* **21**:1333–1340.

Peachey, N. S., Pardue, M. T., Ball, S. L., Hetling, J. R., Chow, V. Y., and Chow A. Y., 2000, Unexpected sensitivity of the mammalian retina to infrared light, *Invest. Ophthalmol. Visual Sci.* **41**:S810.

Peichl, L., and Wassle, H., 1979, Size, scatter, and coverage of ganglion cell receptive field centers in the cat retina, *J. Physiol.* **291**:117–141.

Penn, J. S., Li, S., and Naash, M. I., 2000, Ambient hypoxia reverses retinal vascular attenuation in a transgenic mouse model of autosomal dominant retinitis pigmentosa, *Invest. Ophthalmol. Visual Sci.* **41**:4007–4013.

Pepperberg, D. R., Birch, D. G., Hofmann, K. P., and Hood, D. C., 1996, Recovery kinetics of human rod phototransduction inferred from the two-branched a-wave saturation function, *J. Opt. Soc. Am.* **A13**:586–600.

Pepperberg, D. R., Birch, D. G., and Hood, D. C., 1997, Photoresponses of human rods *in vivo* derived from paired flash electroretinograms, *Visual Neurosci.* **14**:73–82.

Pepperberg, D. R., Birch, D. G., and Hood, D. C., 2000, Electroretinographic determination of human rod flash response *in vivo*, *Methods Enzymol.* **316**:202–223. (Palczewski, K., ed., *Vertebrate Phototransduction and the Visual Cycle*, Academic Press, San Diego.)

Peyman, G., Chow, A. Y., Liang, C., Chow, V. Y., Perlman, J. I., and Peachey, N. S., 1998, Subretinal semiconductor microphotodiode array, *Ophthalmic Surg. Lasers* **29**:234–241.

Pournaras, C. J., 1995, Retinal oxygen distribution. Its role in the physiopathology of vasoproliferative microangiopathies, *Retina* **15**:332–347.

Prince, A. M., and Solomon, I. S., 1996, Automated perimetry diagnostic modalities, in: *Duane's Foundations of Clinical Ophthalmology*, Vol. 2 (W. Tasman and E. A. Jaeger, eds.), Chapt. 109, Lippincott-Raven, Philadelphia, pp. 1–34.

Pugh, E. N., Jr., and Lamb, T. D., 2000, Phototransduction in vertebrate rods and cones: Molecular mechanisms of amplification, recovery and light adaptation, in: *Handbook of Biological Physics*, Vol. 3 (D. G. Stavenga, W. J. DeGrip, and E. N. Pugh Jr., eds.), Elsevier, Amsterdam.

Quigley, H. A., McKinnon, S. J., Zack, D. J., Pease, M. E., Kerrigan-Baumrind, L. A., Kerrigan, D. F., and Mitchell, R. S., 2000, Retrograde axonal transport of BDNF in retinal ganglion cells is blocked by acute IOP elevation in rats, *Invest. Ophthalmol. Visual Sci.* **41**:3460–3466.

Quigley, H. A., Dunkelberger, G. R., and Green, W. R., 1989, Retinal ganglion cell atrophy correlated with automated perimetry in human eyes with glaucoma, *Am. J. Ophthalmol.* **107**:453–464.

Rice, M. E., and Nicholson, C., 1991, Diffusion characteristics and extracellular volume fraction during normoxia and hypoxia in slices of rat neostriatum, *J. Neurophysiol.* **65**:264–272.

Riva, C. E., Grunwald, J. E., and Sinclair, S. H., 1983, Laser doppler velocimetry study of the effect of pure oxygen breathing on retinal blood flow, *Invest. Ophthalmol. Visual Sci.* **24**:47–51.

Rizzo, J. F., and Wyatt, J. L., 1997, Prospects for a visual prosthesis, *Neuroscientist* **3**:251–262.

Rizzo, J. F., and Wyatt, J. L., 1999, Retinal prosthesis, in: *Age-Related Macular Degeneration* (J. W. Berger, S. L. Fine, and M. G. Maguire, eds.), Mosby, St. Louis, Chapt. 25, pp. 413–432.

Rizzo, J. F., Wyatt, J. L., Humayun, M., deJuan, E., Liu, W., Chow, A., Eckmiller, R., Zrenner, E., Yagi, T., and Abrams, G., 2001, Retinal prosthesis: An encouraging first decade with major challenges ahead, *Ophthalmology* **108**:13–14.

Robson, J. G., and Frishman, L. F., 1995, Response linearity and kinetics of the cat retina: The bipolar cell component of the dark-adapted electroretinogram, *Visual Neurosci.* **12**:837–850.

Robson, J. G., and Frishman, L. F., 1996, Photoreceptor and bipolar-cell contributions to the cat electroretinogram: A kinetic model of the early part of the flash response, *J. Opt. Soc. Am.* **A12**:613–622.

Rodieck, R. W., 1965, Quantitative analysis of cat retinal ganglion cell response to visual stimuli, *Vision Res.* **5**:583–601.

Rodieck, R. W., 1973, *The Vertebrate Retina.Principles of Structure and Function*, Freeman and Co., San Francisco.

Rodieck, R. W., 1998, *First Steps in Seeing*, Sinauer Associates, Sunderland, MA.

Rodieck, R. W., and Stone, J., 1965, Analysis of receptive fields of cat retinal ganglion cells, *Visual Neurosci.* **28**:833–849.

Roska, B., Nemeth, E., Orzo, L., and Werblin, F. S., 2000, Three levels of lateral inhibition: A space time study of the retina of the tiger salamander, *J. Neurosci.* **20**:1941–1951.

Rowe, M. H., and Cox, J. F., 1993, Spatial receptive-field structure of cat retinal W cells, *Visual Neurosci.* **10**:765–779.

Saleem, R. A., and Walter, M. A., 2002, The complexities of ocular genetics, *Clin. Genet.* **61**:79–88.

Sassani, J. W., 1996, Glaucoma, in: *Duane's Foundations of Clinical Ophthalmology*, Vol. 3 (W. Tasman and E. A. Jaeger, eds.), Lippincott-Raven, Philadelphia, Chapt. 19, pp. 1–30.

Schiller, P. H., 1992, The ON and OFF channels of the visual system, *Trends Neurosci.* **15**:86–92.

Schneeweis, D. M., and Schnapf, J. L., 1995, Photovoltage of rods and cones in the macaque retina, *Science* **268**:1053–1056.

Schroder, S., Palinski, W., and Schmid-Schoenbein, G., 1991, Activated monocytes and granulocytes, capillary non-perfusion, and neovascularization in diabetic retinopathy, *Am. J. Pathol.* **139**:81–100.

Schwahn, H. N., Gekeler, F., Kohler, K., Kobuch, K., Sachs, H. G., Schulmeyer, F., Jakob, W., Gabel, V.-P., and Zrenner, E., 2001, Studies on the feasibility of a subretinal visual prosthesis: Data from Yucatan minipig, *Graefe's Arch. Clin. Exp. Ophthalmol.* **239**:961–967.

Schnapf, J. L., Nunn, B. J., Meister, M., and Baylor, D. A., 1990, Visual transduction in cones of the monkey *Macaca fascicularis*, *J. Physiol.* **427**:681–713.

Shapley, R. M., and Enroth-Cugell, C., 1984, Visual adaptation and retinal gain controls, *Prog. Retin. Res.* **3**:263–346.

Shapley, R. M., and Lennie, P., 1985, Spatial frequency analysis in the visual system, *Annu. Rev. Neurosci.* **8**:547–583.

Shapley, R., and Perry, V. H., 1986, Cat and monkey retinal ganglion cells and their visual functional roles, *Trends Neurosci.* **9**:229–235.

Shapley, R. M., and Victor, J. D., 1978, The effect of contrast on the transfer properties of cat retinal ganglion cells, *J. Physiol.* **285**:275–298.

Shimazaki, H., and Oakley, B., 1984, Reaccumulation of $[K^+]_o$ in the toad retina during maintained illumination, *J. Gen. Physiol.* **84**:475–504.

Sickel, W., 1972, Retinal metabolism in dark and light, in: *Physiology of Photoreceptor Organs. Handbook of Sensory Physiology* (M. G. F. Fuortes, ed.), Springer, Berlin, pp. 667–727.

Singer, N., 2002, Ambitious MEMS-based retinal prosthesis plan aims to give sight to the blind, *Sandia Lab News* **54**(19):1, 4.

Smith, R. G., Freed, M. A., and Sterling, P., 1986, Microcircuitry of the dark-adapted cat retina: Functional architecture of the rod–cone network, *J. Neurosci.* **6**:3505–3517.

So, Y. T., and Shapley, R. M., 1981, Spatial tuning of cells in and around lateral geniculate nucleus of the cat: X and Y relay cells and perigeniculate neurons, *J. Neurophysiol.* **45**:107–120.

Stamper, R. L., and Sanghvi, S. S., 1996, Intraocular pressure: Measurement, regulation, and flow relationships, in: *Duane's Foundations of Clinical Ophthalmology*, Vol. 2 (W. Tasman and E. A. Jaeger, eds), Lippincott-Raven, Philadelphia, Chapt. 7, pp. 1–31.

Steinberg, R. H., Frishman, L. J., and Sieving, P. A., 1991, Negative components of the electroretinogram from proximal retina and photoreceptor, *Prog. Retin. Res.* **10**:121–160.

Steinberg, R. H., Oakley, B., and Niemeyer, G., 1980, Light-evoked changes in $[K^+]_o$ in retina of intact cat eye, *J. Neurophysiol.* **44**:897–921.

Sterling, P., Freed, M., and Smith, R. G., 1986, Microcircuitry and functional architecture of the cat retina, *Trends Neurosci.* **9**:186–192.

Stone, J., and Fukuda, Y., 1974, Properties of cat retinal ganglion cells: A comparison of W cells with X and Y cells, *J. Neurophysiol.* **37**:722–748.

Sutter, E. E., and Tran, D., 1992, The field topography of ERG components in man. I. The photopic luminance response, *Vision Res.* **32**:433–446.

Tranchina, D., Gordon, J., and Shapley, R., 1983, Spatial and temporal properties of luminosity horizontal cells in the turtle retina, *J. Gen. Physiol.* **82**:573–598.

Troy, J. B., Bohnsack, D. L., and Diller, L. C., 1999, Spatial properties of the cat X-cell receptive field as a function of mean light level, *Visual Neurosci.* **16**:1089–1104.

Troy, J. B., Einstein, G., Schuurmans, R. P., Robson, J. G., and Enroth-Cugell, C., 1989, Responses to sinusoidal gratings of two types of very nonlinear retinal ganglion cells of cat, *Visual Neurosci.* **3**:213–223.

Troy, J. B., Oh, J. K., and Enroth-Cugell, C., 1993, Effect of ambient illumination on the spatial properties of the center and surround of Y-cell receptive fields, *Visual Neurosci.* **9**:535–553.

Troy, J. B., and Shou, T., 2002, The receptive fields of cat retinal ganglion cells in physiological and pathological states: Where we are after half a century of research, *Prog. Retin. Eye Res.* **21**:263–302.

Troy, J. B., Schweitzer-Tong, D. E., and Enroth-Cugell, C., 1995, Receptive field properties of Q retinal ganglion cells in the cat, *Visual Neurosci.* **12**:285–300.

Victor, J. D., Shapley, R. M., and Knight, B. W., 1977, Nonlinear analysis of cat retinal ganglion cells in the frequency domain, *PNAS* **74**:3068–3072.

Victor, J. D., and Shapley, R. M., 1979, Receptive field mechanism of X and Y retinal ganglion cells, *J. Gen. Physiol.* **74**:275–298.

Vo Van Toi and Riva, C. E., 1994, Variations of blood flow at optic nerve head induced by sinusoidal flicker stimulation in cats, *J. Physiol.* **482**:189–202.

Walter, P., Szurman, P., Vobig, M., Berk, H., Ludtke-Handjery, H.-C., Richter, H., Deng, Mittermayer, C., Heimann, K., and Sellhaus, B., 1999, Successful long-tern implantation of inactive epiretinal microelectrode arrays in rabbits, *Retina* **19**:546–552.

Walter, P., and Heimann, K., 2000, Evoked cortical potentials after electrical stimulation of the inner retina in rabbits, *Graefe's Arch. Clin. Exp. Ophthalmol.* **238**:315–318.

Wang, L., Kondo, M., and Bill, A., 1997, Glucose metabolism in cat outer retina, *Invest. Ophthalmol. Visual Sci.* **38**:48–55.

Wangsa-Wirawan, N., and Linsenmeier, R. A., 2003, Retinal oxygen: Fundamental and clinical aspects, *Arch. Ophthalmol.* **121**:547–557.

Wangsa-Wirawan, N., Padnick-Silver, L., Budzynski, E., and Linsenmeier, R. A., 2001, pH regulation in the intact cat outer retina, *Invest. Ophthalmol. Visual Sci.* **42**(4):S367 [ARVO Abstract].

Wassle, H., and Boycott, B. B., 1991, Functional architecture of the mammalian retina, *Physiol. Rev.* **71**:447–480.

Werblin, F., 1991, Synaptic connections, receptive fields, and patterns of activity in the tiger salamander retina, *Invest. Ophthalmol. Visual Sci.* **32**:459–483.

Wolbarsht, M. L., and Landers, M. B., III, 1980, The rationale of photocoagulation therapy for proliferative diabetic retinopathy: A review and a model, *Ophthalmic Surg.* **11**:235–245.

Wise, G. N., Dollery, C. T., and Henkind, P., 1971, *The Retinal Circulation*, Harper and Row, New York.

Wyatt, J., and Rizzo, J., 1996, Ocular implants for the blind, *IEEE Spectrum* **112**:47–53.

Wyatt, J., and Rizzo, J., 1999, Retinal prosthesis, in: *Age-Related Macular Degeneration* (J. W. Berger, S. L. Fine, and M. G. Maguire, eds.), Mosby, St. Louis, pp. 413–432.

Yancey, C. M., and Linsenmeier, R. A., 1989, Oxygen distribution and consumption in the cat retina at increased intraocular pressure, *Invest. Ophthalmol. Visual Sci.* **30**:600–611.

Yamamoto, F., Borgula, G., and Steinberg, R. H., 1992, Effects of light and darkness on pH outside rod photore-ceptors in the cat retina, *Exp. Eye Res.* **54**:685–697.

Yau, K. Y., 1994, Phototransduction mechanism in rods and cones, *Invest. Ophthalmol. Visual Sci.* **35**:9–32.

Young, R. W., 1976, Visual cells and the concept of renewal, *Invest. Ophthalmol.* **15**:700–725.

Yu, D.-Y., and Cringle, S. J., 2001, Oxygen distribution and consumption within the retina in vascularized and avascular retinas and in animal models of retinal disease, *Prog. Retin. Eye Res.* **20**(2):175–208.

Zrenner, E., 2002, Will retinal implants restore vision? *Science* **295**:1022–1025.

Zrenner, E., Miliczek, K.-D., Gabel, V. P., Graf, H. G., Guenther, E., Haemmerle, H., Hoefflinger, B., Kohler, K., Nisch, W., Schubert, M., Stett, A., and Weiss, S., 1997, The development of subretinal microphotodiodes for replacement of degenerated photoreceptors, *Ophthalmic Res.* **29**:269–280.

Zrenner, E., Stett, A., Weiss, S., Aramant, R. B., Guenther, E., Kohler, K., Miliczek, K.-D., Seiler, M. J., and Haemmerle, H., 1999, Can subretinal microphotodiodes successfully replace degenerated photoreceptors, *Vision Res.* **39**:2555–2567.

Zuckerman, R., and Weiter, J. J., 1980, Oxygen transport in the bullfrog retina, *Exp. Eye Res.* **30**:117–127.

SUBJECT INDEX